Chemistry

Neil Jespersen, Ph.D.

Department of Chemistry
St. John's University
Jamaica, New York

BARRON'S

To Dr. Frances S. Sterrett
1913–1996
colleague, mentor, and friend

All inquiries should be addressed to:
Barron's Educational Series, Inc.
250 Wireless Boulevard
Hauppauge, New York 11788

International Standard Book No. 0-8120-9503-0

Library of Congress Catalog Card No. 96-29768

Library of Congress Cataloging-in-Publication Data

Jespersen, Neil D.
Chemistry / Neil Jespersen.
p. cm.—(College review series, Science)
Includes index.
ISBN 0-8120-9503-0
1. Chemistry—Outlines, syllabi, etc. I. Title. II. Series.
QD41.J47 1997
540—dc21 96-29768
CIP

PRINTED IN THE UNITED STATES OF AMERICA
9 8 7 6

CONTENTS

ACKNOWLEDGMENTS

This book would not be possible without the help, encouragement, and advice of my wife, Marilyn Zak Jespersen, and my daughters, Lisa and Kristen. They pursue related careers of their own yet found the time to review the manuscript and offer valuable suggestions.

PREFACE

This book is designed to be used with all currently published textbooks for the first course in college-level chemistry. It is easy to carry with you and can be used any time or place. Marking it with your own notes will make the book especially personal.

Many years of teaching introductory chemistry and the comments of many students taking these chemistry courses led to the development of this book. Surprisingly, these students agreed that the chemistry curriculum must be rigorous and should encourage each student to excel individually. In particular, they wanted:

1. No-nonsense explanations of how to solve their problems.
2. Concise descriptions and applications of fundamental concepts.
3. Clarification of the interrelationships inherent in the various items of chemical information.

There was a need to understand the fundamentals that make chemistry the "central science," and this book is designed to meet this need.

This book is divided into two parts, Principles and Concepts of Chemistry and Techniques and Calculations of Chemistry. These parts are further divided into groups of related topics that will quickly lead to the explanation you need. Principles and Concepts presents basic ideas and theories of chemistry with as little math as possible. Techniques and Calculations provides the details of chemical methods needed to perform calculations and to do laboratory work.

READ ME FIRST!

HOW TO USE THIS BOOK

This book was developed at the suggestion of college students who had either completed their first-year chemistry courses or were currently taking these courses. Although each had a textbook, a study guide, and a faculty to work with, these students **wanted more**—in particular, a no-nonsense volume that would describe each of the individual aspects of chemistry in detail. There was a distinct need to learn how to solve problems. This book meets this important need by providing 12 chapters on specific operations and concepts in chemistry. Where needed, examples are worked out in detail, and most exercises also have detailed solutions.

In this book, the **factor-label method** is used exclusively. Once skill is developed with the factor-label method, the chance of error drops almost to zero. You are encouraged to follow all solutions and to cancel the labels with a colored pencil in each factor-label example or problem. This active participation results in much greater retention of the information learned.

It is important to view this book as an encyclopedia of problem solving. That is precisely what it is. You can find a particular item by using the topical listing of subjects within each grouping in the Table of Contents or by looking up key words in the index at the end of the book. Many items are also described in the Glossary.

Many topics in chemistry require prior knowledge. For example, you cannot balance equations without knowing how to count the number of atoms in a chemical formula. Each group of chapters develops a topic from basic principles, and you can review more basic material in the chapters immediately preceding the one you are reading.

Finally, the current theory of teaching chemistry avoids the memorization of facts and equations. On quizzes and tests, students are given a Periodic Table, tables of data, and often the relevant equations. The student must recognize different situations and problems in order to solve them as quickly as possible with the greatest accuracy. For this reason one valuable study technique is the repetitive solution of problems. The ability to solve problems rapidly and correctly is a great advantage.

HINTS ON MULTIPLE-CHOICE PROBLEMS

Usually, multiple-choice exams are given in general chemistry courses because of the large number of students in these classes. There are some hints that will help you achieve the best scores.

Be sure to provide an answer for each and every question. The odds are usually 1 in 4 or 1 in 5 that a wild guess will be correct.

Second, with a multiple-choice test it is important to quickly read through all questions and answer the easiest ones first. Sometimes the easy questions are at the end, and you should find them and answer them quickly. Answer the medium-difficulty questions second, and save the tough ones for last. In medicine a similar system is called triage, and it works here too.

Third, one of the most important things to learn in chemistry is to estimate a reasonable answer. Of course, a calculator provides an exact answer almost as quickly as you can arrive at an estimate. The two together—your estimate and the calculator—are a powerful force in avoiding gross errors.

In addition, there are some well known generalities that help you recognize reasonable results. For instance, in stoichiometry the mass of a product is rarely more than five times, or less than one-fifth, the mass of the reactant. Also, the vapor pressure of a mixture must be between the vapor pressures of the two pure substances in the mixture. In another case, it is very helpful to know that the mole fraction must be a number between 0 and 1.

HINTS ON WRITTEN EXAMS

If exams consist of written problems, there are strategies that will enhance your scores significantly.

First, always write the fundamental equations needed for the problem. These will include a balanced chemical equation and then the necessary physical equation such as Henry's law or the ideal gas law or the Gibbs free-energy equation. Then list the information given in the problem and convert the units as needed. Finally, substitute the information into the equation and solve the problem.

Very often solutions to chemistry problems involve the use of two or more equations (simultaneous equations). You may solve these symbolically before entering the data. Alternatively, you may enter the data into one equation and solve it to enter the result in the second equation. Either method is valid. To obtain full credit, all algebra steps must be shown clearly.

If you lack the time or knowledge to solve the equations, describe in words how the problem should be solved. For instance, you might say, "Use Henry's law to calculate *k* and then use that value of *k* in Henry's law again to calculate the solubility of the gas."

Avoid, at all costs, saying anything that will detract from your answer. If you are not sure about something, don't write it down.

Finally, always write clearly and neatly so that correct answers will not be misinterpreted.

HINTS ON HOW TO STUDY

The most successful students in chemistry study and learn the material as the course progresses. This means that you must **study each day and keep up with the work** at all times. The surest road to disaster is to cram the night before an exam. Read the assigned chapters before each class, take notes in class, and review the notes (some suggest rewriting them) the same evening. Ask the lecturer about questions as they arise (notice that the best students know how to ask questions and are not hesitant about asking them).

Create a study group. In any group, the real learning occurs with participation. You really have to understand a subject before you can explain it to someone else. Each person in the group should practice explaining problems and concepts to the others. You should not have to study the night before the exam because you have been working all semester. Finally, **relax the night before the exam**. If all of the suggestions above have been followed, you should be fully prepared. The final requirement, then, is physical: Have a relaxing evening and go to bed early. The best help on the day of the exam is a clear, alert mind!

Part 1

PRINCIPLES AND CONCEPTS OF CHEMISTRY

1
ATOMS

ATOMIC THEORY AND ATOMIC MODELS

Atomic Theory

A hypothesis by the ancient Greek philosopher Democritus (ca. 400 B.C.) is the first historical mention of atoms. He reasoned that, if matter was discontinuous like a box of marbles, it could be divided in half repeatedly, until eventually only one marble was left that could not be divided. He called this smallest particle of matter an **atom**. Democritus, like most ancient philosophers, had no experimental proof that atoms really existed.

Roger Bacon, who lived in the thirteenth century, established the concept that science should base its reasoning on experimental evidence, thus converting science from a philosophical to an experimental study of the world. The first chemists were the alchemists who, in the Middle Ages, tried fruitlessly to convert base metals into gold. Despite their shortcomings (secrecy, mysticism, and at times outright fraud), they developed many experimental methods and also built an extensive body of chemical data.

Major milestones in the development of chemistry were reached in 1774, when Antoine Lavoisier performed careful experiments and measurements that led to the **law of conservation of matter**, and in 1799, when Joseph Proust made measurements of chemical reactions and compounds and developed the **law of constant composition**. The first law states that in chemical reactions matter is neither created nor destroyed. The second law states that each pure chemical compound always has the same percentage composition of each element by mass.

These two laws led John Dalton to develop his **atomic theory** over the years from 1803 to 1808.

DALTON'S ATOMIC THEORY

1. All matter is composed of tiny, indivisible particles called atoms that cannot be destroyed or created.
2. Each element consists of atoms that are identical to each other in all of their properties, and these properties are different from the properties of all other atoms.
3. Chemical reactions are simple rearrangements of atoms from one combination to another in small whole-number ratios.

Every scientific theory provides new predictions that can be tested by experiments to support or disprove the theory. (It is important to remember that scientists can never prove a theory to be true. Experiments may be used to support a theory, but not to prove it.)

The atomic theory led John Dalton to propose the law of multiple proportions.

> **LAW OF MULTIPLE PROPORTIONS**
>
> When two elements can be combined to make two different compounds, and if samples of these two compounds are taken so that the masses of one of the elements in the two compounds are the same, then the ratio of the masses of the other element in these compounds will be a ratio of small whole numbers.

This law was used to help verify his atomic theory.

Atomic Models

The first models of matter were devised by Greek philosophers, who developed two logical possibilities. One was that matter was continuous, and a sample, no matter how small, could always be divided into smaller parts. The other was that matter was discontinuous or atomic. In this model there was some point at which a sample would be so small that it could not be divided any further; the atoms are considered solid particles. Since no experimental proof was available to support either of these possibilities, it was a matter of belief in one or the other.

The solid-particle model was put on a firm foundation by Dalton, who in 1803 combined Antoine Lavoisier's law of conservation of mass and Joseph Proust's law of definite proportions to produce his atomic theory.

In the mid 1800s the electron was discovered and investigated. It appeared to be a universal component of all matter. In the prevailing view of the atom negative electrons were floating in a sea of positive charge. Some likened this to the English plum pudding; the negative electrons were seen as similar to the fruit floating in this popular dessert.

The **plum pudding model** came to an abrupt end in 1909, when Ernest Rutherford performed his famous gold foil experiment. He found that the atom is mostly empty space with an extremely dense, positively charged nucleus. The concept of the atom abruptly changed to a **nuclear model**, where the negative electrons surround the very dense, positively charged nucleus.

While the fundamental particles of the atom were being discovered, other physicists were performing experiments that laid the foundation for a revolution in how all matter is viewed. In the mid 1800s physicists were interested in the interaction of light and matter. One of their discoveries was that each element, when heated or sparked with electricity, gives off characteristic colors. A spectroscope was used to show that these colors consist of discrete wavelengths of light (line spectra), not the uniform rainbow observed when white light is separated by a prism. The line spectra of most elements and compounds are very complex. Hydrogen, however, has a seemingly simple series of lines.

In 1885 Johann Balmer found an empirical mathematical relationship between the wavelengths of the lines he observed in the visible region of the spectrum. When similar series of lines were found in the infrared (Paschen series) and ultraviolet (Lyman series) regions, Johannes Rydberg extended Balmer's equation so that all of the wavelengths could be predicted.

In 1913 Niels Bohr completed his theory of how the hydrogen atom is constructed. He assumed, using a **solar system model**, that electrons move around the nucleus in circular orbits. With this model he was able to duplicate the Rydberg equation from fundamental constants already known to physicists. The most important contribution was the concept that electrons exist in only certain "allowed orbits." This concept, along with Max Planck's description, in 1900, of light as packets or quanta of energy called photons, aided Bohr in developing the solar system model of the atom.

ATOMIC MODELS

Solid-Particle Model 400 B.C.
Plum Pudding Model 1909
Nuclear Model Rutherford 1910
Solar System Model Bohr 1913
Wave-Mechanical Model Schrödinger 1927

In 1924 Louis De Broglie suggested that, if light can be considered as particles, then small particles such as electrons may have the characteristics of waves. Three years later. Erwin Schrödinger applied the equations for waves to the electrons in an atom and introduced the **wave-mechanical theory** of the atom. For hydrogen the results are very similar to Bohr's model except that the electron does not follow a precise orbit. The position of the electron in the wave-mechanical model is described by the probability of where it will be located.

Also in the 1920s Werner Heisenberg developed the **uncertainty principle** that bears his name. It states that the position and the momentum of any particle cannot both be known exactly at the same time. As one is known more precisely, the other becomes less certain.

Electrons, Protons, and Neutrons

In 1834 Michael Faraday showed that an electric current can cause chemical reactions to occur, demonstrating the electrical nature of the elements. Four decades later, in the 1870s, Sir William Crookes developed what is known today as the cathode ray tube. He mistakenly thought that the cathode rays were negatively charged molecules, rather than electrons. (He was also the first to suggest the existence of isotopes.) In 1897 J.J. Thomson determined that cathode rays were a fundamental part of matter that he called **electrons**. By measuring the deflection of the cathode rays in the presence of electric and magnetic fields, he also determined their mass-to-charge ratio ($m/e = -1.76 \times 10^8$ coulombs gram^{-1}). Twelve years later, in 1909, Robert Millikan performed his oil drop experiment, which enabled him to calculate the charge of the electron (-1.60×10^{-19} coulomb). With this value, combined with Thomson's charge-to-mass ratio, the mass of the electron was calculated to be 9.11×10^{-28} gram. This information led to the "plum pudding" model of the atom, with electrons bathed in a sea of positive charge, analogous to raisins in the famous English pudding.

At this time Ernest Rutherford was interested in radioactive materials and had identified alpha and beta particles in his research. Along with Hans Geiger and Ernest Marsden, he performed the gold foil experiment wherein heavy alpha particles were aimed at a thin gold foil. While most of the alpha particles went through the foil with no visible effect, a few were deflected from their path and some actually bounced back in the direction from which they came. From these results Rutherford deduced the structure of the atom as an extremely small, dense, and positively charged nucleus surrounded by mostly empty space occupied by electrons. This concept became known as the nuclear model of the atom. Ten years later, in 1919, Rutherford discovered the basic unit of positive charge in the atom and termed it the **proton**. The proton has a positive charge exactly equal to the magnitude of the electron charge, and a mass of 1.67×10^{-24} gram, making it 1,836 times heavier than the electron.

In 1932 James Chadwick discovered a very penetrating form of radiation, which he demonstrated is the third major particle that constitutes the atom. The **neutron**, so named because it is neutral (i.e., it has no charge), has a mass almost equal to that of the proton.

The three fundamental parts of the atom are shown in Table 1.1.

TABLE 1.1. FUNDAMENTAL PARTS OF THE ATOM

Name	Symbol	Absolute Charge (C)	Absolute Mass (g)	Relative Charge	Relative Mass (AMU)
Electron	e or e^-	-1.602×10^{-19}	9.109×10^{-28}	-1	5.486×10^{-4}
Proton	p or p^+	$+1.602 \times 10^{-19}$	1.673×10^{-24}	$+1$	1.0073
Neutron	n	0	1.675×10^{-24}	0	1.0087

The Periodic Table may be used to quickly determine the number of protons and electrons in a particular element. For any given element the number of protons is always equal to the atomic number (Z) of the element:

$$\text{Number of protons} = Z$$

and the number of electrons is also equal to the atomic number:

$$\text{Number of electrons} = Z$$

For an ion of an element, the number of electrons may be calculated as

$$\text{Number of electrons} = Z - \text{charge of the ion}$$

A positive ion (cation) has lost electrons and a negative ion (anion) has gained electrons compared to the element itself.

The number of neutrons in an element depends on the specific isotope that is under consideration. Since the atomic masses listed in the Periodic Table are weighted averages of the masses of all the naturally occurring isotopes, the number of neutrons in an atom cannot normally be determined from the Periodic Table. If, however, mass number for the specific isotope is known, the number of neutrons can be calculated as the difference between the isotope mass and the atomic number:

$$\text{Number of neutrons} = A - Z$$

EXERCISE

Use the Periodic Table and the equations given above to fill in the blanks of the following table:

Symbol	Atomic Number	Isotope Mass	Number of Protons	Number of Electrons	Number of Neutrons
Fe		56			
	60	144			
		102	45	45	
		59			31
Al		27			

Solution:

Symbol	Atomic Number	Isotope Mass	Number of Protons	Number of Electrons	Number of Neutrons
Fe	26	56	26	26	30
Nd	60	144	60	60	84
Rh	45	102	45	45	57
Ni	28	59	28	28	31
Al	13	27	13	13	14

Discovery of the Electron, Proton, and Neutron

Charge-to-Mass Ratio of the Electron

In the 1870s Sir William Crookes designed an evacuated tube with two electrodes, as shown in Figure 1.1. When a high voltage was applied to the electrodes, a glow was noticed between them. When an object was placed in the path of the glow, it blocked part of the rays that produced the glow. This experiment showed that the rays must originate at the negative electrode (cathode) and flow toward the positive electrode (anode).

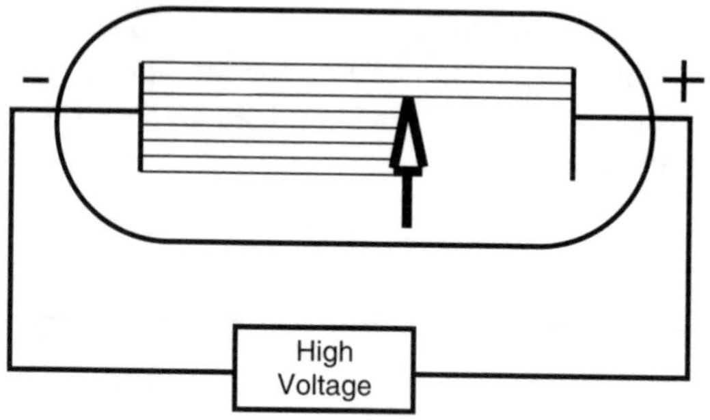

FIGURE 1.1. *Cathode rays blocked by an object, showing direction of flow.*

Thus the rays were called cathode rays since they apparently came from the cathode, or negative terminal. The path of the cathode rays could be deflected by both electric fields and magnetic fields (Figure 1.2). The fact that these rays were attracted toward the positive electric field and repelled by the negative electric field indicated that they had a negative charge.

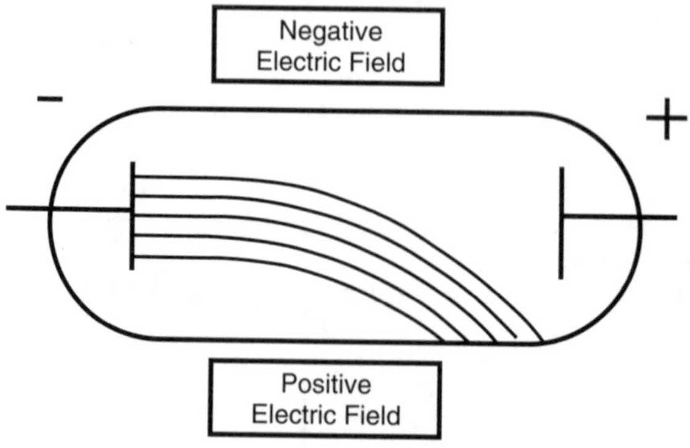

FIGURE 1.2. *Deflection of cathode rays by an electric field.*

In a magnetic field the magnetic force causes a negative particle to move in a circular motion with a radius of r according to this equation:

$$Hev = \frac{mv^2}{r}$$

Where H is the strength of the magnetic field, e is the charge of the electron, m is the mass of the electron, v is its velocity, and r is the radius of curvature caused by the magnetic field. Rearranging this equation yields the e/m ratio:

$$\frac{e}{m} = \frac{v}{rH} \qquad (1.1)$$

The magnetic field (H) and the radius of curvature (r) are easily measured. It is necessary, however, to know the velocity of the electron to calculate the e/m ratio. If an electric field is applied to the beam so that it cancels the deflection of the magnetic field, the force of the electric field must be the same as the force of the magnetic field. Mathematically this is expressed as

$$Ee = Hev$$

where E is the electric field, e is the charge of the electron, and v is the velocity of the electron. Rearranging this equation gives the needed velocity of the electron:

$$v = \frac{E}{H}$$

Substituting this into Equation 1.1 gives

$$\frac{e}{m} = \frac{E}{H^2r}$$

The experiment involves the use of a cathode ray tube with an applied magnetic field having a known strength H; the radius of deflection is measured. An electric field that can be varied is applied to cancel the effect of the magnetic field. The value of the electric field needed to cancel the magnetic field is then used to calculate the e/m ratio. The value obtained was -1.76×10^8 coulombs per gram.

Millikan's Oil Drop Experiment

Robert Millikan set up an apparatus as shown in Figure 1.3, where he could spray oil droplets (from his wife's perfume atomizer) so they would settle into a beam of X rays. The X rays caused the oil droplets to become charged with electrons. Using a small telescope, Millikan could measure the diameter of each droplet. In addition, he applied an electric voltage to the top and bottom of the chamber so that the droplet would stop falling and remain stationary. The positive plate attracted, and the negative plate repelled, the negatively charged droplet. Adjusting the voltage, Millikan was able to make the droplet stand still.

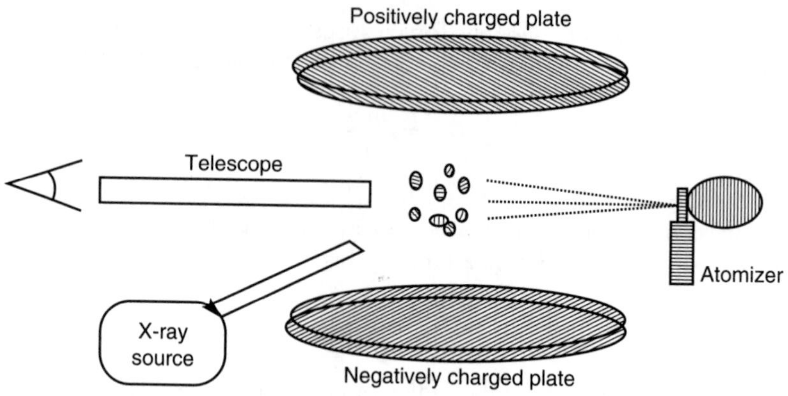

FIGURE 1.3. *Apparatus used by Millikan to determine the charge of the electron.*

Knowing the density of the oil, Millikan could calculate the mass (m) of each oil droplet (volume × density = mass). He knew that the equation for the force of gravitational attraction is

$$\text{Force} = mG$$

where G is the acceleration due to gravity. The equation for electrical force is

$$\text{Force} = Ee$$

where E is the applied voltage and e is the charge of the electron. The gravitational and electrical forces are equal when the voltage is adjusted, so that the oil drop remains stationary and

$$m_{\text{droplet}}G = Ee$$

From this equation the value of the charge (e) of the electron can be easily calculated.

There was one problem: the X rays did not always give an oil droplet with the same negative charge. The result was that Millikan did not get a single value for the charge of the electron. Instead, for four droplets, he got a series of results that may have looked like this:

$$-3.2 \times 10^{-19} \text{ coulomb}$$
$$-6.4 \times 10^{-19} \text{ coulomb}$$
$$-8.0 \times 10^{-19} \text{ coulomb}$$
$$-4.8 \times 10^{-19} \text{ coulomb}$$

From this information, Millikan was able to deduce that the common divisor for each of these values is -1.6×10^{-19} coulomb. He reasoned that the four droplets with the charges described above must have had acquired charges of −2, −4, −5, and −3, respectively, from the X rays. He had the insight to assign -1.6×10^{-19} coulomb as the charge of the electron.

Rutherford's Gold Foil Experiment

Ernest Rutherford devised an elegant experiment carried out by Johannes Geiger and Ernest Marsden. They bombarded thin foils of gold with alpha particles (helium atoms without their electrons) in order to study the structure of the atom (Figure 1.4). At that time the atom was thought to be a uniform ball of positive charge with the negative electrons embedded in it much like the raisins in an English plum pudding.

If the plum pudding model was correct, the alpha particles should have gone straight through the gold foil. To the experimenters' total surprise, a small fraction of the alpha particles were deflected from their original trajectories. Even more surprising was the almost total reflection of a few alpha particles. These results required that the gold atom have a very massive nucleus compared to the alpha particles. In addition, the nucleus had to be positively charged to repel the positively charged alpha particle. These results led Rutherford to postulate the **nuclear model** of the atom, with a dense, positively charged nucleus and a large amount of empty space occupied by the negatively charged electrons.

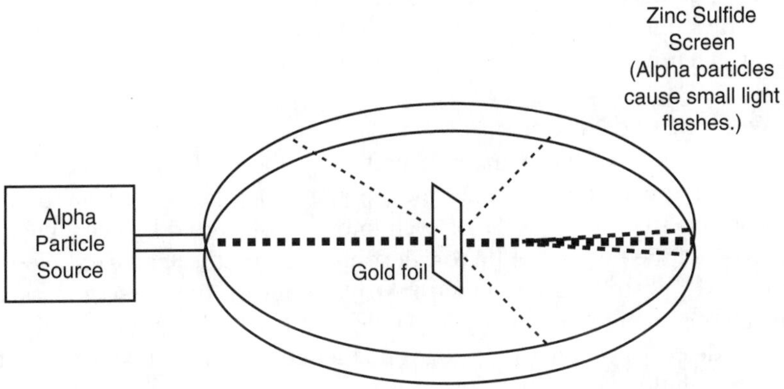

FIGURE 1.4. *Rutherford's gold foil apparatus.*

Discovery of the Proton

Using a cathode ray tube, experimenters drilled holes into the anode of a Crookes tube (Figure 1.5) and discovered rays moving in opposite directions to the electron. Originally called "canal rays," they were recognized as atoms with one or more electrons removed. The measurements described for the *e/m* ratio of the electron were used to determine that the lightest of the canal rays, the **proton**, has a mass 1800 times greater than that of the electron. It also has a charge equal to that of the electron but with the opposite sign (the proton has a positive charge).

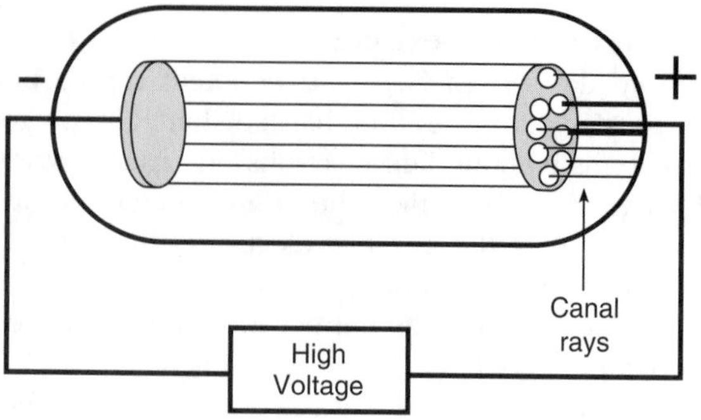

FIGURE 1.5. *Holes drilled in the anode of a Crookes tube reveal atomic ions, the lightest of which is the proton.*

Determination of Atomic Numbers

In 1913, Henry Moseley studied the X rays emitted by various elements in specially designed X-ray tubes. He found that the atomic number (number of protons in the nucleus) was proportional to the square root of the frequency of the X rays emitted by the element. Moseley's brilliant scientific career was cut short when he was killed in World War I.

Discovery of the Neutron

With knowledge of the number of protons in the nucleus of each atom, it soon became apparent that a significant amount of mass was "missing." In fact, only about one-half of the mass of each known atom could be accounted for on the basis of the number of protons it contained. Another particle, called the **neutron**, was postulated by Rutherford to account for the missing mass and was finally discovered in 1932 by Sir James Chadwick. The neutron was very difficult to detect since it has no charge and therefore could not be manipulated with electric and magnetic fields, as was done with the electron and proton.

ELECTRONIC STRUCTURE OF THE ELEMENTS

Structure of the Atom

The atom is assembled in a regular fashion that is not extraordinarily complex. Difficulty often arises, however, because chemists use several different models of this structure. The symbols and terminology employed in these various models are not always consistent; as a result, the subject appears more complex than it really is.

The reason for the different models is that chemists are interested in different features of the atom each of which is best described in its own unique manner. Some chemists desire to describe all of the electrons in an atom. Others are interested only in the outermost electrons, called the **valence electrons**. Yet others are interested only in the one electron that differentiates a particular atom from the immediately preceding atom in the Periodic Table; this electron is called the **differentiating electron**. In another view, some chemists wish to see how an atom is built, adding one electron after another. Others wish to understand how atoms lose electrons to form ions.

The following sections will show the relationships among the ideas and terminology for these different points of view.

Principal Energy Levels (Shells)

The current model of the atom has the positively charged nucleus surrounded by clouds of electrons at different, characteristic, **principal energy levels**. (They are often called principal shells; but since the word *shell* implies a fixed orbit for electrons, it is not the favored terminology.) The principal energy level nearest the nucleus is given the number 1, and each succeeding energy level is numbered with the next consecutive integer. The largest element known with 112 electrons, needs only seven principal energy levels to hold all of its electrons. The number of the principal energy level is given the symbol n and is called the **principal quantum number**.

Principal energy levels near the nucleus can hold fewer electrons than principal energy levels further from the nucleus. Each level can hold a maximum number of electrons equal to $2n^2$. From this fact we can calculate that the first four principal energy levels can hold 2, 8, 18, and 32 electrons, respectively. The last three principal energy levels could hold 50, 72, and 98 electrons, but they are not completely filled.

(The old-style notation used upper-case letters to designate the principal energy levels. The first four were *K*, *L*, *M*, and *N*. This terminology is obsolete but is often used by scientists using and studying X rays.)

Energy Sublevels (Subshells)

Each principal energy level within an atom contains one or more **sublevels**, which may be called subshells. The number of sublevels possible in each principal energy level is equal to the value of n for that level. For instance, the third principal energy level, where $n = 3$, may contain a maximum of three sublevels. For the 112 known elements, only four sublevels are actually *used*. The fifth, sixth, and seventh subsublevels (for $n = 5, 6$, or 7) are theoretically possible but are not currently needed.

Sublevels are numbered with consecutive whole numbers starting with 0. These numbers are called the **azimuthal quantum numbers**, ℓ. The value of ℓ can never be greater than $n - 1$. Numbered sublevels 0, 1, 2, and 3 are also given corresponding letter designations: *s*, *p*, *d*, and *f*.

Table 1.2 shows the sublevels possible for each principal energy level. It is important to remember that a sublevel will not exist unless the atom has enough electrons to occupy at least part of the sublevel.

TABLE 1.2. SUBLEVELS IN THE ATOM

Principal Energy Level Number	Sublevel Number (ℓ)	Sublevel Letter
1	0	*s*
2	0, 1	*s*, *p*
3	0, 1, 2	*s*, *p*, *d*
4	0, 1, 2, 3	*s*, *p*, *d*, *f*
5	0, 1, 2, 3	*s*, *p*, *d*, *f*
6	0, 1, 2	*s*, *p*, *d*
7	0	*s*

As Table 1.2 indicates, each principal energy level has an *s* sublevel, and all except levels 1 and 7 have *p* sublevels. The *d* and *f* sublevels are also found in more than one principal energy level. To distinguish one sublevel from another, chemists usually combine the principal quantum number with the sublevel letter in order to indicate the principal energy level in which the sublevel occurs. For example, the designation 4*p* indicates a *p* sublevel in the fourth principal energy level.

Orbitals

Each sublevel of the atom may contain one or more electron orbitals. An **orbital** is defined as a region of space that has a high electron density. Each orbital may contain a maximum of two electrons. To share an orbital, two electrons must have opposite spins; two electrons that share an orbital are said to be paired. An orbital is designated as *s*, *p*, *d*, or *f* based on the sublevel in which it is found.

The number of orbitals within a sublevel depends on the azimuthal quantum number, ℓ, of the sublevel and is equal to $(2\ell + 1)$. Therefore an *s* sublevel has only one *s* orbital, a *p* sublevel can have three *p* orbitals, a *d* sublevel can have five *d* orbitals, and seven *f* orbitals are possible in an *f* sublevel.

Table 1.3 summarizes this information.

TABLE 1.3. ORBITALS IN THE ATOM

Sublevel Number (ℓ)	Sublevel Letter	Number of Orbitals ($2\ell + 1$)	Number of Electrons (per sublevel)
0	*s*	1	2
1	*p*	3	6
2	*d*	5	10
3	*f*	7	14

This table also shows why the principal energy levels contain 2, 8, 18, and 32 electrons, respectively. The first principal energy level has only one ***s*** orbital and therefore holds only 2 electrons. The second principal energy level has ***s*** and ***p*** orbitals that hold 2 + 6 or 8 electrons for the second level. The third principal energy level holds 2 + 6 + 10 or 18 electrons, and the fourth principal energy level holds 2 + 6 + 10 + 14 or 32 electrons

In addition to the letter designation ***s, p, d,*** or ***f,*** each orbital is given a number called the **magnetic quantum number**, m_ℓ. The possible values of m_ℓ range from $-\ell$ to $+\ell$, including 0, as Table 1.4 shows.

TABLE 1.4. POSSIBLE VALUES OF m_ℓ

Orbital	m_ℓ Values
s	0
p	−1, 0, + 1
d	−2, −1, 0, +1, +2
f	−3, − 2, −1, 0, +1, +2, +3

The designation of a sublevel also tells the chemist the shape of the orbitals within that sublevel. For *s* sublevels the shape of the electron cloud is spherical. For the *s* sublevel in the first principal energy level, the highest electron density is found within a sphere that is 53 picometers from the nucleus, as Bohr predicted. In the second and higher numbered principal energy levels, the *s* sublevel has a high electron density at the expected distance from the nucleus ($n \times 53$ pm) and also has an appreciable electron density at all of the lower energy level radii. Figure 1.6 illustrates this feature of the *s* sublevels.

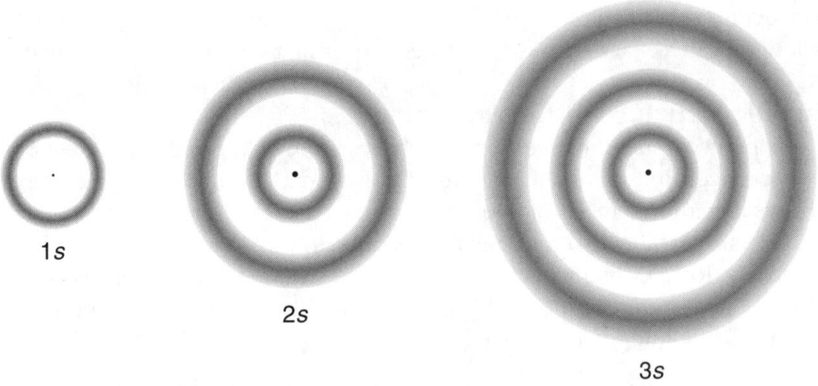

FIGURE 1.6. *Diagrams of the 1s, 2s, and 3s orbitals, showing that electron density for the 2s and 3s orbitals occurs not only at the radius, as expected, but also at intermediate levels.*

A p orbital has a dumbbell shape, with the electron density being greatest in two lobes on either side of the nucleus. There are three p orbitals in each p sublevel. Each is oriented along a different axis, as shown in Figure 1.7. These orbitals may be designated as p_x, p_y, and p_z.

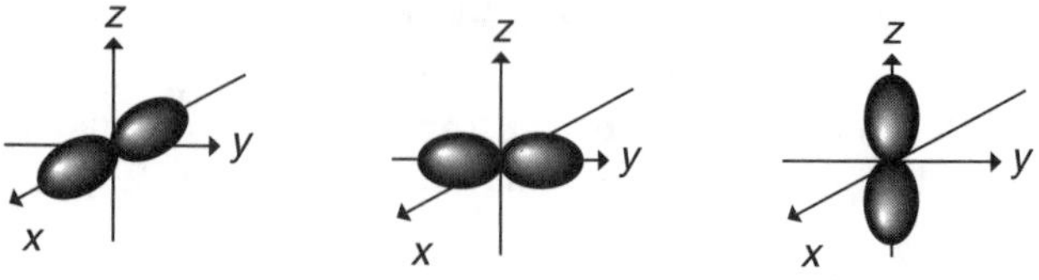

FIGURE 1.7. *Diagrams of the three* p *orbitals aligned with the* x-, y-, *and* z-*axes.*

The five d orbitals in a d sublevel have the shapes shown in Figure 1.8. Often subscripts indicate the general location of the orbitals on the x-, y-, and z-axes.

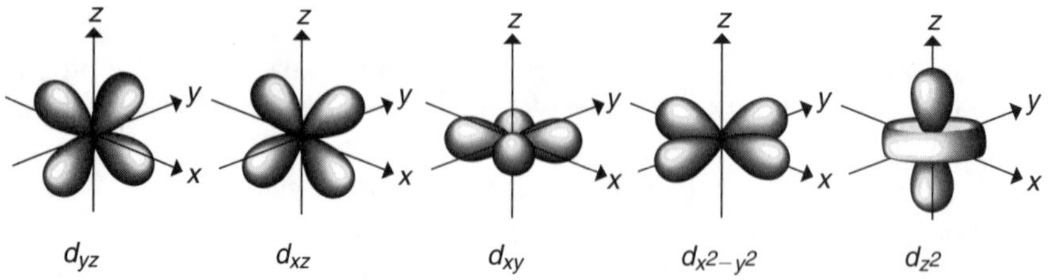

FIGURE 1.8. *Shapes of the five* d *orbitals in the atom. Four of the shapes are identical; the fifth is very distinctive.*

The seven f orbitals have slightly more complex shapes than the d orbitals. Knowledge of the exact shapes is not needed in first-year chemistry courses.

Electronic Structure of the Atom

At this point we need to see how the information in the preceding sections is used to develop a complete picture of the atom. The underlying principle governing the arrangement of the electrons is based on the energy of each orbital. Electrons fill the orbitals that have the lowest energy first, much as water always flows downhill. While the numerical values of the energies of the orbitals are not important here, the order, from lowest to highest energy, is important. The sequence is as follows:

1*s* 2*s* 2*p* 3*s* 3*p* 4*s* 3*d* 4*p* 5*s* 4*d* 5*p* 6*s* 4*f* 5*d* 6*p* 7*s* 5*f* 6*d*

This is known as the **aufbau**, or **energy order**. The electronic configurations of the elements are listed in Appendix 1. Chromium and copper are exceptions to the aufbau order due to the stability of the completely filled and half filled *d* orbitals. Anions are generally formed by completing the *s* and *p* orbitals. Cations are generally formed by removal of the outermost electrons.

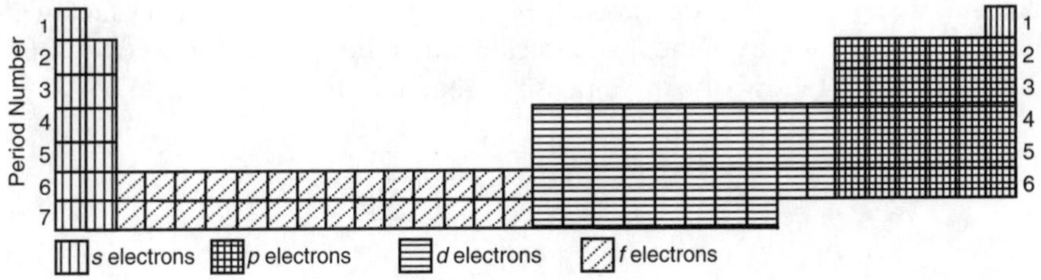

FIGURE 1.9. *Periodic Table divided to show the regions where* s, p, d, *and* f *electrons are the highest energy electrons in each atom.*

Reading across this Periodic Table, we can see that the energy ordering is an integral part of the table. The first period represents the 1*s* electrons. Filling the second period adds the 2*s* and then 2*p* electrons. The third period fills with 3*s* and then 3*p* electrons; the fourth period, with 4*s*, 3*d*, and 4*p* electrons; and the fifth with 5*s*, 4*d*, and 5*p* electrons. Notice that in the fourth and fifth periods the *d* electrons have a number that is 1 less than the period number. The sixth period fills with 6*s*, 4*f*, 5*d*, and 6*p* electrons, and the seventh with 7*s*, 5*f*, and 6*d* electrons. Once again, the *d* electrons have a number that is 1 less than the period number. In addition, the *f* orbitals have numbers that are 2 less than the period number.

Hund's Rule

Hund's rule states that, as a sublevel is filled with electrons, a single electron will occupy each orbital. Once each orbital in a sublevel contains one electron, an additional electron will enter each orbital until every orbital has a pair of electrons that have opposite spins.

Hund's rule makes sense since electrons repel each other strongly because of their negative charges. This repulsion forces each electron into a separate orbital within a sublevel until each orbital has one electron. At this point, additional electrons pair up to fill each orbital with two electrons, until the entire sublevel is filled.

Pairing will occur only if the electrons have opposite spins. A simple view of this is that a spinning electron produces a magnetic field. If the spins are the same, the magnetic fields are aligned and add a magnetic repulsion to the repulsion of the negative charges. If the spins are opposite to each other, the magnetic fields attract, thus reducing the electronic repulsion slightly. The net result is that the opposing-spin configuration is more stable than aligned spins.

Orbital Diagrams

Electron configurations and valence electrons are useful for most purposes in describing the structure of the atom. However, since all of the different orbitals in a sublevel are lumped together, some detail is lost. To show that detail, orbital diagrams are often used. An orbital diagram shows each of the orbitals in the valence shell of the atom as a box or circle. Arrows representing electrons are placed in each orbital. The second electron in each orbital has the arrow facing in the opposite direction from the first, indicating that the spins of the electrons are paired. Figure 1.10 shows the three possible situations for an orbital.

FIGURE 1.10. *Orbital boxes that represent an empty orbital, an unpaired electron, and a pair of electrons, respectively.*

Orbital diagrams are used mainly to describe the valence electrons since all of the inner electrons are paired. At times, however, the *d* electrons are also shown in these diagrams.

The electrons are added to each sublevel, starting with the 1*s* sublevel. The arrow pointing upward traditionally represents the first electron in each orbital, and these arrows are drawn first. Then the downward arrows, representing the electrons of opposite spin, are added to complete the sublevel before a new sublevel starts to fill. Figure 1.11 shows the orbital diagrams of the first ten elements in the Periodic Table.

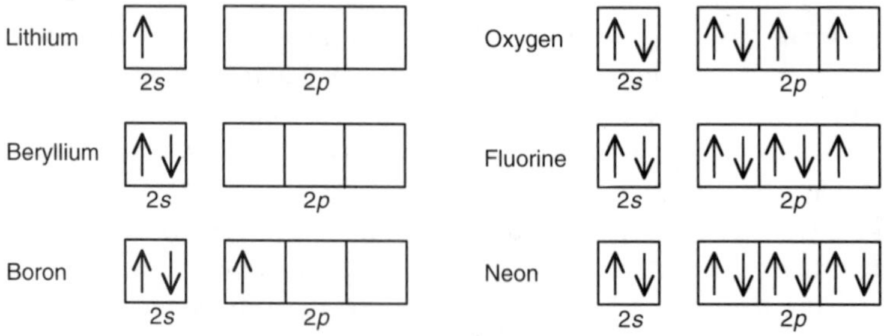

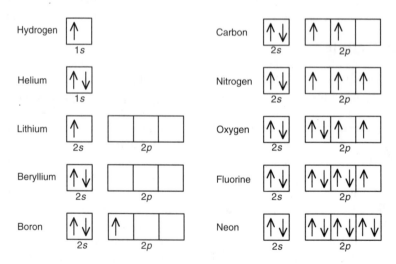

FIGURE 1.11. *Orbital diagrams for the valence electrons of the first ten elements.*

At times we want the orbital diagram to show energy differences between the orbitals. This can be done by positioning the orbitals as shown in Figure 1.12 to indicate that the 2*p* orbitals are higher in energy than the 2*s* orbitals.

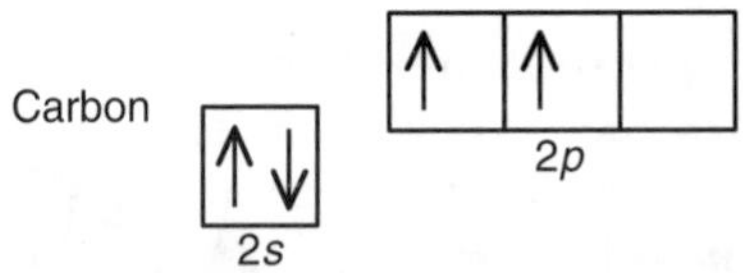

FIGURE 1.12. *Orbital diagram showing that the* 2s *valence electrons in carbon have lower energy than the* 2p *valence electrons.*

Valence Electrons

In many instances chemists are interested in the outermost electrons in an atom. These are located in the principal energy level with the highest number. In a practical sense, these **valence electrons** include only the *s* and *p* electrons of the atom. For any given atom the principal energy level of the outer *d* electrons will always be 1 less than the principal energy level of the *s* and *p* electrons. Similarly for the *f* electrons, the principal energy level is always 2 less than that of the outer *s* and *p* electrons.

Determining the number of valence electrons for any atom involves counting the groups (columns) from the left of the Periodic Table to the element of interest; *d* and *f* electrons are not counted.

Valence electrons are often shown as dots surrounding the symbol of an atom, as in Figure 1.13.

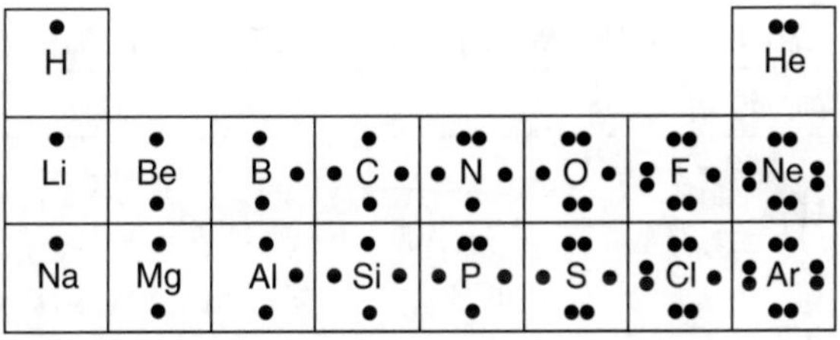

FIGURE 1.13. *Representation of valence electrons as dots around the atomic symbol.*

Quantum Numbers

Schrödinger developed the wave equations that describe the probabilities of where the electrons are located in the atom. These equations require three integers, called **quantum numbers**, to describe each electron. A fourth quantum number was needed to satisfy the Pauli exclusion principle. These four quantum numbers have the following definite rules for their possible values:

1. The **principal quantum number** (n) may have any integer value starting from 1. This represents the principal energy level of the atom in that the electron is located and is related to its average distance from the nucleus.

2. The **azimuthal quantum number** (ℓ) may have any number from 0 up to 1 less than the current value of n ($\ell = 0, \ldots, n - 1$). This designates the sublevel of the electron. It also represents the shape of the orbitals in the sublevel.
3. The **magnetic quantum number** (m_ℓ) may be any integer, including 0, from $-\ell$ to $+\ell$ ($m_\ell = -\ell, \ldots 0, \ldots +\ell$). This quantum number designates the orientation of the orbitals in space.
4. The **spin quantum number** (m_s) may be either $+\frac{1}{2}$ or $-\frac{1}{2}$. This represents the "spin" of an electron. For electrons to pair up within an orbital, one electron must have a $+\frac{1}{2}$ value and the other a $-\frac{1}{2}$ value. This quantum number is not needed for the wave equations, but it is required to satisfy the Pauli exclusion principle.

Finally, the **Pauli exclusion principle** states that no two electrons in the same atom may have the same four quantum numbers.

When designating the quantum numbers for the electrons in an atom, it is common to use the lowest possible values of the first three quantum numbers first. Also, traditionally the positive spin quantum numbers are used before the negative ones. Therefore the hydrogen atom has the lowest possible quantum numbers: 1, 0, 0, $+\frac{1}{2}$. The quantum numbers for the first 20 elements are listed in Table 1.5 It is important to know what a set of quantum numbers appropriately formulated according to the above rules looks like.

TABLE 1.5. QUANTUM NUMBERS FOR THE FIRST 20 ELEMENTS

Element	n	ℓ	m_ℓ	m_s
H	1	0	0	$+\frac{1}{2}$
He	1	0	0	$-\frac{1}{2}$
Li	2	0	0	$+\frac{1}{2}$
Be	2	0	0	$-\frac{1}{2}$
B	2	1	−1	$+\frac{1}{2}$
C	2	1	0	$+\frac{1}{2}$
N	2	1	+1	$+\frac{1}{2}$
O	2	1	−1	$-\frac{1}{2}$
F	2	1	0	$-\frac{1}{2}$
Ne	2	1	+1	$-\frac{1}{2}$
Na	3	0	0	$+\frac{1}{2}$
Mg	3	0	0	$-\frac{1}{2}$
Al	3	1	−1	$+\frac{1}{2}$
Si	3	1	0	$+\frac{1}{2}$
P	3	1	+1	$+\frac{1}{2}$
S	3	1	−1	$-\frac{1}{2}$
Cl	3	1	0	$-\frac{1}{2}$
Ar	3	1	+1	$-\frac{1}{2}$
K	4	0	0	$+\frac{1}{2}$
Ca	4	0	0	$-\frac{1}{2}$

Example

Determine whether each of the following sets of quantum numbers is possible or impossible, and explain your reasoning.

(a) 1, 0, 0, ½
(b) 1, 3, 0, ½
(c) 3, 2, 0, ½
(d) 2, 2, 2, −½
(e) 3, 2, 2, −½
(f) 3, 1, −1, ½
(g) 4, 2, −2, −½
(h) 4, 4, 0, ½
(i) 3, 2, 1, 0
(j) 1, 1, 1, ½
(k) 6, 4, −4, −½
(l) 5, 3, −2, ½
(m) 2, 0, 1, −½
(n) 5, 0, 0, ½
(o) 3, 1, 2, −½

Solution: Sets (a), (c), (e), (f), (g), (k), (l), and (n) are correct. The others disobey one or more of the rules: (b), (d), (h), and (j) violate rule 2; (m) and (o) violate rule 3; (i) violates rule 4.

Relationship of Quantum Numbers to the Periodic Table

It must be remembered that each electron in an atom is described by a set of four quantum numbers. Calcium, for instance, has 20 sets of quantum numbers, one set for each of its 20 electrons. The rules for the sequence in which the quantum numbers are assigned are somewhat arbitrary. Consider, for example, the electron in the hydrogen atom. There is no reason for its spin quantum number to be +½. In fact, since the hydrogen atoms are energetically equal, half of them will have a spin quantum number of +½ and half of −½. Similarly, there is no requirement that the *p* orbitals fill with quantum number $m_\ell = -1$ first. In fact, all three values are equally probable. Energy levels that are the same are said to be degenerate, and all configurations are equally probable.

Referring to pages 16–17 and the discussion of electron configurations, we can see that a knowledge of the first two quantum numbers, n and ℓ, give us the period and sublevel (*s*, *p*, *d*, or *f*) to which the electron belongs. Because the energies of the various possibilities of m_ℓ and m_s are degenerate, they do not help in identifying the atom to which the electron belongs. For example, consider an atom that contains an electron with the quantum numbers 3, 1, 0, −½. All that can be said about this electron is that it may exist in all elements with atomic numbers greater than 13 (Al). If this electron is specified as the last electron in a particular element, we may narrow the possibilities to the elements aluminum to argon, Finally, strictly following the conventional rules above, we would define this electron as the last one to fill the chlorine atom.

Any atom will have electrons that have all of the possible quantum numbers for the completely filled sublevels. If an atom has an incompletely filled sublevel, the electrons in that sublevel can have any of the quantum numbers associated with the sublevel as long as Hund's rule and the Pauli exclusion principle are obeyed. For instance, carbon has completely filled 1*s* and 2*s* subshells, and each carbon atom has electrons with quantum numbers of 1, 0, 0, $+\frac{1}{2}$; 1, 0, 0, $-\frac{1}{2}$; 2, 0, 0, $+\frac{1}{2}$ and 2, 0, 0, $-\frac{1}{2}$. The two additional electrons in the incompletely filled *p* sublevel may be any of the following:

2, 1, −1, $+\frac{1}{2}$	2, 1, −1, $+\frac{1}{2}$
2, 1, 0, $+\frac{1}{2}$	2, 1, 0, $+\frac{1}{2}$
2, 1, 1, $+\frac{1}{2}$	2, 1, 1, $+\frac{1}{2}$

Hund's rule requires that the third quantum number in the selected pair cannot be the same, and the Pauli exclusion principal states that the two electrons cannot have all four quantum numbers the same.

Significance of Quantum Numbers

The four quantum numbers are often regarded as mere items to be memorized, along with the rules for obtaining valid sets. It is important to remember, however, that they represent real physical properties of the atom. These numbers may be summarized in simplified form as follows:

1. The principal quantum number (n) represents the average distance of the electron from the nucleus, or the size of the principal energy level.
2. The azimuthal quantum number (ℓ) represents the shape(s) of the orbitals within the sublevel, as shown in Figures 1.6, 1.7, and 1.8 (pages 15–16).
3. The magnetic quantum number (m_ℓ) represents the orientation of the orbital in space.
4. The spin quantum number (m_s) represents the spin of the electron.

PROPERTIES OF THE ELEMENTS

Atomic Masses

The atomic masses listed in the Periodic Table are all relative to the mass of the carbon-12 isotope of carbon. As such, masses listed in the Periodic Table have no units. Laboratory chemists often assign units of grams to these relative masses for calculation purposes, as is shown in Chapter 9. Other chemists may assign units of kilograms, pounds, or even tons to the masses for large industrial uses.

A second concept to remember is that we may speak of the exact mass of a particular isotope of an atom, or the weighted average of the masses of all naturally occurring isotopes. This **weighted average** is the relative mass listed in the Periodic Table; it usually does not represent the relative mass of any atom of that element since most naturally occurring elements have more than one isotope. Fortunately, the percentage of each isotope is relatively constant throughout the world. As a result, the **measured atomic mass** is the weighted average of the masses of the individual isotopes and their relative abundances.

A weighted average is the sum of the isotope masses multiplied by their natural abundances. The natural abundance of a particular isotope is the fraction of all atoms of an element that have the same number of neutrons.

$$\text{Weighted average} = \sum_{i=1}^{n} (\text{mass of isotope } i)(\text{abundanceof isotope } i)$$

Example Magnesium has three isotopes: ^{24}Mg, ^{25}Mg, and ^{26}Mg. They occur naturally with abundances of 78.6 percent, 10.1 percent, and 11.3 percent respectively. The exact masses of these isotopes are 23.9924, 24.9938, and 25.9898. What is the weighted average of the three isotopic masses?

Solution: The weighted average is the sum of the masses the three isotopes multiplied by their fractions in a natural sample:

$$\begin{aligned}\text{Weighted average} &= (\text{mass of } ^{24}\text{Mg})(\text{abundance of } ^{24}\text{Mg})\\ &+ (\text{mass of } ^{25}\text{Mg})(\text{abundance of } ^{25}\text{Mg})\\ &+ (\text{mass of } ^{26}\text{Mg})(\text{abundance of } ^{26}\text{Mg})\end{aligned}$$

Since the abundances are usually given as percentages, they must be converted to decimal fractions by dividing by 100 before using them in the equation. When the appropriate numbers are substituted, the equation becomes

$$\begin{aligned}\text{Weighted average} &=\\ &(23.9924)(0.786) + (24.9938)(0.101) + (25.9898)(0.113)\\ &= 18.86 + 2.52 + 2.94\\ &= 24.32\end{aligned}$$

You may have noticed that the atomic masses in the Periodic Table may have four, five, six, or seven significant figures. A greater number of significant figures suggests that an atomic mass is known with more certainty than one with fewer significant figures. Experimental difficulties are one source of uncertainty in determining atomic masses. Another source of uncertainty is the variation in natural isotopic abundance. If the natural isotopic abundance is fairly constant, more significant figures may be obtained for atomic masses.

EXERCISE

The atomic mass of bromine is listed as 79.9 in the Periodic Table. There are no isotopes of bromine with a mass of 80. Suggest an explanation for these facts.

Answer

A mass listed in the Periodic Table rarely represents the mass of any specific isotope. The atomic mass of bromine (79.9) can be obtained from many possible combinations of isotope masses and their relative abundances. In fact, bromine has only two natural isotopes (^{79}Br and ^{81}Br), which occur in almost equal amounts in nature.

Ionization Energy

Using the Rydberg equation, we can calculate the energy needed to completely remove an electron from any level in the hydrogen atom. The process of removing an electron from an atom is called **ionization**, and the energy needed to do this is the **ionization energy**. Since moving to higher numbered orbits moves the electron away from the nucleus, moving the electron to n = infinity is the same as removing the electron from the atom. By substituting the number of the starting orbit of the electron for n_1 and infinity for n_2, we can solve the Rydberg equation for the frequency and then calculate the ionization energy using Planck's equation, $E = h\nu$.

Example

What is the ionization energy of a hydrogen atom? What is the value for 1 mole of hydrogen? (1 mole of hydrogen atoms = 6.02 × 10^{23} hydrogen atoms)

Solution: The normal hydrogen atom has its electron in the first orbit, or $n = 1$. When an atom is ionized, the electron is removed. This is equivalent to moving the electron to n = infinity. Solving the Rydberg equation for the energy yields

$$\begin{aligned}\nu &= 3.2881 \times 10^{15}\ \text{s}^{-1}\left(\frac{1}{n_1^2} - \frac{1}{n_2^2}\right)\\ &= 3.2881 \times 10^{15}\ \text{s}^{-1}\left(\frac{1}{1^2} - \frac{1}{\infty^2}\right)\\ &= 3.2881 \times 10^{15}\ \text{s}^{-1}\end{aligned}$$

$$\begin{aligned}E = h\nu &= (6.626 \times 10^{-34}\ \text{J s})(3.2881 \times 10^{15}\ \text{s}^{-1})\\ &= 2.179 \times 10^{-18}\ \text{J}\end{aligned}$$

To obtain the normally quoted ionization energy, it is necessary to multiply by 6.02 × 10^{23} to find the energy needed to ionize 1 mol of hydrogen atoms. The result is 1312 kJ.

The ionization energies of the other elements may be measured by a variety of methods. The energy needed to remove only one electron is called the first ionization energy. The second, third, and higher ionization energies can also be measured for the successive removal of all valence electrons.

Removal of each successive electron from an atom requires an increased amount of energy because the removal of one electron results in a stronger attraction between the nucleus and the remaining electrons, as illustrated in Table 1.6.

TABLE 1.6. IONIZATION ENERGIES (kJ/MOL) OF SELECTED ELEMENTS

Metal	First Electron	Second Electron	Third Electron
Na	496	4563	6913
Mg	737	1450	7731
K	419	3051	4411
Ca	590	1145	4912

In addition, Table 1.6 shows that the Na^+ and K^+ ions have very low first ionization energy and very high second ionization energies. Under normal circumstances the +2 ions of these elements will not form. For Mg^{2+} and Ca^{2+} we see that the first two ionization energies are low and the third much higher. The result is that the +3 ions of calcium and magnesium are not found. A complete tabulation of ionization energies shows similar behavior for other elements.

Finally, the first ionization energies of the elements follow a regular pattern: first, they increase from left to right across a period in the Periodic Table; second, they increase from the bottom to the top of a group. In general, the ionization energies are lowest in the lower left corner of the Periodic Table and increase in going toward the upper right corner, a feature known as a **diagonal relationship**.

The ionization energy is one of the factors involved in determining the electronegativity of an element. Electronegativity is an important concept to master for a complete understanding of chemical properties and reactions.

Electronegativity

To understand bond and molecular polarities, we need some measure of how strongly different atoms attract electrons. Linus Pauling developed just such a measure and called it the **electronegativity** of an element. The Periodic Table in Figure 1.14 gives the electronegativity value for each element. We see that the electronegativity increases from left to right in each period (row). In addition, the electronegativity within any group (column) of the Periodic Table increases from the bottom of the group to the top. This leads to one of the important **diagonal relationships** in the Periodic

H 2.1																
Li 1.0	Be 1.5											B 2.0	C 2.5	N 3.1	O 3.5	F 4.0
Na 1.0	Mg 1.3											Al 1.5	Si 1.8	P 2.1	S 2.4	Cl 2.9
K 0.8	Ca 1.1	Sc 1.2	Ti 1.3	V 1.5	Cr 1.6	Mn 1.6	Fe 1.7	Co 1.7	Ni 1.8	Cu 1.8	Zn 1.7	Ga 1.8	Ge 2.0	As 2.2	Se 2.5	Br 2.8
Rb 0.8	Sr 1.0	Y 1.1	Zr 1.2	Nb 1.3	Mo 1.3	Tc 1.4	Ru 1.4	Rh 1.5	Pd 1.4	Ag 1.4	Cd 1.5	In 1.5	Sn 1.7	Sb 1.8	Te 2.0	I 2.5
Cs 0.7	Ba 0.9	La 1.1	Hf 1.2	Ta 1.4	W 1.4	Re 1.5	Os 1.5	Ir 1.6	Pt 1.5	Au 1.4	Hg 1.5	Tl 1.5	Pb 1.6	Bi 1.7	Po 1.8	At 2.2
Fr 0.7	Ra 0.9	Ac 1.0														

FIGURE 1.14. *Periodic Table showing the electronegativities of the elements.*

Table. The closer an atom is to fluorine, the greater its electronegativity; the closer it is to francium, the smaller its electronegativity. In other words, the electronegativity increases from the lower left corner of the Periodic Table up to the upper right corner.

This diagonal relationship allows the chemist to determine quickly which end of a bond is negative and which end is positive. The element that is closer to fluorine will be the negative end of the bond, and the element further from fluorine will be the positive end. The polarities are indicated with the symbols $\delta+$ and $\delta-$ to indicate partially positive and partially negative atoms, respectively. For example, the B-O bond is written as $^{\delta+}$B- - - - -O$^{\delta-}$ to show that the boron atom is positive compared to the oxygen atom.

EXERCISE

Predict the positive and negative ends of each of the following bonds, using the symbols $\delta+$ and $\delta-$:

(a) S—O

(b) C—N

(c) S—P

(d) C—F

(e) Si—O

(f) H—Br

(g) H—O

Answers

Using the diagonal relationships in the Periodic Table, we can write the following:

(a) $^{\delta+}$S—O$^{\delta-}$

(b) $^{\delta+}$C—N$^{\delta-}$

(c) $^{\delta-}$S—P$^{\delta+}$

(d) $^{\delta+}$C—F$^{\delta-}$

(e) $^{\delta+}$Si—O$^{\delta-}$

(f) $^{\delta+}$H—Br$^{\delta-}$

(g) $^{\delta+}$H—O$^{\delta-}$

Periodic Properties of the Elements

Two outlines of the Periodic Table are shown in Figures 1.15 and 1.16. A row in the Periodic Table is called a period, and a column is termed a group. In Figure 1.15 the Periodic Table is arranged in the conventional manner with the lanthanide and actinide series placed below the body of the table. Figure 1.16 places the lanthanide and actinide elements where they normally belong. This form is rarely used, however, because the lanthanide and actinide elements seldom participate in common chemical reactions and because the boxes become too small to read.

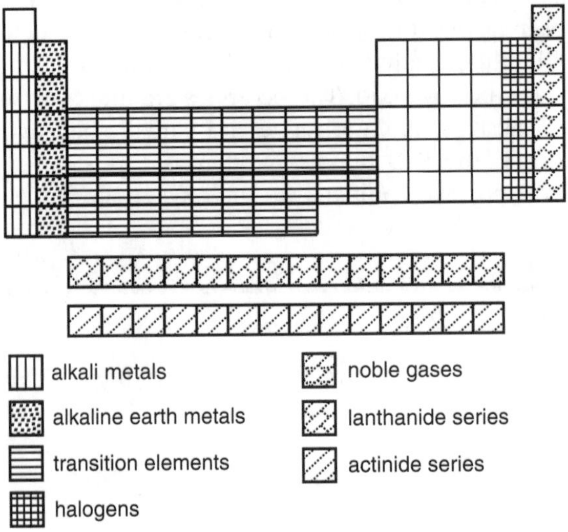

FIGURE 1.15. *Periodic Table showing common classes of elements.*

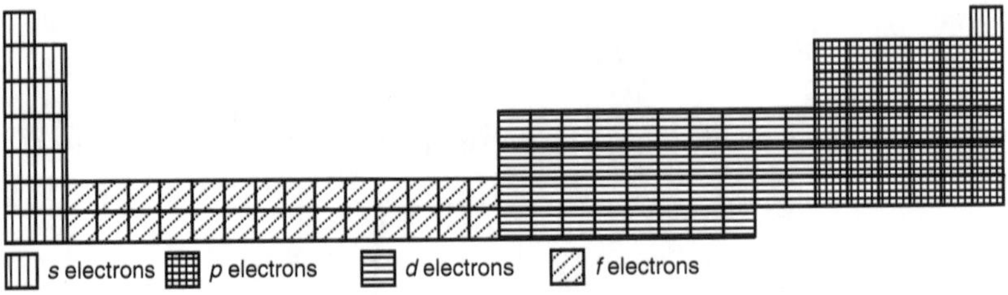

FIGURE 1.16. *Extended form of the Periodic Table. The shading shows the outermost electrons and the way they are grouped.*

Chemists often speak of related classes of elements, such as the alkali metals, halogens, or noble gases. The location of seven of these classifications is shown by the shadings in Figure 1.15. A knowledge of the names of these classes is important; the names are referred to frequently in other chapters.

A review of electronic structure indicates the reasons for the chemical similarities and differences of the elements. First, each column or group has the same number and type of valence electrons, resulting in the chemical similarities of these elements. For instance, the noble gases all have completely filled *s* and *p* sublevels that give them extraordinary stability. The halogens are missing one *p* electron; otherwise they would be electronically the same as the noble gases. The halogens tend to enter reactions that enable them to gain that one *p* electron. Similarly, the alkali metals and the alkaline earth metals have one and two *s* electrons, respectively. They readily lose these electrons to become electronically the same as a noble gas.

Of the 112 elements in the Periodic Table only two, mercury and bromine, are liquids under normal conditions. The noble gases, hydrogen, nitrogen, oxygen, fluorine, and chlorine are gases at room temperature. The remaining elements are solids.

Metals dominate the elements in the Periodic Table. In most periodic tables a heavy line divides the metals from the nonmetals, as shown in Figure 1.17.

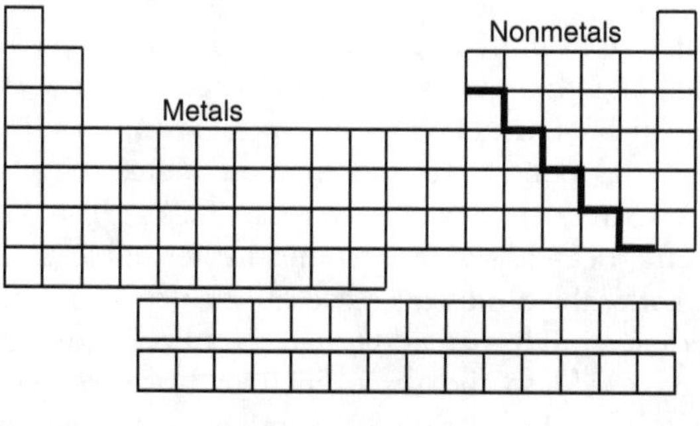

FIGURE 1.17. *Location of metals and nonmetals in the Periodic Table. The heavy line divides the two. Elements along the line have metallic and nonmetallic properties and are called metalloids.*

Elements along this line are often termed metalloids since they exhibit some properties of metals and some properties of nonmetals. The metallic character of the elements increases from the top of the Periodic Table to the bottom. The group headed by nitrogen shows this clearly; nitrogen and phosphorus are nonmetals, arsenic and antimony are metalloids, and bismuth is a metal.

As previously stated, a majority of the elements are metals. Metallic properties depend on loosely held valence (outermost) electrons, and only elements that are close to fluorine in the Periodic Table have electrons that are held strongly enough to cause the elements to be nonmetals.

Finally, most of the elements in the Periodic Table may be considered as individual atoms. A few elements, however, exist naturally as diatomic molecules: H_2, O_2, N_2, and the halogens. Others, notably sulfur and phosphorus, exist in polyatomic units such as S_8 and P_4, but are commonly represented as single atoms, S and P, in chemical reactions.

Other properties of the elements vary regularly throughout the periodic table. In designing the Periodic Table, Mendeleev organized the groups based on similar chemical properties. He then found that many physical and chemical properties varied in a regular manner within these same groups and periods. For instance, as mentioned above, the metallic character of the elements increases from the top to the bottom of a group. The melting and boiling points of metals tend to decrease from the top to the bottom of a group; nonmetals, on the other hand, show an increase in their melting and boiling points. Similar trends in electrical properties, densities, and specific heats are also noted within each group. Many other properties also vary regularly, both horizontally and vertically, in the Periodic Table. Some of these are discussed below.

Atomic Radii

Atoms range in size from hydrogen with a radius of 37 picometers (1 pm = 10^{-12} m) to francium with a radius of 270 picometers. In each period (row) of the Periodic Table, the alkali metal has the largest **atomic radius** and the noble gas at the end of the period has the smallest radius. In periods that have transition elements, there is a slight increase in size for the last transition elements and then a decrease in size to the smaller noble gas.

To understand why the atoms decrease in size across a period, we first see that each succeeding atom has an additional electron and an additional proton. The electrons are added to the same shell and are located at a relatively constant distance from the nucleus. We might expect an increase in size because of electrons repelling each other. However, the increasing nuclear charge is a stronger effect, and it attracts the electron clouds closer to the nucleus, thereby decreasing the overall size.

It is relatively easy to understand why the size of atoms increase as we go from the top to the bottom of a group. From the top each row adds an additional shell to the atom's structure. This added outer shell of electrons must be further from the nucleus than the preceeding shell. In addition, the added distance means that the attractive force of the nucleus is reduced.

Ionization Energy

The **ionization energy** is defined as the energy needed to remove an electron from an atom. Energy is always required to remove electrons. Francium, in the lower left corner of the Periodic Table, has the lowest ionization energy, while helium, in the upper right corner, has the highest. In general, a diagonal line from the lower right to the upper left corner of the Periodic Table defines this relationship. We may also conclude that the ionization energy decreases in going down any group in the Periodic Table and increases from left to right across a period in the table.

Table 1.6 lists the ionization energies for several metals. This table shows that very little energy is needed to remove one electron from sodium and potassium, but much more energy is required to remove a second electron. This confirms the observation that sodium and potassium easily form only Na^+ and K^+ ions. For calcium and magnesium, the table shows that ionization of the first two electrons to form Ca^{2+} and Mg^{2+} ions is relatively easy but removal of a third electron requires too much energy to be feasible. The ionization energies of other metals show similar trends.

Electron Affinity

The **electron affinity** is defined as the energy needed to add an electron to an atom. Some atoms readily attract electrons, and their electron affinities have negative values. Most atoms, however, do not accept additional electrons readily, and their positive electron affinities indicate that energy must be used to add the electron.

Fluorine has the highest electron affinity, and francium the lowest. This is another diagonal relationship; it indicates that atoms close to fluorine tend to accept electrons readily and those close to francium do not.

Electronegativity

The concept of **electronegativity** was developed by Linus Pauling to describe the attraction of electrons by individual atoms. Electronegativity is a combination of ionization energy, electron affinity, and other factors. Electronegativities show the same diagonal trend that is evident with ionization energies and electron affinities: fluorine has the highest electronegativity, and francium the lowest. Differences in electronegativity are used to determine how electrons are distributed in molecules, as shown in later chapters. In Figure 1.14 the electronegativities of the elements illustrate the increasing trend from the lower left corner to the upper right corner of the Periodic Table.

Ionic Radii

There are two types of ions, cations and anions. **Cations** are atoms that have lost one or more electrons and carry a positive charge; **anions** are atoms that have gained one or more electrons and carry a negative charge.

Cations are always smaller than their neutral atoms. Many cations have lost an entire shell of electrons and are only about half the size of the neutral atom. A further decrease in **ionic radius** is due to the fact that cations have more protons in the nucleus than electrons.

Anions are always larger than their neutral atoms; many are almost twice as big. We expect some repulsion from the added electron; but since the electron(s) are added to the same shell, this addition does not fully explain the increase in size observed. Chemists reason that the outer electrons in an atom effectively shield each other from the nuclear charge. This shielding decreases the attractive forces, and the electron cloud expands, resulting in the large anion radius.

Allotropes

Most elements in the Periodic Table (e.g., most metals) exist in nature in only one form. Other elements may have multiple forms, called **allotropes**, which often differ widely in their physical and chemical properties.

Carbon is the central atom in organic compounds. It exists as an element in several different forms, called **allotropes**. These allotropes are graphite, which is the "lead" in pencils; diamond, a precious stone; and buckminsterfullerene.

Graphite consists of layers of sp^2-bonded carbon atoms in six-membered rings that form flat sheets. The unhybridized *p* orbitals allow for delocalization of the *p* electrons. Each layer in graphite is weakly attracted to adjacent layers with electrostatic forces. The loosely bound pi electrons allow electrical conduction along the plane of graphite, but graphite is a nonconductor through the planes. Because of the weak attractive forces between the planes, graphite is an excellent dry lubricant; each plane slips easily by the others. This slippage between planes explains what happens as the graphite in a pencil is transferred to paper.

Diamond consists of tetrahedrally bonded carbon atoms (sp^3 hybrid) in a covalent crystal of extraordinary hardness. In a perfect diamond, each carbon atom is bonded to four other carbon atoms. Imperfections in diamonds represent either incorporated impurities that give the diamond color, specks of graphite that appear as dark spots, or bonding defects whereby a carbon atom is bonded to fewer than four other carbon atoms.

Buckminsterfullerene consists of spheres of 60 carbon atoms covalently bonded together. The existence of C_{60} was inferred from spectral lines observed in astronomical observations of intersteller gas. This allotrope was recently discovered as a natural component of most soot. The structure of buckminsterfullerene is exactly the same as that of the common soccer ball; each seam on the ball is analogous to a carbon-carbon bond, and each intersection to a position of a carbon atom.

Carbon has a total of six electrons, four of which are valence electrons. Carbon always forms four bonds, which may be combinations of single, double, and triple bonds. This element never has a nonbonding pair of electrons. Carbon can form four single bonds with a tetrahedral (sp^3) structure. It can bond to three other atoms in a trigonal planar (sp^2) structure, which involves one double bond and two single bonds. Finally, carbon can bond with two other atoms in a linear (*sp*) structure. These compounds contain either two double bonds, as in CO_2, or one single and one triple bond, as in HCN.

The most important feature of the carbon atom is that it bonds with other carbon atoms to form chains and rings of various sizes and shapes. No other atom in the Periodic Table can form the variety of structures that carbon can.

Arsenic, oxygen, phosphorus, selenium, sulfur, and tin are other elements that have more than one allotropic form. Arsenic in its gray allotrope is metallic in character, while the yellow allotrope of As_4 molecules is nonmetallic. Oxygen exists as O_2 and also as ozone, O_3. Ozone is a pollutant at ground level but in the stratosphere it protects us from ultraviolet radiation. In the same group in the Periodic Table is phosphorus, which has several allotropes; one is white phosphorus (P_4), which burns spontaneously when exposed to air. The red and black allotropes of phosphorus are long chains of atoms that are more stable. Sulfur has many allotropes, one of which is the S_8 molecule; others consist of long chains of sulfur atoms. Selenium, in the same group as sulfur, has an Se_8 allotrope that is nonmetallic and another, semimetallic allotrope. Tin has three allotropes, one that has a distinctly metallic look and two that are white crystals.

THE BOHR ATOM

Electromagnetic Radiation

The Electromagnetic Spectrum

Until the beginning of the twentieth century, the internal structure of the atom was unknown. Nils Bohr advanced a solar system model of electrons revolving around the nucleus, as the planets revolve around the sun. The development of his model depended upon the understanding of electromagnetic radiation and atomic spectra.

Visible light is the most familiar form of electromagnetic radiation. All other electromagnetic radiation is invisible to the human eye but can be detected by a variety of instruments. Table 1.7 lists the various types of electromagnetic radiation, ranging from cosmic rays with very high energy to the lowest energy radio waves. The wavelengths vary from less than a picometer (10^{-12} m) to more than 1 kilometer. Chemists are particularly interested in the ultraviolet, visible, and infrared regions of the spectrum.

TABLE 1.7. REGIONS OF THE ELECTROMAGNETIC SPECTRUM

Common Name	Wavelength Range, λ (nm)	Frequency Range, ν (s^{-1})	Wavelength Range, λ (common units)
Cosmic rays	0.00005	6×10^{21}	50 fm
Gamma rays	0.0005 – 0.14	$6 - 0.02 \times 10^{20}$	0.5 – 140 pm
X rays	0.01 – 10	$300 - 0.3 \times 10^{17}$	10 – 1000 pm
Vacuum ultraviolet	10 – 200	$3 - 0.15 \times 10^{16}$	10–200 nm
Ultraviolet	200 – 350	$1.5 - 0.85 \times 10^{15}$	200–350 nm
Visible	350 – 700	$8.5 - 4 \times 10^{15}$	350–700 nm
Infrared	700 – 50,000	$4000 - 6 \times 10^{12}$	16,000–200 cm^{-1}
Microwaves	$10^6 - 10^7$	$3 - 0.3 \times 10^{11}$	1 – 10 mm
Radio waves	$10^7 - 10^{12}$	$3 - 0.00003 \times 10^{10}$	1 cm – 100 m

Figure 1.18 illustrates the electromagnetic spectrum with a logarithmic scale based on frequency and wavelength. In this representation the visible region is rather small and is surrounded by the ultraviolet and infrared regions. The names of the latter two spectral regions help to remind us that violet (or blue) is at the high-energy end of the visible spectrum and that red is at the low-energy end. The colors of the visible spectrum are violet (highest energy), blue, green, yellow, orange, and red (lowest energy).

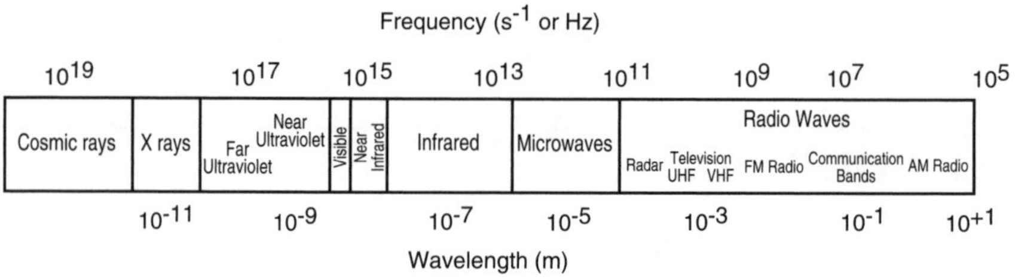

FIGURE 1.18. *Pictorial representation of the electromagnetic spectrum, showing approximate wavelengths and frequencies.*

Wavelength, Frequency, and Energy of Light

All electromagnetic radiation may be considered as waves that are defined by wavelength (λ) and frequency (ν). The **wavelength** (Figure 1.19) is the distance between two repeating points (either two minima or two maxima) on a sine wave. The **frequency** is defined as the number of waves that pass a point in space each second.

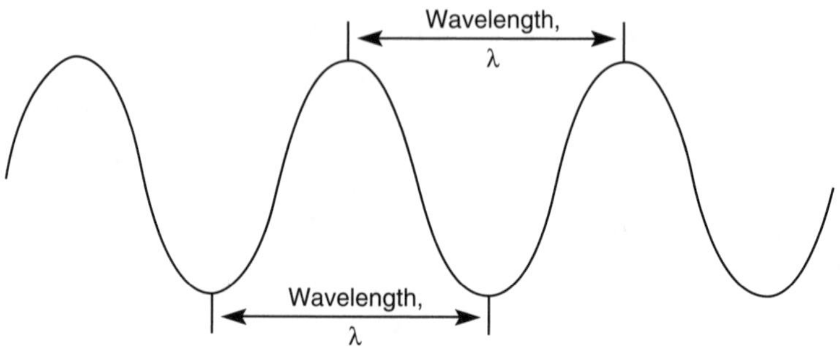

FIGURE 1.19. *Definition of wavelength.*

The wavelength and the frequency of light are inversely proportional to each other and are related by the equation

(Wavelength)(frequency) = speed of light

or

$$\lambda\nu = c$$

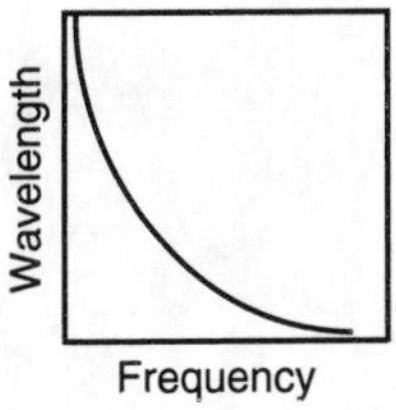

The speed of light is 3.00×10^8 meters per second (m s^{-1}), a number worth remembering. Wavelength has units of meters, often with the appropriate metric prefix (cm, μm, or nm), and frequency has units of reciprocal seconds (s^{-1}), also called Hertz (Hz).

Max Planck found that the energy of electromagnetic waves is proportional to the frequency and inversely proportional to the wavelength. The proportionality constant (*h*) is called Planck's constant; it has a value of 6.62×10^{-34} joule second. (Planck's constant need not be memorized; when required for a problem, it will be given.)

$$E = h\nu = h\frac{c}{\lambda} \qquad (1.1)$$

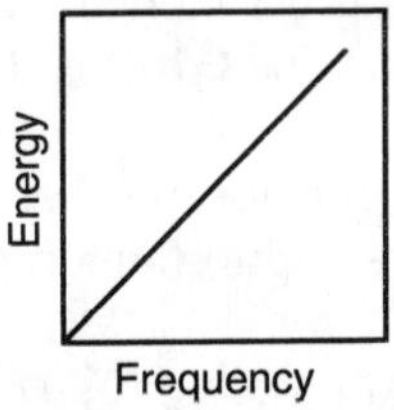

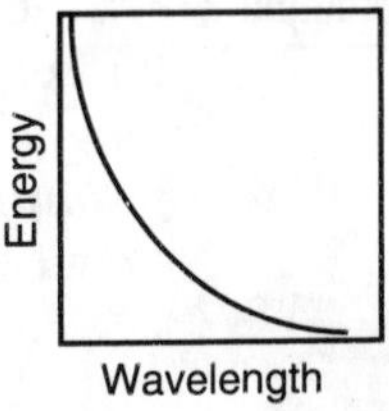

Energy, wavelength and frequency are all related, as shown in the equation above. If the speed of light and Planck's constant are known, only one of these three variables is needed to calculate the others.

Example 1 What are (a) the frequency and (b) the energy of blue light that has a wavelength of 400 nanometers? (Planck's constant = 6.62×10^{-34} J s)

Solution: Substitute the given values into the equation $\lambda\nu = c$:

$$(400 \text{ nm})(\nu) = 3.00 \times 10^8 \text{ m s}^{-1}$$

(a) Substitute 10^{-9} for the prefix *nano-* in the wavelength:

$$(400 \times 10^{-9} \text{ m})(\nu) = 3.00 \times 10^8 \text{ m s}^{-1}$$

Rearrange this equation to solve for the frequency (ν):

$$\nu = \frac{3.00 \times 10^8 \text{ m s}^{-1}}{400 \times 10^{-9} \text{ m}} = 7.50 \times 10^{14} \text{ s}^{-1}$$

(b) The energy of this light can be calculated directly from the frequency or the wavelength by using equation 142.1:

$$\begin{aligned} E &= h\nu = (6.62 \times 10^{-34} \text{ J s})(7.50 \times 10^{14} \text{ s}^{-1}) \\ &= 49.65 \times 10^{-20} \text{ J} \\ &= 4.96 \times 10^{-19} \text{ J} \end{aligned}$$

or

$$E = h\frac{c}{\lambda} = (6.62 \times 10^{-34} \text{ J s})\left(\frac{3.00 \times 10^8 \text{ m s}^{-1}}{400 \times 10^{-9} \text{ m}}\right)$$

$$= 4.96 \times 10^{-19} \text{ J}$$

In solving problems of this type, it is very important that the units cancel properly. The meter units cancel each other since there is one in the numerator and another in the denominator, and the seconds units (s and s^{-1} in the numerator) also cancel.

Example 2 What are (a) the wavelength and (b) the energy of light that has a frequency of 1.50×10^{15} reciprocal seconds (s^{-1})?

Solution: The relationship between the wavelength and the frequency is $\lambda\nu = c$. Substitution of the given values yields

$$\lambda(1.5 \times 10^{15} \text{ s}^{-1}) = 3.00 \times 10^8 \text{ m s}^{-1}$$

(a) Rearrange to solve for the wavelength (λ):

$$\lambda = \frac{3.00 \times 10^8 \text{ m s}^{-1}}{1.5 \times 10^{15} \text{ s}^{-1}}$$

$$= 2.00 \times 10^{-7} \text{ m} = 200 \times 10^{-9} \text{ m}$$

Use the metric prefix *nano-*, 1 nm $= 10^{-9}$ m, to simplify the answer:

$$\lambda = 200 \text{ nm}$$

(b) The energy of this light is calculated from $E = h\nu$. Substituting given data yields

$$E = (6.62 \times 10^{-34} \text{ J s})(1.5 \times 10^{15} \text{ s}^{-1})$$

Canceling units and solving gives the result (remember that s and s^{-1} cancel):

$$E = 9.93 \times 10^{-19}$$

Atomic Spectra of Hydrogen

Visible, ultraviolet, and infrared radiation can be separated into the various wavelengths by focusing the radiation through a triangular prism. The device used for this purpose is called a spectroscope if the light is observed by eye, and a spectrograph if the light is detected by an electronic device and recorded on paper. A spectrometer (or spectrophotometer) is similar to a spectrograph except that the information is read from a meter. Modern instruments often have digital displays and store data in computer memories.

Using spectroscopes, scientists found that when gaseous elements are heated they emit light. The light emitted consists of discrete wavelengths that can be measured with great accuracy. Each element emits a unique pattern of spectral wavelengths. These patterns, called atomic spectra, are used to identify elements. Figure 1.20 illustrates the spectrum of the hydrogen atom in the ultraviolet, visible, and infrared spectral regions. Johann Balmer discovered for the spectrum an empirical mathematical equation that bears his name. When the Lyman and Paschen series were discovered later, Johannes Rydberg extended Balmer's equation to include them. One form of the Rydberg equation is as follows:

$$\nu = 3.2881 \times 10^{15}\ s^{-1} \left(\frac{1}{n_1^2} - \frac{1}{n_2^2}\right)$$

Balmer's and later Rydberg's equations were intriguing. Using different integers for n_1 and n_2, all of the spectral lines of hydrogen could be calculated very accurately. It was not known why this should be true until Neils Bohr demonstrated that n_1 and n_2 represent the energy levels of the electron.

Bohr Model of the Atom

Niels Bohr revolutionized concepts about the atom with his solar system model. This model of the atom required that the electron be confined to specific orbits. The energy (E) of each orbit can be calculated by conventional physics as follows:

$$E = \frac{-2\pi^2\, me^4}{n^2h^2}$$

where m = mass of the electron, e = charge on the electron, h = Planck's constant, and n = orbit number

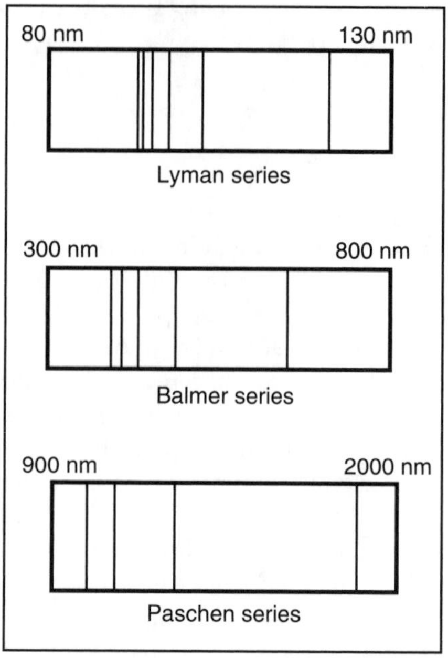

FIGURE 1.20. *Hydrogen spectra that are the basis of the Rydberg equation.*

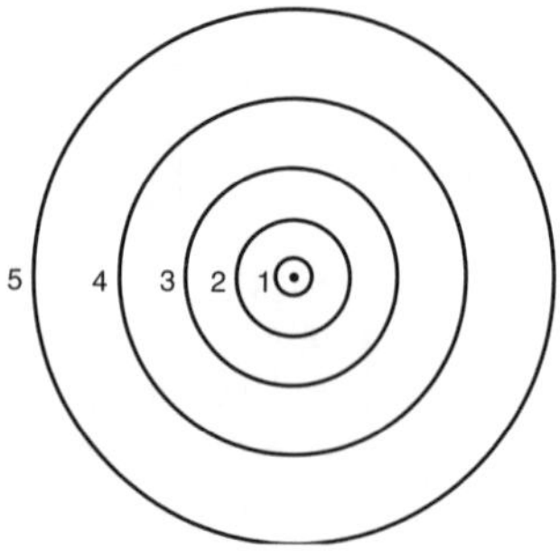

FIGURE 1.21. *Bohr orbits, showing numbering and relative sizes of the electron orbits.*

In Bohr's theory the symbol n represents the number of each orbit, starting with the one closest to the nucleus, as diagrammed in Figure 1.21. This theory was readily accepted because it resulted in an equation identical in form to the empirical equation of Rydberg, and the value of the Rydberg constant calculated by Bohr was almost identical to the value determined by experiment. Bohr's theory also gave meaning to Rydberg's equation.

Energy in the form of light is emitted from an atom when an electron moves from its initial orbit to an orbit that has a lower value of n (Figure 1.22). When an electron is promoted from a low orbit to a higher numbered orbit, energy must be added (Figure 1.23). The energy difference between any two orbits is constant, and the same amount of energy needed to raise an electron from one orbit to another will be released when the electron drops back to the original orbit.

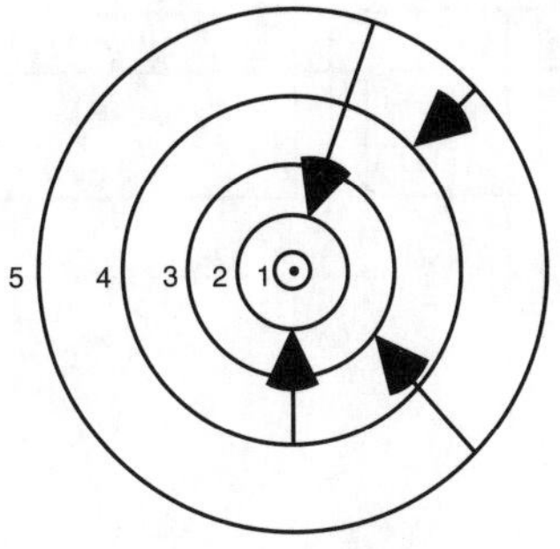

FIGURE 1.22. *Electrons dropping from outer orbits to inner orbits. They emit light energy.*

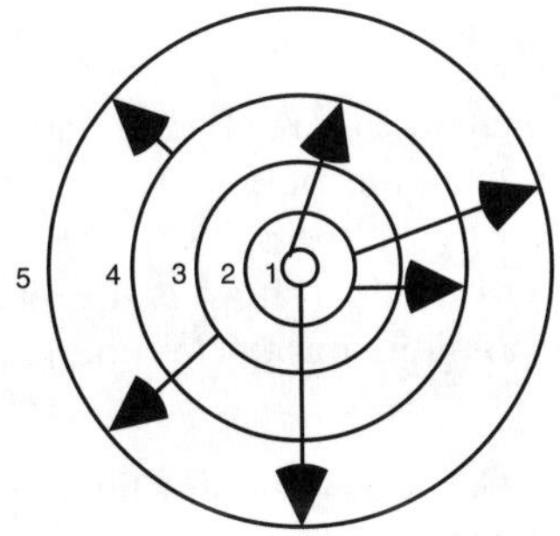

FIGURE 1.23. *Electrons excited from inner to outer orbits. They require added energy.*

Line spectra are due to the emission of light by atoms and represent electrons in excited atoms dropping from high orbits to lower ones. To obtain a positive value for the frequency, the higher number orbit is assigned to n_2 and the lower numbered orbit to n_1 in the Rydberg equation. In the Lyman series n_2 is always equal to 1. The observed lines represent electrons dropping from the second, third, fourth, and fifth orbits down to the first orbit. In the Balmer series n_2 is always 2, and in the Paschen series $n_2 = 3$.

Another way of describing this process is the energy-level diagram (Figure 1.24), where the *y*-axis represents the energy of each orbit. A horizontal line, rather than the circles shown in Figure 1.21, represents the energy level of an orbit. Arrows are drawn from one energy level to another to show where an electron starts and where it ends up. All energy-level diagrams for line spectra show electrons moving from high orbits (energy levels) to lower orbits.

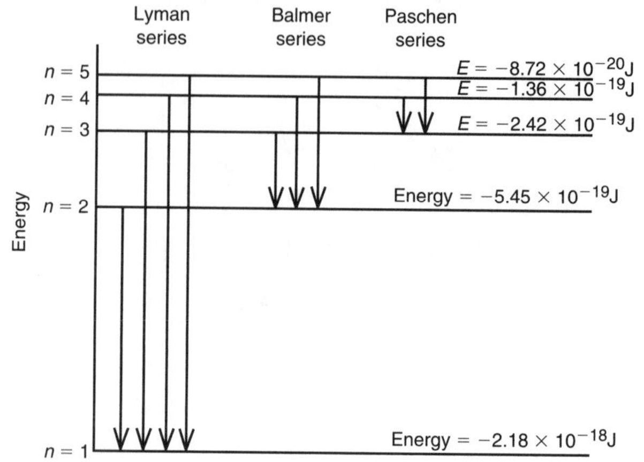

FIGURE 1.24. *Energy-level diagram for the Lyman, Balmer, and Paschen series of lines in the spectra of hydrogen.*

The energy of each level in Figure 1.24 is calculated from the Rydberg equation as shown in the example below. The energies have negative signs since the electron loses energy as it drops towards the $n = 1$ level. The most stable position for the electron in the atom is the first level since the electron has then lost the greatest possible amount of energy.

Example Determine (a) the wavelength of light and (b) the energy needed to promote an electron from the second to the fourth level in the hydrogen atom.

Solution: The Rydberg equation is solved by inserting $n_1 = 2$ and $n_2 = 4$:

$$\nu = 3.2881 \times 10^{15}\ \mathrm{s}^{-1}\left(\frac{1}{n_1^2} - \frac{1}{n_2^2}\right)$$

This becomes

$$\begin{aligned}\nu &= 3.2881 \times 10^{15}\ \mathrm{s}^{-1}(0.25 - 0.00625)\\ &= 3.2881 \times 10^{15}\ \mathrm{s}^{-1}(0.1875)\\ &= 6.165 \times 10^{14}\ \mathrm{s}^{-1}\end{aligned}$$

(a) The wavelength (λ) is calculated as

$$\lambda = \frac{c}{\nu} = \frac{3.00 \times 10^{8}\ \text{m s}^{-1}}{6.165 \times 10^{14}\ \text{s}^{-1}}$$

$$= 4.86 \times 10^{-7}\ \text{m}$$

which is converted to nanometers:

$$\lambda = (4.86 \times 10^{-7}\ \text{m})\left(\frac{1\ \text{nm}}{10^{-9}\ \text{m}}\right)$$

$$= 486\ \text{nm}$$

(b) The energy involved is calculated from

$$E = h\nu$$

$$= (6.62 \times 10^{-34}\ \text{J s})\,(6.165 \times 10^{14}\ \text{s}^{-1})$$

$$= 4.08 \times 10^{-19}\ \text{J}$$

The important principle is that energy must be released (negative sign for energy) when an electron drops from an outer orbit to a lower numbered inner orbit. When an electron is promoted to a higher orbit, energy must be put into the atom and the sign of the energy term will be positive.

Bohr Radius

Bohr decided that the momentum (mass × velocity) of an electron must be related to the size of the electron's orbit. The relationship he used was as follows:

$$mv = \frac{nh}{2\pi r}$$

When Planck's constant (h), the electron mass (m), and the electron velocity (v) are entered into the equation for the first energy level, $n = 1$, a radius of $r = 53$ picometers is calculated. If $n = 2$, the orbit radius is 106 picometers. The value 53 picometers is often called the **Bohr radius** for the hydrogen atom. Whole-number multiples of the Bohr radius give the radii of the other orbits. The Bohr radius gave chemists a theoretical value for the size of a hydrogen atom and confirmed that the atomic sizes determined by experiment are indeed reasonable.

This same reasoning led to the development of an equation similar to the Rydberg equation for the transition of an electron from one orbit to another. Significantly, the empirical Rydberg constant of 109,678 (cm^{-1}) was duplicated almost exactly by Bohr; Bohr's constant was 109,730 cm^{-1}. The agreement between these constants was so remarkable that Bohr's theory was almost immediately accepted.

Development of the wave-mechanical model of the atom changed our view of an atom from the fixed Bohr orbits to the electron cloud picture, where the position of the electron is based on probability factors. One measure that also developed confidence in the wave-mechanical model was that its results duplicated Bohr's. The average distance of an electron from the nucleus in a hydrogen atom was the same 53 picometers that Bohr had calculated. Transitions between energy levels in the hydrogen atom also produced results identical with the Rydberg equation. The Bohr atom was a crucial stepping stone in developing the modern view of the atom, and its electronic structure.

The Wave-Mechanical Model of the Atom

Soon after Bohr's solar system model of the atom (for which he received the Nobel Prize) Louis De Broglie suggested that the electron could behave as a wave as well as a particle. One way of looking at electron behavior involves two equations: the equation for the energy of a wave ($E = h\nu$) and Einstein's well-known equation for the energy of a particle ($E = mc^2$). Since the electron can have only one energy at any given moment, E must be the same in both equations, leading to the equation

$$h\nu = mc^2$$

which indicates the duality between the particle (mass, m) and the wave (frequency, ν) nature of the electron.

Describing the motion of an electron as a wave requires the use of complex "wave equations." Actual use of these wave equations is left for higher level college chemistry courses. It is important, however, to understand the results of using these wave equations, which can be summarized as follows.

1. The wave equations require three numbers in order to reach a solution. These three numbers, called quantum numbers, are the principal quantum number (n), the azimuthal quantum number (ℓ), and the magnetic quantum number (m_ℓ). In addition, to describe an electron completely and uniquely, a fourth quantum number, the spin quantum number, (m_s) is needed. There are specific rules for assigning the four quantum numbers to electrons.
2. The wave equations changed the picture of the atom drastically. In particular, the fixed orbits of the Bohr theory are replaced with a cloud of electrons around the nucleus. The modern orbit is the region of space in which the probability of finding the electron is highest.
3. The simple circular orbits of the Bohr theory have been replaced with spherical electron clouds. The wave equations have shown that the shapes of most electron clouds, although more complex, are still simple geometric shapes.

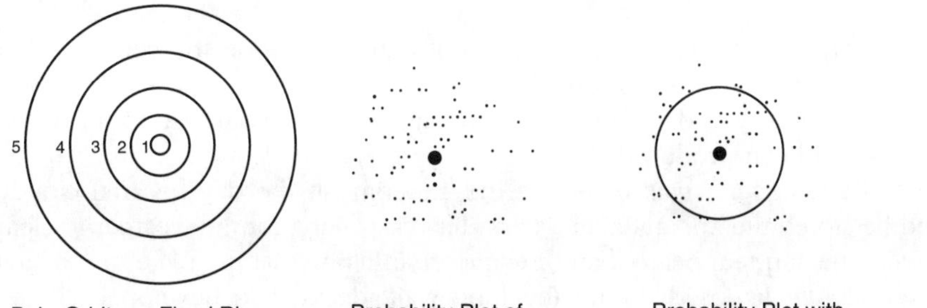

FIGURE 1.25. *Comparison of the Bohr model and the wave-mechanical model of the hydrogen atom.*

4. The arrangement of electrons deduced from wave equations agrees well with the Periodic Table. Many physical and chemical properties of elements and compounds are more fully understood with the knowledge gained in regard to the electronic structure and orbit shape.
5. The results of the wave equations agree completely with the Bohr model. Specifically, the energy change for an electron moving from one electron cloud to another agrees with Bohr's calculations. In addition, the identical 53-picometer radius is found for the electron cloud in the wave-mechanical model of the hydrogen atom.
6. The Heisenberg uncertainty principle is fundamental to the wave-mechanical model of the atom. This principle states that both the position and the momentum of an electron cannot be exactly known at the same time. The more precisely that the position (x) of the electron is known, the more uncertainty there is about its momentum (mv). Heisenberg's uncertainty equation is

$$(\Delta x)(\Delta mv) \leqq \frac{h}{4\pi}$$

which is read as "The uncertainty in the position times the uncertainty of the momentum is equal to Planck's constant divided by four pi."

RADIOACTIVITY

Nuclear Chemistry: A Brief History

The science of nuclear chemistry began with the serendipitous discovery of **radioactivity** by Antoine Becquerel in 1896. Becquerel hypothesized that a **fluorescent** compound, which glows when sunlight strikes it, also emits X rays. To test this idea, he wrapped a photographic plate in black paper, placed a coin on it, and then placed fluorescent uranium ore on the coin. He then exposed this apparatus to sunlight so that the uranium ore would fluoresce. When the plate was developed, a distinct image of the coin demonstrated that X rays had been produced.

While Becquerel was repeating the experiment, the day became cloudy, and he stored the apparatus in a dark drawer, waiting for the weather to clear. Finally the sun reappeared, and Becquerel, thinking that the plate might have been slightly exposed during the long wait, decided to develop it. To his surprise he found that, instead of a slightly exposed plate, the stored plate was very strongly exposed. Becquerel then reasoned that the sun had nothing to do with the success of the experiment and that the uranium compound had emitted the X rays spontaneously. This was the first demonstration of natural radioactivity. In 1903 Becquerel was awarded the Nobel Prize for this discovery.

In 1919 Ernest Rutherford did what alchemists had been trying to do for centuries: he artificially converted one element into another. Rutherford did not prepare gold, however, but converted nitrogen into oxygen. This process of **transmutation** was very expensive and produced extremely small amounts of product that would not have satisfied any alchemist even if gold had been obtained. However, by bombarding nitrogen with alpha particles and producing oxygen, Rutherford demonstrated that the **nucleus** of an atom could be experimentally manipulated.

These two discoveries define the nature of nuclear chemistry. First, radioactive materials spontaneously decay into other substances by emitting energy (X rays and gamma rays) and small particles (alpha particles, beta particles, neutrons, protons, and positrons). Second, by using appropriate experiments, researchers can manipulate nuclear reactions in the laboratory.

No discussion of the early years of nuclear science is complete without mention of the Curies (Marie, Pierre, and their daughter Irene). Their discoveries include radium, polonium, and the positron. They narrowly missed discovering the neutron and nuclear fission. This brilliant family accumulated a total of three Nobel Prizes. Element 96, curium, and one of the basic measures of radiation dosage, the **curie**, are named in their honor.

Radioactivity

Radioactivity is a property of matter whereby an unstable nucleus spontaneously emits small particles and/or energy in order to attain a more stable nuclear state. The process is called **radioactive decay**, and an isotope that contains an unstable nucleus is termed a **radioactive isotope** or **radioisotope**. One radioactive nucleus may decay to another radioactive nucleus, and then to still another. Eventually all radioactive decay results in an isotope with a stable nucleus. Some radioactive isotopes, natural radioactive substances, exist in nature. Artificial radioactive isotopes, on the other hand, are created in the laboratory in nuclear experiments.

Radioactive isotopes, both natural and artificial, emit only a few types of

subatomic particles as they disintegrate. These include the electron (beta particle), neutron, helium nucleus (alpha particle), and positron. When these particles are emitted in a radioactive decay process, the **nuclear mass** (nuclear mass = atomic mass = *A*) and/or **nuclear charge** (nuclear charge = atomic number = *Z*) of the nucleus changes. As a result, one isotope is converted into another with a different identity. Energy may also be released in the form of X rays or gamma rays. The energy released does not affect the identity of the isotope since these rays have neither mass nor nuclear charge. The characteristics of the particles emitted during radioactive decay are described below.

Alpha (α) particles are helium nuclei. They have a mass of 2 and a nuclear charge of +2. The symbol for the alpha particle may take many forms; the two most common are α and ${}^{4}_{2}He$. For example, radon-222 disintegration emits an alpha particle; the reaction may be written as follows:

$${}^{222}_{86}Rn \rightarrow {}^{218}_{84}Po + {}^{4}_{2}He$$

or

$${}^{222}_{86}Rn \rightarrow {}^{218}_{84}Po + \alpha$$

BETA (β) particles are electrons. They have no mass and have a nuclear charge of −1. Like the alpha particle, the beta particle is represented by a variety of symbols, including β and ${}^{0}_{-1}e$. In a decay involving a beta particle, a neutron in the nucleus is converted into a proton, which remains in the nucleus, and the beta particle is emitted. A reaction in which this occurs is the decay of carbon-14 into nitrogen-14.

$${}^{14}_{6}C \rightarrow {}^{14}_{7}N + {}^{0}_{-1}e$$

or

$${}^{14}_{6}C \rightarrow {}^{14}_{7}N + {}^{0}_{-1}\beta$$

Neutrons (n) are often emitted in nuclear reactions. The symbol for a neutron is usually n or ${}^{1}_{0}n$. A neutron has a mass of 1 and a nuclear charge of 0. When scandium-49 decays to calcium-20, a neutron and a beta particle are emitted:

$${}^{21}_{49}Sc \rightarrow {}^{20}_{50}Ca + {}^{1}_{0}n + {}^{0}_{-1}e$$

Positrons are the positive equivalents of electrons or beta particles. The positron has essentially zero mass and a nuclear charge of +1. The symbol is either ${}^{0}_{+1}\beta$ or ${}^{0}_{+1}e$. Carbon-11 is radioactive and emits a positron when it decays to form boron-11:

$${}^{11}_{6}C \rightarrow {}^{11}_{5}B + {}^{0}_{+1}\beta$$

The same overall reaction may occur by **electron capture**, in which we visualize an electron combining with a proton to form a neutron. This reaction does not change the total mass because the electron is very light. However, converting a proton to a neutron results in decreasing the atomic number by 1. The following equation represents an electron capture process:

$$^{11}_{6}C \quad + \quad ^{0}_{-1}\beta \quad \rightarrow \quad ^{11}_{5}B$$

Gamma Rays (γ) are produced when some radioactive decay events occur that leave the nucleus with excess energy. When this excess energy is lost, it is sometimes in the form of gamma rays. Gamma rays have neither charge nor mass and therefore do not change the identity of the nucleus. The symbol for a gamma ray is γ.

X rays represent another form of energy that may be released in radioactive decay. Although X rays and gamma rays may be part of a radioactive decay event, they need not be included in the equations since they do not change the identity of the isotope.

Nuclear Reactions and Equations

Equations for nuclear reactions may be written in the same way as other chemical reactions. In addition to chemical species the nuclear reaction often includes one or more of the following particles. **Beta particles** are energetic electrons ejected from the nucleus and symbolized as $^{0}_{-1}\beta$ or $^{0}_{-1}e^{-}$. Related to the beta particle is the **positron**, or positively charged electron, $^{0}_{+1}e^{+}$. **Alpha particles** are helium nuclei containing two protons and two neutrons and having a mass of 4 and a charge of +2. An alpha particle may be symbolized as $^{4}_{2}He$ or $^{4}_{2}\alpha$; it is the heaviest of the nuclear particles. A **proton**, another common nuclear particle, is simply a hydrogen nucleus, written as $^{1}_{1}H$ or $^{1}_{1}p$. The final particle is the **neutron**, with no charge and a mass of 1, as shown in its symbol, $^{1}_{0}n$.

Gamma rays are often emitted in nuclear reactions. Since these rays have no charge or mass, they are not included in balanced equations. They are very important, however, in determining the amount of energy a nuclear decay produces.

In writing and balancing nuclear equations, careful attention is paid to the total mass and the nuclear charge (atomic number) of each element. For that reason, in nuclear equations all reactants and products have *preceding* superscripts and subscripts to indicate their atomic masses (A) and atomic numbers (Z) respectively.

For example, radon-222 decays spontaneously into polonium-218 by emitting an alpha particle. This information allows us to write the nuclear equation as follows:

$$^{222}_{86}Rn \quad \rightarrow \quad ^{218}_{84}Po \quad + \quad ^{4}_{2}He$$

If necessary, we can use the Periodic Table to determine the atomic number needed to complete the preceding subscript (atomic number or nuclear charge) for each symbol. In the equation above, we see that the superscripts representing the atomic masses add up to 222 on either side of the arrow (218 + 4 = 222), and the subscripts representing the atomic numbers add up to 86 on either side (84 + 2 = 86).

Since the atomic masses and atomic numbers must be equal on both sides of the arrow, it is possible to deduce the identity of a missing isotope or particle in a partial equation. For example, given the partial reaction below, we can identify the particle represented by the question mark:

$$^{238}_{92}\mathrm{U} \rightarrow {}^{234}_{90}\mathrm{Th} + ?$$

Since the total mass on each side must be 238, the question mark must represent a particle that has a mass of 4 (238 − 234 = 4). Also, the atomic number must be 92 on each side, leading to the conclusion that the unknown particle has an atomic number of 2 (92 − 90 = 2). The subatomic particle that has these characteristics is the helium nucleus, also known as the alpha particle ($^{4}_{2}\mathrm{He}$ or $^{4}_{2}\alpha$).

EXERCISES

1. Write the nuclear reaction for each of the following:
   (a) Helium-4 and sodium-23 combine to form aluminum-27.
   (b) An alpha particle combines with nitrogen-14 to produce oxygen-17 and a proton.
   (c) Fluorine-20 decays to neon-20 by emitting a beta particle.
   (d) Uranium-238 decays to thorium-234 by emitting an alpha particle.
   (e) Krypton-87 decays to krypton-86 by emitting a neutron.

2. Predict the particle represented by the question mark in each of the following equations:
   (a) $^{14}_{7}\mathrm{N} \rightarrow ? + {}^{1}_{1}\mathrm{H}$
   (b) $^{54}_{27}\mathrm{Co} \rightarrow {}^{54}_{26}\mathrm{Fe} + ?$
   (c) $^{220}_{86}\mathrm{Rn} \rightarrow {}^{4}_{2}\mathrm{He} + ?$
   (d) $^{51}_{23}\mathrm{V} + {}^{2}_{1}\mathrm{H} \rightarrow ?$
   (e) $^{55}_{25}\mathrm{Mn} + {}^{1}_{1}\mathrm{H} \rightarrow {}^{1}_{0}\mathrm{n} + ?$

Answers

1. (a) $^{4}_{2}\mathrm{He} + {}^{23}_{11}\mathrm{Na} \rightarrow {}^{27}_{13}\mathrm{Al}$
   (b) $^{4}_{2}\mathrm{He} + {}^{14}_{7}\mathrm{N} \rightarrow {}^{17}_{8}\mathrm{O} + {}^{1}_{1}\mathrm{H}$
   (c) $^{20}_{9}\mathrm{F} \rightarrow {}^{20}_{10}\mathrm{Ne} + {}^{0}_{-1}\beta$
   (d) $^{238}_{92}\mathrm{U} \rightarrow {}^{234}_{90}\mathrm{Th} + {}^{4}_{2}\alpha$
   (e) $^{87}_{36}\mathrm{Kr} \rightarrow {}^{86}_{36}\mathrm{Kr} + {}^{1}_{0}\mathrm{n}$

2. (a) $^{13}_{6}C$
   (b) $^{0}_{1}n$
   (c) $^{216}_{84}Po$
   (d) $^{53}_{24}Cr$
   (e) $^{55}_{26}Fe$

Isotopes

Dalton's atomic theory states that all atoms of a given element are identical in all of their properties. It is now known, however, that atoms of an element may have two or more different masses because of differing numbers of neutrons in the nucleus. For instance, a carbon atom may have a mass of 12, 13, or 14 atomic mass units (^{12}C, ^{13}C, or ^{14}C). Although all carbon atoms have six electrons and six protons, the different isotopes have six, seven, and eight neutrons, respectively.

> The number of electrons and the number of protons define an element and its properties.

Isotopes are often associated with nuclear reactions and radioactivity. Many isotopes, however, are not radioactive. Of the carbon isotopes mentioned above, only carbon-14 is radioactive.

In addition, although the chemical and physical properties of the isotopes of an element are virtually identical, there are small differences. The most obvious is the mass; carbon-13 is about 8 percent heavier than carbon-12. In chemical reactions this difference in mass produces subtle effects that can be measured and used to advantage in sophisticated experiments.

Natural and Artificial Radioactive Isotopes

Natural Radioactive Isotopes

Naturally radioactive elements are those from polonium ($Z = 84$) to uranium ($Z = 92$). The 17 transuranium elements with atomic numbers 93–111 are artificially prepared and are also radioactive. In addition to the heavy radioactive elements, a very few of the lighter elements, with atomic numbers less than 83, have naturally occurring radioactive isotopes. These lighter isotopes include potassium-40, vanadium-50, and lanthanum-138. Several interesting naturally radioactive isotopes are described below.

Radon-222 is currently an important potential environmental hazard. It forms, as part of the uranium series, from the decomposition of uranium in many rocks, particularly granite. Radon migrates from the uranium-bearing granite rocks that underlie much of the United States. Often the easiest migration route is through cracks in the basements of many dwellings. Because radon is a dense gas, it collects in the lower, unventilated areas of the house. As a gas, it is readily inhaled and exhaled with the air we breathe. However, its decomposition product, polonium, is a radioactive solid. If the radon atom happens to decay while in the lungs, it first emits an alpha particle, which may do biological damage. Second, it is transmuted into a polonium atom, which is a solid. This solid then completes the decay series within the body. This decay may result in the development of lung disease. Because radon is a noble gas, its unreactive nature makes it very difficult to eliminate from our environment. Other natural radioactive elements may also be implicated in diseases associated with smoking.

Radium was one of the first radioactive elements associated with biological damage. Some radium salts are phosphorescent and glow in the dark. Earlier in this century, these salts were painted on watch dials by workers who licked the paint brushes to get the fine points needed for such work. Many of these workers developed cancers of the mouth. In response to this hazard, radium is no longer used for phosphorescent watch dials. It is also thought that radiation from radium was the cause of the leukemia that killed Marie Curie.

Uranium-238 is implicated in the hazards of radon gas. However, it has been used for a constructive purpose, to estimate the age of the Earth. Uranium eventually decays to lead-206. The half-life, or the time required for half of the element to decay, is 4.5×10^9 years. By measuring the ratio of uranium-238 to lead-206 in rocks, we can estimate how much time has passed since the rocks solidified. The best estimate is that the oldest rocks solidified approximately 4×10^9 years ago. Chapter 12 explains how these calculations are made.

Potassium-40 is one of the few radioactive light elements. One way it decays is by emitting a beta particle to form argon-40. Interestingly, most of the argon in the atmosphere is argon-40, and it is thought to have been produced from the radioactive decay of potassium. This is likely since the amount of argon in the atmosphere is considered to be unusually high for a noble gas.

Artificial Radioactive Isotopes — Transmutation

Today, **artificial radioactive isotopes** are produced by bombarding a target element with nuclei of other elements that have been accelerated to high speeds. This process is carried out in a machine called a **cyclotron**, which accelerates the nuclei toward the target. Another method used to produce artificial radioactive isotopes is to bombard a target with beams of neutrons. Many new and useful radioisotopes have been generated by these methods. Some of these are listed in Table 1.8, which indicates that many of the isotopes have biological applications. Isotopes, whether radioactive or not, retain their

TABLE 1.8. SOME PRACTICAL USES OF RADIOISOTOPES

Isotope Symbol	Isotope Name	Typical Applications
^{3}H	Tritium	Radio-labeled organic compounds and archaeological dating
^{14}C	Carbon-14	Radio-labeled organic compounds and archaeological dating
^{24}Na	Sodium-24	Circulatory system testing for obstruction
^{32}P	Phosphorus-32	Cancer detection, agricultural tracer
^{51}Cr	Chromium-51	Determination of blood volume
^{57}Co	Cobalt-57	*In vivo* vitamin-B^{12} assay
^{59}Fe	Iron-59	Measurements of red blood cell formation and lifetimes
^{60}Co	Cobalt-60	Cancer treatment
^{131}I	Iodine 131	Measurement of thyroid activity and treatment of thyroid disorders
^{153}Gd	Gadolinium-153	Measurement of bone density
^{226}Ra	Radium-226	Cancer treatment
^{235}U	Uranium-235	Nuclear reactors and weapons
^{238}U	Uranium-238	Archaeological dating
^{192}Ir	Iridium-192	Industrial tracer
^{241}Am	Americium-241	Smoke detectors
^{11}C	Carbon-11	Positron emission tomography

characteristic chemical and physical properties, and chemists often use these properties to their advantage.

Chemical properties can be used to predict how an isotope will act in the body and which reactions it will undergo. For instance, the thyroid gland concentrates iodine in the body. To treat thyroid diseases, doctors use iodine-131 as a radioactive source. They know that this isotope, injected into the body, will concentrate exactly where they want it. This property of ^{131}I is of concern when there is an incident at a nuclear reactor and this radioactive substance may be released. When ^{131}I enters the food chain, increased cases of thyroid cancer has been observed.

Another radioactive isotope of concern is strontium-90. We know that strontium and calcium are chemically similar since they are in the same group in the Periodic Table. As a result, ^{90}Sr, which is associated with nuclear explosions, is particularly hazardous. It will become incorporated into the bone structure of all animals. From there, the beta particles it emits can do serious damage.

Radioactive isotopes may be used as tracer materials. For example, we might like to know how the phosphorus in fertilizers is used by plants. By incorporating a small amount of radioactive phosphorus in the fertilizer material, agricultural chemists can measure the time that plants take to utilize the phosphorus. In addition they can track where in the plant the phosphorus goes.

Neutron activation analysis is a sophisticated analysis method using artificial isotopes. A sample is exposed to an intense beam of neutrons that cause many of the elements in the sample to become radioactive. These elements then decay at a rate that is different for each element. When properly calibrated, a neutron activation analysis experiment can identify 20 or more elements in an hour or less. Often very low concentrations can be determined for trace analysis. One advantage of neutron activation analysis is that it is a nondestructive method; in other words, after the induced radiation decays occur, the sample remains unchanged.

The methods used for producing artificial isotopes were applied also to produce the transuranium elements. Many of these 17 elements were discovered during World War II, although secrecy requirements did not allow announcement until the late 1940s and 1950s. The discoverer of an element is allowed to name it. The same is true for the producers of artificial elements. Some of the newer elements have disputed names because more than one group claims credit for their discovery. Although there is some controversy, element 106 was officially named seaborgium in honor of Dr. Glenn Seaborg; its symbol is Sg.

Predicting Radioactivity and Radioactive Decay

It is possible to predict, using a few general principles, whether an isotope will be radioactive. First, bismuth (at. no. = 83) is the last element in the Periodic Table that has a stable, nonradioactive isotope. All elements with atomic numbers greater than 83 are radioactive, and their isotopes decay by emitting alpha particles, beta particles and by electron capture. Alpha particles are emitted by the majority of these isotopes.

Second, the heavy elements often decay to other radioactive elements that, in turn, decay to still other radioactive elements. For instance, uranium-238 decays to lead-206 in a series of 14 steps called a **radioactive disintegration series**. For uranium, this sequence is called the uranium series since uranium is the first element in the series. Two other radioactive disintegration series are the thorium series and the actinium series.

Third, while the heavy isotopes ($Z > 83$) are all radioactive, the lighter isotopes are not always radioactive. If a light isotope is radioactive, it tends to decay by emitting beta particles or positrons. We may also generalize that, if the radioactive isotope of a light element has a mass greater than the average mass of that element, it will undergo a beta emission. If, however, the mass of the isotope is less than the average mass of the element, a positron emission may be expected. For example, the average mass of carbon is 12. Carbon-14 decays by emitting a beta particle, while carbon-10 decays by emitting a positron:

$$^{14}C \rightarrow {}^{0}_{-1}e + {}^{14}N$$

$$^{10}C \rightarrow {}^{0}_{+1}e + {}^{10}B$$

Why some isotopes are radioactive and some are stable is beyond the scope of this book. However, the stability of isotopes with atomic numbers less than 83 seems to depend on the ratio of neutrons to protons (n/p). Up to atomic number 20, the stable isotopes have an n/p ratio very close to 1.00. From atomic number 21 up to atomic number 83, this ratio for stable isotopes increases slowly from 1.00 to 1.5. This stability ratio helps us to understand why carbon-10 and carbon-14 decay as they do. We predicted that ^{14}C (n/p = 1.33) would decay to ^{14}N (n/p = 1.00) by beta emission. The beta-particle emission reduced the n/p ratio to the value expected for a stable isotope. For the decay of ^{10}C (n/p = 0.67), to ^{10}B (n/p = 1.00) by positron emission, the n/p ratio increased to a more favorable value. These examples illustrate that isotopes decay by a process that brings their n/p ratios toward stability.

Instead of memorizing the stable n/p ratio for each element, we can use the relative atomic mass as the average mass of the stable nuclei of a given element. This allows us to estimate the stable n/p ratio for any element. For example, the atomic mass of mercury is 200 and its atomic number is 80; therefore the number of protons is 80 and the number of neutrons is 200 − 80 = 120. The estimated n/p ratio is 120/80 = 1.50. This value is characteristic of the stable isotopes with atomic numbers from 21 to 83. Once we calculate the stable n/p ratios for elements 21 to 83, we can calculate the n/p ratio for a given radioactive isotope and decide whether a positron or a beta particle will be emitted. Emission of a positron will increase the n/p ratio, and emission of a beta particle will decrease the ratio.

EXERCISE

Decide whether each of the following will decay by emitting a positron or a beta particle:

(a) Mn-52

(b) Ni-63

(c) Kr-90

(d) Tc-95

(e) Ce-146

(f) Hf-170

Solutions

If the mass of an isotope is greater than the atomic mass of the element, then the n/p ratio for the isotope is larger than the n/p ratio for the stable element. The result is that a beta particle will most likely be released in a radioactive decay. On the other hand, a positron will most likely be emitted if the isotope mass is less than the atomic mass of the element. As a result we conclude that (a) Mn-52 will emit a positron, (b) Ni-63 will emit a beta particle,

(c) Kr-90 will emit a beta particle, (d) Tc-95 will emit a positron, (e) Ce-146 will emit a beta particle, (f) Hf-170 will emit a positron.

In another measure of stability, it was found that isotopes with even numbers of nucleons (protons plus neutrons) tend to be more stable than isotopes with odd numbers of nucleons. In particular, approximately 60 percent of all stable isotopes have even numbers of protons *and* even numbers of neutrons. Only 1.5 percent of the stable isotopes have odd numbers of neutrons *and* odd numbers of protons.

Measurement of Radioactivity

Radioactivity is a property of chemical substances that is noticeable only with the aid of instrumentation. The first mode of detection was the use of photographic film to detect the X rays emitted in some decay reactions. Since each particle emitted from the nucleus has its own properties, different methods and instruments have been devised to detect the various types. Some of the more common detectors are described below.

A **Geiger-Müller tube** is used for measuring beta and gamma radiation. It consists of a metal tube filled with argon and having a mica window on one end. Inside the tube is a thin wire electrode. When a particle enters through the quartz window, it ionizes the argon gas. These ions are accelerated by the high voltage between the anode and cathode, causing an avalanche of secondary ions. These allow pulses of electricity to flow, and these pulses are counted to determine the number of particles entering per second. If the proper electronics are used, this device can serve also to discriminate among the energies of different types of incoming radiation.

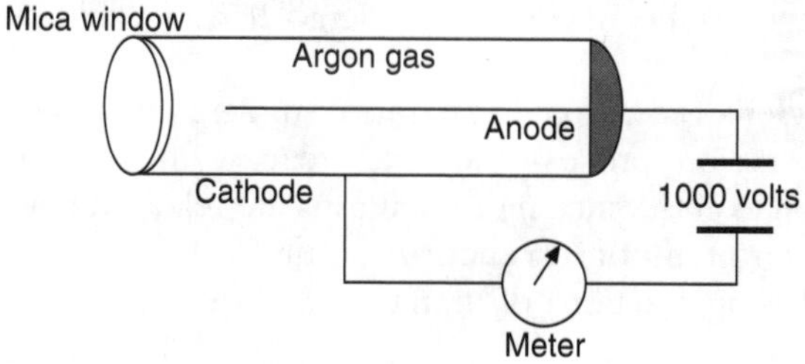

FIGURE 1.26. *Diagram of a Geiger-Müller tube. Ionizing radiation that enters through the mica window ionizes some argon atoms. The ions are accelerated to the cathode and anode, causing an avalanche of secondary ions, which create electrical pulses that are read on the meter.*

Cloud chambers are used to measure charged particles, including alpha and beta particles. A cloud chamber is constructed so that it is saturated with the vapor of a liquid close to its condensation point. Charged particles passing through the chamber condense the vapor into a visible trail that can be photographed and measured. The patterns of these trails can be used to identify different particles. In an electric field the paths will be curved and the charge may be determined from the curvature. When two particles collide, their relative masses can be determined from the angles at which they recoil.

Scintillation counters are used to measure many different types of radiation. The emitting material is mixed with a fluorescent mixture of chemicals. Each time a particle is emitted, it strikes the fluorescent material and emits a flash of light. Phototubes are used to detect and count these flashes (scintillations) as a measure of the number of particles emitted.

Film dosimeters are used mainly as safety devices by people who work with radiation. A small plastic badge encapsulates a piece of special film, which can be made sensitive to different types of radiation. The badge is worn while working with radioactive materials. When developed, the film will tell whether the worker was exposed to excess radiation and will also give the total dose of radiation received.

Nuclear Fission and Fusion

Nuclear Fission

Most nuclear disintegrations involve the emission of small particles, none larger than a helium nucleus. In some nuclear reactions the nucleus of a large atom breaks nearly in half. Any process that yields two nuclei of almost equivalent mass is called **nuclear fission**. Nuclear fission does not occur spontaneously, but requires bombardment of the nucleus with energetic neutrons. Incorporation of a neutron into a fissile nucleus such as $^{235}_{93}U$ or $^{239}_{94}Pu$ causes the nucleus to undergo fission while releasing several more neutrons.

A **fissile nucleus** is one that is capable of undergoing fission. The neutrons emitted in the fission process then can react with other nuclei to cause additional fission events. Since more neutrons are released in each fission than are needed to start it, a **chain reaction** occurs.

One of the fission reactions of uranium-235 is as follows:

$$^{235}_{92}U + ^{1}_{0}n \rightarrow ^{87}_{35}Br + ^{146}_{57}La + 3^{1}_{0}n$$

If uncontrolled, the first reaction produces 3 neutrons, which will result in three more disintegrations and 9 additional neutrons. These 9 neutrons then cause nine more disintegrations and the release of 27 more neutrons. Eventually the reaction is out of control.

There are two ways to keep a fission process from becoming uncontrolled. First, if the sample is small enough, the released neutrons will not hit other ^{235}U nuclei to continue the chain reaction. A **critical mass** of several pounds is needed before the chain reaction will be sustained. Second, excess neutrons can be absorbed by certain materials such as graphite and paraffin. Control rods made of graphite are used in nuclear reactors to adjust the number of available neutrons and therefore the rate of the nuclear reactions. In this use, graphite effectively absorbs some of the neutrons and keeps them from continuing the chain reaction.

Plutonium-239 is another fissile isotope. It is important since it can be produced by bombarding nonfissionable uranium-238 with neutrons. Since only 1 percent of natural uranium is ^{235}U, and most of the rest is ^{238}U, the ability to produce means that ^{239}Pu raw material is more readily available. The reactor that produces plutonium is known as a breeder reactor.

Nuclear reactors are used mainly to generate electricity. The enormous heat generated in fission reactions is used to boil water. The steam is then used to run conventional turbines that produce the electricity. In practice, the nuclear power plant is very complex in order to keep the radioactive materials contained while extracting the desired energy. Uranium reactors require expensive enriched uranium ores. Breeder reactors do not require enriched fuel; however, since they produce weapons-grade plutonium, their use is highly regulated and even discouraged.

Nuclear Fusion

The combination of two nuclei into a larger atom is called **nuclear fusion**. The reactions within the sun are fusion reactions that combine hydrogen nuclei to form a helium atom in a three-step reaction:

$$^{1}_{1}H + ^{1}_{1}H \rightarrow ^{2}_{1}H + ^{0}_{1}e$$

$$^{1}_{1}H + ^{2}_{1}H \rightarrow ^{3}_{2}He$$

$$^{3}_{2}He + ^{3}_{2}He \rightarrow ^{4}_{2}He + ^{1}_{1}H + ^{1}_{1}H$$

Since fusion takes simple, nonradioactive materials and produces helium, it should be a clean and inexpensive source of energy.

Instead of trying to duplicate the three-step reaction that occurs within the sun, current fusion research focuses on this reaction:

$$^{3}_{1}H + ^{2}_{1}H \rightarrow ^{4}_{2}He + ^{1}_{0}n$$

Several designs for fusion reactors are being tested. Although some success has been achieved, it will be some time before this cheap, nonpolluting energy source will become a reality.

2

STATES OF MATTER

GASES

Boyle's Law

In 1660 Robert Boyle discovered the **inverse relationship** between the pressure and volume of a gas that is given in the following equation and graph:

$P \propto \frac{1}{V}$ or $PV = \text{constant}$

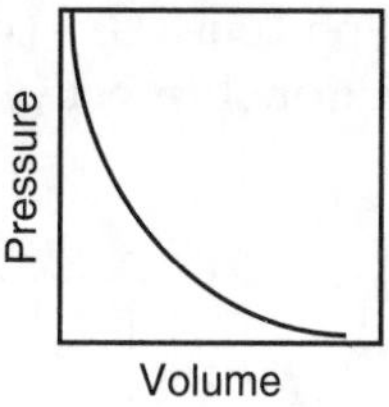

If a sample of gas starts with specific conditions of pressure and volume and an experiment is done to change these conditions (without changing the temperature or the amount of gas), then the relationship

$$P_iV_i = P_fV_f$$

is obtained. The subscripts i and f represent the initial and the final conditions, respectively.

This form of **Boyle's law** shows that, if the pressure on a gas is increased, the volume must correspondingly decrease. Also, if the pressure decreases, the volume will increase.

Charles' Law

In 1787, over 100 years after Boyle's law, Jacques Charles discovered the **direct relationship** between the volume of a gas and its temperature that is given in the following equation and graph:

$V \propto T$ or $\frac{V}{T} = \text{constant}$

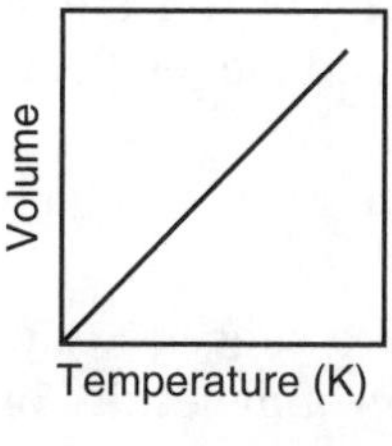

If a sample of gas starts with specific conditions of volume and temperature that are changed to some other, final conditions (while the pressure and the amount of gas do not change), Charles' law may be reformulated as

$$\frac{V_i}{T_i} = \frac{V_f}{T_f}$$

Another way of stating the law is to say that, as the temperature of a gas decreases, its volume will decrease. This equation is applicable ONLY IF the temperature is expressed in Kelvin units.

Gay-Lussac's Law

In the late 1700s, Joseph-Louis Gay-Lussac, along with Jacques Charles, discovered the **direct relationship** between pressure and temperature:

$P \alpha T$ or $\frac{P}{T} = \text{constant}$

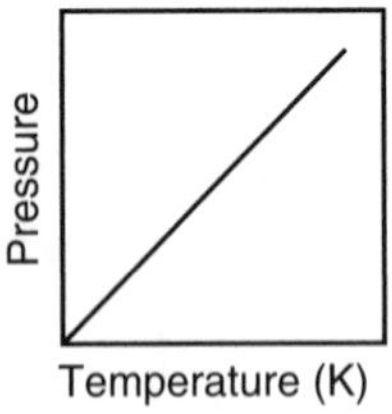

If the initial conditions of pressure and temperature are changed to some final conditions, **Gay-Lussac's law** requires that

$$\frac{P_i}{T_i} = \frac{P_f}{T_f}$$

This equation indicates that the temperature must increase if the pressure increases. Also, if the pressure decreases, the temperature must decrease. This phenomenon can be observed directly with any aerosol spray such as a deodorant, air spray, or perfume (but do not experiment with any potentially hazardous substances such as pesticides). Each of these feels very cold when sprayed directly on the skin because the pressure of the gas is decreased drastically and the temperature undergoes a similar decrease.

Avogadro's Principle

In 1811 Amedeo Avogadro suggested the principle that equal volumes of gases contain equal numbers of molecules or atoms under identical conditions of temperature and pressure. This direct relationship between moles and volume is expressed in the following equation and graph:

$V \alpha n$ or $\frac{n}{V} = \text{constant}$

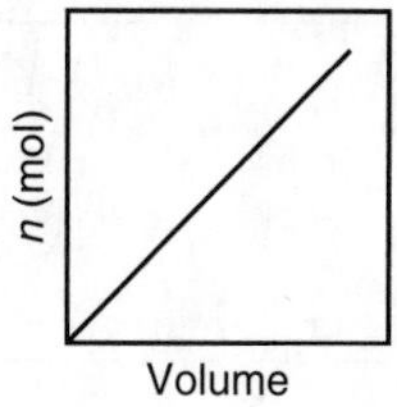

If the conditions of variable n or V are changed to some conditions, **Avogadro's principle** requires that

$$\frac{n_i}{V_i} = \frac{n_f}{V_f}$$

Avogadro's principle can be interpreted as meaning that, at a constant temperature and pressure, the volume of the container must increase as the moles of gas increase.

Ideal Gas Law

The four laws of Boyle, Charles, Gay-Lussac, and Avogadro are combined into the **ideal gas law**:

$$PV = nRT$$

Here P is pressure, V is volume, n is number of moles of gas, and T is temperature in kelvins.

Universal Gas Constant

The constant R, called the **universal gas constant**, is needed to make all of the relationships fit together. Table 2.1 gives some of the possible values and units for R.

This constant appears in many different equations, and the selection of the proper value for R is often critical. This selection is based on the units given in the problem. In many problems not all of the data are given in units compatible with the units of R. For instance, a problem may give the temperature in degrees Celsius, which must *always* be converted to kelvins. In other cases the pressure or volume may need to be converted to the appropriate units. Units must be watched very carefully in gas law calculations.

TABLE 2.1. DIFFERENT FORMULATIONS OF THE UNIVERSAL GAS CONSTANT (*R*)
(These are given with problems and need not be memorized.)

Value	Units
0.0821	L-atm mol^{-1} K^{-1}
8.314	J mol^{-1} K^{-1}
6.24×10^4	L mm Hg mol^{-1} K^{-1}
1.99	cal mol^{-1} K^{-1}
8.314	V C mol^{-1} K^{-1}

Instead of memorizing the four laws developed by Boyle, Charles, Gay-Lussac, and Avogadro, it is more convenient to use only the ideal gas law equation for all problems. There are two ways to use this law. First, a problem may give three of the four variables and ask that the fourth be calculated. This involves direct substitution of data into the ideal gas law equation. Second, a problem may require taking a gas under certain initial conditions of P, V, T, and n and changing to different final conditions of these four variables.

To solve these problems, the ratio of two ideal gas law equations is written:

$$\frac{P_iV_i}{P_fV_f} = \frac{n_iRT_i}{n_fRT_f} \qquad (2.1)$$

Variables that the problem states (or implies) are constants are canceled, along with R, and then the appropriate substitutions are made with the given data to perform the calculation.

Using this approach, we can see that the ideal gas law becomes Boyle's law if n and T are constant (i.e., cancel) and becomes Charles' law if n and P are constant. If V and n are constant, they cancel to give Gay-Lussac's law; if P and T are constant, the ideal gas law becomes Avogadro's principle. Only one equation need be remembered for all ideal gas law calculations!

Derivation of Ideal Gas Law

One method for deriving the ideal gas law uses Avogadro's principle, $n/V = C$, and Gay-Lussac's law, $P/T = C'$. Since C and C' are not the same constants, we multiply Avogadro's principle by the constant R, so that $CR = C'$. Then

$$CR = C'$$

and, substituting for CR and C', we obtain

$$\frac{nR}{V} = \frac{P}{T}$$

Rearranging this equation to eliminate the fractions yields

$$PV = nRT$$

the ideal gas law with R as the universal gas constant.

EXERCISE

A gas occupies 250 milliliters, and its pressure is 550 millimeters of mercury at 25°C.

(a) The gas is expanded to 450 milliliters. What is the pressure of the gas now?

(b) What temperature is needed to increase the pressure of the gas to exactly 1 atmosphere at 250 milliliters?

(c) How many moles of gas are in this sample?

(d) The sample is an element and has a mass of 0.525 gram. What is it?

Solutions

(a) We use the ratio of two ideal gas law equations as shown in equation 2.1:

$$\frac{P_iV_i}{P_fV_f} = \frac{n_iRT_i}{n_fRT_f}$$

All but the *P* and *V* terms can be canceled since no change is specified for the other terms and they are assumed to be constant:

$$\frac{P_iV_i}{P_fV_f} = 1$$

We assign the data to the variables: P_i = 550 mm Hg, V_i = 250 mL, and V_f = 450 mL, enter in the equation, and solve:

$$\frac{(550 \text{ mm Hg})(250 \text{ mL})}{P_f(450 \text{ mL})} = 1$$

$$\frac{(550 \text{ mm Hg})(250 \text{ mL})}{450 \text{ mL}} = P_f$$

$$306 \text{ mm Hg} = P_f$$

(b) We will use the same procedure as in part (a), but cancel *n*, *R*, and *V* since they remain constant:

$$\frac{P_iV_i}{P_fV_f} = \frac{n_iRT_i}{n_fRT_f}$$

$$\frac{P_i}{P_f} = \frac{T_i}{T_f}$$

We assign the data: P_i = 550 mm Hg, T_i = 25 + 273 = 298 K, and P_f = 1 atm = 760 mm Hg. Then:

$$\frac{550 \text{ mm Hg}}{760 \text{ mm Hg}} = \frac{298 \text{ K}}{T_f}$$

$$T_f = 298 \text{ K}\left(\frac{760 \text{ mm Hg}}{550 \text{ mm Hg}}\right)$$

$$= 412 \text{ K} = 139°\text{C}$$

(c) To answer this question, we use the ideal gas law equation

$$PV = nRT$$

and substitute the given values. To use the constant R = 0.0821 L atm mol^{-1} K^{-1}, we must convert the data to the proper units so that P = 550 mm Hg = 0.724 atm, V = 250 mL = 0.250 L, T = 25°C = 298 K. We enter these data and solve:

$$(0.724 \text{ atm})(0.250 \text{ L}) = n(0.0821 \text{ L atm mol}^{-1} \text{ K}^{-1})(298 \text{ K})$$

$$n = \frac{(0.724 \text{ atm})(0.250 \text{ L})}{(0.0821 \text{ L atm mol}^{-1} \text{ K}^{-1})(298 \text{ K})}$$

$$= 7.40 \times 10^{-3} \text{ mol}$$

(d) We can identify a gas by its molar mass:

$$\text{Molar mass} = \frac{\text{mass of sample}}{\text{moles of sample}}$$

$$= \frac{0.525 \text{ g}}{7.40 \times 10^{-3} \text{ mol}}$$

$$= 70.9$$

The element with the molar mass closest to 70.9 is chlorine, Cl_2.

Standard Temperature and Pressure (STP)

The ideal gas law has four variables, P, V, n, and T, along with the constant R. By defining the standard pressure as exactly 1 atmosphere and the standard temperature as exactly 0° Celsius, two of the variables can be stated quickly and easily. Therefore any gas at STP is understood to have P = 1.00 atmosphere and T = 273 K (0°C = 273 K), and only the number of moles, n, and the volume, V, need be stated to describe the system completely.

The acronym STP represents the standard state for gases. Standard states for other experiments may be different. In particular, the standard temperature for a galvanic cell is 25°C, not 0°C.

Molar Volumes and Molar Masses of Gases

In the ideal gas law n represents the number of moles of gas, which is calculated as

$$\frac{\text{grams of gas}}{\text{molar mass}} = n$$

Substituting this into the ideal gas law yields

$$PV = \frac{g}{\text{molar mass}} RT$$

This equation can be used to determine the **molar mass** of a gas if P, V, g, and T are known for a given sample.

Rearranging this equation algebraically yields another useful relationship;

$$P\,(\text{molar mass}) = \frac{g}{V} RT$$

In this equation g/V is the density of the gas in grams per liter. The **density** of a gas can be determined if P, T, and the molar mass are known.

Finally, at standard temperature, 0°C, and pressure, 1.0 atmosphere, the ideal gas law may be solved to calculate the **molar volume**, V/n:

$$\frac{V}{n} = \frac{RT}{P} = \frac{(0.0821 \text{ L atm mol}^{-1} \text{ K}^{-1})(273 \text{ K})}{1.00 \text{ atm}}$$

$$= 22.4 \text{ L mol}^{-1}$$

This result indicates that 1 mole of an ideal gas at STP has a volume of 22.4 liters, a fact that is useful in stoichiometry calculations. Table 2.2 lists the molar volumes of some real gases at STP. These values are all close to 22.4 liters, indicating that the gases behave as ideal gases under these conditions.

TABLE 2.2. MOLAR VOLUMES OF SOME GASES AT STP

Gas	Symbol	Molar Volume (L)
Argon	Ar	22.401
Carbon dioxide	CO_2	22.414
Helium	He	22.398
Hydrogen	H_2	22.410
Nitrogen	N_2	22.413
Oxygen	O_2	22.414

EXERCISE

What are the expected densities of (a) argon, (b) neon, and (c) air at STP?

Solutions

Each of these densities is calculated using the equation

$$\text{Density} = \frac{g}{V} = \frac{(\text{molar mass})\,P}{RT}$$

At STP, $P = 1.00$ atm and $T = 273$ K. Substituting these data gives:

(a) $$\text{Density of argon} = \frac{(39.95\text{g mol}^{-1})(1.00\text{ atm})}{(0.0821\text{ L-atm mol}^{-1}\text{ K}^{-1})(273\text{K})} = 1.78\text{ g L}^{-1}$$

(b) $$\text{Density of neon} = \frac{(20.18\text{g mol}^{-1})(1.00\text{atm})}{(0.0821\text{L-atm mol}^{-1}\text{ K}^{-1})(273\text{ K})} = 0.900\text{g L}^{-1}$$

(c) $$\text{Density of air} = \frac{(28.8\text{g mol}^{-1})(1.00\text{ atm})}{(0.0821\text{L-atm mol}^{-1}\text{ K}^{-1})(273\text{K})} = 1.28\text{g L}^{-1}$$

The molar mass of air is approximated from the fact that air is 80% nitrogen and 20% oxygen. Therefore

$$0.80 \times 28.0 + 0.20 \times 32 = 28.8\text{ g mol}^{-1}$$

We may also conclude from the densities that a balloon full of neon will rise in air but an argon-filled balloon will sink to the floor.

Graham's Law of Effusion

Graham's law of effusion states that the rate (v) at which a gas effuses through a small pinhole is inversely proportional to the square root of the molar mass (m) of the gas. When the rates of effusion of two gases are compared, equation 2.2 may be written:

$$\sqrt{\frac{m_1}{m_2}} = \frac{\bar{v}_2}{\bar{v}_1} \qquad (2.2)$$

The kinetic molecular theory explains Graham's law of effusion. This theory states that the average kinetic energy of gas molecules is directly proportional to their temperature (when temperature is expressed in Kelvin units) as expressed in the following equation:

$$\overline{\text{KE}} = cT = \frac{1}{2}m\bar{v}^2$$

where $\overline{\text{KE}}$ is the average kinetic energy, c is a constant that is the same for all gases, T is temperature in kelvins, m is the mass of the gas molecule, and $\bar{v}$ is the average velocity of the gas. Since cT will be the same for all gases at the same temperature, the average kinetic energy of any two gases at the same temperature will be the same:

$$\overline{KE}_1 = \overline{KE}_2 \qquad (2.3)$$

where subscripts 1 and 2 represent the two different gases. Substituting from equation 2.3, we write

$$\frac{1}{2}m_1\overline{v}_1^2 = \frac{1}{2}m_2\overline{v}_2^2$$

Then, canceling the 1/2's and rearranging, we obtain the equation

$$\sqrt{\frac{m_1}{m_2}} = \frac{\overline{v}_2}{\overline{v}_1}$$

which is **Graham's law of effusion**.

When comparing how rapidly two gases will effuse through a pinhole in a container, we find that the rate for each gas is inversely related to the square root of its molecular mass. Effusion through a pinhole requires that the gas particle hit the pinhole just right in order to pass through it. The more collisions a gas has with the walls of a container, the higher the probability is that the gas molecule will go through the pinhole. We may generalize that a high-velocity gas molecule will have more frequent collisions with the walls than a low-velocity gas molecule. In addition, at the same temperature, heavier molecules must have lower velocities than light molecules.

Equation 2.2 is also the mathematical statement of the relative rates of diffusion of gases. Any gas will expand to fill a container, but some time may be required, depending on the conditions present. Equation 2.2 shows that the velocity of the gas particles is inversely proportional to the square root of the masses of the individual particles. Since this is a ratio and the units cancel, m_1 and m_2 may be either the actual masses of the molecules or the relative molar masses of the gas particles. It is expected that heavier gases will diffuse more slowly than lighter gases, and they do.

EXERCISE

Helium leaks through a very small hole at a rate of 3.22×10^{-5} mole per second. How fast will oxygen effuse through the same hole under the same conditions?

Solution

Graham's law of effusion states that

$$\sqrt{\frac{m_1}{m_2}} = \frac{\overline{v}_2}{\overline{v}_1}$$

and we assign the molar masses as $m_1 = 4$ and $m_2 = 32$. Correspondingly $\bar{v}_1 = 3.22 \times 10^{-5}$ mol s^{-1}, while $\bar{v}_2$ is the unknown. Then

$$\sqrt{\frac{4}{32}} = \frac{\bar{v}_2}{3.22 \times 10^{-5}\ \text{mol s}^{-1}}$$

$$\bar{v}_2 = 1.14 \times 10^{-5}\ \text{mol s}^{-1}$$

One of the most useful applications of Graham's law is the determination of the molar mass of a gas. The rate of effusion of a gas with an unknown molar mass and the rate of effusion of a known gas are measured, and the data are applied to equation 2.2.

Example

Helium effuses through a pinhole at a rate of 5.735×10^{-5} mole per second. An unknown gas, with an empirical formula of C_4H_5, effuses through the same pinhole at a rate of 1.114×10^{-5} mole per second. What is the molecular formula for this compound?

Solution: To obtain the molecular formula, we need the molar mass of the compound. Using equation 2.2, we substitute the data:

$$\sqrt{\frac{m_1}{4\ \text{g mol}^{-1}}} = \frac{5.735 \times 10^{-5}\ \text{mol s}^{-1}}{1.114 \times 10^{-5}\ \text{mol s}^{-1}}$$

Squaring both sides yields

$$\frac{m_1}{4\ \text{g mol}^{-1}} = (5.148)^2$$

$$m_1 = (26.5)(4\ \text{g mol}^{-1})$$

$$= 106\ \text{g mol}^{-1}$$

The mass of the empirical formula unit is 53, and dividing the molar mass obtained above by 53 gives 2. This tells us that there are two empirical formula units in one molecule, and the molecular formula is C_8H_{10}.

Absolute Zero

Absolute zero is the lowest possible temperature: 0 on the Kelvin and −273° on the Celsius temperature scale. In perfect crystals, it is the temperature where all motion ceases. Experimentally, absolute zero has never been achieved, but temperatures of a few tenths of a degree above absolute zero have been attained. At very low temperatures most metals become superconductors with no electrical resistance.

One method of determining absolute zero is to construct a graph of the volume of a gas as its temperature is changed and to extrapolate the data to the temperature that corresponds to zero volume of the gas, as shown in Figure 2.1.

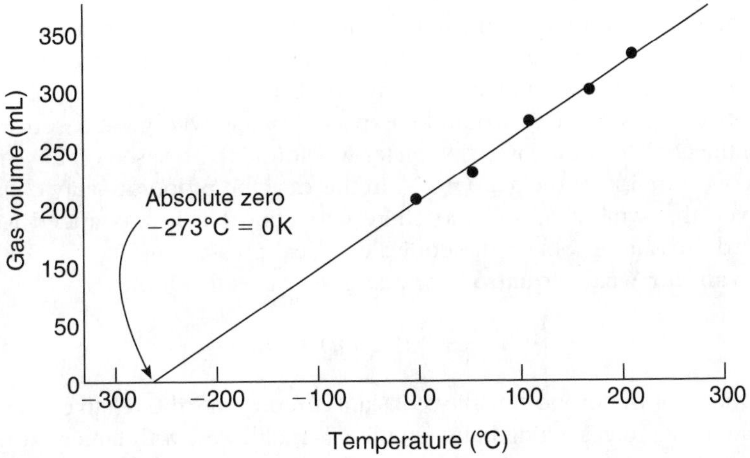

FIGURE 2.1. *Using Charles's law to determine absolute zero. Each dot represents an experimental measurement. The line is the best straight line through the data, and the intercept is at −273°C.*

Real Gases

The ideal gas law works very well for most gases, but not for gases under high pressures or gases at very low temperatures. These conditions are the same ones that are used to condense gases, and therefore it may be generalized that any gas close to its boiling point (condensation point) will deviate significantly from the ideal gas law.

The kinetic molecular theory makes two fundamental assumptions about the properties of gas particles. One assumption is that gases have no volume, and the second is that they exhibit no attractive or repulsive forces. These two assumptions define an **ideal gas**. They work quite well since gas particles are often widely separated and what little volume they have is relatively unimportant. Similarly, because of the large average distance between gas particles the attractive forces and repulsive forces are very weak. As a **real gas** is cooled and/or compressed, however, the distance between the particles decreases dramatically, and these real volumes and forces can no longer be ignored.

Johannes van der Waals developed a modification of the ideal gas law to deal with the nonideal behavior of real gases. He reasoned that, if each of the gas particles occupies some volume, there will be a net decrease in the useful volume of the container. (Imagine a fishbowl filled with marbles; little useful room is left for the fish.) This decrease in volume must be proportional to the moles of gas particles present. With the letter *b* used for the proportionality constant, the volume term in the ideal gas law could be replaced by the term $V - nb$.

In evaluating the effect of intermolecular attractions, it was reasoned that, if gas particles attract each other, they have curved paths and therefore take longer to collide with the container walls. The result will be a decreased collision frequency and a pressure for the real gas that is lower than the pressure of an ideal gas under the same conditions. The frequency of attractions between gas particles would be expected to rise with greater concentrations of the gas. The best correction factor was found to be based on the square of the concentration of the gas, $(n/V)^2$. In this case the proportionality constant was given the symbol a, and the entire correction factor was added to the measured pressure to obtain the equivalent ideal pressure.

The **van der Waals equation** for real gases takes this form:

$$\left(P + \frac{an^2}{V^2}\right)(V - bn) = nRT$$

The value of the proportionality constant a represents the relative strengths of the attractive forces acting between the gas molecules, with larger values of a indicating stronger attractive forces such as dipole-dipole attractions. The value of the constant b represents the relative size of the gas molecule. The larger the value of b, the larger is the size of the molecule.

Size does not vary greatly from one gas molecule to another, and the deviations from the ideal gas law are similar for many molecules. Attractive forces, however, do vary greatly, depending on molecular polarity, and contribute the most to deviations from the ideal gas law.

LIQUIDS AND SOLUTIONS

Physical Properties of Liquids

Intermolecular Attractive Forces

Because molecules in a liquid are much closer together than in a gas, attractive forces between molecules become important. Polar molecules experience **dipole-dipole attractions** with the negative end of one molecule attracted to the positive end of another. Compounds that have hydrogen bonded to fluorine, oxygen, or nitrogen have very large dipole-dipole attractions that are called **hydrogen bonding**. Circulation of electrons in non-polar substances creates momentary charge imbalances that induce dipoles in neighboring molecules. Attractions due to **induced dipoles** are often called **London forces**. Many properties of liquids and solids can be explained on the basis of these three forces.

Surface Tension

Surface tension is due to an increase in the attractive forces between molecules at the surface of a liquid compared to molecules in the center, or bulk, of the liquid. This property causes fluids to minimize their surface areas. As a result, small droplets of liquids tend to form spheres. Surface tension provides a "skin" on a liquid surface that allows small insects to literally stand on water. Also, carefully placed small iron objects such as needles and pins can float on the surface of water even though iron is almost eight times as dense as water.

To visualize surface tension, we can consider the attractive forces that hold a molecule in the liquid state (Figure 2.2). In the interior of the liquid each molecule is attracted to other molecules from every direction (top, bottom, front, rear, left, and right). When a molecule is at the surface of a liquid, it has the same ability to attract neighboring molecules; however, at the surface there are no molecules to attract on its top side. Since the same total attractive force is now divided between fewer adjacent molecules, the result is a stronger attraction of a surface molecule toward the bulk of the liquid. This increased attractive force per molecule is responsible for surface tension.

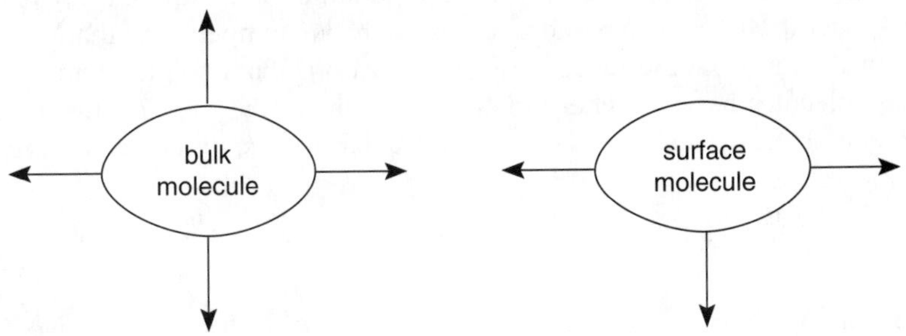

FIGURE 2.2. *Comparison of the attractive forces for molecules in the bulk of a liquid with those at the surface. Arrows indicate attractive forces; front and rear arrows are not shown.*

Surface tension determines whether droplets of a liquid will bead up, as on a freshly waxed car, or spread out when placed on a flat surface. To describe this phenomenon, we define **cohesive forces** as attractions between identical molecules in the liquid, and **adhesive forces** as attractions between different molecules, such as those in the liquid and on a flat surface. If the cohesive forces of the liquid are strong compared to the adhesive forces between the liquid and the flat surface, the liquid will retain its shape and form beads. If the adhesive forces between the flat surface and the liquid are strong enough, the liquid will spread uniformly.

Very clean glass has low adhesive forces, and water beads readily on it. In dishwashers these beads of water evaporate, leaving undesired spots on glasses and dishes. For this reason dishwasher detergents contain chemicals called **surfactants**. A surfactant decreases the cohesive forces, and therefore the surface tension, of liquids. The surfactant in dishwasher detergent lowers the cohesive forces of water so that the adhesive forces between water and glass are small. Then water does not bead up on glassware, and the spotting problem is reduced.

Viscosity

The term **viscosity** refers to a liquid's resistance to flow. A liquid such as water has fairly low viscosity and flows easily. Pancake syrup, on the other hand, has high viscosity, particularly when cold, and flows slowly. The attractive forces within a liquid are responsible for viscosity. For a liquid to flow, the molecules must move past each other. Molecules are able to move more freely in solutions that have relatively low attractive forces. The liquid alkanes (as in gasoline) have lower viscosities than water because alkanes are attracted to each other only by London forces. Water is more viscous because of hydrogen bonding. Syrup is very viscous since its bulky sugar molecules contain many —OH groups, which hydrogen-bond to the water in the mixture and to each other.

Viscosity usually decreases as the temperature of a liquid is increased. Pancake syrup flows much more easily at room temperature than it does when first taken from the refrigerator. The reason is that at higher temperature all the molecules have a higher kinetic energy since $KE_{avg} = cT$. This increase in kinetic energy weakens the intermolecular forces, thus decreasing the viscosity.

Evaporation

Evaporation is a familiar process in which a liquid in an open container is slowly converted into a gas. Some liquids evaporate more rapidly than others. For example, a beaker of gasoline evaporates in a few hours, while a beaker of water may take a day or two. In addition, the rate at which a liquid evaporates increases as the temperature increases.

Evaporation may be explained by considering the attractive forces involved and the kinetic energy needed to overcome these forces. This process is the reverse of condensation. For a molecule to be converted from a liquid to a gas, it must have sufficient kinetic energy to overcome its attractive forces. In any group of molecules the average kinetic energy is proportional to the Kelvin temperature. The actual kinetic energies are distributed as shown in Figure 2.3. Some molecules have low, and others have high, kinetic energies. The escape energy is defined as the minimum kinetic energy needed for a molecule to escape from the liquid into the gas phase. All molecules with kinetic energies greater than the escape energy are capable of evaporating.

The area under the curve in Figure 2.3 represents the total number of molecules. The area of the shaded portion of the graph represents the number of molecules whose kinetic energies are equal to or exceed the escape energy. The ratio of these two areas is constant as long as the temperature is constant. Molecules that are close enough to the surface and are traveling in the correct direction will escape as gas molecules. At a constant temperature the proportion of molecules with enough kinetic energy to escape will remain constant

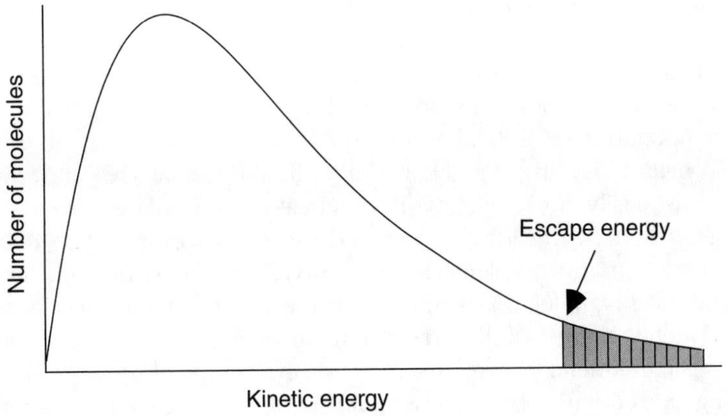

FIGURE 2.3. *Distribution of kinetic energies of a group of molecules at a given temperature. Total area under the curve represents all molecules, and shaded area represents molecules with kinetic energies equal to or greater than the escape energy.*

and the liquid will evaporate at a uniform rate until all molecules have entered the gas phase.

Since only molecules near the surface can escape into the gas phase, the surface area of the liquid is a factor in evaporation (Figure 2.4). Fifty milliliters of a liquid in a narrow test tube will evaporate more slowly than the same 50 milliliters in a beaker. If the liquid is poured from the beaker into an evaporating dish, it will evaporate even more quickly. The reason lies in the increased proportion of molecules that are close enough to the surface to escape readily.

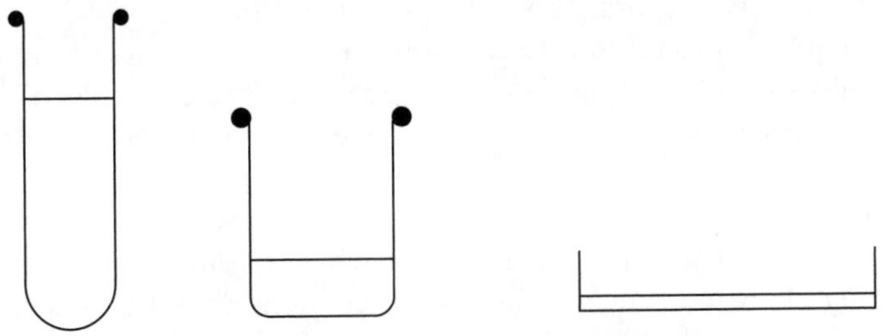

FIGURE 2.4. *Surface areas of a test tube, beaker, and evaporating dish, each holding 10 mL of liquid. The increasing surface area indicates that evaporation will be slowest from the test tube and fastest from the evaporating dish.*

Temperature is another important factor in the evaporation process. An increase in temperature increases the proportion of the molecules that have kinetic energies above the escape energy. Conversely, lowering the temperature decreases the proportion with enough kinetic energy to escape.

The temperature of a beaker of liquid evaporating on a lab bench is usually constant because the beaker and liquid can absorb heat from its surroundings easily and quickly. If the beaker is insulated from the surroundings, however, the liquid cools and the rate of evaporation decreases. These phenomena are explained using curves similar to the one in Figure 2.3. First, when a molecule escapes from the liquid into the gas phase, the average kinetic energy of the remaining molecules decreases. A decrease in the average kinetic energy means that the temperature also decreases, explaining the cooling observed. Second, as the average kinetic energy decreases, the proportion of molecules that have kinetic energies above the escape energy also decreases. Since fewer molecules have enough energy to escape, the rate of evaporation decreases.

Conversely, increasing the temperature of a liquid increases the rate of evaporation. This rise occurs because a greater proportion of all molecules now have kinetic energies greater than the escape energy. When the temperature is increased sufficiently, boiling occurs. Boiling is recognized as the formation, throughout the solution, of gas bubbles that rise to the surface. At the boiling point, the molecules do not have to reach the surface to enter the gas phase. Enough molecules, with the appropriate escape energy, can come together within the solution to form the bubbles we observe.

Vapor Pressure

Vapor pressure is the pressure that develops in the gas phase above a liquid when the liquid is placed in a closed container. Evaporation of molecules from the liquid still occurs in the closed container, but the gas molecules cannot escape to the surroundings. As more molecules enter the gas phase, the pressure increases, finally stopping at a level that is dependent only on the temperature. This final pressure is called the vapor pressure.

When a liquid is placed in a closed container, it starts evaporating just as it would in an open beaker; however, as stated above, the gas molecules cannot escape. As the gas molecules move, they collide with the walls of the container, the liquid at the bottom of the container being one of these "walls." When the gas molecules collide with this liquid, very few bounce off; almost all condense to the liquid state again. Initially the rate at which the molecules evaporate is much greater than the rate at which they condense. As the gas molecules increase in number, however, they collide with the liquid surface more frequently. Eventually the rate at which the liquid molecules evaporate is equal to the rate at which the gas molecules condense. Under these conditions the liquid and the gas are said to be in **equilibrium**. Figure 2.5 illustrates this process.

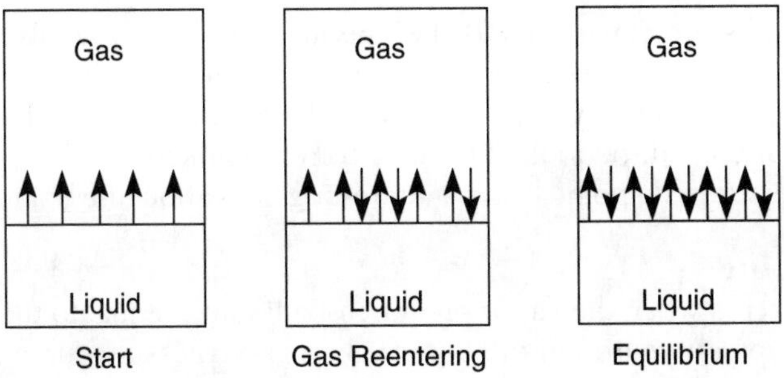

FIGURE 2.5. *Establishment of the equilibrium vapor pressure. At the start, molecules leave the liquid into empty space. The middle diagram shows some gas reentering the liquid but not as rapidly as molecules leave. At equilibrium molecules leave and enter the liquid at the same rate.*

At equilibrium, the rate at which molecules leave the liquid must equal the rate at which they reenter the liquid. The rate at which molecules escape the liquid (evaporate) depends on the temperature. The rate at which they enter the liquid (condense) depends on the frequency at which the gas molecules collide with the liquid "wall" of the container. In the discussion of gases on page 164, it was shown that the frequency of collision is part of the definition of gas pressure. As a result, the vapor pressure depends only on the nature of the liquid (attractive forces) and the temperature (kinetic energy). As the temperature increases, the vapor pressure increases, as shown in Figure 2.6.

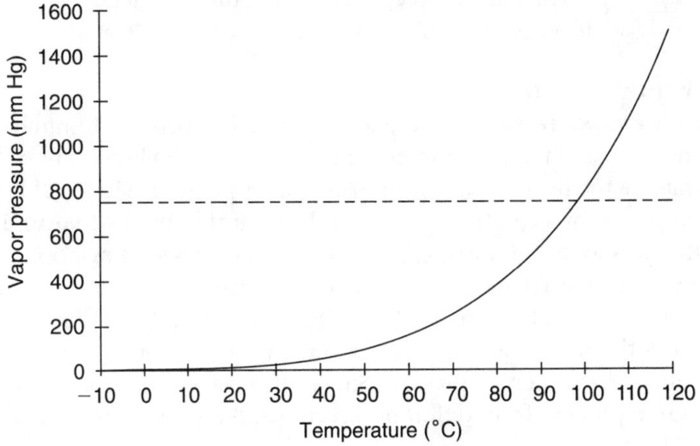

FIGURE 2.6. *Vapor pressure curve for water. Dashed line at 760 mm Hg intersects the curve at 100° C, the normal boiling point of water.*

A vapor pressure curve such as the one in Figure 2.6, or a tabular form of the data, may be used to determine the vapor pressure of a liquid at any temperature. Consequently, chemists do not have to repeatedly determine the vapor pressure; it can be looked up in a convenient source.

Vapor pressure data point to some interesting facts about the boiling process.

Boiling Point

Boiling occurs when the vapor pressure of a liquid is equal to the prevailing atmospheric pressure around the liquid. The vapor pressure curve shows that the temperature at which a liquid boils can vary greatly with changes in the atmospheric pressure. Therefore, the boiling point of a compound is not a constant unless the pressure is also specified. The term **normal boiling point** refers to a boiling point measured when the atmospheric pressure is 760 millimeters of mercury (1.00 atm).

The change in boiling point with pressure has many practical aspects. If, for example, a liquid decomposes at a certain temperature, decomposition can be avoided by reducing the pressure so that boiling occurs at a much lower temperature. This technique is used in vacuum distillation to purify heat-sensitive materials.

As another example, the normal atmospheric pressure in Denver, Colorado, is much lower than 760 millimeters of mercury because of the city's mile-high altitude. The result is that water boils at a lower temperature. At high elevations longer heating times are needed to cook food properly at lower temperatures. For this reason, many people who live at high altitudes use pressure cookers. These covered pots increase the pressure and therefore the boiling point of water. At the increased temperature foods cook faster. The rates of chemical reactions approximately double for every 10°C increase in temperature, and cooking food represents a complex set of chemical reactions carried out simultaneously with pleasing results.

Heat of Vaporization

The **heat of vaporization** is the energy needed to convert 1 gram of liquid into 1 gram of gas at a temperature equal to the normal boiling point of the liquid. The units for the heat of vaporization are joules per gram ($J\ g^{-1}$). If 1 mole of liquid is vaporized, we call the energy used the **molar heat of vaporization** and use the units joule per mole ($J\ mol^{-1}$). In either case the symbol is ΔH_{vap}. Since energy must always be added to a liquid to cause it to vaporize, ΔH_{vap} is always positive. In Chapter 5, pages 175–176, a positive ΔH_{vap} is defined as indicating an endothermic process. In the reverse process, condensation, the gas gives off heat in an exothermic process. Since vaporization and condensation describe the same process from different directions, their heats are related by the equation

$$\Delta H_{vap} = -\Delta H_{cond}$$

Table 2.3 lists some heats of vaporization for compounds with different types of intermolecular forces.

TABLE 2.3. HEATS OF VAPORIZATION OF REPRESENTATIVE COMPOUNDS

Compound	Formula	Heat of Vaporization (kJ mol^{-1})	Attractive Force
Water	H_2O	+ 43.9	Hydrogen bonding
Ammonia	NH_3	+ 21.7	Hydrogen bonding
Hydrogen fluoride	HF	+ 30.2	Hydrogen bonding
Hydrogen chloride	HCl	+ 15.6	Dipole-dipole
Hydrogen sulfide	H_2S	+ 18.8	Dipole-dipole
Fluorine	F_2	+ 5.9	London
Chlorine	Cl_2	+ 10.0	London
Bromine	Br_2	+ 15.0	London
Methane	CH_4	+ 8.2	London
Ethane	C_2H_6	+ 15.1	London
Propane	C_3H_8	+ 16.9	London

The amount of heat required to vaporize a liquid is very large. For example, the heat energy needed to vaporize 1 gram of water could be used to raise the temperature of six times as much water from 0 to 100°C. This fact explains why water can be quickly raised to its boiling point, but a long time is needed to boil away all of the water.

Relationship Between Heat of Vaporization and Intermolecular Attractive Force

There are differences in the heats of vaporization that can be related to the intermolecular attractive forces. For similar-size molecules, hydrogen-bonded substances have the largest ΔH_{vap} values. Polar substances have higher heats of vaporization than similar-size nonpolar substances. Molecules that have the same intermolecular attractive forces also show trends in their heats of vaporization. Water has the highest ΔH_{vap}, and ammonia the lowest, among hydrogen-bonded molecules. Fluorine, chlorine, and bromine show a regular increase in ΔH_{vap}, and methane, ethane, and propane also have increasing ΔH_{vap} values because of increasing London forces.

EXERCISE

For each of the following pairs, predict which compound will have (1) the lower boiling point, (2) the higher heat of vaporization, (3) the higher evaporation rate, and (4) the lower vapor pressure:

(a) C_6H_{14} or C_8H_{18}

(b) C_6H_{14} or $C_6H_{13}OH$

(c) HF or HCl

(d) HBr or HCl

(e) $C_6H_{13}OH$ or $C_3H_7OC_3H_7$

(f) PBr_3 or PBr_5

Solution

For each pair we have to assess the relative attractive forces. The compound with the greater attractive forces will have the higher boiling point, higher heat of vaporization, lower evaporation rate, and lower vapor pressure. For the six pairs the substances with the greater attractive forces are as follows: (a) C_8H_{18} since it has more H atoms for greater London forces; (b) $C_6H_{13}OH$ since it has the $-OH$ group, which forms hydrogen bonds; (c) HF since it forms hydrogen bonds; (d) HBr since its electron cloud is more polarizable; (e) $C_6H_{13}OH$ since it forms hydrogen bonds; (f) PBr_5 since it has more polarizable Br atoms (the polarity of PBr_3 is a minor factor). Once the attractive forces are determined, the answers can be obtained for each pair:

(a) 1. C_6H_{14}
2. C_8H_{18}
3. C_6H_{14}
4. C_8H_{18}

(b) 1. C_6H_{14}
2. $C_6H_{13}OH$
3. C_6H_{14}
4. $C_6H_{13}OH$

(c) 1. HCl
2. HF
3. HCl
4. HF

(d) 1. HCl
2. HBr
3. HCl
4. HBr

(e) 1. $C_3H_7OC_3H_7$
2. $C_6H_{13}OH$
3. $C_3H_7OC_3H_7$
4. $C_6H_{13}OH$

(f) 1. PBr_3
2. PBr_5
3. PBr_3
4. PBr_5

Clausius-Clapeyron Equation

The heat of vaporization and the vapor pressure are both measures of the intermolecular forces that attract molecules together in the liquid state. Because they describe the same forces, there is a mathematical relationship, called the **Clausius-Clapeyron equation**, between the two:

$$\ln P = \frac{-\Delta H_{\text{vap}}}{RT} + C \qquad (2.4)$$

where P represents the vapor pressure, ΔH_{vap} is the heat of vaporization, R is the universal gas law constant, T is the Kelvin temperature, and C is a constant of integration. A plot of the natural logarithm of the vapor pressure versus $1/T$ will produce a straight-line graph in which the slope of the line will be $-\Delta H_{\text{vap}}/R$.

Another form of the Clausius-Clapeyron equation relates the vapor pressures at two temperatures to ΔH_{vap} as shown in equation 2.5.

$$\ln\left(\frac{P_1}{P_2}\right) = \frac{\Delta H_{\text{vap}}}{R}\left(\frac{1}{T_2} - \frac{1}{T_1}\right) \qquad (2.5)$$

This equation is obtained by taking two equations in the form of equation 2.4 and subtracting one from the other. The result is that the constant of integration (*C*) disappears.

Equation 2.5 has five variables: two temperatures, two pressures, and ΔH_{vap}. It can be solved for any one of these if the other four variables are known. For instance, ΔH_{vap} can be determined by measuring the vapor pressures at two different temperatures. Or, if the heat of vaporization is known, the vapor pressure at any temperature can be determined if the vapor pressure is known at any other temperature. Finally, the normal boiling point of a liquid can be determined from the ΔH_{vap} value and a known vapor pressure at any temperature. This calculation is possible since the normal boiling point is always at 760 millimeters of mercury, or 1.00 atmosphere, of pressure.

EXERCISE

What are (a) the heat of vaporization, in kilojoules per mole, and (b) the normal boiling point of a liquid that has a vapor pressure of 254 millimeters of mercury at 25°C and a vapor pressure of 648 millimeters of mercury at 45°C? ($R = 8.314\ \text{J mol}^{-1}\ \text{K}^{-1}$)

Solution

(a) The Clausius-Clapeyron equation is used first to determine ΔH_{vap} and then to determine the normal boiling point at 760 mm Hg. At this point, the necessary units are considered. The temperature must be converted to kelvins because K units are in the units for *R*. *R* has no units for pressure since these units will cancel in the ratio P_1/P_2. The pressures may have any units as long as they are identical. We now use equation 2.5:

$$\ln\left(\frac{P_1}{P_2}\right) = \frac{\Delta H_{vap}}{R}\left(\frac{1}{T_2} - \frac{1}{T_1}\right)$$

In solving this equation, we assign 298 K and 254 mm Hg to T_1 and P_1, respectively. T_2 and P_2 are 318 K and 648 mm Hg. Substituting these values into the equation, we have

$$\ln\left(\frac{254\ \text{mm Hg}}{648\ \text{mm Hg}}\right) = \frac{\Delta H_{vap}}{8.314\ \text{J mol}^{-1}\ \text{K}^{-1}}\left(\frac{1}{318} - \frac{1}{298}\right)$$

Solving yields

$$-0.937 = \Delta H_{vap}\ (-2.53 \times 10^{-5}\ \text{J}^{-1}\ \text{mol})$$

Then

$$\Delta H_{vap} = \frac{-0.937}{-2.53 \times 10^{-5}\ J^{-1}\ mol}$$
$$= +36{,}911\ J\ mol^{-1}$$
$$= +36.9\ kJ\ mol^{-1}$$

(b) Once the heat of vaporization has been determined, one of the two vapor pressure measurements can be combined with it to calculate the normal boiling point. Assigning P_1 = 648 mm Hg, P_2 = 760 mm Hg, and T_1 = 318 K, we enter the data into the equation and calculate T_2, the normal boiling point.

Before actually solving the equation, we can estimate the answer. First, it must be above 318 K since 648 mm Hg is less than 760 mm Hg. Second, the normal boiling point cannot be too much greater than 318 K since 648 mm Hg is not too far from 760 mm Hg.

Now, entering the data into the Clausius-Clapeyron equation, we have

$$\ln\left(\frac{648\ mm\ Hg}{760\ mm\ Hg}\right) = \frac{36{,}900\ J\ mol^{-1}}{8.314\ J\ mol^{-1}\ K^{-1}}\left(\frac{1}{T_2} - \frac{1}{318}\right)$$

We can solve this for T_2, the boiling temperature:

$$-0.159 = 4438\ K\left(\frac{1}{T_2} - 3.14 \times 10^{-3}\ K^{-1}\right)$$

$$= \frac{4438\ K}{T_2} - 13.94$$

$$13.78 = \frac{4438\ K}{T_2}$$

$$T_2 = \frac{4438\ K}{13.78}$$

$$= 322\ K\ (49°C)$$

The 322 K obtained for this part of the problem fits well with the estimate (slightly above 318 K) made beforehand.

Aqueous Solutions

Classification of Solutes

A solution in which water is the solvent is known as an **aqueous solution**. Ionic substances **dissociate** (break apart) completely into ions when dissolved in water. Some molecular compounds, notably hydrogen chloride, HCl, ionize when dissolved in water. Since their solutions conduct electricity very well, these compounds are called **electrolytes**. Other compounds dissociate only slightly when dissolved in water; the solutions conduct electricity, but not well. These compounds are called **weak electrolytes**. The remaining soluble compounds form absolutely no ions when dissolved in water. Since their solutions do not conduct electricity, these compounds are called **nonelectrolytes**.

Many substances dissociate into ions when dissolved in a solvent. Sometimes this process is incorrectly called **ionization**. The term *ionization* should be reserved for processes that form ions from atoms or molecules.

Electrolytes

Electrolytes are ionic compounds, such as sodium chloride, NaCl; potassium bromide, KBr; and magnesium nitrate, $Mg(NO_3)_2$, that are soluble in water. In addition, some electrolytes are molecular (covalent) compounds in the gas phase, but they ionize completely when dissolved in water. These compounds are HCl, HBr, and HI. A typical reaction is as follows:

$$HCl(g) \quad + \quad H_2O(\ell) \quad \rightarrow \quad H_3O^+(aq) \quad + \quad Cl^-(aq)$$

Weak Electrolytes

Weak electrolytes are soluble molecular compounds that dissociate only partially into ions when dissolved in water. Most weak electrolytes dissociate less than 10 percent, meaning that over 90 percent of the molecules remain intact. Weak electrolytes are the weak acids and weak bases. Some weak acids are listed in Table 2.4.

TABLE 2.4. SOME COMMON WEAK ACIDS

Name	Formula	Alternative Formula
Acetic acid	$HC_2H_3O_2$	CH_3COOH
Formic acid	$HCHO_2$	HCOOH
Propionic acid	$HC_3H_5O_2$	CH_3CH_2COOH
Benzoic acid	$HC_7H_5O_2$	C_6H_5COOH
Hypochlorous acid	HClO	HOCl
Hydrosulfuric acid	H_2S	None
Hydrofluoric acid	HF	None
Phosphoric acid	H_3PO_4	None
Water	H_2O	HOH

All weak bases are related to the weak base ammonia. Some weak bases are methylamine, CH_3NH_2, diethylamine, $(CH_3CH_2)_2NH$, and ethylenediamine,

$H_2NCH_2CH_2NH_2$. The reaction of a weak electrolyte with water results in an **equilibrium** that is shown with a double arrow in a reaction:

$$NH_3(g) \quad + \quad H_2O(\ell) \quad \rightleftharpoons \quad NH_4^+(aq) \quad + \quad OH^-(aq)$$

Nonelectrolytes

Nonelectrolytes are molecular compounds that dissolve in a solvent but show absolutely no tendency to form ions. Sugars such as glucose, $C_6H_{12}O_6$, and sucrose, $C_{12}H_{22}O_{11}$ (table sugar), and alcohols such as methanol, CH_3OH, ethanol, CH_3CH_2OH, and propanol, $CH_3CH_2CH_2OH$, are examples of soluble nonelectrolytes.

Solubility and the Solution Process

Whether a substance will dissolve in a solvent is of major importance in chemistry. The ability to separate and purify compounds often depends on the differences in solubility of the compounds. The effectiveness of drugs depends on their solubilities in water and body fats. The ability of plants to use nutrients in the soil depends on the solubility of the nutrients in water.

Since all gases are soluble in each other in all proportions and the consideration of alloys is beyond the scope of this book, we shall focus on liquid-phase solutions. With liquids, the general rule for **solubility** is often stated as "Like dissolves like," meaning that solutions can be made from solutes and solvents with similar polarities but not from solutes and solvents that have very different polarities.

Focusing on the solution process more closely, we find that the dissolution of one substance in another involves three distinct steps, as shown in Figure 2.7. First, the solute molecules are separated so that the solvent can fit between them (Step A: Energy is used to break the attractions between solute molecules). Second, the solvent molecules are separated so that the solute can fit between them (Step B: Energy is used to break the attractions between solvent molecules). Finally, the separated solute and solvent molecules are combined to form the solution (Step C: Energy is released when new attractions are formed between solute and solvent molecules). When more energy is released in Step C than is used in Steps A and B, the solvent will dissolve the solute. This excess energy is released as heat when the solute and solvent are mixed, and an increase in temperature is observed.

In addition to the breaking and making of attractions between molecules, Figure 2.7 illustrates that the randomness of molecules is another factor in the solution process. In diagram form, the solute is shown initially as an ordered crystal but in the solution the solute is randomly arranged. This increase in randomness, more precisely called **entropy**, helps the solution process. In some solutions the energy released in Step C is slightly less than the energy used in Steps A and B, but the increase in randomness provides the extra driving force that makes dissolution possible. In this case, the solution may cool as the solvent and solute are mixed (an example is KCl dissolved in H_2O);

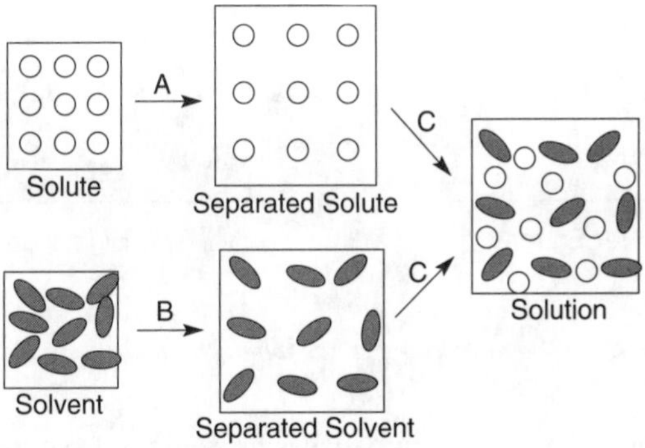

FIGURE 2.7. *Diagram of the three steps involved in forming a solution. Step A separates the solute, Step B separates the solvent, and Step C combines the separated solvent and solute into a solution.*

the shortfall in energy is compensated for by the increased randomness of the solution.

Solutes that are solids have the greatest increase in randomness when dissolved. Liquids have a moderate increase, and gases have virtually no increase.

Examples of the Solution Process

Ionic Solutions: This example describes why ionic compounds dissolve to form aqueous solutions. Ionic compounds have the strongest attractive forces holding the ions in the crystal lattice. The energy needed to disrupt the crystal (Step A of the solution process illustrated in Figure 2.7) is known as the lattice energy. Lattice energies are so high that ionic compounds do not appear to be soluble. However, water, with its high polarity, is strongly attracted to the charged ions in the solution. When many water molecules are attracted to each ion, enough energy is released (Step C of the solution process) so that the dissolution is favored. The increase in entropy due to the breakup of the crystal structure also aids the solution process. Nevertheless, many ionic compounds are insoluble in water because the hydration of the ions cannot offset the lattice energy. The solubility rules for ionic compounds are listed on page 316.

Gas Mixtures: This example demonstrates that gases mix freely with other gases solely on the basis of the entropy increase since there are virtually no attractive forces between gas molecules. When a gas is allowed to expand into a vacuum, it does so because the increased volume means more disorder for the molecules, as shown in Figure 2.8.

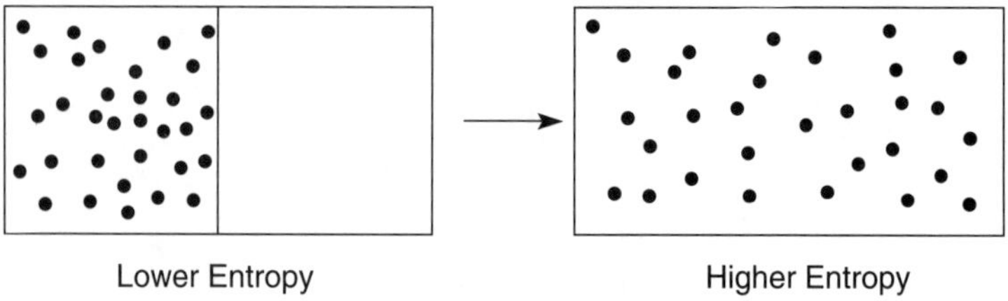

FIGURE 2.8. *Increase in the entropy of a gas when the volume is increased.*

When the barrier between two gases is removed, so that each gas can expand into the volume formerly occupied by the other, the entropy of each gas increases. Since there are no attractive forces, each gas "thinks" it is expanding into a vacuum.

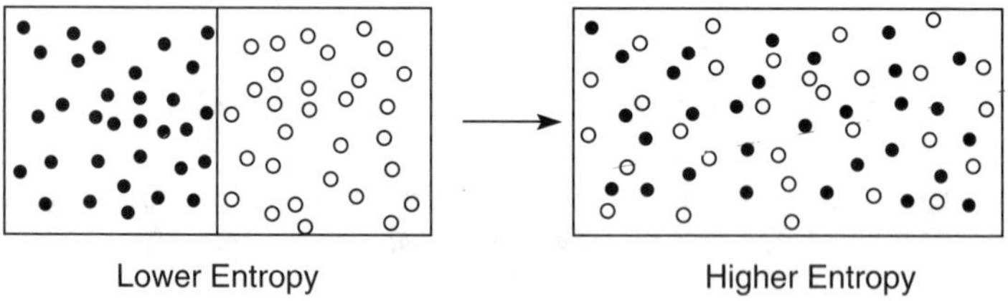

FIGURE 2.9. *Mixing of two gases illustrating the increase in entropy for each.*

Detergents: In this example, the solubility effects of compounds called detergents are described. Detergents have two types of attractive forces. As shown in Figure 2.10, one end of a long-chain carbon molecule contains an ionic group and the other end is a long-chain hydrocarbon. The ionic end causes a detergent molecule to be soluble in water, and the hydrocarbon end can dissolve in nonpolar substances. An aqueous solution of a detergent can remove nonpolar stains from clothes because fats and greases dissolve in the hydrocarbon end of the detergent molecule.

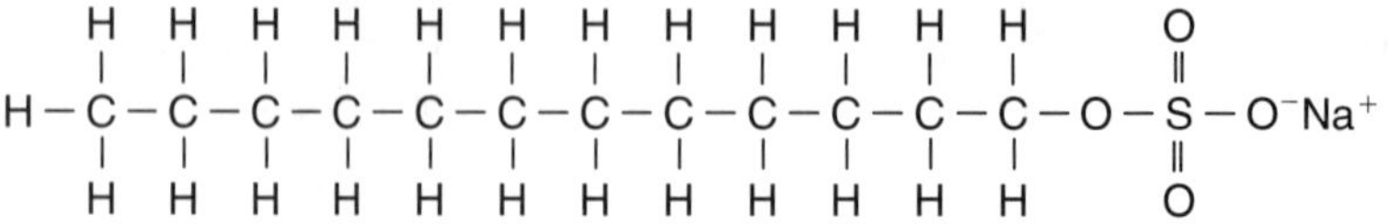

FIGURE 2.10. *Diagram illustrating the structure of the detergent sodium dodecyl sulfate. The ionic end dissolves in water, and the hydrocarbon end dissolves in nonpolar substances such as fats and grease.*

Rates of Dissolution

In addition to the attractive forces that determine if a solute will dissolve, we are interested in the **rate** at which solutes dissolve. Almost intuitively, we heat solutions to increase the speed of dissolution. We also grind chemicals into small particles and stir solutions vigorously to speed the solution process. All three of these operations speed the rate at which solvent molecules can reach the surface of a solid and start the dissolution process. Heating accelerates the motion of the solvent molecules so that they collide with the solid more rapidly. Grinding a solute into fine particles allows more of the solute to be exposed to the solvent. Stirring a solution moves dissolved solute away from the surface of the solute and brings less concentrated solvent in contact with the solute. Although heating, grinding, and stirring increase the rate of dissolution, it must be remembered that the maximum concentration achieved depends only on the temperature of the solution and the chemical identity of the solute itself.

Solubility: Effect of Temperature and Pressure

Effect of Temperature on Solubility

This topic is rather complex, and the details are usually left for more advanced courses. The basic principle of the temperature effect on solubility lies in the randomness (**entropy**) of the molecules. To begin, we recall that gas molecules are always more random than the molecules in liquids, and the molecules in liquids are more randomly arranged than the molecules in highly structured crystals:

Randomness of Molecules
Solids $<<$ liquids $<<$ gases

As a general rule, increasing temperature drives molecules toward the more random phase. Thus an increase in temperature usually increases the solubility of solids and always decreases the solubility of gases. Decreasing temperatures have the opposite effects.

Most solids are more soluble in hot solvents than in cold solvents. Figure 2.11 illustrates the changes in solubility of a few solid compounds as the temperature is increased. A simple explanation is that increasing temperature increases the disorder of all molecules. If two states are present, molecules will tend to move from an ordered to a more disordered state. A more precise explanation is left for more advanced chemistry courses.

In agreement with the general principle stated above, the solubility of gases in liquids always decreases as the temperature is increased. You may have noticed that a warm bottle of soda fizzes more than a cold bottle when opened. In this case the increased disorder due to the warmer temperature favors molecules leaving the liquid phase and entering the gas phase since molecules are always more random in the gas phase than in the liquid phase.

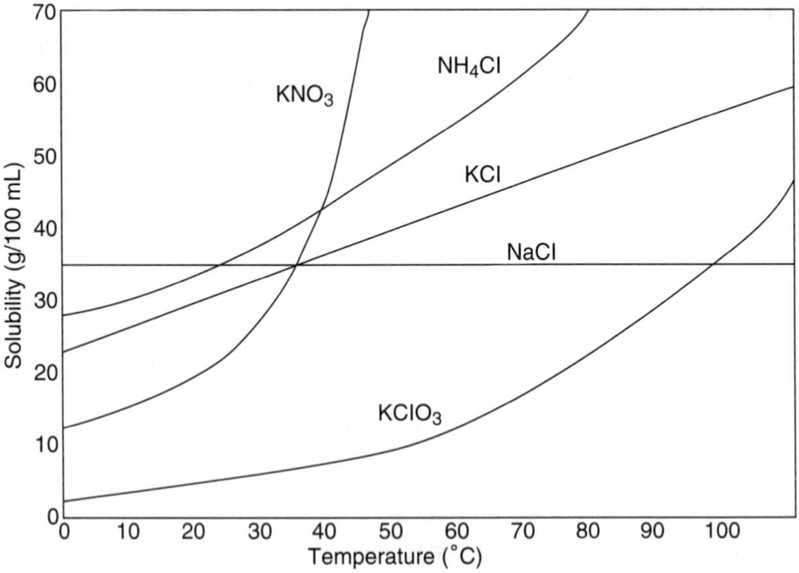

FIGURE 2.11. *Diagram of the temperature dependence of solubility for several representative ionic compounds.*

Effect of Pressure on Solubility

External pressure has no significant effect on the solubility of solids or liquids because solids and liquids are not appreciably compressed when the pressure is increased. Gases, on the other hand, are easily compressed. Compression of a gas increases its concentration and, at the same time, the frequency at which gas molecules hit the liquid phase and enter it, thereby increasing the solubility.

The effect of pressure on the solubility of gases is expressed in Henry's law:

$$\text{Solubility}_{\text{gas}} = kP_{\text{gas}}$$

$\text{Solubility}_{\text{gas}}$ is usually expressed as molarity (M) or weight per volume (wt/vol) concentration units, k is the Henry's law proportionality constant, and P_{gas} is the partial pressure of the gas.

If the solubility of a gas is known at one pressure, Henry's law can be used to determine the solubility at another pressure, as long as the temperature remains the same.

EXERCISE

The solubility of oxygen in water is 1.25×10^{-3} M at 25°C at sea level (1.00 atm). At the same temperature, what will the solubility of oxygen be in Denver, where the atmospheric pressure is 0.800 atmosphere?

Solution

The solution requires two steps. First, the Henry's law constant is determined by inserting the sea-level information into the equation and solving for *k*. Next, the value of *k*, along with the atmospheric pressure in Denver, is used to calculate the solubility.

The value of *k* is calculated as

$$1.25 \times 10^{-3}\ \text{M}\ O_2 = k\,(1.00\ \text{atm})$$
$$k = 1.25 \times 10^{-3}\ \text{M}\ O_2\ \text{atm}^{-1}$$

Then, this value of *k* is used to calculate the solubility of oxygen at Denver's atmospheric pressure:

$$\text{Solubility } O_2 = (1.25 \times 10^{-3}\ \text{M}\ O_2\ \text{atm}^{-1})(0.800\ \text{atm})$$
$$= 1.00 \times 10^{-3}\ \text{M}\ O_2$$

Alternate Method: Another method for solving this type of problem is to write the Henry's law equation for two states, indicated by subscripts 1 and 2:

$$\text{Solubility}_1 = kP_1 \quad \text{and} \quad \text{Solubility}_2 = kP_2$$

The *k* does not need a subscript since it is constant under all conditions. Next, the ratio of these two equations is taken:

$$\frac{\text{Solubility}_1}{\text{Solubility}_2} = \frac{kP_1}{kP_2}$$

The constants *k* cancel, leaving

$$\frac{\text{Solubility}_1}{\text{Solubility}_2} = \frac{P_1}{P_2}$$

For ease of calculation we define solubility_1 as the solubility in Denver, where $P_1 = 0.800$ atm. Then

$$\text{Solubility}_2 = 1.25 \times 10^{-3}\ \text{M} \quad \text{and} \quad P_2 = 1.00\ \text{atm}$$

Entering these data gives

$$\frac{\text{Solubility}_1}{1.25 \times 10^{-3}\ \text{M}} = \frac{0.800\ \text{atm}}{1.00\ \text{atm}}$$

$$\text{Solubility}_1 = \frac{(0.800\ \text{atm})(1.25 \times 10^{-3}\ \text{M})}{1.00\ \text{atm}}$$

$$= 1.00 \times 10^{-3}\ \text{M}\ O_2$$

In this method the algebra is done with the symbolic variables before any data are entered or calculations performed.

SOLIDS

Solids

Properties of Solids

At room temperature and atmospheric pressure many substances exist as solids. In the Periodic Table there are two liquids and 11 gases; the remaining 98 elements are solids. Solids have the property of retaining their shapes, with or without a container, because of their rigid crystal structures. These solid structures may be defined based on the attractive forces that hold them together or on the arrangement of the atoms in the crystals themselves.

The properties of solids are rather limited. We measure the melting and boiling points of substances that are normally solids. A solid may or may not conduct electricity, reflect light, or be hard or soft, brittle or malleable. The properties of solids are related to the type and chemistry of the crystal formed, as discussed below.

Crystal Types Based on Attractive Forces

Metallic Crystals: All metals in the Periodic Table except mercury are solids at 25°C. The **metallic crystal** is visualized as a rigid structure of metal nuclei and inner electrons. The valence electrons are thought to be very mobile in the structure, moving freely from atom to atom. These mobile electrons act to bond metal atoms together with widely varying degrees of force. Metals such as iron, chromium, cobalt, gold, platinum, and copper have melting points above 1000°C. Others, for example, mercury and gallium, have melting points near or below room temperature. Melting points are one measure of the attractive forces since melting disrupts crystal bonding, producing a liquid. The energy needed to disrupt a crystal is called the **lattice energy**.

The mobile valence electrons provide an explanation for the ability of metals to conduct electricity and heat. The electrons can quickly carry charge (electricity) and thermal energy (heat) throughout the metal. Also, the interaction of light with these electrons is responsible for the characteristic metallic luster. Most metals are similar in color to silver or aluminum. Two, copper and gold, are yellow.

Metals such as lead, gold, sodium, and potassium are soft and can be cut with a knife. Other metals, for example, tin and zinc, are somewhat brittle. Most metals are malleable and can be formed into various shapes with a hammer or extruded into thin wires. These properties are due to the metallic crystal structure, which allows the atoms to move from one position to another without a major disruption of the crystal. The softness, hardness, and brittleness of metals can be altered by preparing solutions of one metal dissolved in another. These are known as **alloys**.

The atoms in metallic crystals are arranged in the most compact form possible. As a result, the atoms usually crystallize in either the face-centered cubic structure or the body-centered cubic structure. Both of these are described below.

Ionic Crystals: The attraction of a cation (positive ion) toward an anion (negative ion) is the strongest attractive force known in chemistry. The result is that almost all ionic compounds are solids with rigid crystalline structures (lattices). Because of these strong attractions, a large amount of energy, called the lattice energy, is required to separate the ions. The high lattice energy of ionic compounds gives them very high melting and boiling points compared to molecular compounds of similar size and molar mass. For example, sodium chloride (NaCl, molar mass = 58) melts at 801°C and boils at 1413°C, whereas butane, with almost the same molar mass (C_4H_{10}, molar mass = 58) melts at −135°C and boils at approximately 0°C. The melting points of ionic crystals are consistently high in contrast to the variability in the melting points of metals.

An **ionic crystal** has a regular structure, or lattice, of alternating positive and negative ions. Many of these are cubic structures. Other structures may provide shapes seen in many natural minerals. It is relatively simple to describe the crystal structures of the metals since all of the atoms are the same size. Ionic crystals, however, usually have ions of different sizes, which affect the manner in which they pack. These sizes also may limit the closeness with which the ions approach each other.

Ionic bonding causes these crystals to be rigid and brittle. To understand this property, we visualize a simple ionic substance such as sodium chloride with alternating sodium cations and chloride anions in a crystal lattice. The strong attraction of the positive and negative charges holds the crystal rigidly together. Hitting an ionic crystal with a hammer has a very different result than does hitting a metallic crystal with the same force. In both cases, the atoms can be forced to move. In a metallic crystal, the atoms shift their positions but the metallic bond is not disrupted. In an ionic crystal, however, movement of the atoms by as little as one ionic diameter will cause positive ions to be aligned with positive ions and negative ions to be aligned with negative ions. The repulsion between like-charged ions is so great that the crystal shatters, as diagrammed in Figure 2.12.

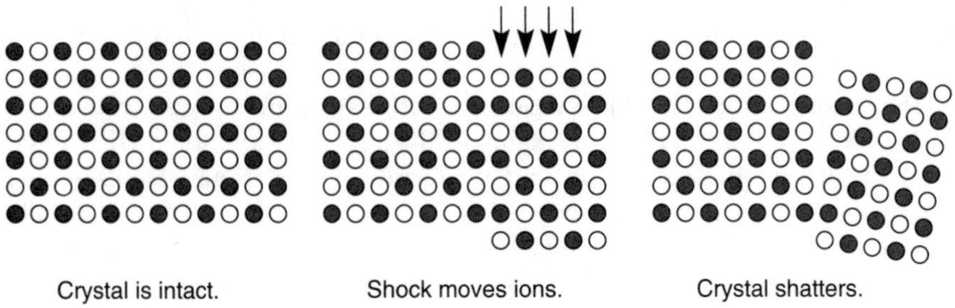
Crystal is intact.
Shock moves ions.
Crystal shatters.

FIGURE 2.12. *Illustration of why ionic crystals shatter. Light circles are cations, and dark circles are anions.*

Molecular Crystals: Molecular crystals may be composed of either atoms of the nonmetals or of covalent molecules. These crystals are held together by London forces, dipole-dipole attractions, hydrogen bonding, or a mixture of these. All of these forces are much weaker than the attractive forces between ions in ionic crystals. As a result, molecular crystals tend to be soft, with low melting points. Some substances that form molecular crystals with London forces holding the crystal together are Ne, Xe, S, F, CH_4, and $C_{10}H_{22}$(decane). Dipole-dipole attractive forces hold crystals of SO_2, NO_2, $CHCl_3$, and other polar molecules together. Hydrogen bonding is responsible for the attractive forces in crystals of H_2O and NH_3.

In many molecules there may be a combination of attractive forces. For example, *n*-decanol has the structure

$$CH_3CH_2CH_2CH_2CH_2CH_2CH_2CH_2CH_2CH_2OH$$

The long carbon chain is responsible for London forces, while the $-OH$ at the end of the molecule provides hydrogen bonding.

Network (Covalent) Crystals: A **network crystal** has a lattice structure in which the atoms are covalently bonded to each other. The result is that the crystal is one large molecule with a continuous network of covalent bonds. A diamond is pure carbon with each carbon atom covalently bonded to four other carbon atoms in a tetrahedral (sp^3) geometry. The totality of this network of covalent bonds makes the diamond the hardest natural substance known. SiO_2 is the empirical formula for sand and quartz. Silicon dioxide forms a covalent crystal with each silicon forming bonds to four oxygen atoms and each oxygen bonding to two silicon atoms with a tetrahedral geometry. Silicon carbide is another network crystal, similar to diamond with alternating tetrahedral silicon and carbon atoms. It is very hard and is used as an industrial substitute for diamonds. Network crystals, like ionic substances, are represented by their empirical formulas.

Graphite is another form of carbon in a covalent crystal. In graphite each carbon atom is covalently bonded to three other carbon atoms in a trigonal planar (sp^2) geometry that gives graphite its structure of flat sheets The extra *p* electron that is not used in the sp^2 bonding holds these sheets together in a manner similar to that in a metallic crystal. The weak bonding of the *p* electrons allows the flat sheets to slide over each other easily and is responsible for the slippery feel of graphite. In addition, these *p* electrons are responsible for the ability of graphite to conduct electricity.

Amorphous (Noncrystalline) Substances: Some materials are **amorphous** and do not form crystals. One characteristic of a noncrystalline substance is that it does not have a precise melting point. Rather, these materials soften gradually over a large temperature range. Ordinary glass is an example. Although glass is composed mainly of SiO_2, the atoms are not arranged in a network crystal as described above. Glass has often been described as a supercooled liquid. Many plastics (polymers) have combined characteristics; they are partially crystalline and partially amorphous.

Crystal Structures

We can determine the positions of atoms in a crystal by using an experimental method called **X-ray diffraction** (Figure 2.13). In an X-ray diffraction experiment a narrow beam of monochromatic (single-wavelength) X rays is aimed at a crystal. On the other side of the crystal is a photographic plate to detect the X rays as they emerge from the crystal. When developed, the photographic plate shows a regular pattern of spots, which can be deciphered to determine the arrangement of the atoms and the distances between them.

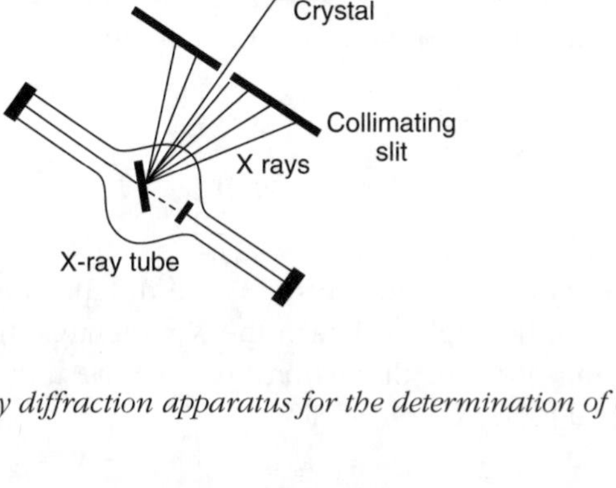

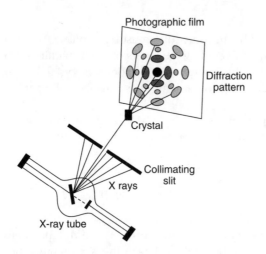

FIGURE 2.13. *X-ray diffraction apparatus for the determination of crystal structures.*

In this technique X rays are diffracted by the layers of atoms within the crystal itself. In diffraction separate light waves may constructively reinforce or destructively reinforce each other when they combine. In Figure 2.14 the two types of reinforcement are diagrammed by representing light as waves, and the combination of two waves as the addition of their amplitudes. If the waves are in phase, with their maxima (peaks) aligned, addition results in a wave with an increased amplitude. If the two waves are out of phase, with maxima aligned with minima, the addition results in zero amplitude or no wave at all.

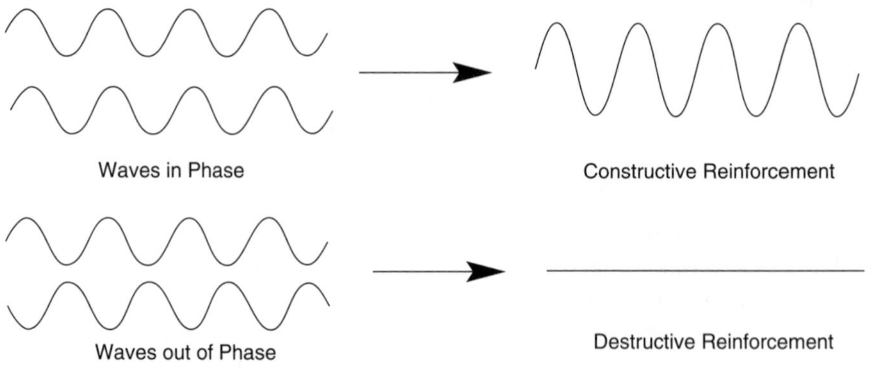

FIGURE 2.14. *How waves in phase constructively reinforce and waves out of phase destructively reinforce.*

X rays generated by an X-ray tube must be in phase. When the waves enter the crystal, they are reflected by the layers of atoms within the crystal as shown in Figure 2.15. When the waves emerge from the crystal, they may or may not be in phase because the lower wave has traveled a longer distance than the upper wave.

The **Bragg equation** (equation 2.6) defines the relationship between the angles and the spacing between the layers in a crystal that results in constructive reinforcement:

$$n\lambda = 2d \sin \theta \tag{2.6}$$

where λ is the wavelength of the X rays, d is the distance between the layers in the crystal, and θ is the angle at which the X ray enters the crystal. Also in this equation n is the order of diffraction and may be any integer value; in most cases n is equal to 1.

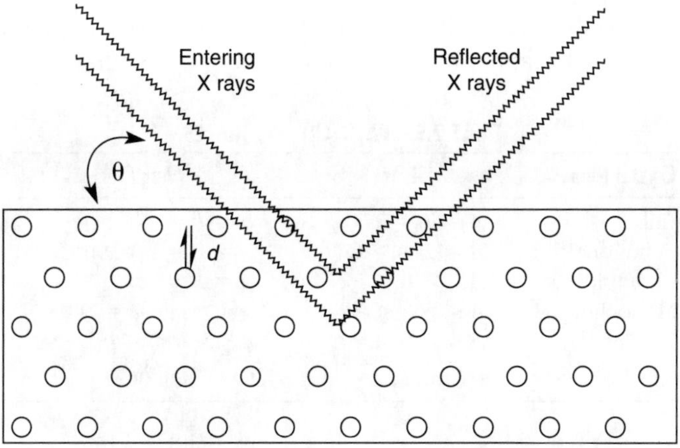

FIGURE 2.15. *Reflection of X rays by the layers of atoms within a crystal.*

EXERCISE

What is the minimum spacing between atoms if X rays with a wavelength of 216 picometers are constructively reinforced when reflected at an angle of 31.2°?

Solution

We have all of the data needed to solve the Bragg equation, equation 2.6, except n. We will solve the equation as far as possible by substituting the data:

$$n\lambda = 2d \sin \theta$$

$$n(216 \text{ pm}) = 2d \sin 31.2$$

$$d = \frac{n(216 \text{ pm})}{2 \sin 31.2}$$

$$= n(208 \text{ pm})$$

Since n may be any integer, starting with 1, we see that the minimum spacing will be 1 × 208 pm, or 208 pm.

The arrangement of the atoms in a solid is called the **crystal lattice**. Since a crystal is made up of repeating **unit cells** stacked together, the crystal observable to the naked eye is an enlarged version of the unit cell itself. Sodium chloride is an example. If we closely examine ordinary table salt, we see many tiny cubes. X-ray diffraction experiments confirm that the atoms are actually arranged in a cubic structure. Basalt, a common rock, forms long crystals with a hexagonal cross section that also describes the shape of the unit cell. Since unit cells must pack together without any spaces, there are only six fundamental crystal shapes as listed in Table 2.5.

TABLE 2.5. BASIC CRYSTAL SHAPES

Crystal Name	Sides	Angles
Cubic	$A = B = C$	$a = b = c = 90°$
Tetragonal	$A = B <> C$	$a = b = c = 90°$
Rhombic	$A <> B <> C$	$a = b = c = 90°$
Monoclinic	$A <> B <> C$	$a <> b = c = 90°$
Triclinic	$A <> B <> C$	$a <> b <> c <> 90°$
Hexagonal	$A <> B$	$a = b = 90°, c = 120°$

We will focus only on the cubic structures, called the **simple cubic**, **body-centered cubic**, and **face-centered cubic**. These three structures are shown in Figures 2.16 and 2.17, which present two different views of the cubic structures.

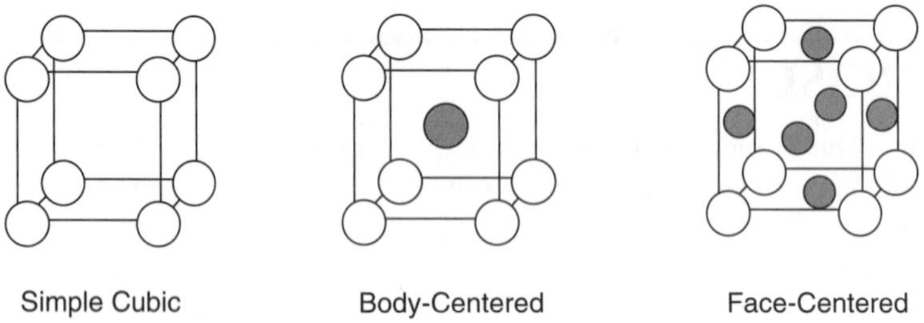

FIGURE 2.16. *Three cubic structures of crystals: simple cubic, body-centered cubic, and face-centered cubic. Diagrams are drawn as "ball and stick" models to show placements of all atoms.*

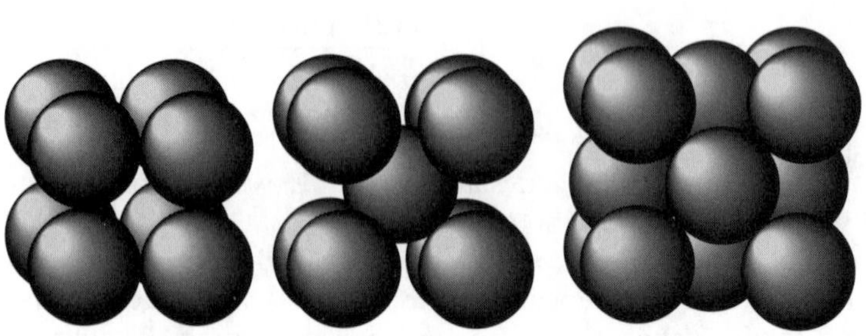

FIGURE 2.17. *Three cubic structures of crystals: simple cubic, body-centered cubic, and face-centered cubic. Diagrams are drawn as "space filling" models to show that atoms actually contact each other.*

In the simple cubic structure, the corner atoms are all in contact with each other. In the body-centered cubic, the corner atoms are not in contact, but all of the corner atoms touch the center atom. In the face-centered cubic, the corner atoms do not touch each other, but they do touch the face atoms.

Another view of the unit cell is obtained by determining the number of atoms that are actually contained within the cell. In Figures 2.16 and 2.17 atoms are placed at the corners and faces of the unit cell. Only parts of these atoms are within the unit cell itself. Figure 2.18 shows the simple cubic structure and the parts of the atoms that are actually within the cell. In this example, only one-eighth of each corner atom is within the unit cell. The other parts of these corner atoms are parts of other unit cells.

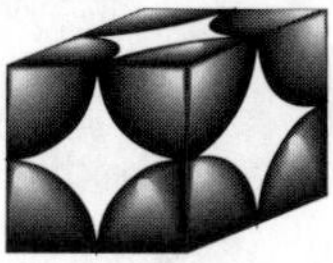

FIGURE 2.18. *Simple cubic unit cell, showing that only one-eighth of each corner atom is actually part of the cell.*

Table 2.6 lists the positions of other atoms and the fractions that are completely inside the unit cell.

TABLE 2.6. FRACTIONAL PARTS OF ATOMS WITHIN A UNIT CELL

Atom Position	Fraction Inside Cell
Body	$\frac{1}{1}$
Face	$\frac{1}{2}$
Edge	$\frac{1}{4}$
Corner	$\frac{1}{8}$

Using the information in this table, we can calculate that there is only one atom inside the simple cubic unit cell (eight corner atoms). For the face-centered cell we have four atoms in the unit cell (eight corner atoms and one-half each of six face atoms). For the body-centered structure there are two atoms inside the unit cell (eighth corner atoms and one body atom).

The length of a side of the cubic cell can be obtained from X-ray diffraction experiments. This measurement, along with the geometry of the atoms in the unit cell, allows us to calculate the radius of an atom. For the simple cubic structure there is a nucleus at each corner of the cube. If the atoms are in contact, the edge of the cube represents two atomic radii. In the face-centered cubic structure, the corner atoms are in contact with the face atom. The diagonal of one side of the cube is equal to four radii (one radius for each corner atom and two radii for the face atom). In the body-centered cubic structure, the atoms are in contact along the diagonal running through the cube from one corner to the opposite corner. Again there are four radii, as in the face-centered cubic. The three equations for determining the radius (r) of the atom from the length of the side (s) of the unit cell are as follows:

$$2r = s \qquad \text{(simple cubic)}$$

$$4r = \sqrt{s^2 + s^2} \qquad \text{(face-centered cubic)}$$

$$4r = \sqrt{s^2 + s^2 + s^2} \qquad \text{(body-centered cubic)}$$

This approach is often used to determine the atomic radii of the elements. The structure of an element is relatively easy to analyze since all of the atoms are identical in size.

The unit cell of a cubic ionic compound may be constructed with one cation at each corner of the cube or, alternatively, with an anion at each corner. Geometrically the structures are the same. In some ionic compounds, for example, lithium bromide, there is sufficient space for the smaller cation to fit between the larger anions. The result is that the anions will be as close as possible to each other. In other compounds, for example, sodium fluoride, the ions are of almost equal size, and the cation will force the anions apart so that they are not in contact with each other.

The size of an ion is more difficult to determine than the size of a metallic atom for two reasons. First, the size of an ion depends on the cation in the compound. Second, because cations and anions are not the same size, it is sometimes difficult to choose the correct geometry for the calculation. Ionic radii are usually average values determined from many compounds.

EXERCISES

1. Calcium metal has a density of 1.55 grams per cubic centimeter. Assuming that it crystallizes into one of the three cubic structures, determine the dimensions of these cubes.
2. Using the data from the previous exercise, estimate the atomic radius of the calcium atom for all three possible structures. If the accepted radius is 197 picometers, what crystal structure does calcium have?

Solutions

1. We can use Avogadro's number, 6.02×10^{23}, along with the gram-atomic mass of calcium, 40.078 g mol^{-1}, to calculate the volume occupied by one atom. We start by deciding on the units for the answer, which are cubic centimeters per Ca atom $\left(\frac{cm^3}{\text{Ca atom}}\right)$, and then we look at the given data to find a starting point. Since our answer requires a ratio of units, we choose the density as the starting point because it also has a ratio of units. Setting up the problem, we see that this calculation involves a conversion from grams to atoms.

$$?\frac{cm^3}{\text{Ca atom}} = \frac{cm^3}{1.55 \text{ g Ca}}$$

Now we convert the denominator from Ca to atoms Ca:

$$?\frac{cm^3}{\text{atom Ca}} = \frac{cm^3}{1.55 \text{ g Ca}}\left(\frac{40.078 \text{ g Ca}}{1 \text{ mol Ca}}\right)\left(\frac{1 \text{ mol Ca}}{6.02 \times 10^{23} \text{ atoms Ca}}\right)$$

We solve this to obtain

$$?\frac{cm^3}{\text{atom Ca}} = 4.30 \times 10^{-23} \text{ cm}^3 \text{ per Ca atom}$$

$$= 4.30 \times 10^{-29} \text{ m}^3 \text{ per Ca atom}$$

Since a simple cubic structure has one atom in its unit cell, the unit cell must have a volume of 4.30×10^{-29} m^3. Taking the cube root of this gives us the length of one side of the cube, or 3.50×10^{-10} m. This can be converted into the more convenient picometer value, 350 pm.

For a body-centered cubic structure there are two atoms in a unit cell, for a volume of 8.60×10^{-29} m^3. Taking the cube root gives us 441 pm.

The face-centered structure has four atoms per unit cell, for a total volume of 1.72×10^{-28} m^3. Taking the cube root gives the length of a side as 556 pm.

2. For the simple cubic structure, $2r = s$, and our value of 350 pm for s gives us a radius of 175 pm.

For the face-centered cubic, $4r = \sqrt{s^2 + s^2}$. Solving this, we obtain $r =$ 197 pm.

For the body-centered cubic, $4r = \sqrt{s^2 + s^2 + s^2}$. Solving this, we obtain $r = 191$ pm.

These data indicate that calcium crystallizes in the face-centered cubic structure.

PHASE CHANGES

Phase Diagrams

A **phase diagram** shows the relationship between pressure and temperature and the three states of matter. A phase diagram of a substance allows us to make many predictions about the physical behavior of a wide variety of materials. Most phase diagrams are similar to the one shown in Figure 2.19. Pressure is plotted on the *y*-axis, and temperature on the *x*-axis.

The diagram is divided into the three physical states by three lines that meet at the triple point. Solids exist at the high pressures and low temperatures in the upper left of the diagram. Liquids exist at the pressures and temperatures in the upper right, and gases exist at low pressures and high temperatures in the lower part of the diagram.

Each line in the diagram represents an **equilibrium** mixture of the two phases on the two sides of that line. At the triple point all three phases are in equilibrium. The liquid-solid line is usually a straight line, while the gas-solid and liquid-gas lines are curved upward as shown.

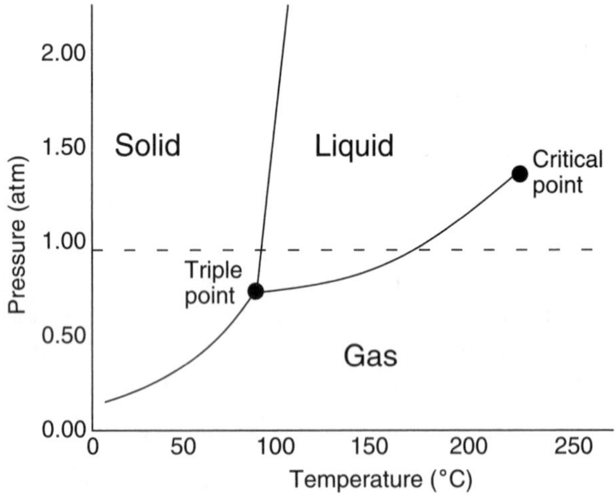

FIGURE 2.19. *A typical phase diagram illustrating the pressure-temperature regions where gas, liquid, and solid phases exist. The lines represent an equilibrium between two phases, and the triple point represents the pressure and temperature at which all three phases are in equilibrium.*

Temperature and pressure are the two variables that define a point in a phase diagram. If we draw a horizontal line representing a selected pressure and a vertical line representing a selected temperature, the intersection of the two lines on the phase diagram will indicate the phase(s) of the substance at that particular combination of temperature and pressure.

The **triple point** is the only point where all three phases are in equilibrium with each other. At this point all the phases have the same temperature and vapor pressure. There is only one temperature and one pressure where this is true for each substance. (Helium is the only substance that does not have a triple point since it has no solid phase.) The triple point of water is at 0.01°C and 4.58 millimeters of mercury.

In the phase diagram three equilibrium lines radiate from the triple point. The pressures and temperatures along these lines indicate the pressures and temperatures where two adjoining phases will be in equilibrium.

The solid-liquid equilibrium line points upward from the triple point:

$$\text{Solid} \rightleftharpoons \text{liquid}$$

This line is nearly vertical because changes in pressure have very little effect on solids and liquids (these two phases are only slightly compressible compared to gases). It leans slightly toward higher temperatures for almost all compounds. For water it slopes in the opposite direction. The slight tilt of this equilibrium line may be explained based on the relative densities of the solid and liquid phases. When the pressure is increased, the equilibrium is shifted toward the denser phase. To counter this shift, the temperature must change to reestablish the equilibrium. Most solids are denser than their liquids. Increasing the pressure forces the reaction toward the formation of solid. To maintain equilibrium, the temperature must be increased; therefore, the line tilts slightly toward the right. For water, the liquid is denser than solid ice. As a result, an increase in pressure favors the formation of liquid water. To maintain equilibrium, the temperature must decrease; therefore, the solid-liquid equilibrium line tilts slightly toward the left. The normal melting point is the point on this line at 760 millimeters of mercury (1 atm) of pressure. We can change the melting point only slightly by changing the pressure.

From the triple point the liquid-gas equilibrium line extends in a curved manner upward toward the higher temperature side:

$$\text{Liquid} \rightleftharpoons \text{gas}$$

This line shows a pronounced curvature because of the compressibility of gases and ends abruptly at the **critical point**, that is, the maximum temperature at which any liquid may exist. Above the critical point the differences between gases and liquids disappear, and the substance is often called a **supercritical fluid**. The normal boiling point of a liquid is the point where the liquid-gas equilibrium line crosses the dashed line, indicating a pressure of 760 millimeters of mercury. Unlike the melting point, the boiling point may vary greatly with pressure.

The solid-gas equilibrium line extends downward toward lower temperatures:

$$\text{Solid} \rightleftharpoons \text{gas}$$

This line is also highly curved because of the compressibility of gases. The process of converting a solid directly into a gas is called **sublimation**. This portion of the diagram is usually at very low pressures not normally encountered in the laboratory. Some substances, such as iodine, have a solid-gas equilibrium line at atmospheric pressure. As a result, iodine readily sublimes from the solid to a vapor without a gas phase.

EXERCISE

Determine the temperatures and pressures for the critical point, triple point, normal boiling point, and normal melting point for the substance depicted in Figure 2.19.

Solution

We may estimate the critical point at 220°C and 1.4 atm of pressure. The triple point is at 95°C and 0.75 atm of pressure. The normal boiling and melting points must be at 1.00 atm of pressure and are at 180°C and 100°C, respectively.

Heating and Cooling Curves

Heating and cooling curves serve as graphical methods for illustrating what happens as a substance is heated or cooled.

A **heating curve** starts with a solid material well below its melting point. As heat is added at a constant rate, the following effects are observed:

1. The temperature of the solid increases at a constant rate until the substance starts to melt.
2. When melting begins, the temperature stops rising and remains constant until all of the solid is converted into a liquid.
3. The temperature of the liquid starts increasing at a constant rate until boiling starts.
4. When boiling begins, the temperature stops rising and remains constant until all of the liquid has been converted into a gas.
5. The temperature of the gas increases at a constant rate.

This process is graphically summarized in the heating curve shown in Figure 2.20.

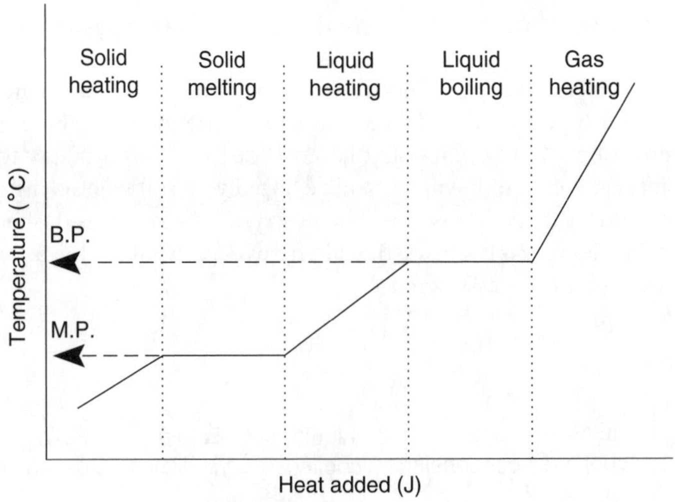

FIGURE 2.20. *Typical heating curve in which a sample is brought from the solid state (left) to the gaseous state (right).*

In the heating curve there are several points of interest. Adding heat to a pure solid, liquid, or gas phase increases its temperature. The heat capacity (J °C^{-1}) of a solid, liquid, or gas is the inverse of the slope of the curve (change in joules divided by a change in temperature) in the regions where the temperature increases as heat is added:

$$\text{Heat capacity} = \frac{1}{\text{slope}}$$

The specific heat, in units of joules per gram per degree Celsius (J g^{-1} °C^{-1}), is the heat capacity divided by the grams of sample used. The plateaus represent the melting and boiling process. The temperature does not change during melting and boiling, and two phases are present. During melting the solid and liquid phases are in equilibrium; during boiling the liquid and gas phases are in equilibrium. The **heat of fusion** (melting) may be determined as the length of the first (solid melting) plateau, which represents the heat added, divided by the grams of sample. The **heat of vaporization** is the length of the second (liquid boiling) plateau divided by the grams of sample. Both have units of joules per gram. (J g^{-1}).

The **cooling curve** is the reverse of the heating curve, as shown in Figure 2.21. All of the features are the same except that we start with a gas and end with a solid as the heat is removed from the sample. The gas condenses to a liquid at the same temperature that the liquid boils, and the liquid crystallizes at the same temperature that the solid melts. The terms *condensation point* and *crystallization point* are sometimes used in place of *boiling point* and *melting point.*

Some liquids exhibit an ability to be supercooled. **Supercooling** is observed when a liquid is cooled to a temperature below its melting point, yet still remains a liquid. A supercooled liquid is described as a metastable stable state. A metastable liquid will crystallize rapidly if sufficiently disturbed by shaking or adding a seed crystal to initiate crystallization. The dashed line on the curve in Figure 2.21 shows the alternative path taken by a liquid that supercools. Gases do not supercool.

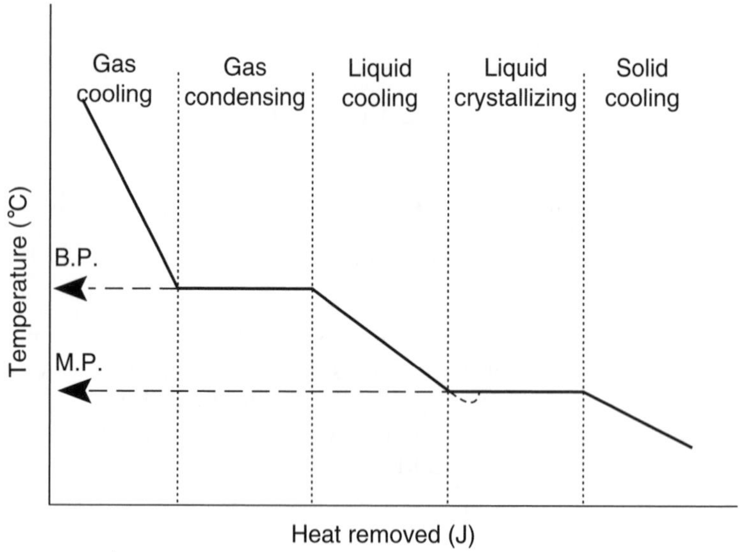

FIGURE 2.21. *A cooling curve showing the effect of removing heat from a gas to form a liquid and then a solid. The dashed line shows an alternative path for a liquid that supercools.*

EXERCISES

1. List all of the information that can be obtained from a heating curve.
2. A liquid is cooled carefully so that it supercools. When a seed crystal is added to the supercooled liquid, it crystallizes within seconds. If we feel the outside of the beaker as the seed crystal is added, what finding can we expect to report?

Answers

1. We can find the following from a heating curve: (1) melting point; (2) boiling point; (3) specific heat and heat capacity of the solid, liquid, and gas; (4) heat of fusion of the solid; (5) heat of vaporization of the liquid.
2. From Figure 2.21 we see that the supercooled liquid returns to the solid form, where the temperature must be higher than the temperature of the supercooled liquid. As a result, we should feel a warming effect as the crystallization process occurs.

3 COMPOUNDS

IONIC AND COVALENT COMPOUNDS

Ionic Formulas

Ionic compounds are formed when cations are attracted to anions because of their opposing charges. The formulas for ionic compounds may be deduced because no compound can have a net charge. The total positive charge of the cations must be exactly canceled by the total negative charge of the anions in the chemical formula. Some chemists refer to this principle as the **law of electroneutrality**. In addition, all ionic compounds have formulas that represent the simplest ratios of the elements needed to obey the law of electroneutrality. This simplest ratio is called the **empirical formula**.

When the anion and cation have the same, but opposite, charges, the formula is written with only one atom of each element. This is another way of saying that the formulas for all ionic compounds are empirical formulas.

Na^+	+	F^-	→	NaF	sodium fluoride
Mg^{2+}	+	O^{2-}	→	MgO	magnesium oxide
Fe^{2+}	+	S^{2-}	→	FeS	iron(II) sulfide
Al^{3+}	+	N^{3-}	→	AlN	aluminum nitride
La^{3+}	+	PO_4^{3-}	→	$LaPO_4$	lanthanum(III) phosphate
NH_4^+	+	NO_3^-	→	NH_4NO_3	ammonium nitrate

When the charges of the anion and cation are not equal and opposite, it is necessary to adjust the numbers of the ions so that the total charge adds up to zero. The most convenient way to do this is to use the charge of the cation for the subscript of the anion and the charge of the anion (without the minus sign) as the subscript for the cation, as shown below. Remember that, when a subscript is 1, it is not written.

Ca^{2+}	+	$2Cl^-$	→	$CaCl_2$	calcium chloride
$2Al^{3+}$	+	$3S^{2-}$	→	Al_2S_3	aluminum sulfide
$3Na^+$	+	PO_4^{3-}	→	Na_3PO_4	sodium phosphate
Pb^{4+}	+	$4Cl^-$	→	$PbCl_4$	lead(IV) chloride

When a subscript must be used with a polyatomic ion, parentheses must be placed around the polyatomic ion first, as shown in the following equations:

Ca^{2+}	+	$2NO_3^-$	→	$Ca(NO_3)_2$	calcium nitrate
$2NH_4^+$	+	SO_4^{2-}	→	$(NH_4)_2SO_4$	ammonium sulfate
$2Al^{3+}$	+	$3SO_4^{2-}$	→	$Al_2(SO_4)_3$	aluminum sulfate

Polyatomic Ions

Many elements combine with oxygen (and sometimes hydrogen and nitrogen) to form a charged group of atoms called a polyatomic ion. (In older textbooks these ions may be called radicals.) **Polyatomic ions** are unusually stable groupings of atoms that tend to act as a single unit in many chemical reactions. The formulas, names, and charges of the common polyatomic ions are listed in Table 3.1 and *should be memorized.* Note that all of these are anions (negatively charged ions) except the ammonium ion, NH_4^+.

TABLE 3.1. COMMON POLYATOMIC IONS

Ion Formula	Ion Name
NH_4^+	Ammonium ion
CO_3^{2-}	Carbonate ion
HCO_3^-	Bicarbonate ion
PO_4^{3-}	Phosphate ion
ClO^-	Hypochlorite ion
ClO_2^-	Chlorite ion
ClO_3^-	Chlorate ion
ClO_4^-	Perchlorate ion
NO_2^-	Nitrite ion
NO_3^-	Nitrate ion
SO_3^{2-}	Sulfite ion
SO_4^{2-}	Sulfate ion
MnO_4^-	Permanganate ion
$Cr_2O_7^{2-}$	Dichromate ion
CrO_4^{2-}	Chromate ion
$S_2O_3^{2-}$	Thiosulfate ion

The atoms in polyatomic ions are bound to each other with covalent bonds (see pages 105–117). Polyatomic ions form ionic compounds by combining with other ions of opposite charge.

When a formula requires more than one polyatomic ion, the polyatomic ion must be placed in parentheses with the appropriate subscript to indicate the number. For example, aluminum nitrate requires three nitrate ions to balance the three positive charges on aluminum. The formula is written as $Al(NO_3)_3$ with parentheses around the NO_3 group.

When a compound containing a polyatomic ion is dissolved in water, the polyatomic ion stays together as a unit and does not break apart into simpler ions. For instance, calcium oxalate, CaC_2O_4, will break apart into the Ca^{2+} and $C_2O_4^{2-}$ ions. The polyatomic oxalate ion, however, does not break apart any further.

Covalent Bonding

The **covalent bond** is the other way in which elements may combine to form compounds and at the same time attain a noble-gas electron configuration. In contrast to ionic bonding, where electrons are transferred from atom to atom, covalent bonding occurs when electrons are shared between two or more atoms. In addition, while ionic bonding is simply the electrical attraction of oppositely charged ions, in covalent compounds the atoms are physically attached to each other. Covalently bonded groups of atoms are called **molecules**.

To conveniently show this sharing of electrons, chemists draw structures of covalent molecules using **Lewis electron-dot structures**. In the Lewis representation the outermost *s* and *p* electrons (valence electrons) are shown as dots arranged around the atomic symbol. Some Lewis structures for the elements are shown in Figure 3.1.

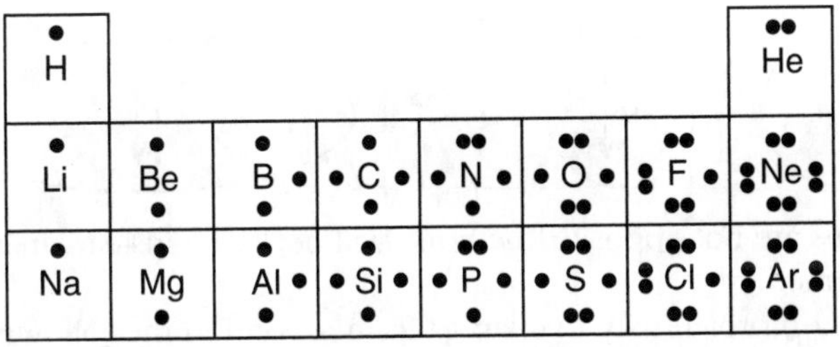

FIGURE 3.1. *Elements normally represented with Lewis electron-dot symbols.*

When two electrons are shared between two atoms, one atom donates one electron and the other atom donates the second electron. The shared pair of electrons represents a covalent bond. If two pairs of electrons are shared between two atoms, a double bond exists. When three pairs of electrons are shared, there is a triple bond.

The sharing of electrons follows the same basic principle that applies to the formation of ions. The atoms are trying to attain a noble-gas electron configuration, in which there are two *s* electrons and six *p* electrons in complete sublevels of the gas. These eight electrons represent the octet of the octet rule that governs covalent compounds. The **octet rule** states that the noble-gas configuration will be achieved if the Lewis structure shows eight electrons around each atom. Hydrogen is an exception; its "octet" consists of two electrons, corresponding to the two outermost electrons in the noble gas helium.

Below are the Lewis electron-dot structures for some diatomic gases. In these structures the electrons of one atom are represented as dots, and the electrons of the other as circles, in order to illustrate that each atom contributes an equal number of electrons to the bond.

FIGURE 3.2. *Lewis structures of diatomic gases.*

In these molecules, each atom regards the shared electrons as its own. Each hydrogen in H_2 "thinks" it has two electrons and the same electron configuration as helium. Nitrogen molecules have triple bonds with three pairs of shared electrons. Fluorine and chlorine look very similar, as expected, since both are halogens.

Another way to view the covalent bond involves the **overlap of orbitals**. The overlap of two *s* orbitals will result in a bond as long as the electrons in the two combining *s* orbitals have opposite spins and can be paired. For hydrogen the process is depicted as follows:

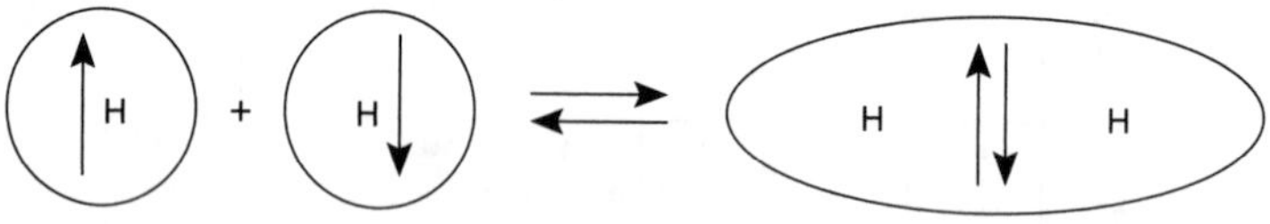

If the spins are not opposite, the bond will not form; instead an antibonding orbital is formed.

An *s* and a *p* orbital may overlap to form a bond in the following manner:

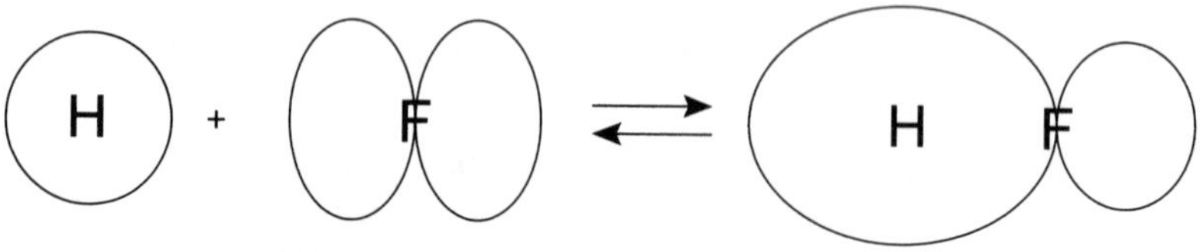

Finally, two *p* orbitals can overlap in a head-to-head manner:

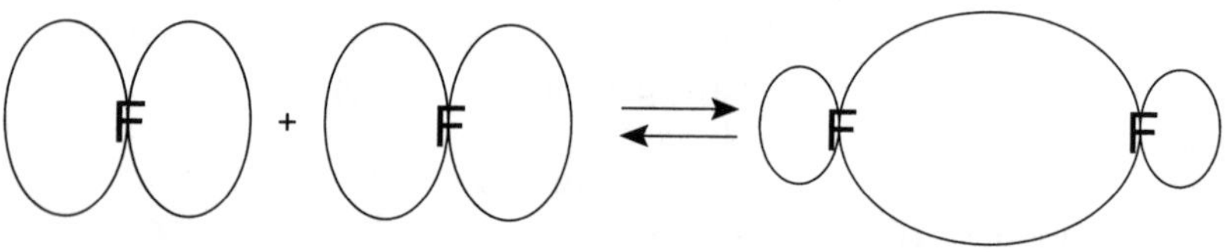

All of the above bonds are designated as sigma bonds. They are characterized by a high electron density along the internuclear axis.

Pi bonds are formed by the sidewise overlap of *p* orbitals:

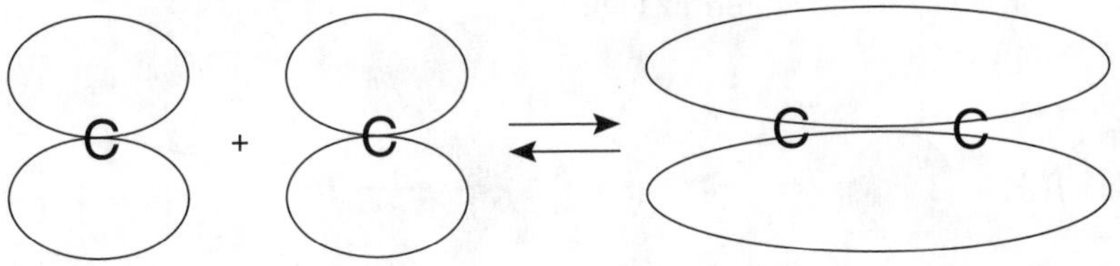

Every covalent bond consists of one sigma bond. In a multiple covalent bond, the additional bonds are pi bonds. A single bond is always a sigma bond. A double bond has one sigma and one pi bond. A triple bond has one sigma bond and two pi bonds.

VALENCE BOND THEORY

Covalent Bond Formation

Orbital Overlap Model (Sigma Bonds)

Up to now we have constructed molecules only according to the octet rule. We now turn to the actual nature of covalent bonds and the way the electrons are shared in these bonds. When an electron is present in an atom, its orbital is called an atomic orbital. When a bond is formed, the shared electrons merge into a new orbital, termed a molecular orbital, where the electrons are paired and localized around two nuclei.

In the formation of hydrogen, H_2, from two hydrogen atoms we visualize the process as shown in Figure 3.3.

FIGURE 3.3. *The combination of two hydrogen atoms to form a hydrogen molecule. The electrons in the hydrogen atoms have opposing spins so that they can pair in the molecular orbital.*

The H's represent the nuclei of the hydrogen atoms, and the shading indicates that the spin of the electron in one hydrogen atom is the opposite of the spin in the other atom. When the electrons with opposite spins are paired, a molecular orbital is formed. The dashed line between the two hydrogen nuclei represents an imaginary line called the internuclear axis. This bond has a high electron density along the internuclear axis.

All bonds with high electron densities along the internuclear axis are called **sigma bonds** (σ bonds). We may view this molecular orbital as the overlap of two *s* orbitals, as illustrated in Figure 3.4.

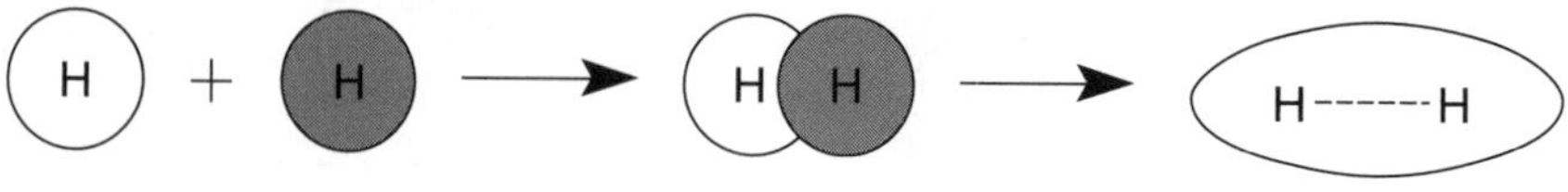

FIGURE 3.4. *The overlap of two* s *atomic orbitals to form a molecular orbital in hydrogen*

The actual molecular orbitals are more complex for compounds beyond H_2. However, the overlap of valence orbitals provides a good picture of bond formation. This method is called Valence Bond Theory. In more complex molecules, sigma bonds may also be formed from the overlap of an *s* orbital and a *p* orbital, as in the formation of HF, illustrated in Figure 3.5, or they may be formed from the overlap of two *p* orbitals, as shown in Figure 3.6.

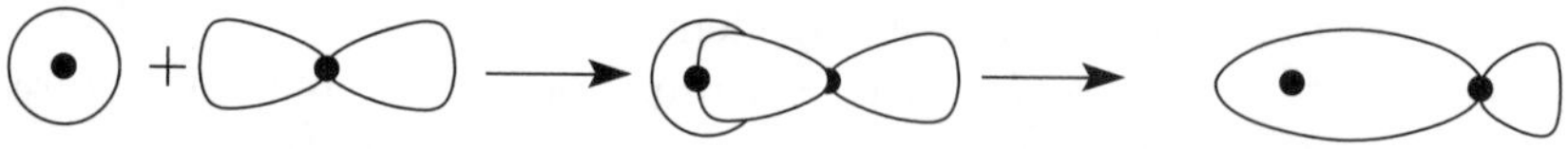

FIGURE 3.5. *An* s *orbital and a* p *orbital overlapping to form a sigma bond in a substance such as HF*

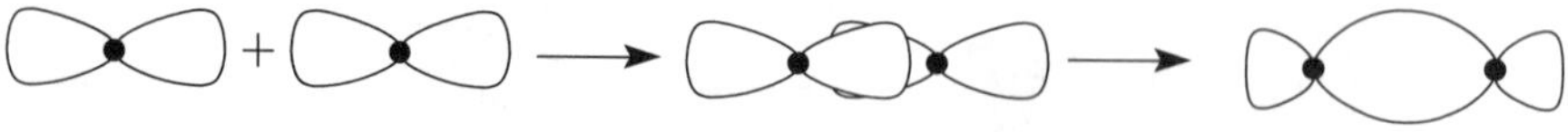

FIGURE 3.6. *Two* p *orbitals overlapping to form a sigma bond in a substance such as* F_2

Orbital Overlap Model (Pi Bonds)

We have described the three important ways in which sigma bonds are formed. Each and every covalent bond has one and only one sigma bond. In a compound that has a double or triple covalent bond, additional overlap of orbitals is needed, and bonds are formed from the sideways overlap of two *p* orbitals, as shown in Figure 3.7. These are called **pi bonds** (π bonds).

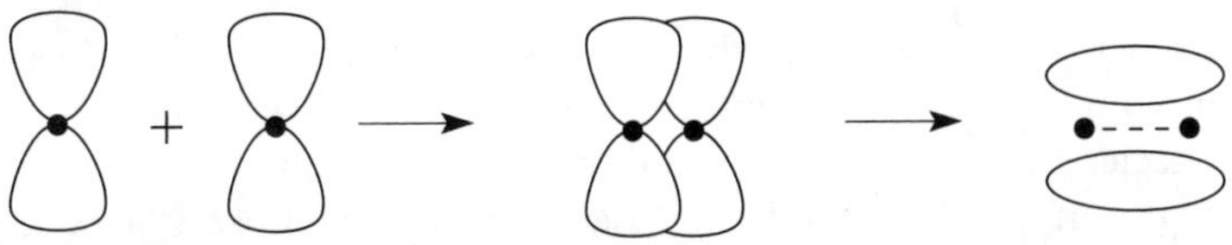

FIGURE 3.7. *The sideways overlap of* p *orbitals to form a pi bond.*

The pi bond has its electrons arranged in two electron clouds, one above and one below the internuclear axis. When arranged in this manner, the electrons of the pi bond do not interfere with the electrons in the sigma bond.

A double bond involves one sigma bond and one pi bond. A triple bond between two atoms may be formed by adding a second pi bond that has its two electron clouds centered behind and in front of the two nuclei.

If we place one atom in front of the other and look down the internuclear axis, the positions of the sigma and pi bonds are as shown in Figure 3.8.

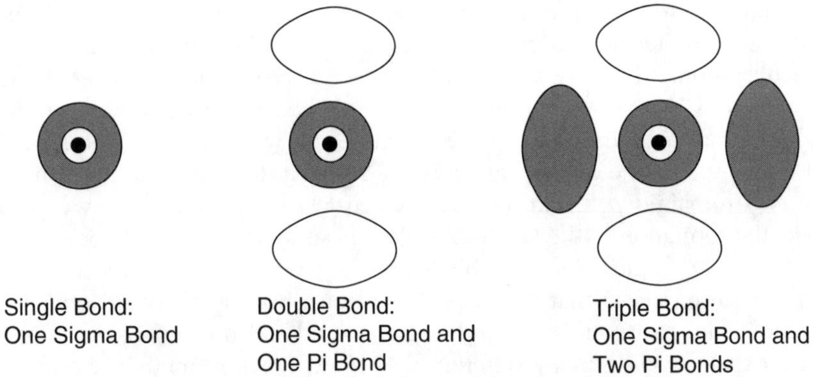

FIGURE 3.8. *End views of single, double, and triple bonds, looking down the internuclear axis. The dot and white circle represent the two nuclei. The shaded circle represents the sigma bond present in all three bonds. In diagrams (2) and (3) the two white ovals represent one of the pi bonds, and in diagram (3) the two shaded ovals represent the second pi bond.*

We see that a single covalent bond is always a sigma bond. A double covalent bond has one sigma and one pi bond, while a triple covalent bond has one sigma and two pi bonds. All of these bonds are arranged so that their electron clouds do not interfere with each other.

The strength of a bond in the valence bond approach is proportional to the amount of overlap of two orbitals. We see that atoms in Period 2 of the Periodic Table form double and triple bonds with the sidewise overlap of *p* orbitals. Atoms in Periods 3–5 do not form multiple bonds since the *p* orbitals cannot effectively overlap. Bonding with hybrid orbitals can be explained by the overlap of hybrid orbitals with each other or with other *s* and *p* orbitals.

HYBRID ORBITALS

Hybridization of Orbitals

The overlap of *s* and *p* orbitals to form sigma and pi bonds works well to describe some features of the covalent bond, and molecules with two and sometimes three atoms. Larger molecules, however, require another model of bond formation.

To understand why a new model is needed, we need to examine the implications of the overlap model. First, *p* orbitals are oriented at 90° from each other and *s* orbitals are spherical, having no directionality. If all covalent compounds were formed from the overlap of these orbitals, we would expect all covalent molecules to have 90° bond angles, but in fact, few of our molecular geometries have angles of 90°. Second, even a simple molecule such as methane, CH_4, cannot be adequately explained using the overlap model. We know that methane is a tetrahedral molecule with four totally equivalent C-H bonds. Using the overlap method, we see that the carbon in methane has only two unpaired *p* electrons (the *s* electrons are paired), and we would expect the formation of the CH_2 molecule. These molecules would have atoms oriented at 90° angles since *p* orbitals are 90° apart. If we allowed the two *s* electrons to unpair so that they could also form bonds, we could obtain the CH_4 molecule. However, the bond angles would still not be correct, and we would expect two distinctly different C-H bond types in methane, one from the overlap of *s* orbitals and the other from the overlap of *p* orbitals. Our new model must be able to explain correctly the molecular geometries and bonding in larger molecules.

The problems posed by the CH_4 molecule require that we develop a better model of sigma bond formation. To construct this model, we postulate the formation of hybrid orbitals. **Hybrid orbitals** may be defined as a set of orbitals with identical properties formed from the combination of two or more different orbitals with different energies.

sp^3 Hybrid Orbitals

The orbital diagram for carbon, is presented in Figure 3.9, along with the conversion of *s* and *p* orbitals into hybrid orbitals called sp^3 orbitals.

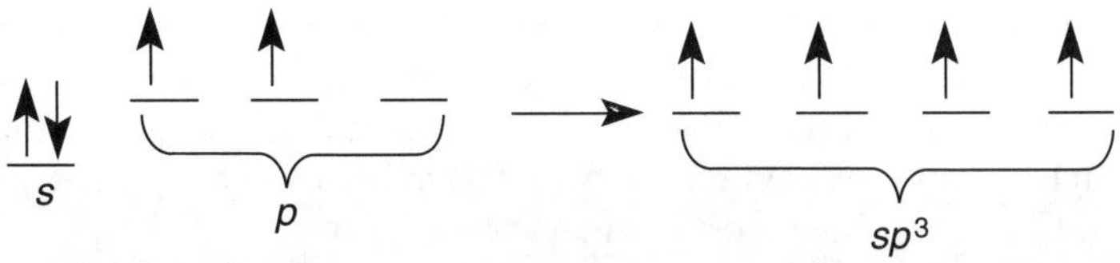

FIGURE 3.9. *The* s *and* p *electrons of the carbon atom and their conversion into* sp^3 *hybrid orbitals used in bonding.*

The designation sp^3 indicates that one *s* and three *p* orbitals have been combined to form the hybrid orbital. In Figure 3.9 the *p* electrons are shown as having a higher energy than the *s* electrons by placing the orbitals at different levels. When carbon forms methane, its electrons reorganize into the four identical sp^3 hybrid orbitals shown on the right. The energy of the electrons in the hybrid orbitals is between the original *s* and *p* energies, as shown. When the sp^3 hybrid orbitals form, their orientation is tetrahedral. Overlap of the four identical sp^3 electrons with electrons from hydrogen atoms form the tetrahedral methane molecule as we know it. Any molecule whose basic structure is the tetrahedron, including the CH_4, NH_3, and H_2O molecules, has sp^3 hybrid orbitals.

sp^2 Hybrid Orbitals

Formaldehyde, CH_2O, is a carbon compound having single bonds to the hydrogen atoms and a double bond to the oxygen atom. Carbon has three sigma bonds and one pi bond in this compound. The structure will be triangular planar because there are no nonbonding electron pairs on carbon. The hybrid orbitals used in this compound are designated as sp^2 hybrid, and their formation is diagrammed in Figure 3.10. We see that three electrons are in three identical sp^2 hybrid orbitals. The remaining electron stays in an unhybridized *p* orbital and is used to overlap with a *p* orbital on the oxygen atom to form the pi bond in the C-O double bond.

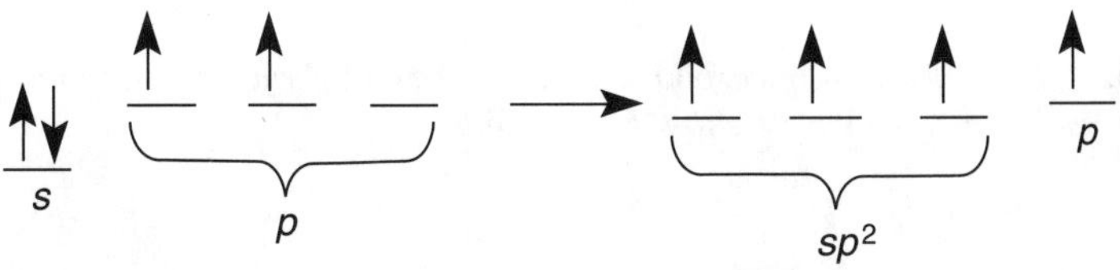

FIGURE 3.10. *The formation of the* sp^2 *hybrid orbitals for carbon.*

The ethylene (CH_2CH_2) and benzene (C_6H_6) molecules are totally flat, and Figure 3.11 shows why. In order for the *p* orbitals to overlap they must be aligned as shown to fix the remaining sp^2 bonds in one plane, resulting in the planar ethylene molecule. When the *p* orbitals in benzene are properly aligned, for pi bonding, it too becomes a planar molecule.

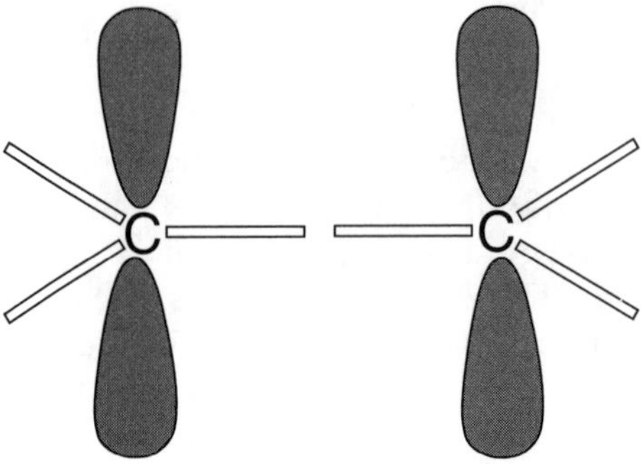

FIGURE 3.11. *Two carbon atoms with* sp^2 *hybridization. The thin lines are the triangular planar* sp^2 *bonding orbitals. The large orbitals are the unhybridized* p *orbitals that overlap to form a pi bond.*

sp Hybrid Orbitals

Carbon dioxide, O=C=O, has two sigma bonds and two pi bonds. The hybridization for this molecule is shown in Figure 3.12.

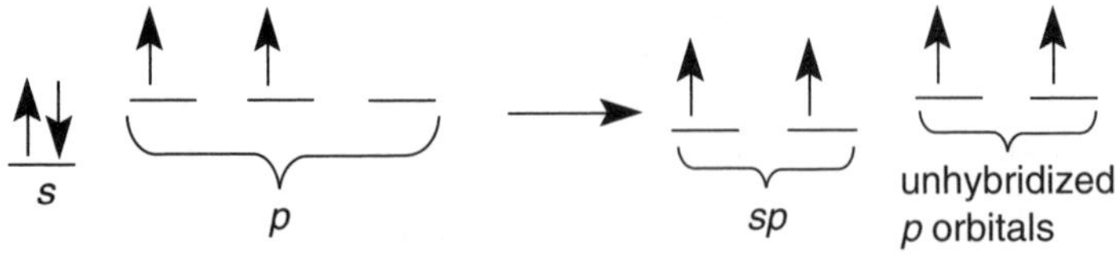

FIGURE 3.12. *Hybridization of carbon to produce the* sp *hybrid orbitals. The two unhybridized* p *electrons are available to form pi bonds.*

When the *sp* hybrid orbitals are formed, two equivalent electrons are obtained that can form sigma bonds. The two remaining, unhybridized *p* electrons can overlap with *p* electrons from the oxygen atoms to form the required double bonds. The hybridized and unhybridized orbitals in the sp^2 hybrid orbital of carbon are pictured in Figure 3.12.

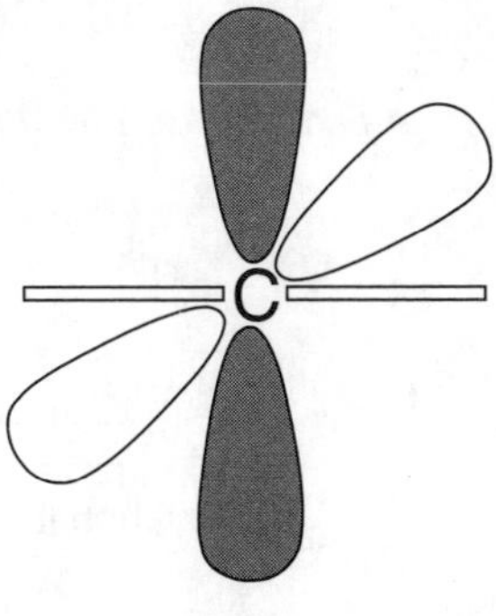

FIGURE 3.13. *Diagram of the* sp *hybrid orbitals of carbon, shown as thin lines. The remaining* p *orbitals (one is shaded and the other is not) are shown as larger lobes. These* p *orbitals overlap with other* p *orbitals to form two pi bonds to the carbon.*

Figure 3.13 shows the *p* orbitals available for pi bonding. Since there are two *p* orbitals, two additional pi bonds can form. These two pi bonds can be directed toward different atoms to form compounds such as carbon dioxide, O=C=O, with two double bonds. They can also be directed toward the same atom to form a triple bond as in the cyanide ion, $C{\equiv}N^-$.

*dsp*³ Hybrid Orbitals

The dsp^3 notation indicates a hybrid that has five equivalent orbitals with at least one electron in each orbital. Since carbon has only four valence electrons, it cannot form a dsp^3 hybrid. In addition, only atoms that have available *d* orbitals can form these hybrids; this requirement means that an element must be in Period 3 or higher. Phosphorus is a typical element that can form the dsp^3 hybrid. It has five valence electrons and is in Period 3.

The orbital diagram for phosphorus is shown in Figure 3.14. The 3*d* orbitals are included even though they are empty before hybridization. When the hybrid orbitals are formed, they may be represented as shown in Figure 3.15.

The five electrons in the dsp^3 hybrid will form five sigma bonds in a covalent compound. The basic structure for the dsp^3 hybrid is the trigonal bipyramid.

Not all phosphorus compounds will be dsp^3 hybrids. When phosphorus forms phosphorus trichloride, PCl_3, for example, the three chlorine atoms can

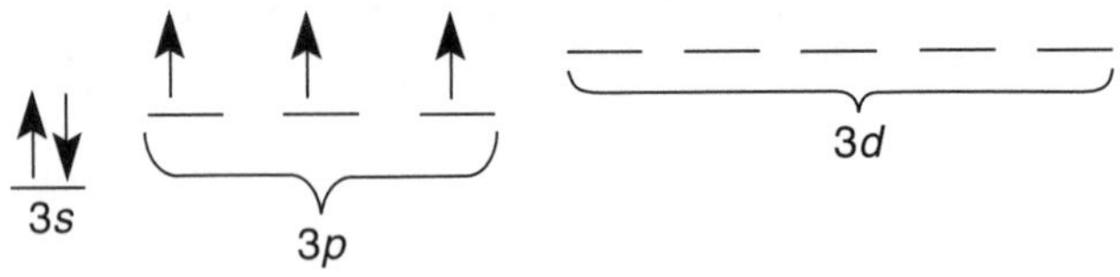

FIGURE 3.14. *Orbital diagram for a phosphorus atom. Energy levels are shown by the position of the orbitals.*

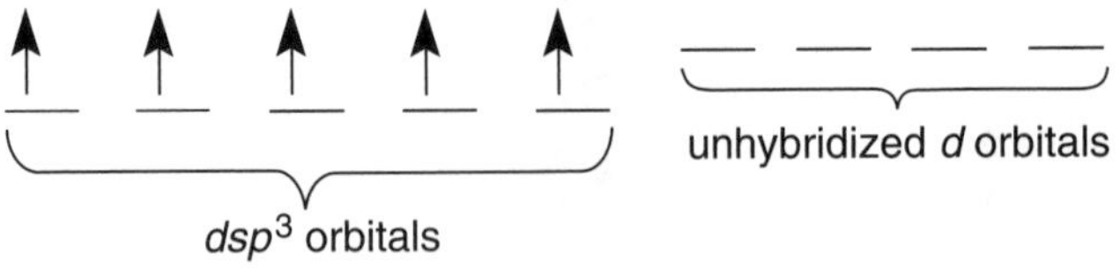

FIGURE 3.15. *Five identical* dsp^3 *hybrid orbitals for phosphorus. Four unoccupied* d *orbitals are not used or hybridized.*

readily combine with the three unpaired electrons in phosphorus without hybridization involving the *d* orbitals. However, since the structure of PCl_3 is tetrahedral, there must be some hybridization. Here the *s* and *p* orbitals hybridize into the sp^3 form shown in Figure 3.16.

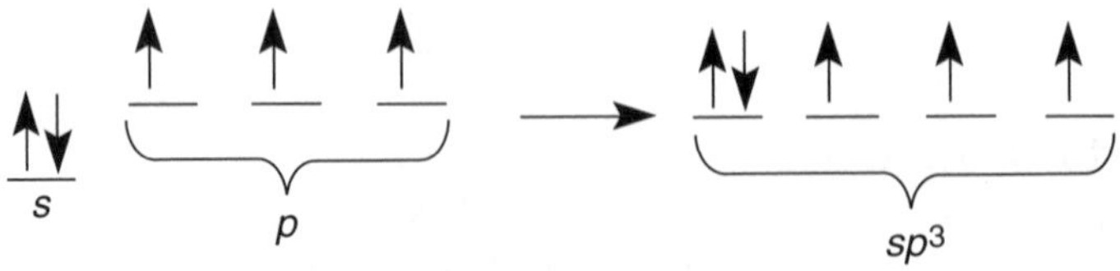

FIGURE 3.16. *The* sp^3 *hybridization for phosphorus.*

Since one of the hybrid orbitals is filled with a pair of electrons, this hybrid of phosphorus will form compounds such as PCl_3 that have only three sigma bonds. PCl_3 has an AX_3E configuration and a triangular pyramid shape (a derived structure in the VSEPR Theory).

d^2sp^3 Hybrid Orbitals

As with dsp^3 hybrids, an element must be in Period 3 or higher to be able to participate in these hybrids, and must also have at least six valence electrons.

Sulfur is one element that forms d^2sp^3 hybrids. Sulfur's orbital configuration and the d^2sp^3 hybrid are shown in Figure 3.17.

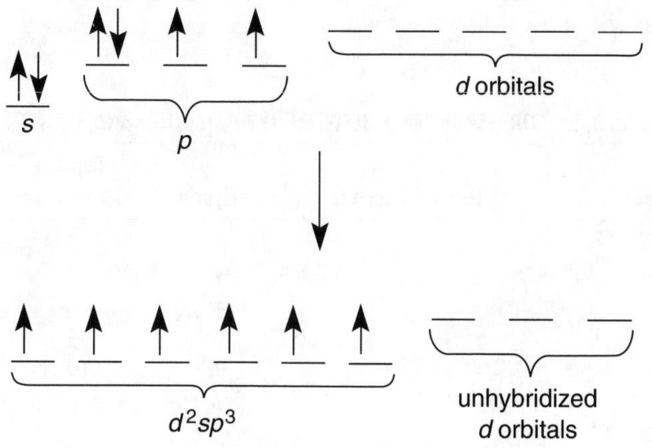

FIGURE 3.17. *Orbital diagram of the valence electrons in sulfur and the conversion to the* d^2sp^3 *hybrid orbitals.*

The d^2sp^3 hybrid allows sulfur to form six covalent bonds with an octahedral structure. One such compound is sulfur hexafluoride SF_6. In addition, a variety of other sulfur compounds are formed when other types of hybridization are used. Sulfur dichloride, SCl_2, uses sp^3 hybridization, as shown in Figure 3.18. In SCl_2, two unpaired electrons form sigma bonds with the chlorine atoms and there are two pairs of nonbonding electrons. This sp^3 hybrid produces an AX_2E_2 structure that also corresponds to a basic tetrahedral configuration. Since the two nonbonding pairs of electrons are not seen, SCl_2 has a bent shape with an angle close to 109.5°. While sulfur can form both d^2sp^3 and sp^3 hybrids, oxygen, in Period 2, does not have available *d* orbitals to form the d^2sp^3 hybrid and can form only compounds with an sp^3 or sp^2 hybrid.

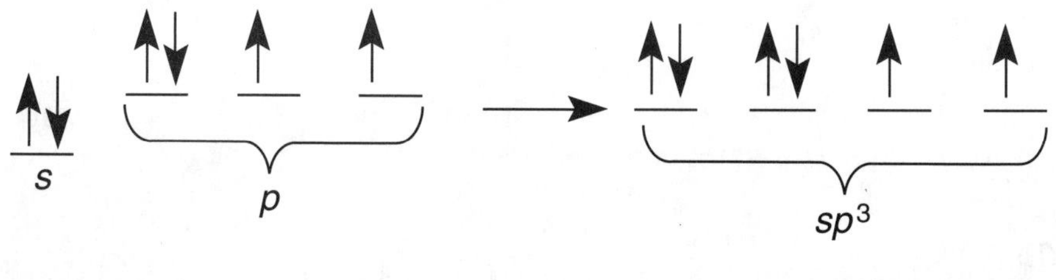

FIGURE 3.18. *Hybridization used to explain the structure of a molecule such as* SCl_2.

There is a direct correspondence between the hybridization and structure, which is shown in Table 3.2. If the basic structure of a molecule is known, the hybridization can be determined; similarly, if the hybridization is known, the structure can be determined. Determination of geometric structures is illustrated in the chapter on Lewis structures.

TABLE 3.2. CORRESPONDENCE BETWEEN HYBRIDIZATION AND STRUCTURE

Basic Structure*	Derived Structure*	Hybrid	Bonding e^- Pairs	Nonbonding e^- Pairs
Linear		*sp*	2	0
Planar triangle		sp^2	3	0
Planar triangle	Bent	sp^2	2	1
Tetrahedron		sp^3	4	0
Tetrahedron	Triangular pyramid	sp^3	3	1
Tetrahedron	Bent	sp^3	2	2
Trigonal bipyramid		dsp^3	5	0
Trigonal bipyramid	Distorted tetrahedron	dsp^3	4	1
Trigonal bipyramid	T-shape	dsp^3	3	2
Trigonal bipyramid	Linear	dsp^3	2	3
Octahedron		d^2sp^3	6	0
Octahedron	Square pyramid	d^2sp^3	5	1
Octahedron	Square planar	d^2sp^3	4	2

EXERCISE

Determine the total number of sigma and pi bonds in each of the following; then, using the simple Lewis structure, determine the hybridization for each:

(a) CH_3Cl
(b) CS_2
(c) PH_3
(d) SiF_4
(e) H_2S
(f) NO_3^-
(g) CO_3^{2-}
(h) PO_4^{3-}
(i) SO_3^{2-}
(j) ClO_4^-

Answers

(a) 4 σ, 0 π; sp^3
(b) 2 σ, 2 π; sp
(c) 3 σ, 0 π; sp^3
(d) 4 σ, 0 π; sp^3
(e) 2 σ, 0 π; sp^3
(f) 3 σ, 1 π; sp^2
(g) 3 σ, 1 π; sp^2
(h) 4 σ, 0 π; sp^3
(i) 3 σ, 0 π; sp^2
(j) 4 σ, 0 π; sp^3

Before leaving the topic of hybrid orbitals, we must recognize that this model is used to explain experimental results. When we find a molecule that has a particular shape, the concept of hybrid orbitals may be used to explain that shape. The reverse, however, is not true. For example, we say that H_2O has sp^3 hybridization because it is a bent structure with a bond angle of 104° that is close to the 109.5° bond angle expected for a tetrahedral structure. Experiments show, however, that the similar molecule H_2S has a bond angle of 90°; in this case the simple overlap of the *p* orbitals of sulfur with the *s* orbitals is sufficient to explain the structure. Here, the concept of hybridization is not needed and apparently is unwarranted.

ORGANIC COMPOUNDS

Carbon-Hydrogen Compounds

The study of carbon-containing compounds is called organic chemistry. Some of the simpler organic compounds contain only carbon and hydrogen. Some of these occur as straight or branched chains of carbon atoms, and others as rings. The most important carbon-hydrogen compounds are described below.

Alkanes

Alkanes, also called paraffins, are compounds with the general formula C_nH_{2n+2}. These compounds contain all single, or sigma, bonds, and each carbon uses sp^3 hybrid orbitals for bonding. At normal temperatures and pressures alkanes containing fewer than 5 carbon atoms are gases; those containing 5–15 carbon atoms are **liquids**; and those with 16 or more carbon atoms are **solids**.

As a group the alkanes are rather unreactive. Their major use is as a fuel in combustion reactions. Methane, propane, and butane are common gaseous fuels used for home heating and cooking; methane is also an industrially important fuel. Gasoline and kerosene are mostly mixtures of liquid alkanes. Liquid alkanes are used also as nonpolar solvents for chemical reactions and for cleaning. The solid alkanes, the paraffins, are often the major components of wax. Solid, firm waxes are used for candles and also in cosmetics. Softer paraffins, called petroleum jelly, are also used in cosmetics as well as in pharmaceuticals.

TABLE 3.3. NAMES AND PHYSICAL PROPERTIES OF SELECTED ALKANES

Name	Formula (C_nH_{2n+2})	Number of Isomers	Melting Point (°C)	Boiling Point (°C)
Methane	CH_4	1	−182.5	−164.0
Ethane	C_2H_6	1	−183.3	−88.6
Propane	C_3H_8	1	−189.7	−42.1
n-Butane	C_4H_{10}	2	−138.3	−0.5
n-Pentane	C_5H_{12}	3	−129.7	36.1
n-Hexane	C_6H_{14}	5	−95.0	68.9
n-Nonane	C_9H_{20}	35	−51.0	150.8
n-Dodecane	$C_{12}H_{26}$	355	−9.6	216.3
n-Icosane	$C_{20}H_{42}$	366,319	36.8	343.0

Table 3.3 lists several alkanes and illustrates the naming system used. The first four alkanes have common names that must be remembered. From pentane on, each name has a prefix that indicates the number of carbon atoms in the longest chain. All the names end in *-ane*.

For the normal or straight-chain alkanes the melting and boiling points increase regularly with the number of carbon atoms in the formula. Branched structural isomers have lower melting and boiling points than normal alkanes with the same numbers of carbon atoms. Increased branching decreases melting and boiling points.

Alkenes

Alkenes are compounds with the general formula C_nH_{2n}, where n must be 2 or larger. Every alkene has a double bond somewhere in its structures. A compound that contains one or more double or triple bonds is said to be **unsaturated**. A **saturated** compound has only carbon-carbon single bonds. The double bond is a reactive site in alkenes and makes them more reactive than alkanes. Alkenes such as ethylene are the major starting reactants, (feed stocks) for many chemical syntheses.

When the double bond can be located in more than one position in the compound, its location is indicated by a number in front of its name. The number represents the lowest numbered carbon atom with the double bond. Compounds that have two double bonds are called dienes, and those with three double bonds are trienes.

Table 3.4 lists, along with other alkenes, the cis and trans isomers for 2-butene. The **cis isomer** has $-CH_3$ groups on the same side of the double bond, and the **trans isomer** has the same groups but on opposing sides of the double bond. Cis-trans isomers are not optically active, and they have different physical and chemical properties. The physical properties of the two isomers indicate that the cis form has stronger London forces.

TABLE 3.4. NAMES AND PROPERTIES OF SELECTED ALKENES

Name	Formula	Melting Point (°C)	Boiling Point (°C)
Ethylene	$CH_2{=}CH_2$	−169	−103.7
Propylene (propene)*	$CH_2{=}CHCH_3$	−185.2	−47.4
1-Butene	$CH_2{=}CHCH_2CH_3$	−185.3	−6.3
cis-2-Butene	$CH_3CH{=}CHCH_3$	−138.9	3.7
trans-2-Butene	$CH_3CH{=}CHCH_3$	−105.5	0.9
1,3-Butadiene	$CH_2{=}CHCH{=}CH_2$	−108.9	−4.4

*Propene is the systemic name, and propylene the common name.

EXERCISE

Newspaper reports sometimes suggest that saturated fats (molecules with long hydrocarbon chains) and trans unsaturated fats (fats with double bonds) may be equally harmful in causing heart attacks. Suggest why this may be so.

Answer

We know that straight-chain molecules have higher melting points than branched-chain molecules, indicating that it is easier for them to be arranged in an orderly crystalline structure. The trans isomers are essentially straight molecules, while a cis isomer has a definite 120° bend. It seems reasonable that trans isomers could deposit themselves almost as easily in artery walls as the straight-chain fats. Deposits of cis fats would be more difficult to form because of the bent structure of these isomers.

Alkynes

Alkynes are compounds with the general formula C_nH_{2n-2}, where n must be 2 or larger. An alkyne has a triple bond somewhere in its structures. Alkynes are very reactive. The most familiar alkyne is acetylene, which is used as a fuel in atomic spectroscopy and welding. A number in front of the name (see Table 3.5) indicates the position of the triple bond.

TABLE 3.5. NAMES AND PROPERTIES OF SELECTED ALKYNES

Name	Formula	Melting Point (°C)	Boiling Point (°C)
Acetylene (ethyne)	$CH{\equiv}CH$	−80.8	−84.0(sublimes)
Propyne	$CH{\equiv}CCH_3$	−101.5	−23.2
1-Butyne	$CH{\equiv}CCH_2CH_3$	−125.7	8.1
2-Butyne	$CH_3C{\equiv}CCH_3$	−32.2	27

Since a carbon atom with triple bonds has only one additional bond, the alkynes do not form cis-trans isomers.

Ring Compounds

In addition to straight chains, carbon atoms can form rings. Rings of six carbon atoms are the most common since the bond angles in these rings are very close to the bond angles in sp^3 or sp^2 hybridized carbon atoms. Five-membered rings are also common, particularly in biochemical molecules.

Cyclohexane is a ring of six carbon atoms with no double bonds. It has two possible structures, the chair and the boat, shown in Figure 3.19, which can be converted from one into the other.

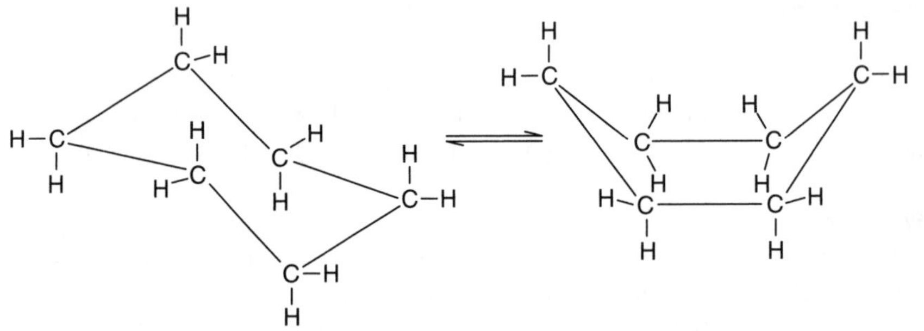

FIGURE 3.19. *Diagrams of the chair (left) and boat (right) forms of cyclohexane.*

The shape of the cyclohexane ring is dictated by the sp^3 hybridization of the carbon atoms, which prefer 109° bond angles. Both the chair and boat forms have these bond angles. Hydrogen atoms on the cyclohexane ring may be replaced by other functional groups to form compounds of added complexity.

Another type of ring is the **benzene** ring. In benzene the carbons have sp^2 hybridization with 120° bond angles. The benzene ring may be visualized as a ring of carbon atoms with alternating double bonds. These are actually resonance structures, as described on pages 278–279, and chemists recognize this fact by drawing a circle in the center of the ring as shown in Figure 3.20.

The benzene ring is unusually stable because the double bonds are conjugated. In a **conjugated double bond** every other bond is a double bond. The electrons in the pi orbitals are delocalized, stabilizing the structure. This stability makes it very difficult to break the carbon-carbon bonds in benzene. The typical reactions of benzene involve replacing the hydrogen atoms with other functional groups.

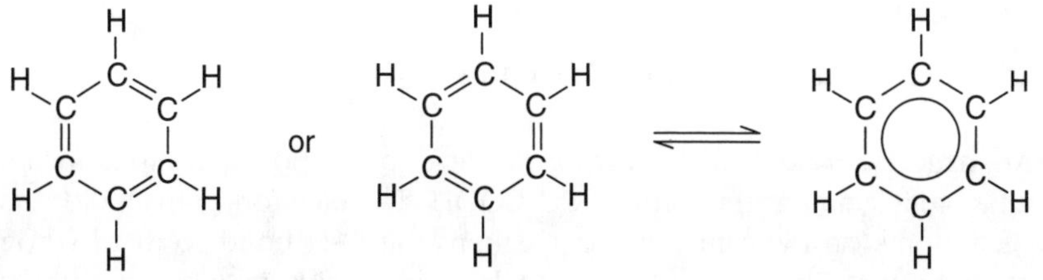

FIGURE 3.20. *Structures of benzene. The two structures on the left are the resonance structures; the structure on the right symbolizes that resonance. These figures are generally drawn without the symbols for carbon and hydrogen.*

Organic Functional Groups and Side Chains

Organic molecules can be considered to be structures developed from just a few essential building blocks, called **functional groups** or **side chains**. Some of these building blocks, such as chains and rings, give the molecule its general shape. Other functional groups are responsible for the molecule's characteristic physical properties and chemical reactivity.

Alkyl Side Chains

The alkanes, without a terminal hydrogen, can be considered to be functional groups. Methane, CH_4, with one hydrogen removed is termed the methyl group, $-CH_3$. Ethane without a hydrogen is the ethyl group, $-CH_2CH_3$. In general formulas the **alkyl group** is given the symbol R. RH represents an alkane, ROH an alcohol, and R=O an aldehyde.

Aryl Side Chains

Benzene, C_6H_6, and related substances are aromatic compounds. The term *aryl* is derived from the word *aromatic*. The functional **aryl group** for benzene is $-C_6H_5$, and it is called the phenyl group.

Alcohol Functional Groups (−OH)

Every **alcohol** has the −OH functional group attached to a carbon, as in ethanol, CH_3CH_2OH. Since the C−O−H bonds are rather strong, the alcohol functional group does not dissociate as the OH^- ion or the H^+ ion. This functional group does hydrogen-bond to other alcohols and water. A primary alcohol has the −OH group bonded to a carbon that has only one other carbon bonded to it. In a secondary alcohol the −OH group is attached to a carbon that is bonded to two other carbon atoms, and a tertiary alcohol has the −OH bonded to a carbon that is bonded to three other carbon atoms.

Acid Functional Groups ($-COOH$ or $-C(=O)-O-H$)

An **organic acid** has the carboxylic acid or carboxyl group, functional group. Acetic acid has the formula CH_3COOH. The electronegativity of the two oxygen atoms on the same carbon weakens the O-H bond so that hydrogen ions dissociate. Organic acids tend to be weak acids in which only a few percent of all molecules dissociate. Carboxyl groups must be located on the terminal carbon of a molecule. Some organic acids have two or even three carboxyl groups.

Aldehyde Functional Groups ($-C=O$)

Compounds with a terminal $-C=O$ group are called **aldehydes**. Although this group is polar, only the smaller aldehydes are soluble in water. The aldehydes often have pleasant odors. They are used in perfumes and are found in many beverages such as coffee, tea, beer, and liquor. Formaldehyde, although thought to be carcinogenic, is a major component of urea-formaldehyde foam insulation. Its structure is shown in Figure 3.21.

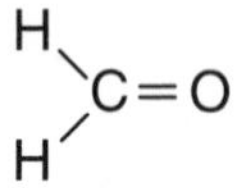

FIGURE 3.21. *Structure of formaldehyde, also called methanal.*

Ketone Functional Groups ($\overset{O}{\overset{\|}{C}}-C-C$)

Ketones are similar to aldehydes except that the double-bonded oxygen is located on a nonterminal carbon atom. Acetone, a common solvent and nail-polish remover, is a ketone; its structure is shown in Figure 3.22. Ketones are also found in many natural materials.

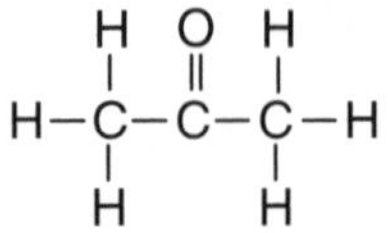

FIGURE 3.22. *Structure of acetone (dimethylketone), commonly used as a solvent. Its systematic name is propanone.*

Amine Functional Groups (−NH_2)

Amines are related to ammonia, NH_3. Replacing one or more of the hydrogen atoms of ammonia with a carbon chain produces an amine. Replacing one hydrogen with a carbon chain yields a primary amine such as ethylamine, $CH_3CH_2NH_2$. Replacing two hydrogen atoms results in a secondary amine, and replacing three hydrogen atoms yields a tertiary amine. Primary and secondary amines can form hydrogen bonds and are generally soluble in water. Amines are the organic version of bases.

Ether Functional Groups (−C−O−C−)

A compound with a carbon-oxygen-carbon unit is known as an **ether**. Diethyl ether, $CH_3CH_2OCH_2CH_3$, was used as the first anesthetic in medical procedures. With the oxygen in the center of a carbon chain, some of its polarity is diffused in the two separate bonds it forms. As a result ethers are much less polar than ketones and aldehydes of similar size. This lesser polarity translates into lower boiling points and higher vapor pressures than those of aldehydes and ketones. There is a slight polarity, however, and ethers have higher boiling points than similar alkanes.

Because the oxygen is bonded to two different carbon atoms, ethers are generally less reactive than other oxygen-containing organic compounds. Because of their high vapor pressures, ethers are more flammable than other organic compounds.

Halides (−X)

The halogens, fluorine, chlorine, bromine, and iodine, are common organic functional groups. Along with hydrogen, the **halides** are terminal atoms and do not appear in the center of carbon chains. A halide is often represented by the symbol X. Thus the formula CH_3CH_2X can represent fluoroethane, chloroethane, bromoethane, or iodoethane.

Isomers

Two compounds may have exactly the same atoms in their formulas but exhibit different properties. Such compounds are called **isomers**. There are three main types of isomers: structural isomers, cis-trans isomers, and stereoisomers.

Structural Isomers

Structural isomers are two or more different compounds that have the same atoms in their formulas. However, the atoms are bonded to each other in different configurations. The hydrocarbon C_5H_{12} has three different structural isomers based on the arrangement of the carbon skeleton, as shown in Figure 3.23. A compound with a single straight chain of carbon atoms is designated as **normal** with the letter *n* preceding its name, as in *n*-pentane.

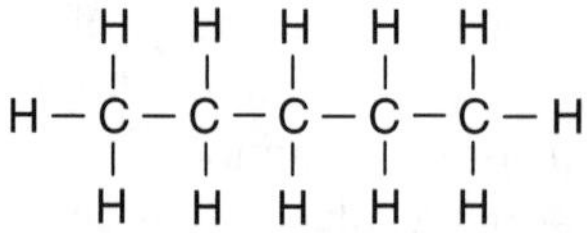

n-pentane
(b.p. = 35.2°C)

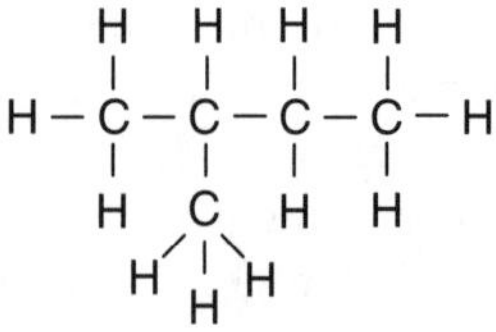

2-methylbutane
(b.p. = 28°C)

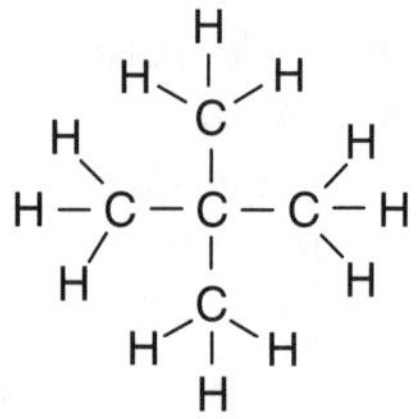

2,2-dimethylpropane
(b.p. = 9.5°C)

FIGURE 3.23. *Structural isomers each having the formula C_5H_{12}.*

Each of these three structural isomers has different chemical and physical properties. Figure 3.23 shows the different boiling points of the isomers. To explain the difference in boiling points, we see that the *n*-pentane molecules can line up side by side to form many instantaneous dipoles. Because it has the largest London forces, *n*-pentane has the highest boiling point. The 2,2-dimethylpropane molecules can interact at only a few points, and this compound has the lowest boiling point because of smaller London forces.

Some of the structural isomers having the formula $C_4H_{10}O$ are shown in Figure 3.24.

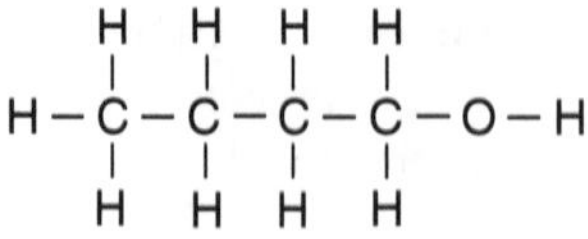

n-butanol

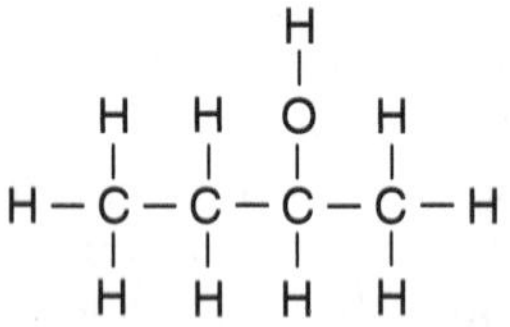

2-butanol

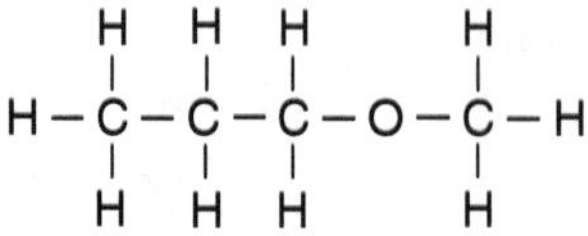

propyl ethyl ether

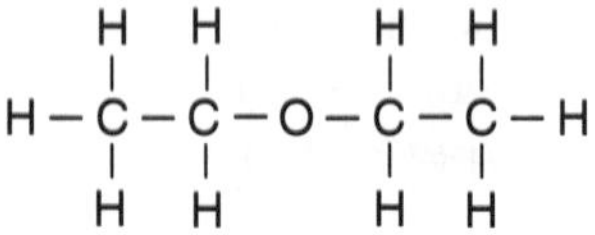

diethyl ether

FIGURE 3.24. *Some possible isomers with the formula $C_4H_{10}O$.*

Each of these compounds has distinctly different physical and chemical properties. The alcohols, with −OH groups, hydrogen-bond extensively and have high boiling points. The ethers, with the C−O−C bonds, have much lower boiling points because the forces of attraction are mainly London forces.

Cis Isomers and Trans Isomers

A carbon-carbon single bond allows the carbon atoms at each end to rotate freely. A carbon-carbon double bond, however, is a rigid structure since rotation would require constant breaking and reforming of the pi bond. This rigidity means that the atoms bound to double-bonded carbon atoms will have fixed orientations. *Cis*-means "on the same side"; *trans*-means "on opposite sides" and refers to groups attached to the double-bonded carbon atoms. Figure 3.25 illustrates the structures of *cis*- and *trans*-2-butene. The two $-CH_3$ groups are on the same side of the double bond in the cis form, and on opposite sides in the trans form.

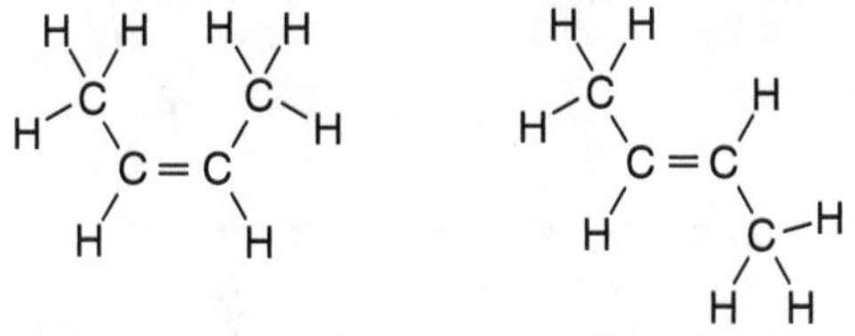

cis-2-butene *trans*-2-butene

FIGURE 3.25. *The difference between the cis and trans isomers of butene.*

Since the double-bonded carbon atoms cannot rotate, these two structures are distinctly different. The molecules have different chemical and physical properties as a result of their structures.

Stereoisomers

Stereoisomers have the same formula, and in both structures all the atoms are bonded to the same atoms. Stereoisomers are not structural isomers. The unique feature is that the arrangement of atoms around a single carbon atom can produce two different molecules. Figure 3.26 shows a two-dimensional representation of stereoisomers. Four different groups, designated as W, X, Y, and Z, are attached to a central carbon atom. These structures are not superimposable. No matter how they are rotated, it is impossible to line up the W, X, Y, and Z of one structure with the W, X, Y, and Z of the other without lifting them from the plane of the paper.

Stereoisomers are often called **mirror images**, and the dashed line in Figure 3.26 can be regarded as a mirror reflecting the two shapes. Three-dimensional molecules have the same ability to form nonsuperimposable mirror images. Your left and right hands are three-dimensional mirror images of each other. If you hold your right hand up to a mirror, the exact image of your left hand is produced—try it and see!

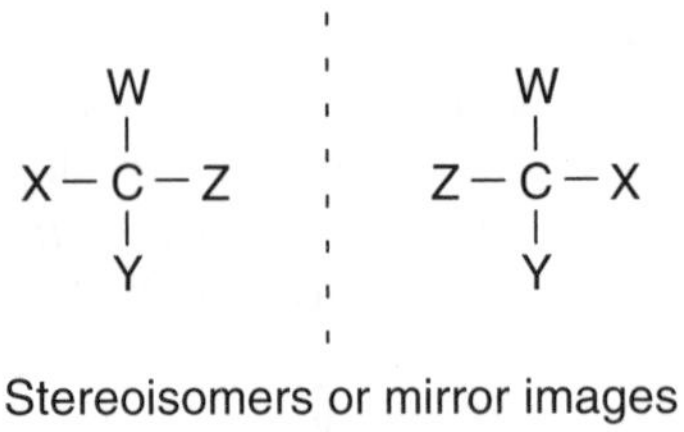

FIGURE 3.26. *Two carbon compounds that have four different groups, W, X, Y, and Z, attached to the central carbon atom. These compounds are not superimposible by rotation. They are the reflections of each other, or mirror images.*

Stereoisomers have identical physical properties such as melting and boiling points. The chemical reactivity of two stereoisomers is, for the most part, identical also. Stereoisomers often have different chemical reactivities in biological systems, where the overall shape of the molecule is important. Most amino acids have two stereoisomers.

Stereoisomers that rotate polarized light are called **optical isomers** because of this property. One isomer rotates polarized light to the right (**dextrorotatory**) and the other rotates it to the left (**levorotatory**) by an equal amount. One style of nomenclature uses D to indicate the dextrorotatory isomer and L to designate the levorotatory isomer. In all known living matter the stereoisomers are the L type.

A substance can be a stereoisomer only if it has at least one carbon atom that has four different groups bound to it. Organic synthesis methods may be **stereospecific**, resulting exclusively in one stereoisomer or the other. Other synthetic methods, however, are not steriospecific and result in a 50:50 mixture of D and L isomers. Such a mixture is called a **racemic mixture**. Racemic mixtures do not rotate polarized light because the rotation due to the D isomers is canceled by the equal and opposite rotation of the L isomers.

Organic Structures: Three-Dimensional Drawings

Geometries of more complex molecules are constructed by determining the geometry around each atom in sequence and then stringing these geometries together. Organic, or carbon-based, compounds are often **complex structures** for which these geometries are very important. The three-dimensional drawings that show these structures often help to define the chemical, physical, and biological properties of complex molecules.

Carbon, with its four valence electrons, can form a maximum of four covalent bonds with four other atoms (an sp^3 hybrid). It can also bond to three atoms as long as one of the bonds is a double bond (an sp^2 hybrid). In bonding to two atoms, carbon will form either two double bonds, as in CO_2 (O=C=O), or one single bond and one triple bond, as in hydrogen cyanide, HCN (H−C≡N); both of these are *sp* hybrids. In no instance does carbon have a nonbonding pair of electrons. As a result, a carbon bonded to four atoms is tetrahedral; when bonded to three atoms, it is trigonal planar; and when bonded to two atoms, it is linear.

For instance, the ethylene molecule, CH_2CH_2, has the Lewis structure shown in Figure 3.27.

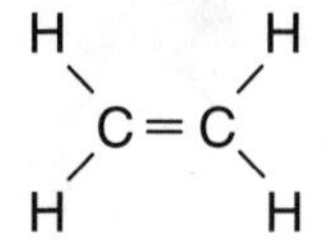

FIGURE 3.27. *The ethylene molecule.*

Since each carbon atom is bonded to only three atoms (two hydrogen and one carbon), each must have a trigonal planar geometry. For reasons discussed below, both planes are lined up so that this molecule is perfectly flat. We can also predict that the benzene ring discussed on page 112 must also be flat since each of its six carbon atoms is trigonal planar.

For a molecule such as butane, $CH_3CH_2CH_2CH_3$, we can write the Lewis structure as shown in Figure 3.28.

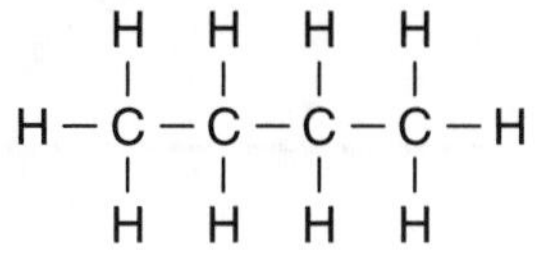

FIGURE 3.28. *Line structure of butane, C_4H_{10}.*

Each carbon atom is bonded to four other atoms; therefore, the geometries of all the carbon atoms are tetrahedral. Placing the four tetrahedral structures together, we obtain the three-dimensional structure shown in Figure 3.29. Since three-dimensional structures are difficult to draw on paper, organic chemists often find it convenient to use models to build three-dimensional structures so that they can inspect their features better.

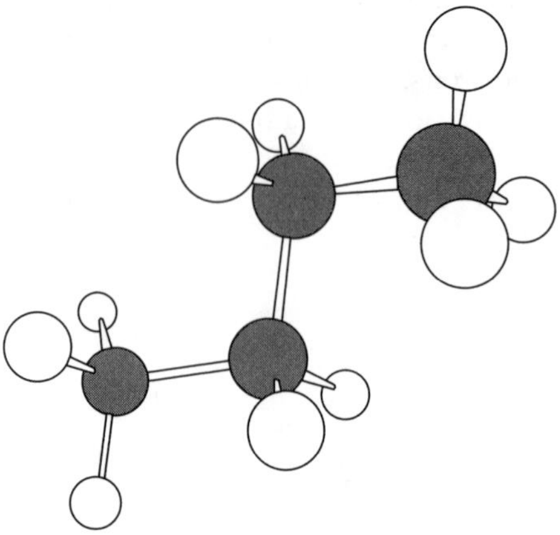

FIGURE 3.29. *Three-dimensional computer-generated structure for butane, showing the tetrahedral arrangement of atoms around each carbon (shaded circles).*

An oxygen atom in an organic compound always has two nonbonding pairs of electrons. An oxygen bonded to two atoms has a bent structure (sp^3), and a double-bonded oxygen is bonded to only a single atom for linear structure.

A nitrogen atom in an organic compound has one nonbonding pair of electrons. It has a triangular pyramid structure if bonded to three other atoms. When bonded to only two atoms, one with a double bond, a nitrogen atom has a bent structure. These structures are reviewed on page 116.

EXERCISE

Predict the geometry around each of the carbon atoms in the following molecule:

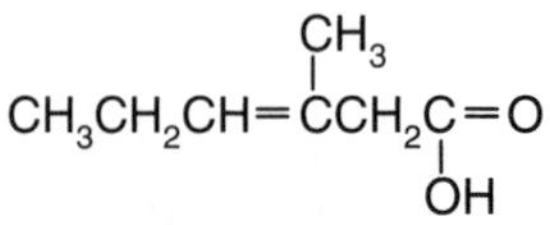

Answer

From left to right along the main chain, the geometries are tetrahedral, tetrahedral, triangular planar, triangular planar, tetrahedral, triangular planar. The CH_3 above the molecule is tetrahedral. With this information, a more realistic drawing or a molecular model of the compound can be made.

Polymers

Many people consider polymers (plastics) to be the chemical compounds most important in today's society, and most chemists work in industries involved in producing or fabricating polymers. Economically, polymers represent a large proportion of the chemicals bought and sold.

Polymers are long chains of repeating structural units, called **monomers**. One of the simplest polymers is polyethylene. Its structure is shown in Figure 3.30.

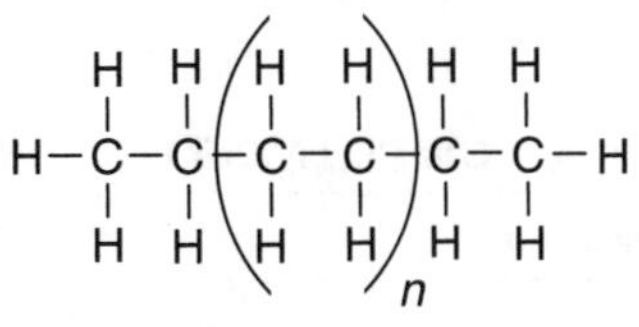

polyethylene

FIGURE 3.30. *Polyethylene, showing the general form used to draw the structure of a polymer molecule.*

The number of repeating ethylene units ($-CH_2CH_2-$) in the complete molecule is *n* plus the two end groups. The value of *n* is usually greater than 100 and often exceeds 1000. As the name suggests, the molecule is made from ethylene, $CH_2{=}CH_2$. Ethylene is the monomer from which the polyethylene polymer shown in Figure 3.30 is constructed. Polymers may be linear or branched.

Addition Polymers

A monomer that contains a double bond often reacts in a process called an **addition reaction**. Special catalysts start the reaction by forming free radicals, and more free radicals are then produced in a chain reaction. The net result is that the double bond opens (one bond in the double bond breaks) and then reforms with two adjacent molecules, as illustrated in Figure 3.31.

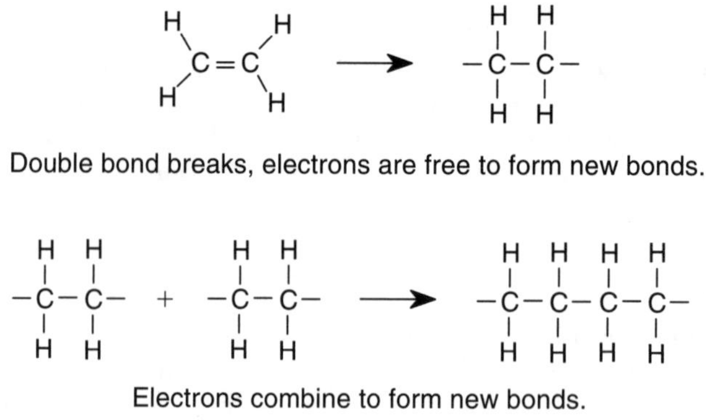
Double bond breaks, electrons are free to form new bonds.
Electrons combine to form new bonds.

FIGURE 3.31. *The initiation (top) and propagation (bottom) steps of an addition polymer reaction.*

Some familiar polymers and the monomers from which they are made are listed in Table 3.6.

TABLE 3.6. COMMON ADDITION POLYMERS

Name	Monomer	Typical Uses
Polyethylene	$CH_2{=}CH_2$	Bottles, coatings
Polypropylene	$CH_2{=}CHCH_3$	Bottles, coatings
Polyvinyl chloride	$CH_2{=}CHCl$	Credit cards, pipes
Polystyrene	$CH_2{=}CHC_6H_5$	Foamed beads and blocks for packaging
Teflon	$CF_2{=}CF_2$	Nonstick surfaces

Condensation Polymers

Condensation reactions are similar to the reactions that form esters and peptide bonds. Each reactant must have two functional groups; one is an acid and the other is a base, either $-OH$ or $-NH_2$. (We can remember that condensation reactions produce water as a product by recalling that condensation on a cold window or glass of ice water also produces water.) In the following reactions R represents a hydrocarbon group:

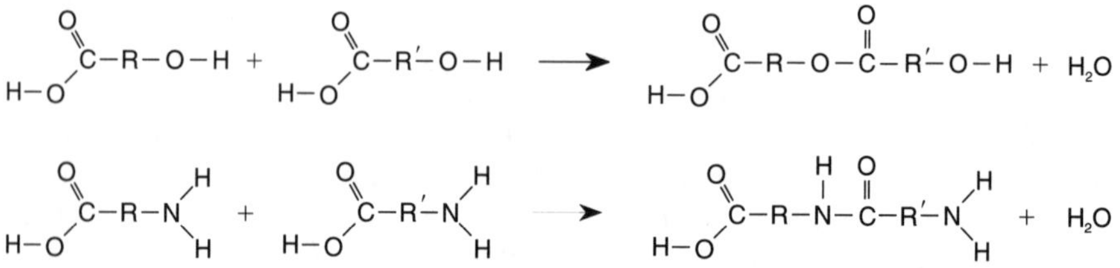

Each monomer may contain an acid and a basic functional group, as shown previously, in which case there is only one reactant. Another method of producing a condensation polymer involves using one monomer that contains two acid groups and another that contains two basic groups. An equal molar mixture of these two monomers is used to form the polymer:

$$HOOC{-}R{-}COOH \quad + \quad H_2N{-}R'{-}NH_2 \quad \rightarrow$$
$$HOOC{-}R{-}CO{-}NH{-}R'{-}NH_2 \quad + \quad H_2O$$

Starch and cellulose are both condensation polymers of glucose. The different physical and chemical properties are due to the different arrangement of the bonds in these two polymers.

Some common condensation polymers and their uses are listed in Table 3.7.

TABLE 3.7. COMMON CONDENSATION POLYMERS

Name	Monomer(s)	Typical Uses
Dacron	Terephthalic acid, ethylene glycol	Fibers, fabrics
Nylon	1,6-Hexanedicarboxylic acid, 1,6-diaminohexane	Fibers, fabrics
Proteins	Amino acids	Biological reactants, food
Disaccharides	Two sugar molecules	Biological energy source
Starch	Many glucose molecules	Food source
Cellulose	Many glucose molecules	Biological structural material
DNA	Nucleotides	Genetic code

Polymer Properties

We may deduce some of the properties of polymers from fundamental chemical principles developed in other chapters.

Polymers with polar, hydrogen-bonding functional groups may be soluble in water despite their tremendous size. This fact explains the solubility of proteins and polyvinyl alcohols. Most other polymers, however, are insoluble. The length of these polymers allows them to align with each other, thereby generating substantial London forces of attraction.

Although addition polymers are prepared from monomers with double bonds, the finished polymer has no double bonds. As a result, these polymers are fairly inert. The most inert is Teflon, in which the very strong C—F bond cannot be broken easily.

Polymers, also have distinct properties related to the absorption of water and gases. Many of these properties depend on the method used to manufacture the polymer. A polymer can be manufactured or treated to increase or decrease its porosity. Branched-chain polymers have different properties than linear polymers. In designing consumer products, chemists take advantage of the specific properties of polymers.

ACIDS AND BASES

Acid-Base Theories

Arrhenius Theory

Svante Arrhenius considered an acid to be any substance that increases the concentration of hydrogen ions, H^+, in aqueous solution. When an acid is dissolved in water, the reaction may be written in two different forms.

For example, when gaseous hydrogen chloride is dissolved in water, the equation may be written as

$$HCl(g) \xrightarrow{H_2O} H^+(aq) + Cl^-(aq)$$

Modern chemistry recognizes, however, that the hydrogen ion is unlikely to exist in aqueous solution. Instead, the hydrogen ion is hydrated or bound to one or more water molecules. Therefore, H_3O^+ is used to represent the hydration of the hydrogen ion, and the reaction of HCl with water is expressed as

$$HCl(g) + H_2O(\ell) \rightarrow H_3O^+(aq) + Cl^-(aq)$$

Common examples of acids are HCl, HBr, H_2SO_4, H_3PO_4, and $HC_2H_3O_2$.

The **Arrhenius theory** considers a base to be any substance that increases the concentration of hydroxide ions, OH^-, when dissolved in water. A typical reaction is

$$KOH(s) \xrightarrow{H_2O} K^+(aq) + OH^-(aq)$$

Common bases are NaOH, KOH, $Ca(OH)_2$, and $Al(OH)_3$.

Brönsted-Lowry Theory

For a long time it was known that ammonia, NH_3, when dissolved in water increased the hydroxide ion concentration. In fact, ammonia solutions were often called ammonium hydroxide, with the formula NH_4OH. This designation fit the Arrhenius theory, but not the facts since NH_4OH does not really exist. The **Brönsted-Lowry theory** solves the problem by defining acids as proton donors, as the Arrhenius theory does, but defining bases as proton acceptors. Ammonia is a base since it accepts protons from water molecules in the reaction

$$H_2O(\ell) + NH_3(g) \rightleftharpoons NH_4^+(aq) + OH^-(aq)$$

Ethylamine, $CH_3CH_2NH_2$, and dimethylamine, $(CH_3)_2NH$, are both bases related to ammonia. In ethylamine the ethyl group, CH_3CH_2-, replaces one hydrogen in ammonia. Two methyl groups, CH_3-, in dimethylamine replace two of the ammonia hydrogens.

Lewis Theory

To completely generalize acid-base theory and to account for the formation of complex ions, G. N. Lewis proposed that acids are substances that accept electron pairs from other atoms, ions, or molecules, and bases are substances that donate electron pairs, in forming chemical bonds. The reaction between boron trichloride, BCl_3, an electron-deficient compound, and ammonia, NH_3, which has a nonbonding pair of electrons, is an acid-base reaction according to the Lewis definitions. The ammonia is an electron-pair donor and is a base, while the boron trichloride is the electron-pair acceptor and is an acid.

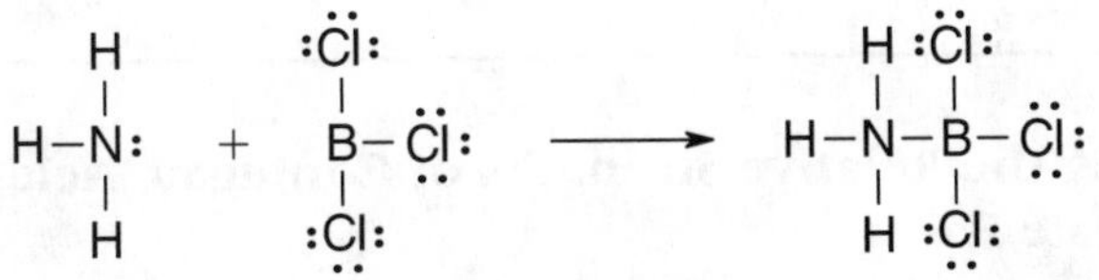

The **Lewis theory** is used mainly to explain the formation of substances called **complexes**, while the Arrhenius and Brönsted-Lowry theories serve to explain the more traditional acid-base reactions.

Acid Formulas

An acid, according to both the Arrhenius and Brönsted-Lowry theories, is any compound in which one or more hydrogen atoms are weakly bound to the rest of the molecule. When dissolved in water, hydrogen ions ionize from the rest of the molecule. The formulas of most acids are written with hydrogen as the first element. The formulas of organic acids can be written with hydrogen as the first element or with the $-COOH$ unit, which is the functional group for organic acids. The following three formulas all represent acetic acid:

$$HC_2H_3O_2 \qquad CH_3COOH \qquad CH_3\overset{\overset{\displaystyle OH}{|}}{C}{=}O$$

Often in beginning chemistry courses the formulas for all acids are written with hydrogen as the first element. This is done to make it easy for the student to identify acids.

Brönsted-Lowry Theory: Conjugate Acid-Base Pairs

The Brönsted-Lowry theory of acids and bases not only redefined acids and bases but also introduced the concept of **conjugate acid-base pairs**. In the equilibrium reaction of acetic acid with hydroxide ions:

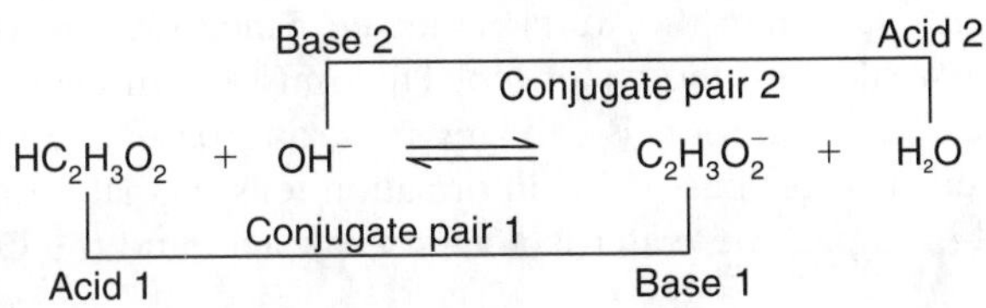

$HC_2H_3O_2$ is an acid that reacts with the base OH^- in the forward reaction. The products, however, are also acids and bases. $C_2H_3O_2^-$ is a base that can accept an H^+ from the water, while water is an acid since it donates a proton in the reverse reaction. $HC_2H_3O_2$ and $C_2H_3O_2^-$ are called a conjugate acid-base pair. In a similar way, H_2O and OH^- are another conjugate acid-base pair in this equation. Conjugate acid-base pairs always have formulas that differ by only one H^+.

$$\text{Conjugate acid} \rightleftharpoons \text{conjugate base} + H^+$$

Determining the Relative Strengths of Conjugate Acids and Conjugate Bases

In a conjugate acid-base pair the relative strengths of the conjugate acid and conjugate base are determined by the position of the equilibrium in equation 3.1. If the position of equilibrium results in more products than reactants, the conjugate acid is stronger than the conjugate base. If, however, there are more reactants than products at equilibrium, then the conjugate base is stronger.

Acetic acid dissociates slightly in the reaction

$$HC_2H_3O_2 \rightleftharpoons C_2H_3O_2^- + H^+ \qquad (3.1)$$

and we conclude that acetic acid is a weak acid and the acetate ion is a stronger base.

The conjugate acid-base pair for ammonia may be written as

$$NH_4^+ \rightleftharpoons NH_3 + H^+$$

Since an aqueous solution of ammonia has a much higher concentration of ammonia than of the ammonium ion, the equilibrium lies to the right of the double arrow. Therefore we conclude that the NH_4^+ ion is a stronger conjugate acid than NH_3 is a conjugate base.

Using similar logic, we can determine the relative strengths of conjugate acids and bases in a complete chemical reaction. For example, when equal numbers of moles of acetic acid and base are mixed, the following reaction goes virtually to completion, leaving little $HC_2H_3O_2$ and OH^-:

$$HC_2H_3O_2 + OH^- \rightarrow C_2H_3O_2^- + H_2O$$

From the position of the equilibrium, chemists deduce that the OH^- ion is a stronger base than the acetate ion, $C_2H_3O_2^-$. At the same time, the acetic acid is also a stronger acid than the water molecule. Since we already know that acetic acid is a weak acid, the water molecule must be an even weaker acid. Similarly, since the acetate ion is a relatively strong base, the hydroxide ion must be an even stronger base. This information tells us that in the H_2O/OH^- conjugate acid-base pair the water is a very weak acid and the OH^- is a very strong base.

Another reaction takes place when acetic acid is dissolved in distilled water:

$$HC_2H_3O_2 \quad + \quad H_2O \quad \rightleftharpoons \quad C_2H_3O_2^- \quad + \quad H_3O^+$$

We know that acetic acid is a weak acid, so by definition most of it remains in the molecular form. Since the reactants are favored in this equilibrium, we conclude that H_3O^+ is a stronger acid than $HC_2H_3O_2$. Also, the acetate ion is a stronger base than water. Knowing the position of equilibrium gives us another method to determine the relative strengths of conjugate acids and bases.

In aqueous solution the strongest acid is the H_3O^+ ion (often written simply as the H^+ ion) and the strongest base is the OH^- ion. Acids that are stronger than water react with water to produce the H_3O^+ ion:

$$HCl \quad + \quad H_2O \quad \rightarrow \quad Cl^- \quad + \quad H_3O^+$$

Because HCl is a very strong acid, this reaction goes all the way to completion, forming the H_3O^+ ion and leaving no HCl molecules. Bases that are stronger than water produce the OH^- ion when dissolved in water. In aqueous solutions, all strong acids react completely with the weak base water, and all strong bases also react completely with the water, which acts as a weak acid. This phenomenon is known as the **leveling effect** of water.

In discussing the strengths of acids and bases, it was mentioned that the acetate ion is a stronger base than water. This results in the concept that the anion of a weak acid may be considered as a base. Very strong acids such as HCl and HNO_3 have anions that are extremely weak conjugate bases. On the other hand, very weak acids have anions that are strong conjugate bases. Carbonic acid, H_2CO_3, is a very weak acid, and the carbonate ion, CO_3^{2-} is a strong conjugate base. Carbonate salts such as sodium carbonate and calcium carbonate (limestone) are rather strong bases and are used in many industrial processes instead of the more expensive hydroxide bases such as NaOH and KOH.

In Chapter 10 we will see that the strengths of conjugate acids and bases can be expressed numerically as K_a and K_b values. These two values are related to the constant K_w in the following equation:

$$K_a K_b = K_w$$

This equation indicates that K_a is inversely proportional to K_b and that strong conjugate acids have weak conjugate bases and vice versa.

Amphiprotic (Amphoteric) Substances

The words ***amphiprotic*** and ***amphoteric*** describe the same phenomenon, in which a substance may act as both a conjugate acid and a conjugate base. An amphiprotic salt is an anion and must have at least one proton so that it can act as a proton donor. It must also be able to accept a proton and thereby act as a base. The anions of partially neutralized polyprotic acids are always amphiprotic.

The common amphiprotic anions are the hydrogen carbonate ion, HCO_3^-, the hydrogen sulfate ion, HSO_4^-, the hydrogen sulfite ion, HSO_3^-, the dihydrogen phosphate ion, $H_2PO_4^-$, and the monohydrogen phosphate ion, HPO_4^{2-}. Each of these ions can accept a proton and act as a base, and can also lose a proton and act as an acid.

Water is an amphiprotic solvent that can act both as an acid and as a base, as shown in this reaction:

$$H_2O \quad + \quad H_2O \quad \rightleftharpoons \quad H_3O^+ \quad + \quad OH^-$$

Here one water molecule donates a proton and is an acid while the other accepts a proton and is a base. In either case water is an extremely weak acid and an extremely weak base.

Lewis Acids and Bases; Complexation Reactions

Complexation Reactions

The primary use of the Lewis theory is to explain complexation reactions that are not covered in the discussions of ionic reactions or covalent reactions. For instance, we know that silver chloride, AgCl, is an insoluble salt. It is not a hydroxide or a basic anhydride that can be dissolved in acids. However, this salt does dissolve in ammonia solutions. Analysis of this process shows that the reaction is as follows:

$$AgCl(s) \quad + \quad 2\ NH_3 \quad \rightarrow \quad Ag(NH_3)_2^+ \quad + \quad Cl^-$$

The same reaction occurs, without the Cl^- ions, when silver ions in solution react with ammonia. This reaction is a **complexation reaction**.

The Lewis theory explains why such ions as $Ag(NH_3)_2^+$ form. In this reaction the silver ion is a Lewis acid, accepting pairs of electrons, and the ammonia, acting as a Lewis base, donates a pair of electrons to the bond. The Lewis structure shows the nonbonding pair of electrons that the ammonia molecule donates:

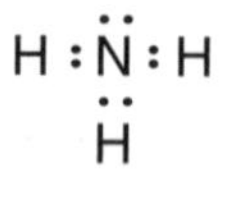

The silver ion starts as a silver atom with the electron configuration [Kr] $5s^1$, $4d^{10}$. In forming the Ag^+ ion, it loses the $5s^1$ electron, resulting in the [Kr] $4d^{10}$ electron configuration. Therefore it has empty $5s$ and $5p$ orbitals that may accept pairs of electrons. In fact, silver ions accept only two pairs of electrons from two ammonia molecules. In a similar fashion, all metal ions have available orbitals that may accept electron pairs. Metal ions are generally Lewis acids.

In complexation reactions Lewis bases have many names. They are called **ligands**, **complexing agents**, **chelates**, and **sequestering agents**. Most ligands have one pair of electrons to donate, as ammonia does. Some ligands have two pairs of electrons, and some have up to six pairs. A ligand that provides more than one electron pair in forming a complex must be a large, flexible molecule so that each pair of electrons can be oriented properly to form a bond. The chloride ion has four pairs of electrons but forms only one bond because the remaining six electrons cannot be aligned properly to form additional bonds.

Complexation reactions can be written generally as follows:

$$M^{n+} \quad + \quad xL^{m-} \quad \rightleftharpoons \quad ML_x^{n-mx}$$

Silver tends to accept two electron pairs, and copper accepts four. The rest of the metal ions tend to accept six electron pairs in complexes. With this information we can write most complexation reactions accurately once the number of electron pairs that a ligand can donate has been determined.

Ligands that form only one bond are called monodentate ligands. Halide ions are monodentate ligands. Even though they have four available electron pairs, once one electron pair is donated, the other electron pairs are not in the proper positions to form additional bonds. Cyanide ions, CN^-, thiocyanate ions, SCN^-, and anions of weak acids are other common monodentate ligands. Ammonia is a neutral molecule that is a monodentate ligand; water and carbon monoxide are other monodentate molecular ligands.

Ligands that can form two bonds are called bidentate ligands. The most common bidentate ligands are the diamines, such as ethylene diamine, $H_2NCH_2CH_2NH_2$, and the anions of diprotic organic acids, such as the oxalate ion, $^-OOCCOO^-$ or $C_2O_4^{2-}$.

One special ligand is ethylene diaminetetraacetic acid (EDTA), shown below:

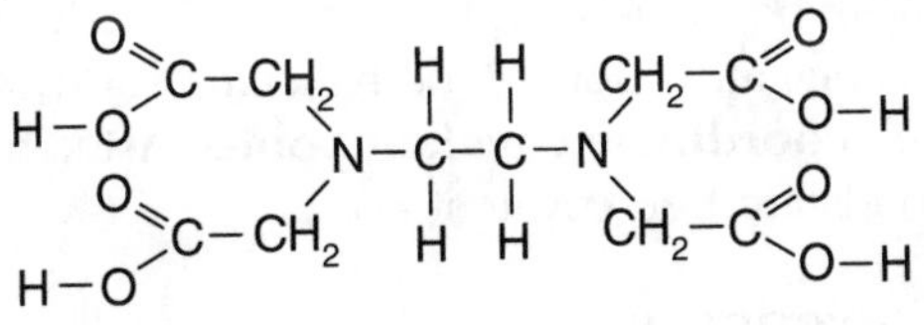

It has six pairs of electrons to donate, and the molecule is flexible enough to allow each of the six pairs to bond with a metal ion. One molecule of EDTA always reacts with only one metal cation, as shown in the following general equation, where Y^{4-} is the EDTA molecule without its four outer protons and M^{n+} represents any metal ion:

$$M^{n+} \quad + \quad Y^{4-} \quad \rightarrow \quad MY^{n-4}$$

EDTA is an important molecule for the chemical analysis of metal ions using simple titration methods. EDTA is also found in many consumer products. Cosmetics, drugs, and even foods contain EDTA because it acts as a preservative by forming complexes with metal ions. The same metal ions, if uncomplexed, could act as catalysts to promote oxidation. Thus complexation reduces or eliminates catalytic activity and increases the shelf life of these products.

Water has been mentioned as a monodentate ligand. In fact, when chemists write an ion with the symbol (aq) after it, they are recognizing the fact that all ions are actually complexed to water in aqueous solution. Therefore the $Fe^{2+}(aq)$ ion is actually the $Fe(H_2O)_6^{2+}$ complex ion. Complexes with other ligands simply replace the water molecules with other electron-pair donors.

The formulas of complexes and complex ions are written in the same way as any other chemical formula. For any complex ion the total charge on the ion is the sum of the charges on all the ions in the complex. The complex made of Fe^{3+} and six chloride ions, $FeCl_6^{3-}$, has a charge of -3:

$$Fe^{3+} \quad + \quad 6Cl^- \quad \rightleftharpoons \quad FeCl_6^{3-}$$

Molecular complexing agents do not affect the charge since they are neutral molecules. For example, Cu^{2+} still has a $+2$ charge when complexed with four molecules of ammonia in $Cu(NH_3)_4^{2+}$:

$$Cu^{2+} \quad + \quad 4NH_3 \quad \rightleftharpoons \quad Cu(NH_3)_4^{2+}$$

One interesting application of complexation reactions involves gold mining. In modern gold mining, dilute cyanide solutions, along with air, are sprayed on gold-bearing ore. The gold, oxidized to Au^+, complexes with two CN^- ions to form the soluble complex $Au(CN)_2^-$. The liquid containing the gold complex is then treated to recover the concentrated gold.

Coordinate Covalent Bonds

The bonds formed in complexation reactions are **covalent bonds**. All of the properties of covalent bonds, as well as molecular geometry, discussed on pages 269–279 and 279–284, also apply to the bonds in complexes. However, the formation of these bonds is unique. Instead of each atom donating one electron to the bond, one atom donates both electrons. Covalent bonds formed in this way are called **coordinate covalent bonds**. Aside from the method of formation, these bonds are true covalent bonds.

Acid and Base Strengths

Acid Strengths

It is possible to use the concepts of electronegativity and bond polarity to determine the relative strengths of acids based on their chemical formulas. The strength of an acid is inversely proportional to the strength of the bond between the hydrogen and the rest of the molecule. Hydrogen is bonded very weakly in strong acids and more strongly in weak acids. To determine the relative strengths of acids, an estimate of the bond strengths is needed.

Binary Acid Strengths: **Binary acids** are composed of hydrogen and one other atom. Experimental evidence shows that these acids increase in strength according to the position of this other atom from left to right in a period (row) of the Periodic Table. For example, it is observed that PH_3 is weaker than H_2S, which is weaker than HCl. Across a period the acid strength parallels the electronegativity of the anion. This statement is reasonable because, as the anion attracts the electrons more strongly, the bond with hydrogen becomes weaker while the size of the anions remains relatively constant.

Experimental evidence also shows that binary acids increase in strength from the top of a group to the bottom. Thus HF is weaker than HCl, which in turn is weaker than HBr.

Electronegativity cannot explain this sequence of binary acid strengths. Instead, the size of the anion is important. An increase in anion size requires a corresponding increase in bond length, the distance between two nuclei. Longer bond lengths imply weaker bonds. Acids that have weaker bonds with hydrogen are stronger acids.

Oxoacid Strengths: Most mineral acids that are not binary acids are **oxoacids**, that is, acids that contain hydrogen, oxygen, and another element. In all oxoacids the oxygen atoms are bound to the central atom and the hydrogen atoms are bound to the oxygen atoms. A typical structure is represented by sulfuric acid:

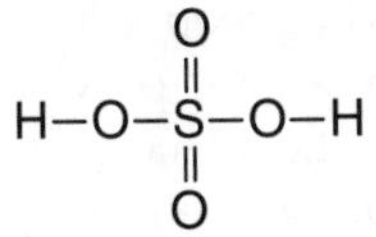

As for binary acids, the strength of an oxoacid depends on the relative strength of the oxygen-hydrogen bond. In oxoacids the strength of this bond depends on (1) the number of oxygen atoms per hydrogen in the formula and (2) the electronegativity of the central atom in the formula.

Oxoacids that have the same central atom and the same number of hydrogen atoms increase in strength as the number of oxygen atoms increases. For example:

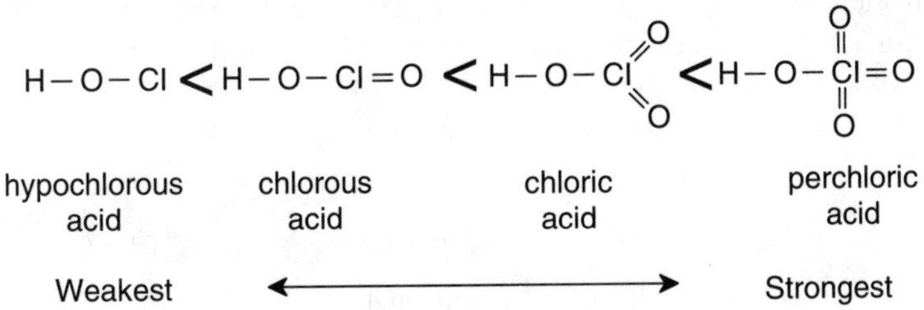

The reason for this increase is that each extra oxygen withdraws additional electron density from the O-H bond, thereby weakening it. The more oxygen atoms per hydrogen, the stronger is the acid.

The strengths of oxoacids that have the same number of hydrogen and oxygen atoms but a different central atom are affected by the electronegativity of the central atom. For example:

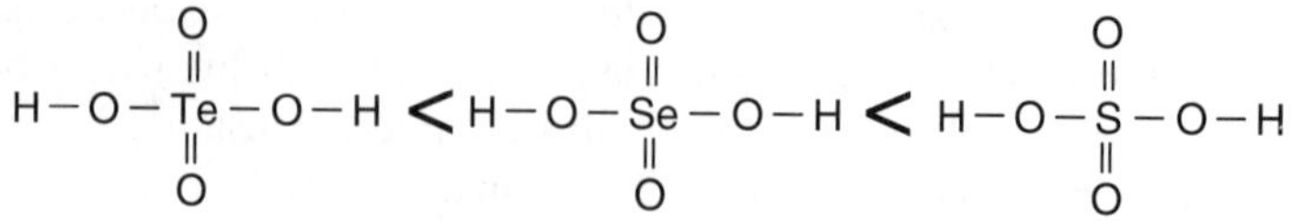

telluric acid　　selenic acid　　sulfuric acid

Weakest ⟷ Strongest

The electronegativity of the central atom increases from tellurium to selenium to sulfur, weakening the O-H bond and resulting in increasingly stronger acids.

Oxoacids that have the same number of oxygen atoms but differing central atoms and differing numbers of hydrogens can also be compared. For instance, experiments show that $H_3PO_4 < H_2SO_4 < HClO_4$. In this sequence the electronegativity of the central atom increases as the strength increases; at the same time the number of oxygen atoms per hydrogen atom is also increasing.

A correct understanding of the effects of oxygen involves two considerations:

1. Double-bonded oxygen atoms (=O) attract electrons more strongly than do oxygen atoms bonded to hydrogen atoms (−O−H), and the effects of double-bonded oxygen atoms are additive.
2. Double-bonded oxygen atoms provide a pi electronic structure to delocalize and stabilize the electron that remains behind when H^+ ionizes. The more double-bonded oxygens present, the more stabilization of the anion results.

These two principles lead to the general rule that more oxygen atoms per hydrogen atom result in a stronger acid.

Organic Acid Strengths: It is difficult to compare **organic acids** with very different structures. However, the principles of electronegativity may be used when considering similar acids. We can consider the effect when an organic acid is modified by replacing hydrogen atoms with other atoms in the carbon chain.

Electronegative atoms such as fluorine, chlorine, bromine, iodine, oxygen, and sulfur on nearby carbon atoms will withdraw electron density from the O-H bond and increase the strength of the organic acid. The effect of these electronegative atoms is approximately additive:

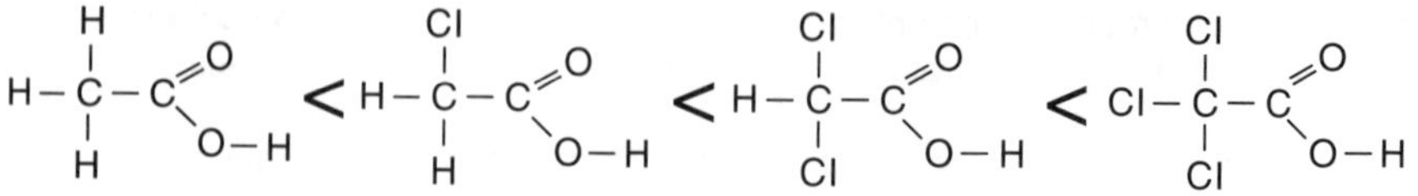

acetic acid　　chloroacetic acid　　dichloroacetic acid　　trichloroacetic acid

Weakest ⟷ Strongest

Base Strengths

Metal Hydroxides: All **metal hydroxides** are strong bases; that is, the hydroxide ion dissociates completely when the compound is dissolved in water. Most metal hydroxides, however, are also very slightly soluble. Only the hydroxides of Group IA metals, strontium and barium, have appreciable solubility; calcium hydroxide is moderately soluble. Soluble hydroxides may cause severe skin burns. Insoluble hydroxides are much less harmful; for instance, $Mg(OH)_2$ can safely be swallowed as an antacid to neutralize excess stomach acid.

Nitrogen Bases: All **nitrogen bases** are related to ammonia and are weak bases. The organic compounds related to ammonia are called amines and have carbon-containing groups replacing one or more of the hydrogen atoms of ammonia, NH_3. Two such bases, ethylamine and dimethylamine, are described on page 132.

The relative strengths of the weak bases may be evaluated based on the electronegativities of the organic functional groups that replace the hydrogen atoms of ammonia. Electronegative substituents such as chlorine increase the strength of organic acids; the reverse is true of organic bases. For example, chloromethylamine is a weaker base than methylamine.

Numerical Evaluation of Acid and Base Strengths

The preceding sections described how the relative strengths of acids and bases can be determined from their chemical structures. Another way to compare the strengths of weak acids and weak bases is to compare the numerical values of their respective ionization constants (K_a and K_b).

Stronger weak acids have larger K_a values than weaker weak acids. Since most K_a values have negative exponents of 10, it is important to remember that the smaller the negative exponent, the larger the number. For example, 10^{-9} is larger than 10^{-11}. Similarly, for weak bases, the stronger weak bases have larger K_b values.

Strong and Weak Acids

Strong Acids

Strong acids are acids that dissociate completely into ions when dissolved in water. The six most important strong acids are as follows:

Hydrochloric acid	HCl	$\rightarrow$	H^+	+	Cl^-
Hydrobromic acid	HBr	$\rightarrow$	H^+	+	Br^-
Hydroiodic acid	HI	$\rightarrow$	H^+	+	I^-
Perchloric acid	$HClO_4$	$\rightarrow$	H^+	+	ClO_4^-

Nitric acid	HNO_3	$\rightarrow$	H^+	+	NO_3^-
Sulfuric acid	H_2SO_4	$\rightarrow$	H^+	+	HSO_4^-

For sulfuric acid, only the dissociation of the first hydrogen ion is considered strong; the second hydrogen ion dissociates only slightly.

These strong acids are also called mineral acids, as opposed to organic acids. It is worthwhile to remember this short list of strong acids.

Weak Acids

Almost all of the other acids are weak, meaning that, when they are dissolved in water, only a small percentage of the molecules dissociate into ions. Most of the **weak acids** are organic acids. The most common weak acids are hydrofluoric acid (HF), carbonic acid (H_2CO_3), sulfurous acid (H_2SO_3), phosphoric acid (H_3PO_4), and hypochlorous acid (HClO). Appendix 3 lists many additional weak acids.

Neutralization Reactions

The reactions between acids and bases, called **neutralization reactions**, are often double-replacement reactions, and the products can be predicted using the methods described on pages 314–323. Most neutralization reactions can be described as the mixing of an acid with a base to form a salt and water, as in this example:

$$\underset{\text{(acid)}}{HBr(aq)} + \underset{\text{(base)}}{KOH(aq)} \rightarrow \underset{\text{(salt)}}{KBr(aq)} + \underset{\text{(water)}}{H_2O(\ell)}$$

A neutralization reaction may be written as a molecular, ionic, or net ionic equation, depending on the type of information the chemist is interested in communicating. For a typical neutralization reaction with soluble acids and bases the three types of reactions are as follows:

$$HCl(aq) + NaOH(aq) \rightarrow NaCl(aq) + H_2O(\ell) \quad \text{(molecular equation)} \quad (3.2)$$

$$H^+(aq) + Cl^-(aq) + Na^+(aq) + OH^-(aq) \rightarrow Na^+(aq) + Cl^-(aq) + H_2O(\ell) \quad \text{(ionic equation)} \quad (3.3)$$

$$H^+(aq) + OH^-(aq) \rightarrow H_2O(\ell) \quad \text{(net ionic equation)} \quad (3.3)$$

If all of the reactants are soluble strong acids and bases, the net ionic equation of a neutralization will always be the one shown in equation 3.3.

Neutralization of Oxides

Acids are often used to dissolve insoluble hydroxides or oxides of metals. For instance, lanthanum ions serve to suppress interferences in atomic absorption spectroscopy. Dissolution of lanthanum oxide, La_2O_3, is achieved by reacting it with either HCl or HNO_3:

$$6HCl + La_2O_3 \rightarrow 2LaCl_3 + 3H_2O$$

or

$$6HNO_3 + La_2O_3 \rightarrow 2La(NO_3)_3 + 3H_2O$$

These are reactions between an acid and a basic anhydride. In both reactions the chloride and nitrate salts of lanthanum are soluble.

Acid anhydrides such as SO_3 and CO_2 react with bases as in the following reactions:

$$2NaOH(aq) + SO_3(g) \rightarrow Na_2SO_4(aq) + H_2O(\ell)$$
$$Ca(OH)_2(aq) + CO_2(g) \rightarrow CaCO_3(s) + H_2O(\ell)$$

Acid and basic anhydrides may react with each other without any water present. Lime can react with sulfur trioxide in this reaction:

$$SO_3(g) + CaO(s) \rightarrow CaSO_4(s)$$

Neutralization of Polyprotic Acids

Acids that contain more than one ionizing hydrogen are called **polyprotic acids**. Sulfuric acid, H_2SO_4, and phosphoric acid, H_3PO_4, are two examples. Although acetic acid, $HC_2H_3O_2$, has four hydrogen atoms in its formula, it is not a polyprotic acid since it has only one ionizable hydrogen.

All polyprotic acids are weak acids except sulfuric acid, which is unique in that its first proton dissociates completely but the second proton does not. One property of polyprotic acids is that the protons dissociate and react in a stepwise manner; the first proton dissociates or reacts before the second proton. The stepwise dissociation of phosphoric acid is written as follows:

$$H_3PO_4 \rightleftharpoons H_2PO_4^- + H^+$$
$$H_2PO_4^- \rightleftharpoons H_2PO_4^{2-} + H^+$$
$$H_2PO_4^{2-} \rightleftharpoons PO_4^{3-} + H^+$$

The product(s) formed in the neutralization of a polyprotic acid depend on the amount of base used. When 1 mole of hydroxide ions per mole of phosphoric acid reacts, the equation is

$$H_3PO_4 + OH^- \rightarrow H_2PO_4^- + H_2O$$

When 2 moles of hydroxide ions per mole of phosphoric acid react, the equation is

$$H_3PO_4 + 2OH^- \rightarrow H_2PO_4^{2-} + 2HO$$

When 3 moles of hydroxide ions per mole of phosphoric acid react, the equation is

$$H_3PO_4 \quad + \quad 3OH^- \quad \rightarrow \quad PO_4^{3-} \quad + \quad 3H_2O$$

Salts containing $H_2PO_4^-$, HPO_4^{2-}, or PO_4^{3-} ions can be prepared by accurately adjusting the amount of base added to the phosphoric acid.

If exactly 1, 2, or 3 moles of hydroxide ions per mole of phosphoric acid are not reacted, a mixture of phosphate salts will form. For example, if 2.25 moles of OH^- are added per mole of phosphoric acid, we may deduce that the first 2 moles of OH^- convert the phosphoric acid to HPO_4^{2-}. The additional 0.25 mole of OH^- converts only some of the HPO_4^{2-} to PO_4^{3-}. Consequently the final mixture will be a combination of HPO_4^{2-} and PO_4^{3-} salts.

EXERCISES

1. Predict the salt(s) that will be formed in each of the following cases:
   (a) 2.30 mol H_3PO_4 and 4.60 mol KOH
   (b) 0.250 mol $H_3C_6H_5O_7$ and 0.750 mol NaOH
   (c) 0.345 mol H_2SO_3 and 0.500 mol KOH
   (d) 1.67 mol H_4Y (EDTA) and 4.00 mol NaOH
2. Write the complete set of ionization reactions for (a) H_2TeO_3 and (b) H_3AsO_4.

Answers

1. (a) K_2HPO_4; exactly 2 mol of base for each mole of acid
   (b) $Na_3C_6H_5O_7$; exactly 3 mol of base for each mole of acid
   (c) $KHSO_3$ and K_2SO_3; more than 1 but less than 2 mol of base for each mole of acid
   (d) Na_2H_2Y and Na_3HY; more than 2 but less than 3 mol of base for each mole of acid
2. (a) $H_2TeO_3 \rightleftharpoons H^+ + HTeO_3^-$
   $HTeO_3^- \rightleftharpoons H^+ + TeO_3^{2-}$
   (b) $H_3AsO_4 \rightleftharpoons H^+ + H_2AsO_4^-$
   $H_2AsO_4^- \rightleftharpoons H^+ + HAsO_4^{2-}$
   $HAsO_4^{2-} \rightleftharpoons H^+ + AsO_4^{3-}$

Anhydrides of Acids and Bases

The word ***anhydride*** means "without water," and the acidic and basic anhydrides are compounds that, when added to water, become common acids and bases. **Acid anhydrides** are the oxides of nonmetals; the following are examples of their reactions with water:

$$SO_2 + H_2O \rightarrow H_2SO_3 \quad \text{(sulfurous acid)}$$
$$SO_3 + H_2O \rightarrow H_2SO_4 \quad \text{(sulfuric acid)}$$
$$CO_2 + H_2O \rightarrow H_2CO_3 \quad \text{(carbonic acid)}$$
$$P_2O_5 + 3\,H_2O \rightarrow 2\,H_3PO_4 \quad \text{(phosphoric acid)}$$

Basic anhydrides are the oxides of metals. Two typical reactions with water are these:

$$K_2O + H_2O \rightarrow 2KOH$$
$$CaO + H_2O \rightarrow Ca(OH)_2$$

In industrial applications CaO is called lime and $Ca(OH)_2$ is termed slaked lime. Lime is the important ingredient in the cement and concrete used in construction. The production of lime is a major industry that involves mining naturally occurring limestone, $CaCO_3$, and heating it to very high temperatures to drive off carbon dioxide in this reaction:

$$CaCO_3 \xrightarrow{\text{heat}} CaO + CO_2$$

Anhydrides react as acids and bases without water present. Sulfur dioxide, SO_2, reacts with lime as follows:

$$SO_2 + CaO \rightarrow CaSO_3$$

Limestone kilns (ovens) react with sulfur oxides, SO_2 and SO_3, produced in the combustion of wood in these reactions:

$$SO_2 + CaCO_3 \rightarrow CaSO_3 + CO_2$$
$$SO_3 + CaCO_3 \rightarrow CaSO_4 + CO_2$$

While limestone is reasonably hard, $CaSO_3$ and $CaSO_4$ are powders. The result is the slow destruction of the kiln.

Marble is another form of limestone. Many marble sculptures have been destroyed by a similar reaction with acidic oxides or by acidic oxides dissolved in rain (i.e., acid rain).

4
REACTIONS

CLASSIFICATION OF REACTIONS

Classification of Chemical Reactions

Many chemical reactions fall into distinct groups with definite similarities. By classifying chemical reactions, it is possible to compare the properties of the reactants and products. In addition, classification often serves as a shorthand substitute for writing a complete chemical reaction. The combustion of propane offers one example. When the reaction is described as a combustion process, the knowledgeable chemist knows that the other reactant is oxygen and that the products are carbon dioxide and water. Some types of reactions are described below.

COMBUSTION REACTIONS

In a **combustion reaction**, an organic (carbon-containing) compound reacts with oxygen to form carbon dioxide and water. If the organic compound contains elements other than carbon, hydrogen, and oxygen, it is assumed that those elements end up in the elemental state as products. Here is a typical combustion reaction:

$$C_5H_{12} + 8O_2 \rightarrow 5CO_2 + 6H_2O$$

SINGLE-REPLACEMENT REACTIONS

In some reactions, an element may react with a compound to produce a different element and a new compound. Here is a typical reaction of this sort:

$$2AgNO_3 + Zn \rightarrow 2Ag + Zn(NO_3)_2$$

In this reaction zinc replaces (or displaces) the silver in the silver nitrate. This type of reaction is known as a **single-displacement reaction**.

DOUBLE-REPLACEMENT REACTIONS

In a **double-replacement reaction**, when two compounds react, the metal in one compound replaces the other and vice versa. Here is an example of a double-replacement reaction:

$$MgSO_4 \quad + \quad BaCl_2 \quad \rightarrow \quad MgCl_2 \quad + \quad BaSO_4$$

In this reaction the magnesium replaces the barium and the barium replaces the magnesium—thus the term *double replacement.*

NEUTRALIZATION REACTIONS

A **neutralization** reaction is a special type of double-replacement reaction in which one reactant is an acid and the other is a base. The products are a salt and water. Here is a typical neutralization reaction:

$$HCl \quad + \quad NaOH \quad \rightarrow \quad NaCl \quad + \quad H_2O$$

SYNTHESIS REACTIONS

Reactions of the elements to form a compound are often called **synthesis reactions**. One such reaction is the formation of rust, Fe_2O_3:

$$4Fe \quad + \quad 3O_2 \quad \rightarrow \quad 2Fe_2O_3$$

FORMATION REACTIONS

A **formation reaction** is the same as a synthesis reaction except that the product must have a coefficient of 1. The reactants are the elements in their normal state at room temperature and atmospheric pressure. The formation reaction for $Fe(NH_4)_2(SO_4)_2$ is as follows:

$$Fe \quad + \quad N_2 \quad + \quad 4H_2 \quad + \quad 2S \quad + \quad 4O_2 \quad \rightarrow \quad Fe(NH_4)_2(SO_4)_2$$

If necessary, the use of fractional coefficients for the reactants is permitted in a formation reaction.

ADDITION REACTIONS

In an **addition reaction** a simple molecule or element is added to another molecule, as in the addition of HCl to pentene, C_5H_{10}:

$$HCl \quad + \quad C_5H_{10} \quad \rightarrow \quad C_5H_{11}Cl$$

DECOMPOSITION REACTIONS

In a **decomposition reaction** a large molecule decomposes into elements or smaller molecules. When sucrose is heated strongly, this reaction occurs:

$$C_{12}H_{22}O_{11} \rightarrow 12C + 11H_2O$$

NET IONIC REACTIONS

When ionic compounds react in aqueous solution, usually only one ion from each compound reacts. The other ions are "spectator ions" and do not react. The use of **net ionic reactions** focuses attention on the actual reactions and allows the chemist to find substitute reactants to achieve the same reaction. For instance, the reaction of silver nitrate with sodium chloride produces a precipitate of silver chloride in this molecular equation:

$$NaCl + AgNO_3 \rightarrow AgCl(ppt) + NaNO_3$$

Written as a net ionic equation, this becomes

$$Cl^- + Ag^+ \rightarrow AgCl(ppt)$$

From this equation the chemist knows that any soluble chloride salt (KCl, $MgCl_2$, etc.) and any soluble silver salt ($AgClO_4$, $AgSO_4$, etc.) will also give AgCl as the product. In reactions with ions the charges must balance, as well as the atoms.

HALF-REACTIONS

Half-reactions are used extensively with oxidation-reduction equations and in describing electrochemical processes on pages 305–314 and 431–436. The half-reaction is a reduction reaction if electrons are on the reactant side and an oxidation half-reaction if the electrons are products:

$$I_2 + 2e^- \rightarrow 2I \qquad \text{(reduction half-reaction)}$$

$$Fe^{2+} \rightarrow Fe^{3+} + e^- \qquad \text{(oxidation half-reaction)}$$

Half-reactions may be combined to make a complete oxidation-reduction reaction as long as the electrons all cancel.

OXIDATION-REDUCTION REACTIONS

Oxidation-reduction reactions involve the loss of electrons by one compound or ion and the subsequent gain of the same electrons by another compound or ion. The two half-reactions in the preceding section may be added (after multiplying the second reaction by 2 and canceling the electrons) to obtain the oxidation-reduction reaction:

$$I_2 + 2Fe^{2+} \rightarrow 2Fe^{3+} + 2I^-$$

The single-replacement reactions discussed previously are also oxidation-reduction reactions.

EXERCISE

Classify each of the reactions below as one of the reaction types described in this chapter.

(a) $C_6H_6 + O_2 \rightarrow CO_2 + H_2O$
(b) $MgCl_2 + AgNO_3 \rightarrow AgCl + Mg(NO_3)_3$
(c) $Al + O_2 \rightarrow Al_2O_3$
(d) $CaO + H_2SO_4 \rightarrow H_2O + CaSO_4$
(e) $Al + Fe_3O_4 \rightarrow Fe + Al_2O_3$
(f) $NO_2 + O_2 \rightarrow N_2O_5$
(g) $HCl + CaCO_3 \rightarrow CaCl_2 + H_2O + CO_2$
(h) $O_2 + C_4H_9NH_2 \rightarrow CO_2 + H_2O + N_2$
(i) $Mg + HCl \rightarrow H_2 + MgCl_2$
(j) $Zn + Cu(NO_3)_2 \rightarrow Cu + Zn(NO_3)_2$
(k) $CoCl_3 + Ba(OH)_2 \rightarrow BaCl_2 + Co(OH)_3$
(l) $Ba(OH)_2 + H_3PO_4 \rightarrow H_2O + Ba_3PO_4$
(m) $C_6H_8 + H_2 \rightarrow C_6H_{12}$
(n) $SO_2 + O_2 \rightarrow SO_3$
(o) $C_4H_{10} + O_2 \rightarrow CO_2 + H_2O$
(p) $H_2S + AuCl_3 \rightarrow Au_2S_3 + HCl$

Answers

(a) Combustion
(b) Double replacement
(c) Synthesis
(d) Double replacement
(e) Single replacement
(f) Combustion or addition
(g) Double replacement
(h) Decomposition
(i) Single replacement
(j) Single replacemen
(k) Double replacement
(l) Neutralization
(m) Addition
(n) Addition or combustion
(o) Combustion
(p) Double replacement

Organic Reactions

There are literally thousands of different organic reactions. Descriptions of several of these major types follow.

Combustion

In a **combustion** reaction, organic molecules may be burned in excess oxygen to produce carbon dioxide, CO_2, and water. When the oxygen is limited, the carbon dioxide is replaced by carbon monoxide, CO. If there is a severe depletion of oxygen soot, elemental carbon is formed instead of CO_2 or CO. When a piece of paper is burned, all three products, CO_2, CO, and C, are formed.

Hydrogenation

Alkenes and alkynes have double and triple bonds. Hydrogen may be added to these bonds in a **hydrogenation** reaction that is usually catalyzed by platinum. For example, 1-butene is hydrogenated to butane in this reaction:

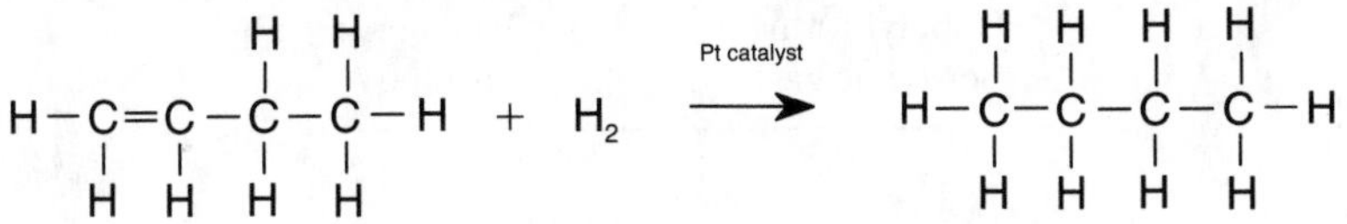

Compounds containing double or triple bonds are said to be unsaturated. A compound that has no double or triple bonds is termed saturated since it has the maximum number of hydrogen atoms possible.

Halogenation and Hydrohalogenation

Halogenation is a reaction similar to the hydrogenation process. Instead of H_2, a halogen such as F_2, Cl_2, or Br_2 may be used, as in this reaction:

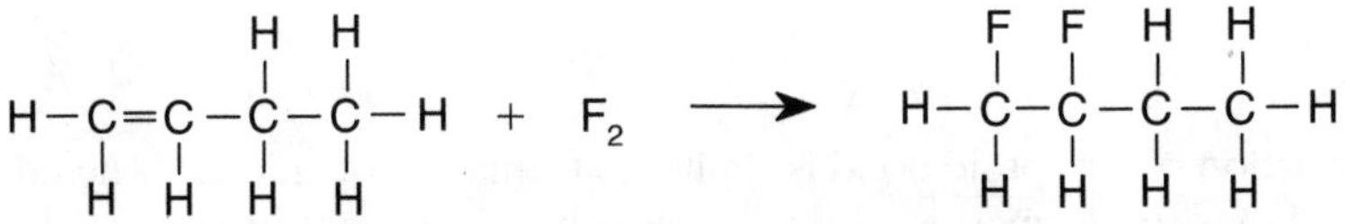

Halogenation reactions are generally more vigorous than hydrogenations and do not need catalysts.

Hydrohalogenation is the corresponding reaction in which compounds such as HCl and HBr add one hydrogen and one halogen atom to a double bond.

Esterification

In **esterification** an alcohol reacts with an organic acid to produce an ester and water. The formation of propyl acetate is shown below:

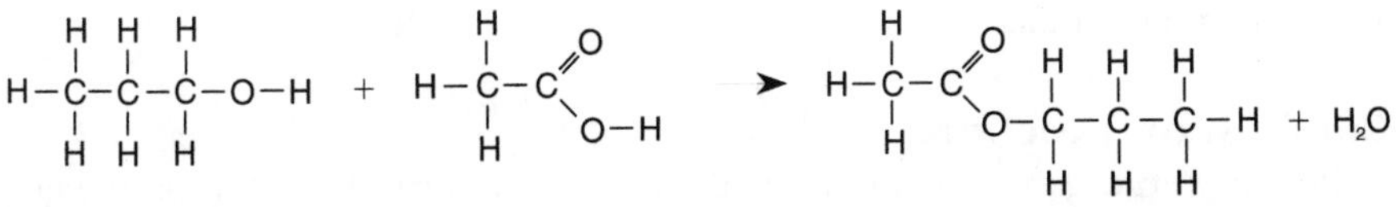

1-propanol acetic acid propyl acetate

In the name of an ester, the name of the alcohol comes first and the name of the acid second. For example, butyl propionate is made from butyl alcohol and propionic acid. Most esters have sweet, fruity aromas. Some of them are used as substitutes for natural products. Some common esters and their aromas are listed in Table 4.1.

TABLE 4.1. ESTERS AND THEIR AROMAS

Ester Name	Aroma
Octyl acetate	Orange
Ethyl formate	Rum
Methyl butyrate	Apple
Ethyl butyrate	Pineapple
Isopentyl acetate	Banana
Pentyl propionate	Apricot
Isobutyl formate	Raspberry
Benzyl acetate	Jasmine

Peptide Bond Synthesis

Amino acids contain two functional groups, $-NH_2$ and $-COOH$. Reactions between amino acids form a bond called the **peptide bond**:

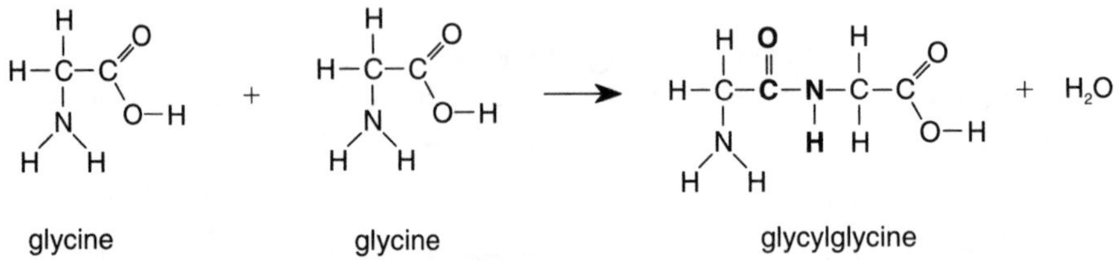

Formation of a peptide bond is similar to formation of an ester. The equation above shows the formation of one peptide bond (in bold print), which leaves an NH_2 on one end and a COOH on the other end of the molecule for the formation of additional peptide bonds.

Long chains of amino acids joined by peptide bonds are known as proteins. Twenty-one naturally occurring amino acids produce thousands of different proteins in the body.

Oxidation-Reduction Reactions

The main groups of **oxidation-reduction**, or **redox**, reactions are presented here as a unit.

Combustion Reactions

Reacting an organic compound with oxygen often results in the production of a large amount of heat along with a flame that is characteristic of the combustion process. The products of combustion reactions are usually carbon dioxide and water as in the combustion of glucose:

$$C_6H_{12}O_6 \quad + \quad 6O_2 \quad \rightarrow \quad 6CO_2 \quad + \quad 6H_2O$$

When the amount of oxygen is limited, the products of the reaction may include carbon monoxide or elemental carbon (soot):

$$C_6H_{12}O_6 + 3O_2 \rightarrow 6CO + 6H_2O \quad \text{(limited } O_2\text{)}$$
$$C_6H_{12}O_6 \rightarrow 6C + 6H_2O \quad \text{(very limited } O_2\text{)}$$

The same products, CO_2 and H_2O, are formed when glucose is metabolized in the body. However, since the process is slow and does not produce a flame, it is not called a combustion reaction. In metabolism the body uses the heat and energy produced in a more efficient manner than is the case in combustion. Even when the amount of oxygen is limited, CO and C are not produced in metabolic reactions.

Oxidation of Metals

Metallic elements have differing affinities toward oxygen. The very unreactive elements, such as platinum, gold, silver, and copper, do not react with O_2. Silver tarnishes by reacting with small amounts of hydrogen sulfide in the air. Copper reacts with water and carbon dioxide to form a carbonate compound. Some metals, particularly the alkali and alkaline earth metals, react readily with oxygen and are completely converted into oxides if exposed long enough.

Magnesium burns with a bright white flame in oxygen. The bright light is used in magnesium flares by the military, in flashbulbs for photography, and in fireworks. The reaction is so energetic that magnesium will continue burning even in an atmosphere of carbon dioxide:

$$2Mg + CO_2 \rightarrow 2MgO + C$$

When finely divided into a powder, many metals burn in oxygen. Steel wool burns when placed in a flame, and powdered metals such as aluminum are classified as highly combustible or even explosive.

Aluminum reacts very well with oxygen, as mentioned above. In larger sheets or bars, however, the aluminum oxide formed produces an impervious coating, so that complete oxidation of aluminum does not occur. This oxide layer is only a few molecules thick and is not visible to the eye.

Iron and steel react poorly with gaseous oxygen. The formation of rust requires the presence of water for oxidation to occur; iron does not rust in pure oxygen or in water that contains no oxygen. In this complex electrochemical process, the iron actually acts as the anode of a chemical reaction when moisture is present and as the cathode when it is not. Corrosion and its prevention are major concerns of approximately 30 percent of working chemists and chemical engineers. Each year, damage due to corrosion of buildings, bridges, and even computer circuits costs billions of dollars to correct and repair.

Single-Replacement (Displacement) Reactions

In single-replacement reactions an element replaces an atom in a compound, producing another element and a new compound. For example, the element zinc replaces the hydrogen in hydrochloric acid, HCl, forming the element H_2 and the new compound zinc chloride:

$$Zn + 2HCl \rightarrow ZnCl_2 + H_2$$

Very Active Metals: Very active metals, which have the lowest ionization energies, are Li, Na, K, Rb, Cs, Ca, Sr, and Ba. All these elements react with water in single-displacement reactions to form hydrogen. Here is one example:

$$2Na + 2H_2O \rightarrow 2NaOH + H_2$$

Many of these reactions also produce so much heat that the hydrogen ignites.

Active Metals: Active metals do not react with water, but will react with acids in a single-replacement reaction. An example is

$$Mg + 2H^+ \rightarrow Mg^{2+} + H_2$$

The common active metals are Mg, Zn, Pb, Ni, Al, Ti, Cr, Fe, Cd, Sn, and Co.

Inactive Metals: Inactive metals do not undergo simple single-replacement reactions with water or acids. The most common inactive metals are Ag, Pt, Au, and Cu. Copper and silver react with concentrated nitric acid in a reaction that produces nitrogen oxides but not hydrogen. Gold reacts with a mixture, called aqua regia, of three parts concentrated HCl and one part concentrated HNO_3.

The reactions of very active, active, and inactive metals with water and acid are summarized in Table 4.2.

TABLE 4.2. SUMMARY OF METAL REACTIONS WITH WATER AND ACID

Type of Metal	Examples	Comments
Very active metal	Li, Na, K, Rb, Cs, Ca, Sn, Ba	React with H_2O to produce H_2, which may ignite
Active metal	Mg, Zn, Pb, Ni, Al, Ti, Cr, Fe, Cd, Sn, Co	React with acids to form H_2, but not with H_2O
Inactive metal	Ag, Au, Cu, Pt	Do not form H_2 with acids; may react with conc. oxidizing acids HNO_3 and H_2SO_4

In addition to replacing hydrogen, metals will displace less active metal ions from their compounds. For instance, copper metal will react with silver nitrate in the single-replacement reaction

$$Cu + 2AgNO_3 \rightarrow Cu(NO_3)_2 + 2Ag$$

The net ionic equation is written as

$$Cu + 2Ag^+ \rightarrow Cu^{2+} + 2Ag$$

In these two reactions, copper is more active than silver and the reaction does occur. The reverse reaction:

$$Cu^2 + 2Ag \rightarrow Cu + 2Ag^+$$

does not occur, however, since silver is less active than copper.

An **activity series** is a listing of metals in the order of their activities. We can use an activity series to determine whether a certain metal will displace another metal ion from its compounds. A table of standard reduction potentials, such as the one given on page 211, contains the same information as an activity series. The metal having the lower, or more negative, standard reduction potential is the more active metal.

Only the active and inactive metals will displace each other from compounds. The very active metals listed in Table 4.2 do not displace other metals. Although the very active metals are certainly reactive enough to result in such displacements, their high activity causes them to react preferentially with water instead.

Non-metals: Nonmetals such as the halogens also participate in single-displacement reactions. The activity series for the halogens is $F_2 > Cl_2 > Br_2 > I_2$. As a result, adding chlorine to a solution of potassium bromide results in this reaction:

$$Cl_2 + 2KBr \rightarrow Br_2 + 2KCl$$

The net ionic reaction is as follows:

$$Cl_2 + 2\,Br^- \rightarrow Br_2 + 2Cl^-$$

Industrially important single-replacement reactions use carbon to displace metals from metal oxides in the refining process. For example, Fe_2O_3 is refined into iron in the reaction

$$2Fe_2O_3 + 3C \rightarrow 4Fe + 3CO_2$$

Reactions of the Permanganate Ion

Permanganate solutions have a very deep violet color, and the permanganate ion, MnO_4^-, is a very versatile oxidizing agent. It reacts differently in acidic, neutral, and basic solutions.

In acidic solutions the half-reaction is

$$MnO_4^- + 8H^+ + 5e^- \rightarrow Mn^{2+} + 4H_2O$$

Five electrons are used in this reduction, and the soluble Mn^{2+} ion is colorless. In addition, this reaction is slow and the presence of Mn^{2+} ions catalyzes the process.

In neutral or slightly acidic solutions the permanganate half-reaction is

$$MnO_4^- \quad + \quad 4H^+ \quad + \quad 3e^- \quad \rightarrow \quad MnO_2 \quad + \quad 2H_2O$$

The hydrogen ions in this three-electron half-reaction come from the dissociation of water molecules. The product MnO_2 is an insoluble black precipitate.

In basic solutions the reaction involves a one-electron transfer:

$$MnO_4^- \quad + \quad e^- \quad \rightarrow \quad MnO_4^{2-}$$

The reaction mixture changes color from the deep violet of the MnO_4^- ion to green for the MnO_4^{2-} ion.

Reactions of Chromium(VI)

When chromium is in the +6 oxidation state, it forms different compounds, depending on the pH of the solution. In neutral or basic solutions the chromate ion, CrO_4^{2-}, predominates.

In acidic solutions the dichromate ion predominates:

$$2CrO_4^{2-} \quad + \quad 2H^+ \quad \rightleftharpoons \quad Cr_2O_7^{2-} \quad + \quad H_2O$$

In very acid solutions the dichromate ion becomes protonated to form chromic acid:

$$Cr_2O_7^{2-} \quad + \quad 2H^+ \quad + \quad H_2O \quad \rightleftharpoons \quad 2H_2CrO_4$$

As the acidity of a solution increases, the strength of chromium(VI) as an oxidizing agent increases. Dichromate salts are dissolved in concentrated sulfuric acid to produce chromic acid, H_2CrO_4, which is a very effective cleaning agent because it oxidizes most organic materials. Dichromate salts dissolved in approximately 1 molar acid solutions are used for chemical analysis. For example, the reaction of dichromate with ethyl alcohol is used to test the sobriety of drivers suspected of driving while intoxicated (DWI):

$$8H^+ + Cr_2O_7^{2-} + 3CH_3CH_2OH \rightarrow 2Cr^{3+} + 3CH_3CHO + 7H_2O$$

Since the dichromate ion is orange, and the Cr^{3+} ion is green, the change in color from orange to green is a measure of the amount of alcohol present in the breath of a DWI suspect.

Iodine, Hydrogen Peroxide, and Thiosulfate Reactions

Iodine is a weak oxidizing agent. A dilute solution of iodine in alcohol, known as tincture of iodine, is an effective antiseptic for minor wounds. Because it is a weak oxidizing agent, iodine can be used in reaction mixtures when chemists want to oxidize only the more active reducing agents that are present.

Hydrogen peroxide, H_2O_2, is another weak oxidizing agent. Its mode of action is to decompose into water and atomic oxygen (O, not O_2). Atomic oxygen normally combines to form molecular O_2. However, in the brief time that it is available, atomic oxygen acts as a very good oxidizing agent if it encounters a suitable reactant. Hydrogen peroxide is used, in 3 percent solutions, as a household disinfectant and hair bleach. In higher concentrations, 30 percent, it is a very powerful oxidizing agent that must be handled with care. In fact, higher concentrations of H_2O_2 have been used as rocket fuel because of the oxygen this compound supplies.

The thiosulfate ion, $S_2O_3^{2-}$, is a reducing agent, one of very few that is stable in air because it reacts slowly with O_2. Thiosulfate solutions are used as reducing agents in chemical analysis, often with iodine as the oxidizing agent. Active metals such as zinc and magnesium can also serve as reducing agents in chemical reactions.

Redox Principles

Chemical reactions in which electrons are transferred from one atom to another, or those in which oxidation numbers change, are called **oxidation-reduction** reactions. As a group, more reactions may be classified as oxidation-reduction reactions than as acid-base, double-replacement, or complexation reactions combined.

Oxidation is the loss of electrons, and **reduction** is the gain of electrons. When an atom of barium reacts with an atom of sulfur, the barium loses its two valence electrons and is oxidized while the sulfur gains those two electrons and is reduced:

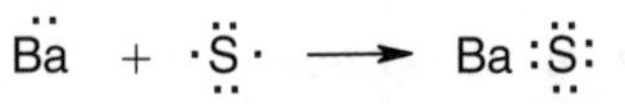

This reaction may be written as two **half-reactions** that show the individual oxidation and reduction steps:

$$Ba \rightarrow Ba^{2+} + 2e^- \quad \text{(oxidation)}$$

and

$$S + 2e^- \rightarrow S^{2-} \quad \text{(reduction)}$$

Although we may write separate half-reactions, neither of the two steps can exist without the other. Perhaps the word ***redox*** was coined to emphasize this point.

In the oxidation-reduction reaction above, the barium atom loses its electrons to the sulfur atom, causing the sulfur atom to be reduced. Consequently, barium is the **reducing agent**. The opposite view is that the sulfur atom gains electrons from the barium atom, causing the barium atom to be oxidized. Consequently, sulfur is the **oxidizing agent**. In any redox reaction, the substance oxidized is the reducing agent, and the substance reduced is the oxidizing agent.

LE CHÂTELIER'S PRINCIPLE

Equilibrium Dynamics

In 1888 Henry Le Châtelier proposed his fundamental principle of chemical equilibrium. He observed that chemical systems react until they reach a state of equilibrium. He also observed that, if chemicals in a state of equilibrium are disturbed in some manner, they will start reacting again and will continue until equilibrium is reestablished.

> **LE CHÂTELIER'S PRINCIPLE**
>
> Whenever a system in dynamic equilibrium is disrupted by changes in chemical concentrations or physical conditions, the system will respond with internal physical and chemical changes to reestablish the equilibrium state.

Chemical changes involve the addition or removal of one or more of the products or reactants. Physical changes to a system include changes in temperature, pressure, and volume. Understanding how these factors affect chemical equilibria allows chemists to adjust experimental conditions to maximize the desired products and to minimize waste.

Effects of Concentration Changes

Changing the concentration of any reactant or product in a chemical reaction will alter the concentrations of the other chemicals present as the system reacts to reestablish equilibrium. Figure 4.1 illustrates how this works. A two-compartment container is set up. The two compartment's are divided by a very porous barrier such as a window screen (dashed line). When a liquid is added to the container, the fluid easily flows to the same height in both compartments to establish Equilibrium 1. Then, if more liquid is added to the reactant, R, compartment, the equilibrium is momentarily disturbed, as shown in the middle diagram. However, the liquid flows rapidly through the barrier and establishes a new equilibrium condition, Equilibrium 2, in the last diagram.

This diagram clearly illustrates the action of a chemical system. Adding a reactant to an equilibrium system disturbs it by raising the reactant concentration. This disturbance causes the reaction to produce more products and reduce the amount of reactant in reestablishing equilibrium. Also important is that the final equilibrium has different amounts of reactants and products compared to the initial equilibrium state.

Illustrations similar to Figure 4.1 can be used to visualize other possible concentration changes and their effects. Increasing the concentration of a product will cause an increase in reactant formation (reverse reaction). Decreasing the concentration of a reactant will cause more reactant to be formed

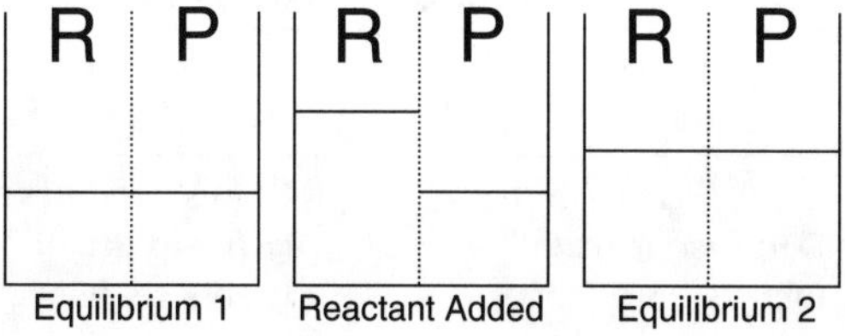

FIGURE 4.1. *Diagrams illustrating an initial equilibrium of two liquids with a porous barrier, a disturbance to the equilibrium by adding liquid to one side and finally the reestablishment of equilibrium.*

(reverse reaction) and decreasing the concentration of a product will cause more product to form (forward reaction).

Increasing the concentration of a reactant or product is a simple experimental process of adding more chemicals to the reaction mixture. Decreasing the concentration of a product or reactant is experimentally more difficult. The next paragraph describes some techniques that are used to remove products or reactants from a reaction mixture.

In the reaction to form ammonia

$$N_2(g) \quad + \quad 3H_2(g) \quad \rightleftharpoons \quad 2NH_3(g)$$

the ammonia gas produced is very soluble in water $NH_3(g) \xrightarrow{H_2O} NH_3(aq)$ so that a little water in the reaction system effectively removes the product NH_3. Another method used to remove a gas from a reaction is to condense it into a pure liquid.

Substances in solution may be removed by causing them to precipitate as solids. Additional chemicals may be added to a reaction mixture in order to cause precipitation.

Formation of a complex is an effective method for removing metal ions from solution. Complexing a metal ion changes it into a distinctly different substance, the complex. For example, the dissolution of silver chloride with ammonia is a complexation reaction. The $Ag(NH_3)_2^+$ complex is very soluble, and it removes Ag^+ from the solution so that more AgCl may dissolve. The two separate steps of the reaction are as follows:

$$AgCl(s) \quad \rightleftharpoons \quad Ag^+(aq) \quad + \quad Cl^-(aq)$$

and

$$Ag^+(aq) \quad + \quad NH_3(aq) \quad \rightleftharpoons \quad Ag(NH_3)_2^+(aq)$$

They add up to

$$AgCl(s) \quad + \quad 2NH_3(aq) \quad \rightleftharpoons \quad Ag(NH_3)_2^+(aq) \quad + \quad Cl^-(aq)$$

The effects of concentration changes are summarized in Table 4.3.

TABLE 4.3. EFFECTS OF CONCENTRATION CHANGES

Concentration Change	Observed Effect
Increase reactant	Favors products
Decrease reactant	Favors reactants
Increase product	Favors reactants
Decrease product	Favors products

Effects of Pressure Changes

An increase in pressure easily compresses gases but has little effect on solids and liquids. Increasing the pressure of a gas increases its molar concentration. Changing the pressure of individual gaseous reactants by adding or removing a gas follows the same principles as changing the concentrations, discussed in the preceding section. Changing the pressure of a system by adding an inert gas has no effect since the gases originally present still have the same partial pressures and concentrations.

Increasing the pressure of a gaseous reaction system by decreasing its volume will have an effect on the equilibrium only if Δn_g, the difference between the moles of gaseous products and gaseous reactants is not zero. If $\Delta n_g = 0$, there will be no shift in the equilibrium since the number of moles of gaseous products is the same as the number of moles of gaseous reactants. When $\Delta n_g > 0$, the reaction will be forced toward the reactant side because the larger number of moles of product will be compressed to a higher concentration than the reactants. Similarly, if $\Delta n_g < 0$, there will be more moles of gaseous reactant and the reaction will be forced toward producing more product. Decreasing the pressure will have the opposite effects.

The effects of pressure changes are summarized in Table 4.4.

TABLE 4.4. EFFECTS OF PRESSURE CHANGES

Value of Δn_g	Increasing Pressure	Decreasing Pressure
Positive	Favors reactants	Favors products
Zero	Has no effect	Has no effect
Negative	Favors products	Favors reactants

Effects of Temperature Changes

> The only experimental variable that has any effect on the value of the equilibrium constant is the temperature.

For some reactions the equilibrium constant increases as the temperature increases; for others, the equilibrium constant decreases. The direction in which the equilibrium constant moves depends on whether the reaction is exothermic (ΔH is negative) or endothermic (ΔH is positive). An exothermic reaction gives off heat to the surroundings, and an endothermic reaction absorbs heat from the surroundings.

An exothermic reaction may be represented as one in which one of the products is heat:

$$\text{Reactants} \rightleftharpoons \text{products} + \text{heat}$$

Raising the temperature for an exothermic reaction is similar to increasing the concentration of the products. Then, as with a chemical change, the increase in product will move the reaction toward the left, or reactant, side.

For an endothermic reaction, increasing the temperature is equivalent to increasing the concentrations of the reactants. The result is to move the equilibrium toward the product side:

$$\text{Heat} + \text{reactants} \rightleftharpoons \text{products}$$

and represents an increase in the equilibrium constant.

TABLE 4.5. EFFECTS OF TEMPERATURE CHANGES

Temperature Change	Reaction Type	Effect on Reaction	Effect on K
Increase	Exothermic	Favors reactants	Decrease
Increase	Endothermic	Favors products	Increase
Decrease	Exothermic	Favors products	Increase
Decrease	Endothermic	Favors reactants	Decrease

The effects of temperature changes are summarized in Table 4.5. The last column indicates that the actual effect of a change in temperature is a change in the value of the equilibrium constant. In addition to the direction of change, the amount of increase or decrease in the equilibrium constant is related to the magnitude of the heat of reaction.

REACTION RATES

Rates of Chemical Reactions

Most people are aware of the factors that influence reaction rates, perhaps without even realizing that they have this knowledge.

Concentration of Reactants

To make a fire burn more fiercely, we add wood to it. A car travels faster when we press on the accelerator pedal to give the engine more gas. In any chemical reaction the **concentration of reactants** is an important factor in the observed rate.

Temperature

To cook food faster, we turn up the heat in the oven. To decrease the spoilage of foods, we refrigerate or even better, freeze them. **Temperature** is an important factor in the rates of these chemical reactions. As a consequence, all rate experiments must have their temperatures carefully controlled. It is important to remember that increasing the temperature always increases reaction rates and decreasing the temperature always decreases reaction rates.

Catalysts

A **catalyst** will increase the rate of reaction. A catalyst is a substance that participates in a chemical reaction but does not appear in the balanced chemical equation. Perhaps the most familiar catalysts are those in the catalytic converters in automobiles. The platinum in a catalytic converter provides a surface on which reactants meet and react more efficiently. Although the platinum promotes the effective reaction of two chemicals, it is not included in the balanced reaction. An important feature of catalysts is that they increase proportionately the rates of both the forward and the reverse reactions. Therefore a catalyst does not alter the equilibrium constant of the reaction.

Other experimental factors, as long as they do not affect the concentration, the temperature, or the catalysts, will have no effect on the rate of a chemical reaction.

Collision Theory

The **collision theory** states that the reaction rate is equal to the frequency of effective collisions between reactants. For a collision to be effective, the molecules must collide with sufficient energy and in the proper orientation so that products can form.

The minimum energy needed for a reaction is the activation energy, (E_a). If two molecules collide head-on, they will stop at some point and all of the kinetic energy will be converted into potential energy. If the molecules strike each other with a glancing blow, only part of the kinetic energy will be converted into potential energy. As long as the increase in potential energy is greater than the activation energy, a reaction is possible. The fraction of all collisions that have the minimum energy needed for reaction can be calculated using the kinetic molecular theory of gases discussed on pages 163–164. The fraction of collisions with this minimum energy increases with increasing temperature since the average kinetic energy of molecules increases with temperature.

In addition to the energy requirement, an effective collision requires that the molecules collide in the proper orientation. An example is the reaction of hydrogen iodide molecules with chlorine atoms. In this reaction the chlorine replaces the iodine in the molecule:

$$HI \quad + \quad Cl \quad \rightarrow \quad HCl + I.$$

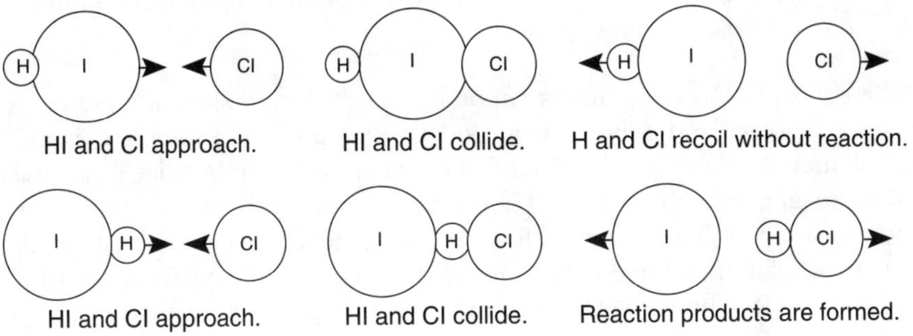
H
I
Cl
HI and Cl approach.
HI and Cl collide.
H and Cl recoil without reaction.
HI and Cl approach.
HI and Cl collide.
Reaction products are formed.

FIGURE 4.2. *Diagrams illustrating how the orientation of colliding molecules determines if a reaction can occur.*

Figure 4.2 shows hydrogen iodide colliding with a chlorine atom from two directions. When chlorine collides with the iodide end of HI, the reactants recoil from the collision without a reaction occurring. In the other sequence the chlorine atom collides with the hydrogen end of HI. This collision can cause the iodine atom to be released while the chlorine bonds with the hydrogen. Thus the products are HCl and I.

The overall reaction rate predicted by the collision theory can be summarized by the equation

$$\text{reaction rate} = Nf_e f_o$$

where N is the number of collisions per second, which depends on the temperature and concentration of the reactants, f_e is the fraction of the collisions that have the minimum energy, and f_o is the fraction of collisions with the correct orientation. The fraction f_e will increase as temperature increases, but the fraction f_o remains constant for a given reaction.

Kinetic Molecular Theory

The ideal gas law describes the relationships between the variables P, V, T, and n for ideal gases. The **kinetic molecular theory** describes gases at the level of individual gas particles. This theory was developed largely by Ludwig

Boltzmann, Rudolf Clausius, and James Maxwell between 1850 and 1880. It is often stated as five postulates:

1. Gases consist of molecules or atoms in continuous random motion.
2. Collisions between these molecules and/or atoms in a gas are elastic.
3. The volume occupied by the atoms and/or molecules in a gas is negligibly small.
4. Attractive forces between the atoms and/or molecules in a gas are negligible.
5. The average kinetic energy of a molecule or an atom in a gas is directly proportional to temperature.

The concept of gas pressure is important to understand since it is central to the kinetic molecular theory. Pressure is defined in physics as the force exerted per unit area. In English units, pounds per square inch are familiar. For gases, the force is generated by collisions of the gas particles with the container walls. Each collision will have a certain force, which is related to the velocity of the gas particle. The total force is the sum of the forces of all collisions occurring each second per unit area. Thus the pressure is dependent on the velocity of the gas particles and the collision frequency. The collision frequency depends, in turn, on the velocity of the gas particles and the distance to the container walls.

Changing the temperature will change the force of the collisions as well as the frequency of collision. Altering the size of the container will change the frequency of collision but not the force of the collisions.

With this understanding of pressure, the molecular meaning of the various gas laws can be appreciated as follows.

Boyle's law states the inverse relationship between pressure and volume ($P \alpha 1/V$). The kinetic molecular theory agrees with this observation. If the volume of a gas is decreased, the gas particles will strike the walls of the container more frequently. Increasing the frequency of these collisions increases the observed pressure of the gas.

Gay-Lussac's law expresses a direct relationship between temperature and pressure ($P \alpha T$). In this case, an increase in temperature increases the kinetic energy of the gas particles with two effects: first, the energy of each collision is greater; second, since the average velocity of the gas particles increases with temperature, the frequency of collisions with the container walls also increases. Thus the kinetic molecular theory predicts an increase in gas pressure as the temperature increases.

Charles's law predicts a direct relationship between the temperature and volume of a gas ($V \alpha T$). An increase in temperature increases both the force of each collision and the frequency of collisions. Both of these effects increase the pressure. If the pressure of the gas is to remain constant, the volume must increase to correspondingly decrease the frequency of collisions with the walls of the container.

In addition to explaining the gas laws, the kinetic molecular theory explains **Graham's law of effusion**, which is summarized by the equation

$$\sqrt{\frac{m_1}{m_2}} = \frac{\bar{v}_2}{\bar{v}_1}$$

where m_1 and m_2 are the relative masses of two gases, and $\bar{v}_1$ and $\bar{v}_2$ are the respective rates of effusion. This law can be directly derived from the fifth postulate of the kinetic molecular theory, as shown on page 164.

Finally, the kinetic molecular theory allows us to develop a fundamental explanation of reaction rates. This is done, using collision theory, discussed on pages 162–163.

Transition State Theory

The **transition state theory** attempts to describe in detail the molecular configurations and energies as a collision of reactants occurs.

The transition state theory recognizes that, as molecules approach on a collision course, they do not act like billiard balls simply bouncing off each other. Instead, as the molecules get closer, their orbitals interact and distort each other. This distortion weakens bonds within the molecules so that at the moment of collision some bonds are so weak that they break and new bonds may form.

Using the diagram in Figure 4.3, we can visualize the effective collision in this sequence. First, the chlorine approaches the hydrogen end of the HI molecule. As the chlorine and HI get closer, the very electronegative chlorine starts attracting the electrons that the hydrogen shares with the iodine atom, thus weakening the H-I bond and beginning the formation of a H-Cl bond. At the moment of collision, the H-I bond is approximately half broken and the H-Cl bond is approximately half formed. The I-H-Cl group of atoms is often called the **activated complex**. When the atoms recoil, the activated complex breaks apart. The result may be a successful reaction giving new products, or an unsuccessful collision with the original reactants remaining intact.

$$\text{HI} + \text{Cl} \rightarrow \text{HCl} + \text{I}$$

In the transition state theory the energies of the reactants during a collision are described by a reaction profile. As the molecules approach, interact, and become distorted, their kinetic energy is converted into potential energy. As a result the potential energy increases. In other words, as the molecules approach a collision, they slow down because their kinetic energy is converted to potential energy. The **reaction profile** plots the increase in potential energy of the reactants as they approach, reaching a maximum at the moment of collision, and then the decrease in potential energy as the products recoil. A reaction profile is shown in Figure 4.4.

In the reaction profile the minimum amount of kinetic energy that must be converted into potential energy in order to form products is called the activation energy (E_a). It is often referred to as an energy barrier between the reactants and products.

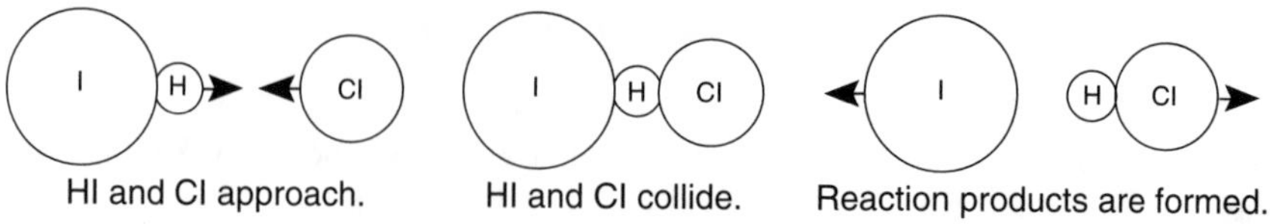

FIGURE 4.3. *Explanation of Figure 4.2 in terms of activated complexes. The reaction of HI with Cl. On the left the reactants approach. In the center is the activated complex. The result may be either products (right) or reactants (left).*

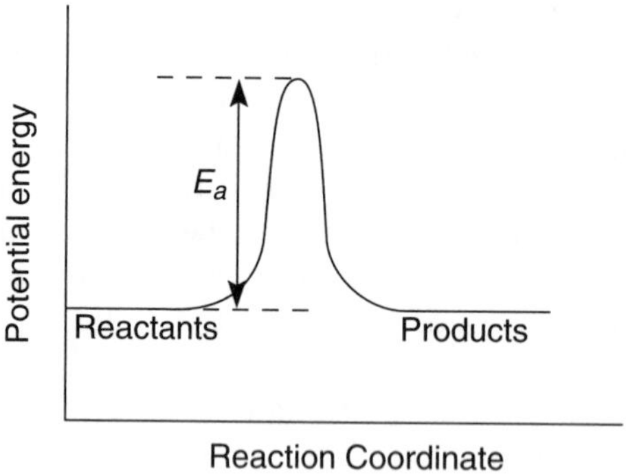

FIGURE 4.4. *Reaction profile illustrating the energy barrier between reactants and products.*

Comparison of the Collision and Transition State Theories

The transition state theory is based on the same considerations as the collision theory. If reactants collide with enough energy to overcome the energy barrier, a reaction may occur. The other requirement is that the activated complex formed at the moment of collision (top of the energy barrier) must have the proper structure, so that it can proceed to fall apart into products. If it has the wrong structure, products cannot form and the molecules recoil as the original reactants. These are the same energy and orientation factors that are important in the collision theory.

The major difference in the two theories is that the collision theory views reactions as collisions between hard spheres, similar to the collisions of billiard balls. The transition state theory views the collisions as interactions between sponge-like balls that are deformed in the collision process. Compared to the collision theory, the transition state theory involves more details about the energy and shapes of the molecules.

Interpretation of Reaction Profiles

Reaction profiles provide a rich source of information about the rates of chemical reactions. These profiles serve also as a graphical view of the conversion of reactants into products and are often more informative than words alone in describing features of the reaction process.

In a reaction profile, the rate constant, and therefore the rate of a chemical reaction, is inversely related to the height of the energy barrier (E_a). When the activation energy is low, a large proportion of the collisions will have sufficient energy for a reaction to occur. A high activation energy, on the other hand, indicates that few collisions will have enough energy to convert reactants into products.

Reaction profiles can be used to determine whether a reaction is endothermic or exothermic. This can be done because the potential energy difference between the products and reactants is equal to the heat of reaction (ΔH):

$$\Delta H = PE_{\text{products}} - PE_{\text{reactants}}$$

When heat is absorbed from the surroundings, the reaction is endothermic and ΔH has a positive sign. For **endothermic reactions** the potential energy of the products is greater than the potential energy of the reactants. The reverse is true for **exothermic reactions**. The reaction profiles for these two types of reactions are shown in Figure 4.5.

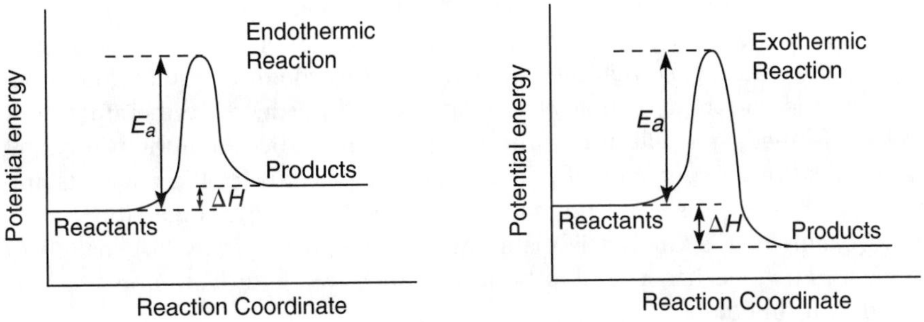

FIGURE 4.5. *Reaction profiles illustrating the difference between an exothermic reaction and an endothermic reaction.*

Another advantage of the reaction profile is that it allows the chemist to explain the reverse as well as the forward chemical reaction. To visualize what occurs in the reverse process, we begin our examination of the reaction profile on the product side and proceed toward the reactant side. When the products are reacting to form reactants, the energy of activation and heat of reaction are different. The activation energy for the reverse reaction is the difference between the products' potential energy and the maximum of the curve. For the heat of reaction the sign of ΔH will be opposite to that for the forward direction. Figure 4.6 is the same as Figure 4.5 except that it shows the activation energies and heats of reaction for the reverse reactions.

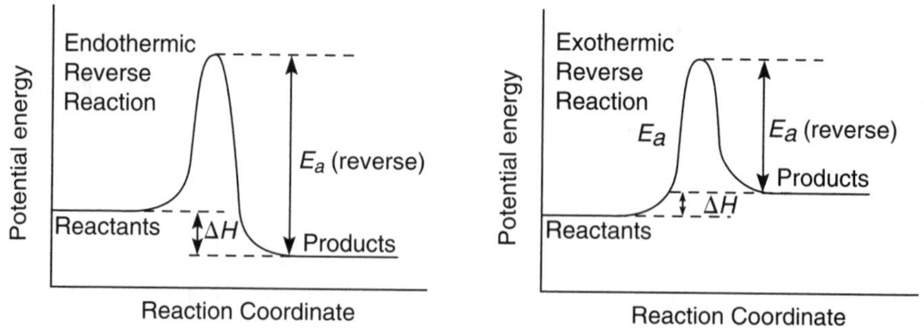

FIGURE 4.6. *Reaction profiles illustrating the activation energy of reverse reactions and the fact that the sign of ΔH is the opposite of the sign of ΔH for the forward reaction.*

Reaction profiles also allow us to explain the action of **catalysts**. A catalyst is a substance that increases the rate of a chemical reaction without itself being reacted. It does this by providing an alternative reaction pathway that has a lower energy barrier in the reaction profile. As a result the energies of activation of the forward and reverse reactions are simultaneously decreased by the same amount, so that the reaction comes to chemical equilibrium more quickly. It is important to note that a catalyst will not increase the amount of product formed, nor will it alter the composition of the equilibrium mixture, as indicated by the unchanged potential energy plateaus for the reactants and products. Figure 4.7 shows the reaction profile of a catalyzed reaction.

An example of a catalyst is platinum metal, used in the hydrogenation of ethylene. Ethylene has a double bond to which two hydrogen atoms can be added to form ethane:

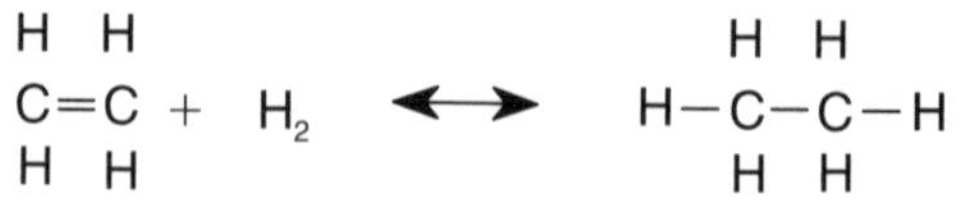

A mixture of hydrogen and ethylene at room temperature does not show appreciable reaction. The apparent reason is that the hydrogen molecule is fairly stable, and the energy required to break the hydrogen atoms apart results in a high activation energy. Addition of a small amount of finely divided platinum, however, catalyzes a very rapid reaction. Platinum adsorbs hydrogen readily on its surface, and the hydrogen molecule separates into

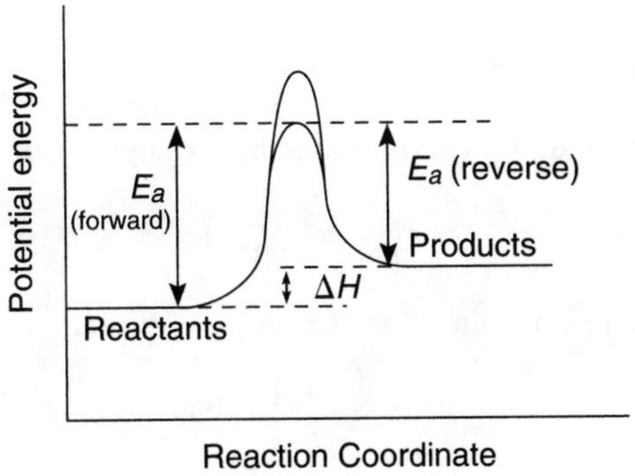

FIGURE 4.7. *Reaction profile of a catalyzed reaction illustrating that the forward and reverse activation energies are both reduced. The heat of reaction is not affected, nor is the position of equilibrium, because the potential energies of the reactants and products are not affected.*

individual hydrogen atoms. These hydrogen atoms then react readily with the ethylene. In the end, the original platinum may be recovered and used again.

The discovery of new, more effective, and more specific catalysts is a major objective of many industrial chemists. Many catalysts are naturally occurring minerals whose catalytic properties are often discovered by trial and error. Others are synthetic compounds designed by studying natural catalysts and using chemical methods to improve upon them.

Enzymes are another class of natural catalysts found in living organisms. They catalyze very specific reactions, and most enzyme reactions have an optimum temperature around 36°C. Generally the enzyme, E, and the reactant(s), called the substrate, S, react to form an enzyme-substrate, ES, complex. The **enzyme-substrate complex** then decomposes into the product(s), P, and the original enzyme. This process is written as follows:

$$E + S \rightleftharpoons ES \rightleftharpoons P + E$$

Enzymes are specific for certain reactants because they recognize specific chemical structures by their three-dimensional shape as well as their chemical properties. The way this occurs is often called the **lock and key** model of enzyme action. Small changes in the shape of a key will make it useless in a lock that the key once was able to open. Similarly, small differences in the shapes of molecules will determine whether or not they form effective enzyme-substrate complexes.

The rate of reaction of a simple enzyme reaction is given by the **Michaelis-Menton equation**:

$$\text{Rate} = \frac{k[\text{E}][\text{S}]}{C + [\text{S}]}$$

In this equation k is the rate at which the enzyme-substrate complex reacts to form products and the constant C is a combination of the other rate constants in the process. When the substrate concentration is very large, this equation becomes

$$\text{Rate} = k[\text{E}]$$

When the substrate concentration is very low, the equation becomes

$$\text{Rate} = \left(\frac{k}{C}\right)[\text{E}]\,[\text{S}]$$

To understand these equations, it is helpful to visualize the rates at which cars pass through a toll booth on an expressway. When there is a large volume of traffic, [S], the rate at which cars go through the booths depends on the number of booths, [E]. When there is little traffic, the rate depends on the frequency at which cars arrive at the toll booths, [S]. At low substrate concentrations the rate depends on [S], while at high substrate concentrations the rate depends on [E], as shown in Figure 4.8.

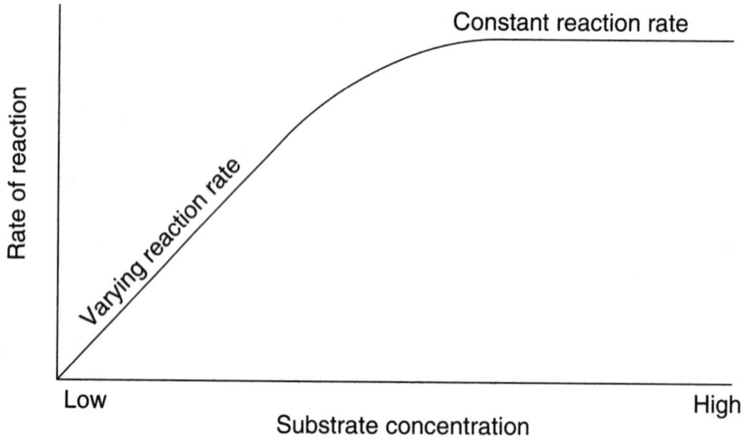

FIGURE 4.8. *Dependence of the reaction rate of enzyme-catalyzed reactions on substrate concentration [S]. At low substrate concentration the rate is proportional to [S]; at high substrate concentrations the rate is independent of [S]. The enzyme concentration, [E], is assumed to be constant.*

5
PHYSICAL CHEMISTRY

ENERGY

Forms of Energy

Energy takes many forms. Heat and light, along with chemical, nuclear, electrical, and mechanical energy, are some common types of energy. Any one of these forms of energy can be converted into any other form. In addition, the law of conservation of energy states that energy is never created or destroyed. As a result of these properties, all forms of energy can be converted into heat energy, which can be measured in a calorimeter as described below.

Energy can also be categorized into just two groups, either kinetic energy (KE) or potential energy (PE). **Kinetic energy** is the energy that matter possesses due to its motion and is described by one equation:

$$\text{KE} = \tfrac{1}{2}\,\text{mv}^2$$

When the mass (m) is expressed in kilograms and the velocity (v) in meters per second, the energy units are joules.

Potential energy is stored energy that may be released under the appropriate conditions, as in a nuclear reaction. There are several forms of potential energy, such as gravitational energy and the energy of electrostatic attraction between oppositely charged ions. Each form of potential energy is described by its own equation, but all these equations are similar. For example:

$$\text{PE}_{\text{grav}} = K_{\text{grav}}\left(\frac{m_1 m_2}{r}\right) \quad \text{and} \quad \text{PE}_{\text{elect}} = K_{\text{elect}}\left(\frac{q_1 q_2}{r}\right)$$

In these equations the two masses (m) in gravitational attraction and the two charges (q) in electrostatic attraction are both separated by a distance (r). K is a proportionality constant that is different for each type of potential energy.

The total energy of a substance is the sum of its kinetic and potential energies:

$$\text{Energy } (E) = \text{potential energy (PE)} + \text{kinetic energy (KE)}$$

In chemical substances, the kinetic energy is the motion of the molecules. The potential energy of a chemical is the sum of all attractions, including all covalent bonds, ionic bonds, or electrostatic attractions in the substance.

Kinetic Energy and Temperature

Experimentation has shown that the average kinetic energy of a gas is directly proportional to the temperature. Some gas molecules have kinetic energies above the average, and some have kinetic energies below the average.

Figure 5.1 illustrates the distribution of the kinetic energies of gas particles at two different temperatures. The figure shows the position of the average kinetic energy. The maximum of each curve is the most probable kinetic energy. It can be seen that the average and most probable kinetic energies are not the same. The reason is that the curves in Figure 5.1 are not symmetrical, and the point at which half of the molecules have higher, and half have lower, kinetic energies lies to the right of the peak, as shown.

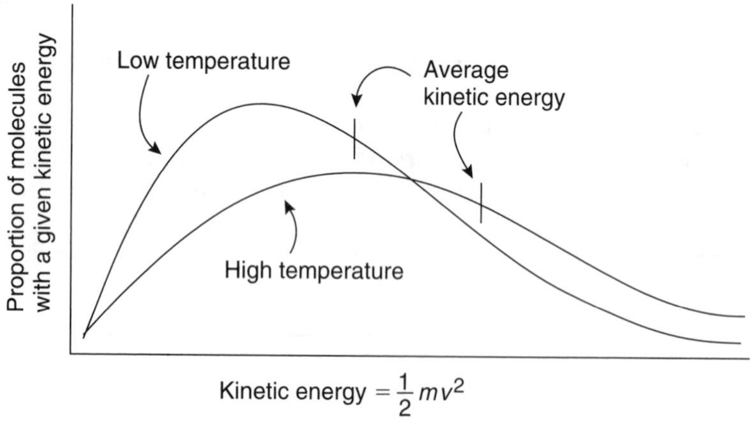

FIGURE 5.1. *Kinetic energy distribution diagrams for gas particles at two different temperatures. Note that average kinetic energy is not at the curve maximum.*

Using the equation KE = $\frac{1}{2}mv^2$, we may draw a similar graph with the molecular velocity on the x-axis. In this case too, some molecules have velocities above the average and some have velocities below the average. We would also see that heavier molecules have lower average velocities than light molecules, at the same temperature.

Laws of Conservation of Matter and Energy

These are the two most important laws of chemistry.

> **LAW OF CONSERVATION OF MATTER**
> Matter cannot be created or destroyed.

The law of conservation of matter governs all chemical reactions. The total mass of reactants we start with must be the total mass of the products we end with. If the reaction is very efficient and well planned, all reactants will be converted into products, but the mass of the products can never exceed the mass of the reactants. This law serves as an important—and quick—check for a reasonable answer in any mass-to-mass stoichiometry calculation.

LAW OF CONSERVATION OF ENERGY

Energy cannot be created or destroyed, but it can be converted readily from one form to another.

The law of conservation of energy allows energy to be converted from one form to another. Chemists prefer to define all energy as either potential energy (PE) or kinetic energy (KE). **Potential energy** is due to the positions of two objects that may be repelling or attracting each other; **kinetic energy** is present only in moving objects.

Potential energy is easily converted into kinetic energy, and kinetic energy is easily converted into potential energy. The KE + PE sum, however, must remain constant for any isolated system. The traditional example is the pendulum. At its highest points, at the end of each swing, all of the kinetic energy has been converted into potential energy since the pendulum stops momentarily to reverse the swing. At the lowest point in the swing, the pendulum is moving its fastest and has the most kinetic energy, which was derived from the potential energy.

We may look at various forms of energy and see whether the chemist classifies each as kinetic or potential. *Heat energy* is due to rapidly moving atoms; the greater the speed, the higher the heat content. Heat energy is kinetic energy. *Nuclear energy* is the energy released when radioactive materials disintegrate. Until the actual disintegration occurs, this energy is potential energy. When disintegration occurs, it converts the potential energy into kinetic energy such as a speeding alpha particle. *Chemical energy*, or the energy in a chemical bond, is also potential energy. When a chemical reacts, however, the rearrangement of the bonds can release heat, which is kinetic energy.

Einstein's famous equation shows the relationship that allows for the conversion of matter to mass or mass to matter:

$$E = mc^2$$

A very small amount of matter produces a large amount of energy. For the small energy changes involved in chemistry, the corresponding mass changes are so small (less than 1 nanogram) they cannot be measured even with the best instruments.

EXERCISE

The combustion of one mole of propane, C_2H_6, produces 890 kJ of heat energy. Using Einstein's equation, $E = mc^2$, how many grams of matter was converted into energy,

Answer

Since $E = 890 \times 10^3$ J and $c = 3.0 \times 10^8 \text{m s}^{-1}$ we obtain

$$m = \frac{890 \times 10^3 \text{ J}}{(3.0 \times 10^8 \text{ m s}^{-1})^2}$$

The metric units for Joules are

$$\frac{\text{kg m}^2}{\text{s}^2}$$

and our answer will be in kg.

$$m = \frac{890 \times 10^3 \text{ kg m}^2 \text{ s}^{-2}}{9 \times 10^{16} \text{ m}^2 \text{ s}^{-2}} = 98 \times 10^{-13} \text{ kg}$$

$$m = 9.8 \times 10^{-9} \text{ g} = 0.0000000098 \text{ g}$$

This is too small a mass for easy measurement.

LAWS OF THERMODYNAMICS

Thermodynamic Definitions

In the discussion of thermodynamics precise terminology is used to avoid confusion and to ensure that experimental results can be compared from one laboratory to another. Some of the important terms described below are *system, state function, standard state*, and the *exo*- and *endo*- prefixes.

Systems

A **system** is that part of the universe that is under study. Everything else in the universe is called the **surroundings**. When hydrogen and oxygen are placed in a bomb calorimeter to study the formation of water, all of the hydrogen and oxygen atoms and the calorimeter are the system. The surrounding laboratory, the building, the city, and so on are all parts of the surroundings.

There are several types of systems. **Open systems** can transfer both energy and matter to and from the surroundings. An open bottle of perfume is an example of an open system. In a **closed system** energy can be transferred to the surroundings but matter cannot. A well-stoppered bottle of perfume is a closed system. In an **isolated system** there is no transfer of energy or matter to or from the surroundings. A thermos bottle is almost an isolated system since it minimizes the transfer of heat energy to the surroundings. A calorimeter is designed to be as nearly as possible an isolated system.

State Function

In thermodynamics we encounter the terms *enthalpy change*, (ΔH), *entropy change*, (ΔS), *free-energy change*, (ΔG), and *energy change* (ΔE), which designate **state functions**. The numerical values and mathematical signs of these state functions depend only on the difference between the final state and the initial state of the system. The state of a system is defined by the mass and phase (solid, liquid, or gas) of the matter in the system, as well as by the temperature and pressure of the system. In describing the melting of 1 mole of ice, the initial state is described as 18 grams of $H_2O(s)$ at 1.00 atmosphere of pressure and 273 K. The final state will be 18 grams of $H_2O(l)$ at 1.00 atmosphere of pressure and 273 K. With this precise description all scientists should, within experimental error, obtain identical values of ΔH, ΔS, ΔG, and ΔE for the melting of ice.

Two quantities that are not state functions are heat (q) and work (w). The values for these quantities depend on the experimental methods used to transform matter from the initial state to the final state.

Standard State

The thermodynamic quantities ΔH, ΔS, ΔG, and ΔE are **extensive properties** of matter, meaning that they change as the amount of sample changes. To make these quantities **intensive properties** of matter, we must define precisely the temperature, pressure, mass, and physical state of the substance. A system is in the **standard state** when the pressure is 1 atmosphere, the temperature is 25°C, and 1 mole of compound is present. When the thermodynamic quantities are determined at standard state, they are intensive properties and a superscript ° is added to their symbols: $\Delta H°$, $\Delta S°$, $\Delta G°$, and $\Delta E°$. For a chemical reaction, the standard state involves the number of moles designated by the stoichiometric coefficients in the simplest balanced chemical equation.

Exo- and *Endo-* Prefixes and Sign Conventions

The prefix ***exo-***, as in *exothermic*, indicates that heat energy is being lost from the system to the surroundings. Mathematically, *exo-* corresponds to a negative sign for numerical thermodynamic quantities. In an exothermic reaction the heat of reaction (ΔH), is a negative number.

The prefix ***endo-*** indicates that energy is gained from the surroundings. An endothermic reaction absorbs heat energy from the surroundings. If heat is not transferred rapidly, however, the reaction mixture may appear to cool as the reaction progresses. Also, if a beaker containing such a mixture is touched, it will feel increasingly cold as heat is transferred from the hand to the reaction. In an endothermic reaction the heat of reaction (ΔH) is a positive number.

First Law of Thermodynamics

The **first law of thermodynamics** states that energy is always conserved. In chemistry this law means that the measurable quantities heat (q) and work (w) must add up to the total energy change (ΔE) in a system:

$$\Delta E = q + w$$

The value of q has a positive sign if heat is added to the system. If a beaker cools during a reaction, it is absorbing heat from the surroundings and q is positive. Conversely, q is a negative value if heat is released from the system. When the beaker feels warm, heat is being released and q is negative.

The value of w is positive if work is done *on* the system and is negative if work is done *by* the system. Work is equal to the pressure times the change in volume, ($P\,\Delta V$). If the volume increases, work is done by the system and w has a negative sign. If the volume decreases, work is done on the system and w has a positive sign.

The change in energy of a system, at constant temperature, is also the difference in potential energy (PE) between the final and initial states of the system:

$$\Delta E = \text{PE}_{\text{final}} - \text{PE}_{\text{initial}}$$

As shown on page 178, ΔE is mainly heat energy. Therefore a system that increases its potential energy is often said to be **endothermic**, while a decrease in potential energy indicates an **exothermic** process.

Work

The energy change is the sum of the heat and work, as shown above. Heat is measured using a calorimeter.

Work is defined as the force applied to an object as it moves a certain distance. The negative sign is necessary because of the definition above.

$$\text{Work} = -\text{force} \times \text{distance moved}$$

Force can be defined as the pressure exerted over a given area, so

$$\text{Work} = -\text{pressure} \times \text{area} \times \text{distance moved}$$

Multiplying the area by the distance results in volume units or an overall volume change:

$$\text{Work} = -\text{pressure} \times \text{volume change}$$

$$\text{Work} = -P\,\Delta V$$

Thus work is the product of the pressure and the change in volume that occurs during a chemical reaction and so is easily measured.

EXERCISES

1. Demonstrate that work is not a state function by calculating the work involved in expanding a gas from an initial volume of 1.00 liter and pressure of 10.0 atmospheres to (a) 10.0 liters and 1.0 atmosphere and (b) 5.00 liters and 2.00 atmospheres and then to 10.0 liters and 1.00 atmosphere.

   The change in volume (ΔV) is calculated as $V_{\text{final}} - V_{\text{initial}}$. The pressure is the final pressure for the change stated. In fact, the statement of this problem assumes that the initial pressure is instantaneously changed to the final pressure and then the gas is allowed to expand against this final pressure.
2. The preceding Exercise uses units of liter atmospheres. Convert the results into joules.

Solutions

1. (a)
$$\begin{aligned} w &= -P\,\Delta V \\ &= -(1.00\text{ atm})(10.0\text{ L} - 1.00\text{ L}) \\ &= -9.00\text{ L atm} \end{aligned}$$

   (b)
$$\begin{aligned} w &= -P\,\Delta V \\ &= -(2.00\text{ atm})(5.0\text{ L} - 1.00\text{ L}) - (1.00\text{ atm})\,(10.0\text{ L} - 5.00\text{ L}) \\ &= -13.0\text{ L atm} \end{aligned}$$

   In both parts of this exercise the sample starts in the same state (1.00 L and 10.0 atm) and ends in the same state (10.0 L and 1.00 atm). However, the work, in units of liter atmospheres, is different. This can occur only if w is not a state function.

2. We can construct the necessary conversion factor from the defined values of R, the universal gas law constant. Two definitions of R are

$$R_1 = 0.08206\text{ L atm mol}^{-1}\text{ K}^{-1}$$

   and

$$R_2 = 8.315\text{ J mol}^{-1}\text{ K}^{-1}$$

   Dividing R_1 into R_2, we obtain

$$\frac{R_2}{R_1} = \frac{8.315 \text{ J mol}^{-1} \text{ K}^{-1}}{0.08206 \text{ L atm mol}^{-1} \text{ K}^{-1}}$$

$$= \frac{101.3 \text{ J}}{\text{L atm}}$$

This is the necessary factor label for converting the liter atmospheres into joules. Applying this factor label to the answers for Exercise 4.1, we obtain (a) −912 J and (b) −1317 J of energy.

Definitions of q_p, q_v, ΔE, and ΔH

The first law of thermodynamics may be rewritten by substituting $-P\,\Delta V$ for w:

$$\Delta E = q_p - P\,\Delta V$$

The minus sign is needed because an increase in volume means that the system does work on the surroundings and this has been defined as a negative quantity. The heat term (q) is given the subscript p to indicate that the pressure must be constant.

When the heat energy is measured in a calorimeter that does not allow the volume to change, $P\,\Delta V$ must be zero. As a result, $\Delta E = q_v$, where the subscript v indicates that the volume is held constant. Calorimeters that do not allow the volume to change are called bomb calorimeters because the heavy stainless steel reaction vessel has been known to explode if not correctly used.

For most real reactions chemists are interested in the heat generated at constant pressure (q_p). The symbol q_p represents the enthalpy change and is given the symbol ΔH. Enthalpy (H) is the heat content of a chemical, and ΔH is the difference in the heat contents of the products and reactants:

$$\Delta H = H_{\text{products}} - H_{\text{reactants}}$$

The relationship between ΔE and ΔH can be expressed as follows:

$$\Delta E = \Delta H - P\,\Delta V$$

For many reactions the value of ΔH is very large and the value of $P\,\Delta V$ is relatively small, so that ΔE and ΔH are approximately equal.

Standard Enthalpy Changes; the Standard Heat of Reaction (ΔH°)

The heat energy at constant pressure, or the enthalpy change (ΔH), produced by a chemical reaction is an extensive property, since reacting a larger amount of chemicals produces a larger amount of heat. For example, when propane is burned according to this equation:

$$CH_3CH_2CH_3(g) + 5O_2(g) \rightarrow 3CO_2(g) + 4H_2O(g) \quad (5.1)$$

more heat is generated as more propane is burned. To make the heat produced by a reaction an intensive property, the amount of chemical that reacts must be specified. The standard heat of a reaction (ΔH°) is generally defined as the heat produced when the number of moles specified in the balanced chemical equation reacts. For the reaction of propane the heat of reaction (ΔH°_{react}) is equal to -2044 kilojoule when 1 mole of propane reacts with 5 moles of oxygen as shown in Equation 5.1. The negative sign indicates that a large amount of heat is released in this reaction, making propane an excellent fuel for cooking.

Entropy and the Second Law of Thermodynamics

Entropy

The degree of randomness in a sample of matter is called its **entropy** (S). The more randomly that the atoms and molecules move or are arranged, the greater is the entropy of a given substance. We can immediately visualize that ice has a very ordered structure compared to liquid water; therefore, water has greater entropy than ice. Similarly, since the molecules of water are much more randomly arranged in steam than in liquid water, the gas has a higher entropy.

The units for entropy are joules per Kelvin (J K^{-1}), thus making entropy an extensive property of matter. Chemists transform it into an intensive property by calculating the entropy of 1 mole of matter; in this case the units become joules per mole per Kelvin (J mol^{-1} K^{-1}) and the symbol is S°. The entropy (S°) values for selected elements and compounds are given in the table in Appendix 2. It should be noted that the entropy values of the elements are not zero.

The change in entropy is a state function, just as ΔH and ΔE are state functions. However, unlike other thermodynamic quantities such as energy (E) and enthalpy (H), the actual value for the entropy (S) of a substance can be determined. From fundamental principles, a perfect crystal at absolute zero (0 K or -273.16°C) has zero entropy since all motion ceases and there is perfect order. As the temperature of 1 mole of a chemical is increased from absolute zero, however, the entropy increases and the standard entropy is defined as

$$S^\circ = \frac{q_{rev}}{T}$$

In this equation T represents the temperature in kelvins, and q_{rev} represents the heat added to raise the temperature very slowly from absolute zero to T. Heat (q) is not a state function, and it appears that S should not be a state function. However, if heat is always added in a carefully defined manner, the results will always be the same. This carefully defined path is called a reversible process. A **reversible process** is defined as one that occurs in infinitesimally small steps from the initial to the final state.

Entropy changes due to a chemical process are calculated in the same fashion as the heats of reaction. Like heat of formation (ΔH°_f) values, absolute entropy (S_f°) values are tabulated for 1 mole of substance, making S_f° an intensive physical property. We can then calculate the entropy change for a reaction in a manner similar to that for heats of reaction by using this equation:

$$\Delta S^\circ = \Sigma(S^\circ \times coeff)_{\text{products}} - \Sigma(S^\circ \times coeff)_{\text{reactants}}$$

For example, we can calculate the entropy change for the combustion of propane using the S° values in Appendix 2:

$$\begin{aligned}\Delta S^\circ &= \left[\left(\frac{213.6\ \text{J}}{\text{mol K}}\right)(3\ \text{mol}) + \left(\frac{188.7\ \text{J}}{\text{mol K}}\right)(4\ \text{mol})\right] - \left[\left(\frac{205.0\ \text{J}}{\text{mol K}}\right)(5\ \text{mol}) + \left(\frac{270.2\ \text{J}}{\text{mol K}}\right)(1\ \text{mol})\right] \\ &= 1395.6\ \text{J K}^{-1} - 1295.2\ \text{J K}^{-1} \\ &= +100.4\ \text{J K}^{-1}\end{aligned}$$

This result represents an increase in entropy. We think that this would be true for this reaction since there are 7 moles of gaseous products and only 6 moles of gaseous reactants. The increase in the number of moles of gas in this reaction is 1 ($\Delta n_g = 1$), indicating that the entropy change is expected to be positive.

Estimating ΔS° for Chemical Reactions

In addition to calculating the entropy change from tabulated data, we can estimate the sign, and to some degree the magnitude, of an entropy change for a chemical process. In making such an estimate, the following principles are important:

1. Formation of a gas increases the entropy greatly. The greater the value of Δn_g, the greater is the entropy increase. The reverse is true for a decrease in entropy.
2. If Δn_g, = 0, then changes from the solid to the liquid phase are the next largest contributors to increasing entropy. A solid that melts and a solute that dissolves in a solvent both exhibit an increase in entropy. Formation of solids always results in a decrease in entropy.
3. A rise in temperature increases the entropy of a system, and a temperature decrease lowers the entropy.

Gibbs Free Energy (ΔG)

Gibbs free energy, a thermodynamic quantity, was named to honor J. Willard Gibbs, the preeminent physical chemist who developed the concept. The standard **free-energy change**, represented by the symbol ΔG°, is the maximum amount of energy available from any chemical reaction. Two forces drive chemical reactions. The first is the enthalpy (ΔH°), which represents the change in the internal potential energy of the atoms. The second is the standard entropy change, *($\Delta S°$)*, which represents the change in randomness of the system. In calculating the Gibbs free energy, $\Delta S°$ is multiplied by the temperature, *(T)*, in Kelvin units. If the enthalpy is negative, it means that the internal potential energy of the system is decreased, thereby favoring a spontaneous reaction. If the entropy increases, a spontaneous reaction is also favored.

The combination of the two driving forces is represented in equation form as follows:

$$\Delta G° = \Delta H° - T\,\Delta S°$$

The Gibbs free-energy equation is derived directly from the second law of thermodynamics, which states that any physical or chemical change must result in an increase in entropy of the universe. In addition a $\Delta G°$ that is negative indicates a spontaneous reaction.

Since $\Delta G°$ is a combination of $\Delta H°$ and $T\Delta S°$, we can make some generalizations about the spontaneity of a reaction based only on the signs of these quantities. Temperature may be an important factor, as shown in Table 5.1.

TABLE 5.1. RELATIONSHIPS OF THE SIGNS OF $\Delta H°$ AND $\Delta S°$ TO THE SIGN OF $\Delta G°$

$\Delta H°$	$\Delta S°$	$\Delta G°$	Comment
Negative	Positive	Negative	Always spontaneous
Positive	Negative	Positive	Never spontaneous
Negative	Negative	Positive or negative	Decrease temperature to make spontaneous
Positive	Positive	Positive or negative	Increase temperature to make spontaneous

When $\Delta H°$ is negative and $\Delta S°$ is positive, the only value possible for $\Delta G°$ is a negative one. In this case the reaction will be spontaneous at all temperatures. Similarly, when $\Delta H°$ is positive and $\Delta S°$ is negative, $\Delta G°$ must be positive, indicating a nonspontaneous reaction at all temperatures.

When $\Delta H°$ is negative and $\Delta S°$ is negative, $\Delta G°$ may be either negative or positive, depending on the relative magnitudes of $\Delta H°$ and $\Delta S°$. However, the $-T\Delta S°$ term will always be positive. It will have a larger magnitude at high temperatures and a smaller magnitude at low temperatures. This fact suggests that lowering the temperature may make the positive magnitude of $T\Delta S°$ small enough so that, when it is combined with the negative $\Delta H°$, the resulting $\Delta G°$ will be negative and the reaction will be spontaneous.

The reverse is true when $\Delta H°$ is positive and $\Delta S°$ is positive. Once again,

$\Delta G°$ may be either positive or negative. The same reasoning leads to the conclusion that increasing the temperature will eventually cause the reaction to be spontaneous with a negative $\Delta G°$.

Spontaneity of Reactions

> When $\Delta G°$ is negative, a reaction is spontaneous. When $\Delta G°$ is positive a reaction is nonspontaneous.
>
> When the signs of $\Delta S°$ and $\Delta H°$ are different, the reaction will be either spontaneous or nonspontaneous
>
> When the signs of $\Delta S°$ and $\Delta H°$ are the same, the reaction may be either spontaneous or nonspontaneous, depending on the temperature.

As is true of $\Delta H°$ and $\Delta S°$ calculations, we can calculate the value of $\Delta G°_{RXN}$ from values of the free energies of formation *($\Delta G°$),* which are listed in the table in Appendix 2. Since temperature is an important variable that affects the value of the free energy, the temperature must be specified. It is most common to list $\Delta G°$ values for room temperature of 25°C or 298 K. The symbol for free energy incorporates the temperature, as in $\Delta G°_{298}$.

In using the free-energy table in Appendix 2 we subtract the $\Delta G°_{298}$ values of the reactants from the $\Delta G°_{298}$ values of the products in the equation

$$\Delta G°_{298} = \Sigma(\Delta G°_{298} \times \text{coeff})_{\text{products}} - \Sigma(\Delta G°_{298} \times \text{coeff})_{\text{reactants}}$$

For the combustion of propane, for example, we obtain

$$C_3H_8 + 5O_2 \rightarrow 3CO_2 + 4H_2O$$

$$\Delta G°_{298} = \left[\left(\frac{-394.4\ \text{J}}{\text{mol}}\right)(3\ \text{mol}) + \left(\frac{-228.6\ \text{J}}{\text{mol}}\right)(4\ \text{mol})\right] - \left[\left(\frac{-23.5\ \text{J}}{\text{mol}}\right)(1\ \text{mol})\right]$$

$$= (-1183.2\ \text{J} - 914.4\ \text{J}) - (-23.5\ \text{J})$$

$$= -2074.1\ \text{J}$$

The negative value indicates that the reaction is spontaneous, as anyone who has used a barbecue grill or propane torch already knows.

We have calculated $\Delta H°$ and $\Delta S°$ *f*or this reaction in other chapters. Using those values, along with a temperature of 298 K, we have a second way to determine the value of $\Delta G°_{298}$:

$$\Delta^\circ_{298} = \Delta H^\circ \quad - T\,\Delta S^\circ$$

$$= -2044 \text{ kJ} - (298 \text{ K})\ (100.4 \text{ J K}^{-1})$$

$$= -2044 \text{ kJ} - 29.9 \text{ kJ}$$

$$= -\ 2074 \text{ kJ}$$

Free Energy at Temperatures other than 298 K

Using the standard free energies of formation given in Appendix 2, we can calculate the standard free-energy change at 298 K. The standard free-energy changes at other temperatures may also be calculated. For this purpose we need to know the standard heat of reaction (ΔH°_{RXN}) and the standard entropy change (ΔS°_{RXN}) for the reaction. These values are correct for 298 K, but we may assume that they do not change significantly with temperature. Using these values in the free-energy equation with a temperature other than 298 K gives the free-energy change at that different temperature.

EXERCISES

1. For a certain reaction, ΔH°_{react} = +2.98 kilojoules and ΔS°_{react} = +12.3 joules per kelvin. What is ΔG°_{RXN} at 298 K, 200 K, and 400 K?
2. For a certain reaction $\Delta H^\circ_{react} = -13.65$ kJ and $\Delta S^\circ_{react} = -75.8 \text{ J K}^{-1}$. 1. What is ΔG°_{react} at 298 K? 2. Will increasing or decreasing the temperature make the reaction spontaneous? 3. If so, at what temperature will the reaction become spontaneous?

Solutions

1. The equation to be solved is

$$\Delta G^\circ_{react} = \Delta H^\circ_{react} - T\,\Delta S^\circ_{react}$$

Substituting the values in the problem yields, for a temperature of 298 K,

$$\Delta G^\circ_{react} = 2.98 \text{ kJ} - 298(12.3 \text{ J K}^{-1})$$

$$= 2.98 \text{ kJ} - 3665 \text{ J}$$

To complete the problem, − 3665 J must be converted into − 3.67 kJ:

$$\Delta G^\circ_{react} = -0.69 \text{ kJ} \quad \text{(a spontaneous reaction)}$$

For 200 K and 400 K the answers are

$$\Delta G^\circ_{react} = 2.98 \text{ kJ} - 200 \text{ K}(12.3 \text{ J K}^{-1}) = +\ 0.52 \text{ kJ}$$

and

$$\Delta G^\circ_{\text{react}} = 2.98 \text{ kJ} - 400 \text{ K}(12.3 \text{ J K}^{-1}) = -1.94 \text{ kJ}$$

In this exercise we see that the reaction is not spontaneous at 200 K but is spontaneous at 298 and 400 K.

2. 1. At 298 K the free energy is

$$\Delta G^\circ_{\text{react}} = -13.65 \text{ kJ} - 298 \text{ K}(-75.8 \text{ JK}^{-1}) = +8.94 \text{ kJ}$$

2. The reaction is not spontaneous at 298 K. Since $\Delta H^\circ_{\text{RXN}}$ and $\Delta S^\circ_{\text{RXN}}$ both have the same sign, the free energy will change from positive to negative at some temperature. We have the information to calculate the temperature where $\Delta G^\circ_{\text{RXN}} = 0.00$ kJ. This temperature will be the dividing line between spontaneous and nonspontaneous reactions. Consequently, the free-energy equation is set up as

$$\Delta G^\circ_{\text{react}} = \Delta H^\circ_{\text{react}} - T\Delta S^\circ_{\text{react}}$$

$$0.00 \text{ kJ} = -13.65 \text{ kJ} - T(-75.8 \text{ J K}^{-1})$$

$$T = \frac{13.65 \text{ kJ}}{0.0758 \text{ kJ K}^{-1}}$$

$$= 180 \text{ K}$$

3. Since we know that the reaction is not spontaneous at 298 K, we may conclude that it is not spontaneous at all temperatures above 180 K. Below the dividing-line temperature of 180 K the reaction must be spontaneous with a negative value for $\Delta G^\circ_{\text{react}}$. From this fact we predict that the reaction will be spontaneous below 180 K but will not be spontaneous above 180 K. The condensation of a gas and the crystallization of a liquid are two physical processes that are spontaneous at low temperatures and nonspontaneous at higher temperatures.

Free Energy and Equilibrium

When a system is not at standard state, the free-energy change is represented by ΔG, not ΔG°. Equation 5.2 shows the relationship between ΔG and ΔG°.

$$\Delta G = \Delta G^\circ + RT \ln Q \qquad (5.2)$$

Later in this chapter the concept of Q, the reaction quotient, is developed. If the value of Q is not equal to the equilibrium constant, further reaction occurs until the system reaches equilibrium.

Using Equation 5.2, we find that when a system is at standard state all concentrations are equal to 1 and $Q = 1$. The natural logarithm of 1 is zero ($\ln 1 = 0$), and consequently $\Delta G = \Delta G^\circ$.

The value of ΔG (without the superscript) tells us whether the reaction will continue and, if so, in which direction it will go. When ΔG is negative, the reaction will proceed in the forward direction. When ΔG is positive, the reaction proceeds in the reverse direction. If ΔG is zero, the reaction is at equilibrium and no further reaction occurs. For the equilibrium condition we find that

$$\Delta G^\circ = -RT \ln K$$

by setting $\Delta G = 0$, substituting the equilibrium constant (K) for the reaction quotient (Q), in Equation 5.2, and rearranging. This shows that measurements or tabulated values of $\Delta G^\circ_{\text{react}}$ will allow us to calculate the equilibrium constant, for many reactions.

The relationships between ΔG and ΔG° are illustrated in Figure 5.2. The standard free energies (G°), are shown for the reactants on the left side and the products on the right side of each graph. The difference between the two is the ΔG° value for the reaction. The curved line connecting the two G° values represents the values of G for the reaction mixture. The slope of the curved line is ΔG, and at the minimum, where the slope $= 0$, the reaction is in equilibrium.

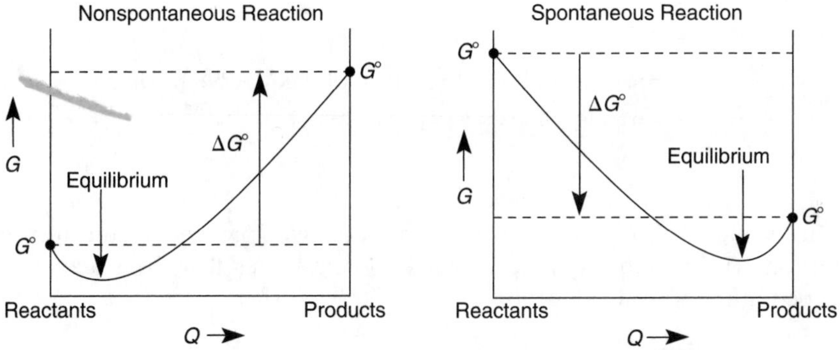

FIGURE 5.2. *Free-energy diagrams for a nonspontaneous reaction (left) and a spontaneous reaction (right). The difference between the G° points is ΔG°. The slope of the curved line is ΔG, and the equilibrium point is at the minimum of the curve.*

The curves in Figure 5.2 illustrate the fact that even in a nonspontaneous reaction a small amount of reactants are converted to products at equilibrium (minimum of curved line). In a spontaneous reaction, on the other hand, most of the reactants are converted into products because the minimum is closer toward the product side. The slope of the curved line is negative to the left of the equilibrium point, so that ΔG is negative and the reaction moves toward the products and also toward the minimum. On the right side of the equilibrium point the slope representing ΔG is positive. The result is that a mixture of products will form more reactants, moving toward the minimum of the curve.

EXERCISE

The value of the equilibrium constant (K) is 45 at 298 K. At that same temperature Q is 35. Determine the value of $\Delta G°$ for the reaction at 298 K and show that the value of ΔG indicates the same direction for the reaction as Q predicts.

Solution

The value of $\Delta G°$ is calculated as follows:

$$\Delta G° = -RT \ln K$$

$$= -(8.314 \text{ J mol}^{-1} \text{ K}^{-1})(298 \text{ K})(\ln 45)$$

$$= -9.43 \text{ kJ mol}^{-1}$$

The value of ΔG is calculated as follows:

$$\Delta G = \Delta G° + RT \ln Q$$

$$= -9.43 \text{ kJ mol}^{-1} + (8.314 \text{ J mol}^{-1} \text{ K}^{-1})(298 \text{ K})(\ln 35)$$

$$= -9.43 \text{ kJ mol}^{-1} + 8.81 \text{ kJ mol}^{-1}$$

$$= -0.62 \text{ kJ mol}^{-1}$$

The value of ΔG predicts that the reaction will proceed in the forward direction. The fact that Q is less than K also indicates that the reaction will proceed in the forward direction.

CHEMICAL EQUILIBRIUM

Dynamic Equilibrium

The concept of a dynamic equilibrium is central to many aspects of chemistry. In a dynamic equilibrium, chemicals are reacting rapidly at the molecular scale while their concentrations remain constant on the macroscopic scale. Many physical and chemical processes participate in dynamic equilibria. For example, (1) in a closed container gas molecules condense just as fast as liquid molecules evaporate; (2) a saturated solution involves a solid dissolving at the same rate as solute crystallizes; and (3) in a weak electrolyte solution the ions recombine into molecules just as fast as other molecules dissociate into ions. In all of these chemical and physical processes the rate in one direction is exactly equal to the rate in the other direction. It is important to note that, although reactions never stop in a dynamic equilibrium, the overall concentrations remain constant.

Figure 5.3 shows that a chemical reaction has two well-defined regions in time, the kinetic and equilibrium regions, that are studied and measured in very different ways. When compounds are first mixed in a chemical reaction, they interact to form different compounds. During the reaction process the concentration of the reactants decreases and the concentration of the products increases. While the concentrations are changing, the reaction is studied using the principles of chemical kinetics that are reviewed in Chapter 12. At some point in time the concentrations of the reactants and products stop changing. Although reactions do not stop at the molecular level, at the macroscopic level the concentrations of compounds in a dynamic equilibrium remain constant. At this point the compounds are in a dynamic chemical equilibrium with each other, and they are studied and described using the concepts of chemical equilibrium.

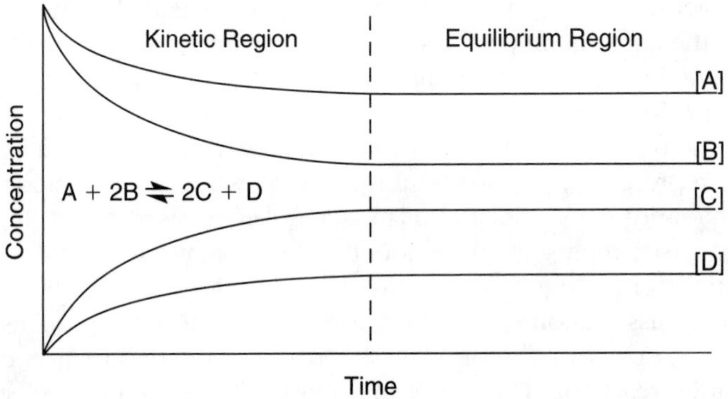

FIGURE 5.3. *This diagram illustrates the two regions of chemical reactions. On the left, in the kinetic region, the concentrations are changing with time. On the right, in the equilibrium region, the concentrations, on a macroscopic or laboratory scale, no longer appear to change.*

EXERCISE

It is observed that large crystals will grow from a precipitate that originally contains only very small crystals. How can the concept of dynamic equilibrium be used to explain this phenomenon?

Answer

In a dynamic equilibrium, the crystals are continuously dissolving while solute is continuously crystallizing at the same rate. The crystallization process depends on the concentration of solute, and the rate at which crystallization occurs will be the same on all crystal surfaces. The dissolution process depends on the surface area of the crystals, and 1 g of a large crystal has a much smaller surface area than 1 g of a very small crystal. The result is that small crystals dissolve at a faster rate than large crystals. The net result is an increase in size for large crystals at the expense of small crystals.

Equilibrium Law

In a chemical reaction the actual concentrations of the reactants and products, at equilibrium, are dependent on the initial concentrations in the reacting mixture. When several experiments with different initial concentrations of reactants are performed, they result in equilibrium mixtures with different concentrations. While all of these mixtures may be different, they all obey the **equilibrium law**.

The equilibrium law states that the concentrations of all of the products of a reaction multiplied together, divided by the concentrations of all the reactants multiplied together, will be equal to a number called the equilibrium constant K. The value of the equilibrium constant depends only on the specific reaction and the temperature of the reaction mixture when equilibrium is reached.

The upper-case letter K serves as the symbol for the equilibrium constant. To describe the type of equilibrium constant, a subscript is often used after the K. Thus K_c represents the equilibrium constant when concentration is expressed in molarity units (mol L^{-1}). The symbol K_p is used when the partial pressures of gases represent the amounts of reactants and products. Other special symbols for the equilibrium constant are K_{sp} for the solubility product, K_a for the acid dissociation constant, K_b for the base dissociation constant, K_f for the formation constant of complexes, and K_d for the dissociation constant in complexation reactions. These special forms of the equilibrium constant are described in Chapter 12.

The equilibrium law depends on the chemical equation for the reaction under study. A general equilibrium reaction may be written as

$$a\text{A} \quad + \quad b\text{B} \quad \rightleftharpoons \quad p\text{P} \quad + \quad n\text{N}$$

and its equilibrium law is expressed as

$$K_c = \frac{[\text{P}]^p[\text{N}]^n}{[\text{A}]^a[\text{B}]^b}$$

Whenever possible, the chemical reaction is balanced with the smallest possible whole-number coefficients. All tabulated values for equilibrium constants refer to equations with the simplest coefficients.

A more specific example of the formulation of the equilibrium law can be shown using the reaction between hydrogen and chlorine to form hydrogen chloride:

$$H_2(g) \quad + \quad Cl_2(g) \quad \rightleftharpoons \quad HCl\,(g) \quad + \quad HCl(g) \tag{5.3}$$

or

$$H_2(g) \quad + \quad Cl_2(g) \quad \rightleftharpoons \quad 2HCl\,(g) \tag{5.4}$$

The equilibrium law for this reaction is written as

$$K_c = \frac{[HCl][HCl]}{[H_2][Cl_2]} = \frac{[HCl]^2}{[H_2][Cl_2]}$$

Writing the chemical reaction with two separate HCl molecules as in Equation 5.3 illustrates that the concentration of HCl, written with square brackets as [HCl], should be multiplied by itself in the equilibrium law. Writing the chemical equation in the form of Equation 5.4 illustrates that the coefficient in a chemical equation will become an exponent in the equilibrium law.

For the combustion of propane, C_3H_8, the reaction is

$$C_3H_8(g) \quad + \quad 5O_2(g) \quad \rightleftharpoons \quad 3CO_2(g) \quad + \quad 4H_2O(g)$$

and the equilibrium law for this equation is written as

$$K_c = \frac{[CO_2]^3[H_2O]^4}{[C_3H_8][O_2]^5}$$

For this reaction, the coefficients of the reactants and products are written as exponents for their concentrations in the equilibrium law.

Any substance that has a constant concentration during a reaction, including all solids and all pure liquids, is not written as part of the equilibrium law. In dilute solutions, the solvent concentration is also constant and is not written in the equilibrium law. Three examples are a reaction where water is a pure liquid:

$$CH_4(g) \quad + \quad 2O_2(g) \quad \rightleftharpoons \quad CO_2(g) \quad + \quad 2H_2O(\ell)$$

$$K_c = \frac{[CO_2]}{[CH_4][O_2]^2}$$

a reaction where AgCl is a solid:

$$AgCl(s) \quad \rightleftharpoons \quad Ag^+(aq) \quad + \quad Cl^-(aq)$$

$$K_c = [Ag^+][Cl^-]$$

and a reaction where water is the solvent for dilute solutions:

$$NH_3(g) \quad + \quad H_2O(\ell) \quad \rightleftharpoons \quad NH_4^+(aq) \quad + \quad OH^-(aq)$$

$$K_c = \frac{[NH_4^+][OH^-]}{[NH_3]}$$

In most cases a solution is considered to be dilute when the concentration of the solute is less than 1 mole per liter.

EXERCISE

Write the equilibrium law for each of the following chemical equations:

(a) $HF(aq) + H_2O(\ell) \rightleftharpoons F^-(aq) + H_3O^+(aq)$

(b) $2NH_3(aq) + 3I_2(s) \rightleftharpoons 2NI_3(s) + 3H_2(g)$

(c) $CO(g) + H_2O(g) \rightleftharpoons CO_2(g) + H_2(g)$

(d) $Ba^{2+}(aq) + SO_4^{2-}(aq) \rightleftharpoons BaSO_4(s)$

(e) $C_2H_4(g) + 3O_2(g) \rightleftharpoons 2CO_2(g) + 2H_2O(g)$

Answers

(a) $K_c = \dfrac{[F^-][H_3O^+]}{[HF]}$ (H_2O is a pure liquid and is not included.)

(b) $K_c = \dfrac{[H_2]^3}{[NH_3]^2}$ (The two solids are not included.)

(c) $K_c = \dfrac{[CO_2][H_2]}{[CO][H_2O]}$ (All compounds are included.)

(d) $K_c = \dfrac{1}{[Ba^{2+}][SO_4^{2-}]}$ ($BaSO_4$ is a solid and is not included.)

(e) $K_c = \dfrac{[CO_2]^2[H_2O]^2}{[C_2H_4][O_2]^3}$ (All compounds are included.)

> It is extremely important to remember that the numerical value of the equilibrium constant will change only if there is a change in temperature. It will not change under any other experimental conditions.

EQUILIBRIUM CONSTANTS

The Equilibrium Law and How to Use It

The equilibrium law for a reaction is written directly from the balanced chemical equation. On paper, chemical equations are easily manipulated. We can reverse the direction of a reaction by writing the reactants as products and the products as reactants. Equations can be added and subtracted. The coefficients of an equation can all be multiplied or divided by a constant factor.

Each of these operations results in a different equilibrium law and a different value for the equilibrium constant.

Reversing a Chemical Equation

The chemical equation

$$O_2(g) + 2SO_2(g) \rightleftharpoons 2SO_3(g) \qquad \left(K_c = \frac{[SO_3]^2}{[O_2][SO_2]^2}\right)$$

can be written in the reverse direction as

$$2SO_3(g) \rightleftharpoons O_2(g)\ 2SO_2(g) \qquad \left(K_c' = \frac{[O_2][SO_2]^2}{[SO_3]^2}\right)$$

The two equilibrium constants (K_c and K_c') are inversely related to each other

$$K_c = \frac{1}{K_c'}$$

If a chemical reaction is reversed, the value of the new equilibrium constant will be the inverse of the original equilibrium constant.

Multiplying or Dividing Coefficients by a Constant

Taking the same reaction of sulfur dioxide with oxygen:

$$O_2(g) + 2SO_2(g) \rightleftharpoons 2SO_3(g) \qquad \left(K_c = \frac{[SO_3]^2}{[O_2][SO_2]^2}\right)$$

we may multiply each of the coefficients by 3 to obtain another balanced equation:

$$3O_2(g) + 6SO_2(g) \rightleftharpoons 6SO_3(g) \qquad \left(K_c' = \frac{[SO_3]^6}{[O_2]^3\,[SO_2]^6}\right)$$

The relationship between K_c and K_c' can be shown since

$$\frac{[SO_3]^6}{[O_2]^3\,[SO_2]^6} = \left(\frac{[SO_3]^2}{[O_2][SO_2]^2}\right)\left(\frac{[SO_3]^2}{[O_2][SO_2]^2}\right)\left(\frac{[SO_3]^2}{[O_2][SO_2]^2}\right) = \left(\frac{[SO_3]^2}{[O_2][SO_2]^2}\right)^3$$

$$K_c' = K_c K_c K_c = K_c^3$$

We see that the original equilibrium constant is raised to the power equal to the factor used in the multiplication.

Dividing an equation by 2 is the same as multiplying the equation by 1/2. Therefore, when an equation is divided by 2, the new equilibrium constant (K_c') is the square root of the original K_c:

$$K_c' = K_c^{\frac{1}{2}} = \sqrt{K_c}$$

Finally, reversing a reaction is mathematically the same as multiplying it by −1. As a result

$$K'_c = K_c^{-1} = \frac{1}{K_c}$$

Adding Chemical Reactions

Two chemical reactions are added by adding all the reactants and all the products in the equations and writing them as one equation. For example:

$$2S(g) + 2SO_2(g) \rightleftharpoons 2SO_2(g) \qquad \left(K_1 = \frac{[SO_2]^2}{[S]^2\,[O_2]^2}\right)$$

$$O_2(g) + 2SO_2(g) \rightleftharpoons 2SO_3(g) \qquad \left(K_2 = \frac{[SO_3]^2}{[O_2]\,[SO_2]^2}\right)$$

Adding these two equations and canceling the $2SO_2(g)$ that are identical on both sides yields

$$2S(g) + 3O_2(g) \rightleftharpoons 2SO_3(g) \qquad \left(K_{overall} = \frac{[SO_3]^2}{[S]^2\,[O_2]^3}\right)$$

Mathematically we find that

$$K_{overall} = K_1 \times K_2$$

When equations are added, the overall equilibrium constant is the product of the equilibrium constants of the reactions that were added.

EXERCISE

Given these two reactions and their equilibrium constants;

$$Ag^+(aq) + Cl^-(aq) \rightleftharpoons AgCl(s) \qquad (K_c = 1.0 \times 10^{10})$$

$$Ag^+(aq) + 2NH_3(aq) \rightleftharpoons Ag(NH_3)_2^+(aq) \qquad (K_c = 1.6 \times 10^7)$$

Calculate the equilibrium constant of the reaction

$$AgCl(s) + 2NH_3(aq) \rightleftharpoons Ag(NH_3)_2^+(aq) + Cl^-(aq) \qquad (K_{overall} = ?)$$

and show how the given reactions are added to obtain this reaction.

Solution

To add the given equations, it is necessary to reverse the first equation to make AgCl(s) a reactant and Cl^- a product as required in the overall reaction:

$$AgCl(s) \rightleftharpoons Ag^+(aq) + Cl^-(aq) \qquad \left(K_c = \frac{1}{1.0 \times 10^{10}} = 1.0 \times 10^{-10}\right)$$

The equilibrium constant is inverted, as shown when a reaction is reversed. The new equation is added to the second equation, and the $Ag^+(aq)$ ions canceled. When reactions are added, the equilibrium constants are multiplied:

$$K_{overall} = (1.0 \times 10^{-10})(1.6 \times 10^7) = 1.6 \times 10^{-3}$$

We recall that Hess's law states that, when reactions are added, the heats of reaction are added, and when reactions are multiplied by a factor, the heats are multiplied by the same factor. There is a distinct difference, however, when equilibrium constants are combined. It is important to note that the two procedures cannot be interchanged.

Equilibrium Constants K_p and K_c

Equilibrium Constant K_c

The equilibrium law requires a balanced chemical equation. A general equilibrium reaction may be written as

$$a\text{A} \quad + \quad b\text{B} \quad \rightleftharpoons \quad p\text{P} \quad + \quad n\text{N}$$

and its equilibrium law, with the products in the numerator and reactants in the denominator, as

$$K_c = \frac{[P]^p[N]^n}{[A]^a[B]^b}$$

In the equilibrium law, the square brackets indicate the molarity of each of the reactants and products used in the ratio, and the coefficients in the balanced equation become exponents for the concentrations. Whenever possible, the chemical reaction is balanced with the smallest possible whole-number coefficients. All tabulated values for equilibrium constant K_c refer to equations with the simplest coefficients.

A more specific example of the formulation of the equilibrium law involves the reaction between hydrogen and chlorine to form hydrogen chloride:

$$H_2(g) \quad + \quad Cl_2(g) \quad \rightleftharpoons \quad HCl(g) \quad + \quad HCl(g) \qquad (5.5)$$

or

$$H_2(g) \quad + \quad Cl_2(g) \quad \rightleftharpoons \quad 2HCl(g) \qquad (5.6)$$

The equilibrium law for this reaction is written as

$$K_c = \frac{[HCl] \times [HCl]}{[H_2] \times [Cl_2]} = \frac{[HCl]^2}{[H_2] \times [Cl_2]}$$

Writing the chemical reaction with two separate HCl molecules as in Equation 5.5 illustrates that the concentration of HCl, written with square brackets as [HCl], should be multiplied by itself in the equilibrium law. Writing the chemical reaction in the form of Equation 5.6 illustrates that the coefficient in a chemical equation will become an exponent in the equilibrium law.

Equilibrium Constant K_p

In the preceding section, the equilibrium law was written in terms of concentrations and the equilibrium constant was given the subscript *c* to indicate that fact. In some situations, however, it is more convenient to formulate the equilibrium law using pressures rather than concentrations. In such cases the equilibrium law may be written using the partial pressures of the gaseous reactants and products. Under these conditions the equilibrium constant is given the symbol K_p. Usually only reactions that are entirely in the gas phase are written in this way.

One gas phase reaction is written as

$$H_2(g) \quad + \quad I_2(g) \quad \rightleftharpoons \quad 2\ HI(g)$$

and its equilibrium law is

$$K_p = \frac{p_{HI}^2}{p_{H_2} p_{Br_2}}$$

where *p* represents the partial pressure of each gas. When the equilibrium constant is K_p, all calculations and procedures are done in exactly the same manner as when the equilibrium constant is K_c.

Example For the reaction of gaseous sulfur with oxygen at high temperatures:

$$2S(g) \quad + \quad 3O_2(g) \quad \rightleftharpoons \quad 2SO_3(g)$$

when the system reaches equilibrium, the partial pressures are measured as p_S = 0.0035 atmosphere, P_{SO_3} = 0.0050 atmosphere, and P_{O_2} + 0.0021 atmosphere. What is the value of K_p under these conditions?

Solution: This problem is solved by writing the correct equilibrium law. Since all of the chemicals in the reaction are gases and are measured in partial pressures, the equilibrium law should be written in terms of K_p.

$$K_p = \frac{p_{SO_3}^2}{p_S^2 p_{O_2}^3}$$

With this equilibrium law, the values for the partial pressures are entered and the solution calculated:

$$K_p = \frac{[0.0050]^2}{[0.0021]^3\,[0.0035]^2}$$

$$= 2.2 \times 10^8$$

Example

With the known value $K_p = 2.2 \times 10^8$ from the previous example, determine whether each of the following systems is at equilibrium. If a system is not at equilibrium, determine in which direction the reaction will proceed.

System (a):

$P_{SO_3} = 1.25$ atm, $p_{O_2} = 0.256$ atm, $p_S = 0.0112$ atm

System (b):

$P_{SO_3} = 0.00677$atm, $p_{O_2} = 0.122$ atm, $p_S = 0.212$ atm

System (c):

$P_{SO_3} = 0.123$ atm, $p_{O_2} = 0.00145$ atm, $p_S = 0.0332$ atm

Solutions: The values for the pressures are entered into the equilibrium law equation to calculate the reaction quotient *(Q)*, which is then compared to the known value of K_p.

System (a):

$$Q = \frac{p^2_{SO_3}}{p^2_S\, p^3_{O_2}} = \frac{[1.25]^2}{[0.256]^3\,[0.0112]^2} = 7.4 \times 10^5$$

System (b):

$$Q = \frac{p^2_{SO_3}}{p^2_S\, p^3_{O_2}} = \frac{[0.00677]^2}{[0.122]^3\,[0.212]^2} = 5.6 \times 10^{-1}$$

System (c):

$$Q = \frac{p^2_{SO_3}}{p^2_S\, p^3_{O_2}} = \frac{[0.123]^2}{[0.00145]^3\,[0.0332]^2} = 4.5 \times 10^9$$

None of the three systems has a value of Q equal to the known K_p of 2.2×10^8, indicating that none of the systems is at equilibrium. For System (a) and System (b), Q is smaller than the known K_p. Q must increase as these systems approach equilibrium, and the numerator of the ratio will increase while the denominator decreases. Since the numerator contains the products and the denominator contains the reactants, the products must increase and the reactants decrease. Therefore the reaction must proceed in the forward direction. System (c) has a value of Q that is greater than the known K_p. The ratio must decrease for the reaction to reach equilibrium. Products must be used to form more reactants, and the reaction must proceed in the reverse direction.

Relationship between K_P and K_C

The ideal gas law, $PV = nRT$, can be rearranged to read

$$\frac{n}{V} = \frac{P}{RT} \tag{5.7}$$

Since the n/V term has units of concentration of moles per liter, the pressure of a gas, at a constant temperature, is directly proportional to its concentration.

An equilibrium constant for a gas-phase reaction can be written as either K_p or K_c, and the two can be converted from one into the other. Equation 5.7 shows that (P/RT) can be substituted for the molar concentrations in the equilibrium law, resulting in the relationship

$$K_p = K_c(RT)^{\Delta n_g} \tag{5.8}$$

In this equation R is the universal gas law constant (0.0821 L atm mol^{-1} K^{-1}), T is the Kelvin temperature, and Δn_g is the change in the moles of gas in the balanced reaction.

Δn_g = moles of gas products − moles of gas reactants

The reaction

$$I_2(g) + H_2(g) \rightleftharpoons 2\ HI(g)$$

has $K_c = 49$. The equivalent K_p at 100°C is calculated by determining Δn_g:

$$\begin{aligned}\Delta n_g &= (2\text{ mol HI}) - (1\text{ mol }I_2 + 1\text{ mol }H_2)\\ &= 0\end{aligned}$$

If $\Delta n_g = 0$, then $K_p = K_c$ for this reaction.

In the reaction of sulfur with oxygen, however, the value of Δn_g is not 0:

$$\begin{aligned}\Delta n_g &= (2\text{ mol }SO_3) - (2\text{ mol S} - 3\text{ mol }O_2)\\ &= -3\end{aligned}$$

With $K_p = 2.2 \times 10^8$ at 300°C we can calculate K_c by first converting 300°C to 573 K and then substituting the information into equation 5.8:

$$\begin{aligned} 2.2 \times 10^8 &= K_c[(0.0821)(573)]^{-3} \\ K_c &= (2.2 \times 10^8)[(0.0821)(573)]^3 \\ &= 2.3 \times 10^{13} \end{aligned}$$

Example The value of K_c for the following reaction is 5.6×10^{12} at 290 K. What is the value of K_p?

$$2NO(g) \quad + \quad O_2(g) \quad \rightleftharpoons \quad 2NO_2(g)$$

Solution: To answer this question, we need to use equation 5.7. Since K_p is the unknown, we need values for K_c, R, T, and Δn_g. All of these are given except Δn_g. We see that this reaction has 3 moles of gases as reactants and 2 moles of gases as products. Since this is a decrease of 1 mole of gas in going from reactants to products, Δn_g is -1. Entering the numbers in to the equation gives

$$K_p = 5.6 \times 10^{12}[(0.081)(290)]^{-1}$$

Rearranging the equation, we calculate the value of K_p as

$$K_p = \frac{5.6 \times 10^{12}}{(0.081)(290)} = 2.4 \times 10^{11}$$

The Equilibrium Constant and Its Meaning

Extent of Reaction and Spontaneous Reactions

The value of the equilibrium constant indicates the extent to which reactants are converted into products in a chemical reaction. If the equilibrium constant is large, it indicates that much more product is present at equilibrium than reactants. When K is very large, greater than 10^{10}, for example, for all intents and purposes the reaction goes to completion. On the other hand, when K is very small, less than 10^{-10}, very little product is formed and virtually no visible reaction occurs. If $K = 1$, the equilibrium mixture contains approximately equal amounts of reactants and products. These general ideas allow us to quickly estimate the composition of an equilibrium mixture.

A **spontaneous reaction** is one in which, when reactants are mixed, products are formed without any additional assistance. Chemists generally define a spontaneous reaction as one with an equilibrium constant greater than 1.00. A reaction with an equilibrium constant less than 1.00 is a **nonspontaneous reaction**. The fact that $K' = 1/K_c$ when a reaction is reversed tells the chemist that a reaction that is nonspontaneous in one direction is spontaneous if written in the opposite direction.

EXERCISE

1. Which of the following reactions is (are) spontaneous?
2. List the reactions in order from the greatest extent of reaction to the lowest.

(a) $H_2(g)$	$+ I_2(g)$	$\rightleftharpoons$ 2 HI	($K_c = 49$)
(b) Br_2	$+ Cl_2$	$\rightleftharpoons$ 2 BrCl	($K_c = 6.9$)
(c) HF(aq)	$+ H_2O(\ell)$	$\rightleftharpoons F^-(aq) + H_3O^+(aq)$	($K_c = 6.8 \times 10^{-4}$)
(d) $2H_2(g)$	$+ O_2(g)$	$\rightleftharpoons 2\ H_2O(g)$	($K_c = 9.1 \times 10^{80}$)
(e) $2N_2O(g)$	$+ O_2(g)$	$\rightleftharpoons 2N_2(g) +)_2(g)$	($K_c = 7.0 \times 10^{34}$)

Answers

1. All the reactions except for (c) have equilibrium constants greater than 1 and are spontaneous.
2. The correct order of extent of reaction is (d) > (e) > (a) > (b) > (c), based on the magnitude of the equilibrium constants.

The Reaction Quotient; Predicting the Direction of a Reaction

The **reaction quotient** (Q) is defined as the number obtained by entering all of the required concentrations into the equilibrium law and calculating the result. For example:

$$O_2(g) + 2SO_2(g) \rightleftharpoons 2SO_3(g)$$

$$K_c = \frac{[SO_3]^2}{[O_2][SO_2]^2} \quad \text{(equilibrium law)}$$

$$Q = \frac{[SO_3]^2}{[O_2][SO_2]^2} \quad \text{(reaction quotient)}$$

The equilibrium constant is the numerical value of K_c when the reaction is at equilibrium. If the chemicals in the reaction are not in equilibrium, the numerical value of the equilibrium law is called the reaction quotient (Q). This equation has exactly the same form as the equilibrium law except that the K_c is replaced by Q.

Four principles may be ascribed to the value of Q.

1. If Q does not change with time, the reaction is in a state of equilibrium and $Q = K_c$.
2. If $Q = K_c$, the reaction is in a state of equilibrium.
3. If $Q < K_c$, the reaction will move in the forward direction (to the right) in order to reach equilibrium.
4. If $Q > K_c$, the reaction will move in the reverse direction (to the left) in order to reach equilibrium.

The first principle tells us how to determine whether a chemical reaction has reached equilibrium. It is necessary to measure the concentrations of reactants and products at different times—for example 1 hour, 2 hours, and 5 hours after the reaction has started. If the value of Q does not change, the system is in equilibrium and $Q = K_c$.

The second principle tells us that, if K_c is known from a prior experiment, the determination of Q will tell us whether the reactants and products have reached equilibrium. In particular, if $Q = K_c$, the system is in equilibrium.

The third and fourth principles tell us what will happen if Q is not equal to K_c. If Q is less than K_c, the numerator of the equilibrium law must increase, while the denominator decreases, to raise the value of Q to the level of K_c. Since the numerator represents the concentrations of the products, the reaction must proceed toward the product side of the reaction, or in the forward direction.

A Q that is larger than K_c indicates that the numerator must decrease, while the denominator increases, to reach equilibrium. Again, the numerator represents the concentrations of products, and a decrease in products indicates that the reaction must proceed in the reverse direction.

If the value of K_c is known, we can tell whether a reaction is at equilibrium by determining Q and comparing it to K_c. If the reaction is not at equilibrium, we may predict the direction in which it will go to reach equilibrium.

EXERCISE

1. Using the equilibrium constants and reactions below, calculate the values of Q and determine whether each of the following systems is or is not in equilibrium.
2. If the system is not in equilibrium, predict whether it will proceed in the forward or the reverse direction.

(a) $H_2(g) + I_2(g) \rightleftharpoons 2\ HI(g)$ ($K_c = 49$)
$[H_2] = 0.10$ M; $[I_2] = 0.10$ M; $[HI] = 0.70$ M

(b) $Br_2 + Cl_2 \rightleftharpoons 2\ BrCl$ ($K_c = 6.9$)
$[Br_2] = 0.10$ M; $[Cl_2] = 0.20$ M; $[BrCl] = 0.45$ M

(c) $HF(aq) + H_2O(\ell) \rightleftharpoons F^-(aq) + H_3O^+(aq)$ ($K_c = 6.8 \times 10^{-4}$)
$[HF] = 0.20$M; $[F^-] = 2.0 \times 10^{-4}$ M;
$[H_3O^+] = 2.0 \times 10^{-4}$ M

(d) $2H_2(g) + O_2(g) \rightleftharpoons 2\ H_2O(g)$ ($K_c = 9.1 \times 10^{80}$)
$[H_2] = 3.0 \times 10^{-30}$ M; $[O_2] = 2.2 \times 10^{-24}$ M;
$[H_2O] = 0.0180$ M

(e) $2N_2O(g) \rightleftharpoons 2N_2(g) + O_2(g)$ ($K_c = 7.0 \times 10^{34}$)
$[N_2O] = 2.4 \times 10^{-18}$ M; $[N_2] = 0.0360$ M;
$[O_2] = 0.0090$ M

Answers

1. Values of Q: (a) 49 (b) 10.1 (c) 2×10^{-7} (d) 9.1×10^{80} (e) 2.0×10^{30} Reactions (a) and (d) are in equilibrium since $Q = K_c$.
2. For reaction (b), $Q > K_c$, indicating that the reaction must go in the reverse direction to reach equilibrium. For reactions (c) and (e), $Q < K_c$ so the reaction must go forward to attain equilibrium.

Equilibrium Constant Units

In the exact derivation of equilibrium constants there are no units, and we say that the equilibrium constants are dimensionless quantities.

At times, however, it is convenient to assign dimensions to the equilibrium constant. To assign units to K_p or K_c, we determine the value of Δn as

$$\Delta n = \text{moles of products} - \text{moles of reactants}$$

that is, the sum of the coefficients of the reactants subtracted from the sum of the coefficients of the products in a balanced chemical reaction.

With Δn determined, the units are assigned as follows:

$$\text{units for } K_p = (\text{atm})^{\Delta n}$$

and

$$\text{units for } K_c = (\text{M})^{\Delta n} = \left(\frac{\text{mol}}{\text{L}}\right)^{\Delta n}$$

It must be realized that these units are artificial. However, the assignment of units often helps to avoid calculation errors when the factor-label method is used.

Special Equilibrium Constants

Earlier in this chapter we saw that the equilibrium law is based on a balanced chemical reaction. The equilibrium constant is formulated as shown on pages 188–190. Although all equilibrium constant equations are constructed in the same way, chemists often name equilibrium constants for the general reactions they describe, as shown in the following examples.

Solubility Product

The rules for determining solubility given on page 316 are used to determine whether an ionic compound will dissolve to an appreciable extent in water. In quantitative terms, it is loosely agreed that a salt is soluble if at least 0.1 mole of it will dissolve in 1 liter of water (0.1 M solution). Saturated solutions of insoluble salts have concentrations that are less than 0.1 molar. However, insoluble salts do dissolve to a small extent.

The solution process may be written in a form similar to a chemical reaction. For solid $Fe_2(OH_3)$ the equation is

$$Fe(OH)_3(s) \rightleftharpoons Fe^{3+}(aq) + 3\ OH^-(aq)$$

The equilibrium law for this process is written as

$$K_c = [Fe^{3+}][OH^-]^3$$

$Fe(OH)_3(s)$ does not appear in the denominator of the equilibrium law because it is a solid. In the special case of the solubility of slightly soluble compounds, the equilibrium law is always the product of the ions produced when the compound dissolves. The equilibrium constant is called the **solubility product constant** and is given the symbol K_{sp}.

$$K_{sp} = [Fe^{3+}][OH^-]^3$$

Weak-Acid and Weak-Base Equilibria

Weak acids and weak bases are compounds that dissociate only slightly when dissolved in water, as shown for acetic acid and ammonia in the following equations:

$$CH_3COOH(aq) + H_2O(\ell) \rightleftharpoons CH_3COO^-(aq) + H_3O^+(aq)$$

$$NH_3(aq) + H_2O(\ell) \rightleftharpoons NH_4^+(aq) + OH^-(aq)$$

The equilibrium laws for these two reactions are as follows:

$$K_c = K_a = \frac{[CH_3COO^-][H_3O^+]}{[CH_3COOH]}$$

$$K_c = K_b = \frac{[NH_4^+][OH^-]}{[NH_3]}$$

For weak acids the equilibrium constant is called the **acid dissociation constant** and given the symbol K_a. Weak bases have a corresponding **base dissociation constant** (K_b). Since every weak acid dissociates to form an anion and the hydronium ion, H_3O^+, all weak-acid dissociation reactions have the same form as the reaction for acetic acid, and their K_a expressions are similar. Likewise, weak bases all dissociate in the same way as ammonia, and their base dissociation expressions are similar to that for NH_3.

Acid- and base-equilibrium problems are approached in exactly the same manner as other equilibrium problems. These same equilibrium constants are used also for solving buffer problems and problems involving the hydrolysis of salts.

Formation Constants

Metal ions can react with anions and molecules to form chemical species called **complexes**. Ammonia complexes with copper ions in solution, turning the color from a light blue to a much darker blue.

$$Cu^{2+}(aq) \quad + \quad 4NH_3(aq) \quad \rightleftharpoons \quad Cu(NH_3)_4^{2+}$$

$Cu(NH_3)_4^{2+}$ is a complex ion of copper and ammonia. Since this reaction is written as an equilibrium, its equilibrium law is as follows:

$$K_c = K_f = \frac{[Cu(NH_3)_4^{2+}]}{[Cu^{2+}][NH_3]^4}$$

The reaction represents the formation of the complex, and the equilibrium constant is known as the formation constant (K_f).

When the complexation reaction is reversed, it represents the dissociation of the complex into its parts. The equilibrium constant for the dissociation is often called the **dissociation constant** K_d. Constants K_f and K_d are inversely proportional to each other:

$$K_f = \frac{1}{K_d}$$

Problems involving complexation equilibria are solved in exactly the same way as other equilibrium problems.

ELECTROCHEMISTRY

Electrolysis

In an electrolysis experiment a nonspontaneous chemical reaction is forced to occur when two electrodes are immersed in an electrically conductive sample and the electrical voltage applied to the two electrodes is increased until the reaction starts. At the electrode supplying the electrons, reduction reactions occur. This electrode is called the **cathode**. The other electrode, where oxidation reactions occur, is termed the **anode**.

Applications of Electrolysis

One type of sample that may be electrolyzed is a molten salt. Salts are composed of ions, and in a salt that is a solid the ions are immobile in the crystal lattice. Heating the salt until it melts frees the ions, and their mobility in the molten salt makes the salt electrically conductive. In the electrolysis of a molten salt that does not contain any polyatomic ions, the cation of the salt is reduced at the cathode and the anion of the salt is oxidized at the anode.

For example, sodium chloride, NaCl, may be melted at 801°C; the reactions are as follows:

$$\text{Cathode: } Na^+ + e^- \rightarrow Na$$
$$\text{Anode: } 2Cl^- \rightarrow Cl_2 + 2e^-$$

Electrolysis of molten salts containing polyatomic ions is much more complex. These reactions are covered in advanced chemistry courses.

Aqueous solutions of salts are also electrically conductive and may be electrolyzed. In the case of solutions, two additional reactions are possible:

$$\text{Cathode: } 2H_2O + 2e^- \rightarrow H_2 + 2OH^- \quad (5.9)$$

$$\text{Anode: } 2H_2O \rightarrow O_2 + 4H^+ + 4e^- \quad (5.10)$$

There are two possible reactions at each electrode. At the cathode either the reduction of water, as in Equation 5.9, or the reduction of a cation will occur. At the anode either the oxidation of water, as in Equation 5.10, or the oxidation of the salt's anion will take place.

The following principles can be used to decide which reaction takes place at each electrode:

1. *Cathode:* If the metal ion is a very active metal, water will be reduced. If the metal ion is an inactive metal or an active metal, the metal ion will be reduced.
2. *Anode:* If the anion is a polyatomic ion, it generally will not be oxidized. In particular, the sulfate, nitrate, and perchlorate polyatomic anions are not oxidized in aqueous solution. Chloride, bromide, and iodide ions, however, are oxidized at the anode in aqueous solution. If an anion in one salt is oxidized in an aqueous electrolysis, the same anion in any other salt will also be oxidized. For example, since a solution of sodium bromide, NaBr, results in Br^- being oxidized to Br_2, we may predict that solutions of KBr, $CaBr_2$, NH_4Br, and $AlBr_3$ will all produce Br_2 at the anode.

Electrolysis is important in the industrial production of several chemical materials. Because of the cost of electricity, however, most operations that use electrolysis are found in areas where electricity is inexpensive. Two of these are located in the Pacific Northwest and in Niagara Falls, close to large hydroelectric generators.

The electrolysis of concentrated aqueous NaCl solutions, called brine, produces hydrogen and hydroxide ions at the cathode, as shown in Equation 5.9. At the anode chlorine gas is produced. If the electrodes are separated by a porous membrane, the products are H_2, NaOH, and Cl_2. In another arrangement, if the solution is stirred, the chlorine gas reacts with the sodium

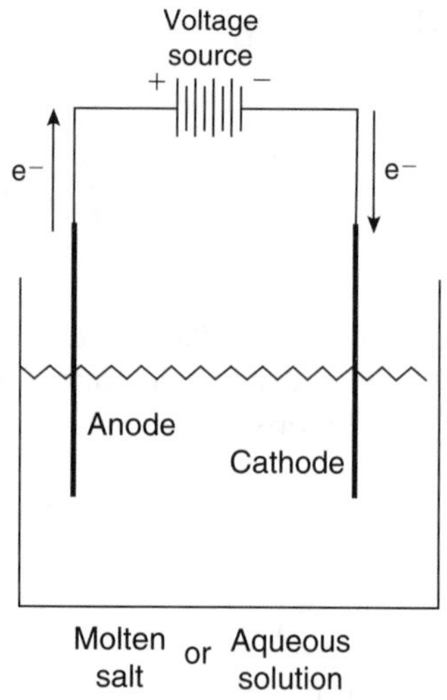

FIGURE 5.4. *Diagram of the setup of an electrolysis (electrolytic) cell as described above.*

hydroxide to form a sodium hypochlorite, NaOCl, solution, better known as bleach:

$$2NaOH + Cl_2 \rightarrow NaOCl + NaCl + H_2O$$

Aluminum oxide, Al_2O_3, has a melting point above 2000°C. In 1886 a college student, Charles Hall, discovered that the melting point of Al_2O_3 can be effectively decreased to around 1000°C when this compound is mixed with cryolite, Na_3AlF_6. When electrolyzed, this mixture produces aluminum metal and oxygen. The more active sodium and fluorine will be electrolyzed only when all of the aluminum and oxygen are used up.

$$2Al_2O_3(\ell) \rightarrow 4Al(\ell) + 3O_2(g)$$

Copper, the metal most used for electrical wiring, is refined by electrolysis. Copper ore is often a sulfide of copper that is converted into an impure form of copper by roasting:

$$CuS + O_2 \rightarrow Cu + SO_2$$

The impure copper is then refined by using a bar of crude copper as the anode of an electrolytic cell. The cathode is a small strip of pure copper. During electrolysis, the copper is oxidized to Cu^{2+} at the anode and then reduced back to copper metal at the cathode. In the process, impurities such as silver and gold drop to the bottom of the vessel as sludge. The value of the sludge comes close to paying for the cost of the electricity used in the refining.

In addition to producing industrial quantities of chemicals, electrolysis is used to electroplate thin layers of a decorative metal onto a less expensive metal. Silver and gold are electroplated onto iron utensils for decoration. Gold is electroplated onto electrical contacts on computer circuit boards to decrease corrosion failure. Chromium is electroplated on automobile parts for decoration and improved resistance to corrosion.

Quantitative Electrochemistry

Every balanced half-reaction specifies the number of electrons lost or gained. Consequently, stoichiometric calculations can be used to convert between moles of electrons and of all other substances in the half-reaction. It is important that, if the number of electrons flowing through the electrolysis cell is measured, the quantity of material reacted can be calculated. Michael Faraday discovered the relationship between the mole and electric current. He started with the definition of the coulomb (C), which is the number of electrons that flow past a given point in a wire in exactly 1 second when the current is exactly 1 ampere:

$$1 \text{ coulomb} = 1 \text{ ampere} \times 1 \text{ second} \tag{5.11}$$

Faraday found that 96,485 coulombs are equal to 1 mole of electrons:

$$1 \text{ mole of } e^- = 96{,}485 \text{ coulombs} \tag{5.12}$$

Faraday's constant ($\mathcal{F}$), is the factor label (96,485 C/mol e^-). Using equations 5.11 and 5.12, we can determine the number of moles of electrons by measuring the current and the time that the current flows:

$$\text{Moles of } e^- = \frac{I \times t}{96{,}485} \tag{5.13}$$

The ampere, which is the measure of electric current (I), has units of coulombs per second ($C\ s^{-1}$); time has units of seconds, and Faraday's constant is equal to 96,485 coulombs per mole of electrons.

Once the moles of electrons are calculated, factor-label stoichiometric calculations are used to determine other quantities. For example, the half-reaction for the reduction of iodine is

$$I_2 + 2e^- \rightarrow 2I^-$$

The number of moles of I^- produced at an electrode with a current of 0.500 ampere for 90 minutes can be calculated by first determining the moles of electrons as

$$\text{mol } e^- = \frac{(0.500 \text{ C s}^{-1})(90 \text{ min})(60 \text{ s min}^{-1})}{96485 \text{ C mol}^{-1}}$$

$$= 0.0280 \text{ mol } e^-$$

Using the moles of electrons as the starting point, we can write the stoichiometric calculation as

$$? \text{ mol } I^- = 0.0280 \text{ mol } e^- \left(\frac{2 \text{ mol } I^-}{2 \text{ mol } e^-}\right) = 0.0280 \text{ mol } I^-$$

Chemists often bypass the stoichiometric step by incorporating the stoichiometry into Equation 5.13:

$$x \text{ moles } X = \frac{It}{nf} \tag{5.14}$$

In this equation x is the stoichiometric coefficient of substance X, and n is the number of electrons in the balanced half-reaction.

EXERCISE

A current of 2.34 amperes is delivered to an electrolytic cell for 85 minutes. How many grams of (a) gold from $AuCl_3$, (b) silver from $AgNO_3$, and (c) copper from $CuCl_2$ will be obtained?

Solutions

(a) $Au^{3+} + 3e^- \rightarrow Au$

$$? \text{ mol Au} = \frac{(2.34 \text{ C s}^{-1})(85 \text{ min})(60 \text{ s min}^{-1})}{(3)(96{,}485 \text{ C mol}^{-1})}$$

$$= 4.12 \times 10^{-2} \text{ mol Au}$$

$$? \text{ g Au} = 4.12 \times 10^{-2} \text{ mol Au} \left(\frac{197 \text{ g Au}}{1 \text{ mol Au}}\right)$$

$$= 8.12 \text{ g Au}$$

(b) $Ag^+ + e^- \rightarrow Ag$

$$? \text{ mol Ag} = \frac{(2.34 \text{ C s}^{-1})(85 \text{ min})(60 \text{ s min}^{-1})}{(1)(96{,}485 \text{ C mol}^{-1})}$$

$$= 0.124 \text{ mol Ag}$$

$$? \text{ g Ag} = 0.124 \text{ mol Ag} \left(\frac{108 \text{ g Ag}}{1 \text{ mol Ag}}\right)$$

$$= 13.4 \text{ g Ag}$$

(c) $Cu^{2+} + 2e^- \rightarrow Cu$

$$? \text{ mol Cu} = \frac{(2.34 \text{ C s}^{-1})(85 \text{ min})(60 \text{ s min}^{-1})}{(2)(96{,}485 \text{ C mol}^{-1})}$$

$$= 6.18 \times 10^{-2} \text{ mol Cu}$$

$$? \text{ g Cu} = 6.18 \times 10^{-2} \text{ mol Cu} \left(\frac{63.5 \text{ g Cu}}{1 \text{ mol Cu}}\right)$$

$$= 3.92 \text{ g Cu}$$

Galvanic Cells

Galvanic cells are used to harness the energy of spontaneous redox reactions. This is done by physically separating the chemicals of two half-reactions so that the electrons generated by the oxidation half-reaction must flow through an electrical conductor before they can be used in the reduction half-reaction. The electrons may be diverted through meters, motors, light bulbs, and other devices to perform useful work before they reach their destination.

A galvanic cell is constructed as shown in Figure 5.5. All of the reactants in the oxidation half-reaction are placed in the left beaker, and all of the reactants in the reduction half-reaction in the right beaker. If a half-reaction is written with a metal, the metal is used as the electrode for that beaker; otherwise, an inert electrode made of platinum, silver, or gold is used. Electrodes are connected to each other with a metal wire, usually copper, and a device such as a meter (voltmeter or ammeter), motor, or light may be inserted in the electrical circuit. If a voltmeter is used, the positive side of the voltmeter is connected to the cathode, and the negative or ground is connected to the anode. To complete the circuit, a salt bridge is needed. The flow of charge is carried by electrons in the wires and by ions through the solutions and salt bridge.

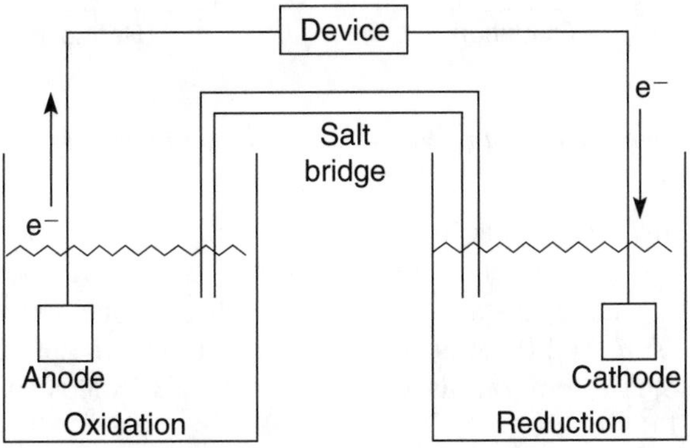

FIGURE 5.5. *Diagram of the galvanic cell.*

In setting up a galvanic cell, we start with a balanced redox reaction. For the spontaneous reaction of permanganate with iron(II) in acid solution the equation is as follows:

$$5Fe^{2+} + MnO_4^- + 8H^+ \rightleftharpoons 5Fe^{3+} + Mn^{2+} + 4H_2O \qquad (5.15)$$

To identify where the chemicals should be placed, the half-reactions are written:

$$Fe^{2+} \rightarrow Fe^{3+} + e^-$$

$$5e^- + MnO_4^- + 8H^+ \rightleftharpoons Mn^{2+} + 4H_2O$$

The first equation is the oxidation half-reaction, and Fe^{2+} and Fe^{3+} are placed in the left beaker. Since neither Fe^{2+} nor Fe^{3+} is a metal, a platinum electrode will be used. MnO_4^-, H^+, and Mn^{2+} must be placed in the right beaker since they are the components of the reduction half-reaction. Once again, a platinum electrode will be used. A voltmeter will serve to measure the tendency of electrons to flow. The complete cell is shown in Figure 5.6.

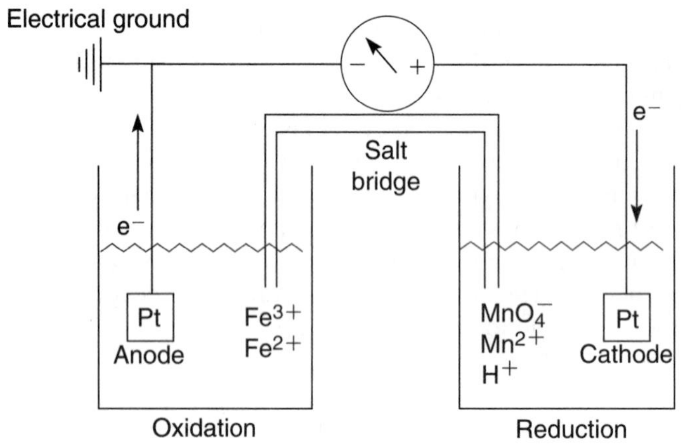

FIGURE 5.6. *The galvanic cell set up to study the reaction in equation 5.15.*

To obtain consistent results from a galvanic cell, the variables of temperature, pressure, and concentration must be controlled. For this purpose electrochemists define a standard state for galvanic cell experiments; a temperature of 298 K, a pressure of 1.00 atmosphere for all gases and 1.00 molar concentrations for all soluble compounds. Concentrations of all solids and pure liquids are defined as 1.00 molar.

At standard state, the galvanic cell diagram assumes that reaction 5.15 is spontaneous and that oxidation and the anode are on the left while reduction and the cathode are on the right. Since this reaction is known to be spontaneous, the cell in Figure 5.6 is correct and a positive voltmeter reading will be obtained. This voltmeter reading is called the standard cell voltage (E°_{cell}) or the electromotive force, ($\mathscr{EMF}$).

In reality, we often do not know whether a reaction is spontaneous or nonspontaneous. In such a situation, a chemical equation is written and the galvanic cell is constructed for that reaction. When E°_{cell} is measured, it will be either positive or negative. If E°_{cell} is positive, the reaction is spontaneous as written. If E°_{cell} turns out to be negative, the reaction is nonspontaneous as written and should be reversed to construct the galvanic cell. A nonspontaneous reaction may be converted into a spontaneous reaction by writing the chemical equation in the reverse directions (reactants become products, and products become reactants). Then the steps outlined above are used again to determine the new and correct setup of the galvanic cell.

The chemical reaction and the setup for the galvanic cell are directly related to each other. Knowing one of them allows us to determine the other. The chemicals present in the cell with the anode are written as an oxidation half-reaction; those present in the cell with the cathode, as the reduction half-reaction.

A cell diagram is a shorthand method of drawing a galvanic cell. The cell diagram "reads" the galvanic cell from left to right, starting with the anode and ending with the cathode. For the standard-state reaction of permanganate with iron(II) the cell diagram is as follows:

$$\mathrm{Pt}\,|\,\mathrm{Fe^{2+}(1\ M),\ Fe^{3+}(1\ M)}\,\|\,\mathrm{MnO_4^-(1\ M),\ Mn^{2+}(1\ M),\ H^+(1\ M)}\,|\,\mathrm{Pt}$$

The single vertical lines near the beginning and the end represent phase changes between the solid platinum electrodes and the solutions. The double vertical lines represent the salt bridge. As shown, concentrations are entered in parentheses if known. At standard state all concentrations are 1 molar. The order in which the chemicals are written, within each compartment of the cell, is not important.

Standard Reduction Potentials

The standard cell voltage is the difference between the standard reduction potentials (voltages) of the cathode and the anode:

$$E^\circ_{cell} = E^\circ_{cathode} - E^\circ_{anode} \qquad (5.16)$$

$E^\circ_{cathode}$ is the standard reduction potential for the reaction occurring at the cathode and represents its tendency to remove electrons from the electrode surface. E°_{anode} is the standard reduction potential for the reaction occurring at the anode and represents its tendency to remove electrons from the anode. It is impossible to measure independently the electromagnetic force ($\mathscr{EMF}$) values of the cathode and anode. The only measurement possible is the combined E°_{cell}.

If it were possible to measure $E^\circ_{cathode}$ or E°_{anode} for any reaction, the standard electrode potentials of all other half reactions could be determined using equation 5.15. As an alternative, chemists define the reduction of hydrogen, at standard state (M = 1.00, 25°C, and 1 atm pressure), as having a reduction potential of exactly 0.00 volt:

$$H^+(aq) + e^- \rightarrow \tfrac{1}{2}H_2(g); \quad E^\circ = 0.00 \text{ V}$$

This definition is used to determine the potentials of other half-reactions. For example, the reaction of

$$Zn(s) + 2H^+(aq) \rightarrow Zn^{2+}(aq) + H_2(g)$$

has a standard cell voltage of $E^\circ_{cell} = +0.76$ volt. Since the zinc oxidation occurs at the anode and the reduction of hydrogen ions occurs at the cathode, we may write

$$E^\circ_{cell} = +0.76 \text{ V} = E^\circ_{H^+} - E^\circ_{Zn}$$

E°_{H+} has been defined as 0.00 volt, and entering that value into the preceding equation gives

$$+0.76 \text{ V} = 0.00 \text{ V} - E^\circ_{Zn}$$

Rearranging this equation yields

$$E^\circ_{Zn} = -0.76 \text{ V}$$

This is the standard reduction potential for zinc.

Once the standard reduction potential for zinc is known, we may then make measurements involving the reaction

$$Zn(s) + 2Ag^+(aq) \rightarrow Zn^{2+}(aq) + 2Ag(s)$$

The standard cell voltage for this reaction is measured as +1.56 volts, so

$$E^\circ_{cell} = +1.56 \text{ V} = E^\circ_{Ag+} - E^\circ_{Zn}$$

Substituting the value for the standard reduction potential of zinc into the above equation yields

$$+1.56 \text{ V} = E^\circ_{Ag+} - (-0.76 \text{ V})$$

Rearranging and solving yields the standard reduction potential of silver ions:

$$E^\circ_{Ag+} = +1.56 \text{ V} - 0.76 \text{ V} = +0.80 \text{ V}$$

In a similar manner the standard reduction potentials of all half-reactions are determined. The standard reduction potentials for most half-reactions are known, and some of these are listed in Table 5.2.

With a table of standard reduction potentials, E°_{cell} for any redox reaction may be predicted. If E°_{cell} is positive, the reaction is spontaneous; if it is negative, the reaction is not spontaneous (i.e., the reverse reaction is spontaneous).

TABLE 5.2. STANDARD REDUCTION POTENTIALS (25°C)

Half-Reaction							$E°$ (V)
$F_2(g)$	+			$2e^-$	→	$2F^-$	2.87
Co^{3+}	+			e^-	→	Co^{2+}	1.82
Au^{3+}	+			$3e^-$	→	Au	1.50
$Cl_2(g)$	+			$2e^-$	→	$2Cl^-$	1.36
$O_2(g)$	+	$4H^+$	+	$4e^-$	→	$2H_2O$	1.23
$Br_2(g)$	+			$2e^-$	→	$2Br$	1.07
$2Hg^{2+}$	+			$2e^-$	→	Hg_2^{2+}	0.92
Ag^+	+			e^-	→	Ag	0.80
Hg_2^{2+}	+			$2e^-$	→	Hg	0.79
Fe^{3+}	+			e^-	→	Fe^{2+}	0.77
I_2	+			$2e^-$	→	$2I^-$	0.53
Cu^+	+			e^-	→	Cu	0.52
Cu^{2+}	+			$2e^-$	→	Cu	0.34
Cu^{2+}	+			e^-	→	Cu^+	0.15
Sn^{4+}	+			$2e^-$	→	Sn^{2+}	0.15
S	+	$2H^+$	+	$2e^-$	→	H_2S	0.14
$2H^+$	+			$2e^-$	→	H_2	0.00
Pb^{2+}	+			$2e^-$	→	Pb	−0.13
Sn^{2+}	+			$2e^-$	→	Sn	−0.14
Ni^{2+}	+			$2e^-$	→	Ni	−0.25
Co^{2+}	+			$2e^-$	→	Co	−0.28
Tl^+	+			e^-	→	Tl	−0.34
Cd^{2+}	+			$2e^-$	→	Cd	−0.40
Cr^{3+}	+			e^-	→	Cr^{2+}	−0.41
Fe^{2+}	+			$2e^-$	→	Fe	−0.44
Cr^{3+}	+			$3e^-$	→	Cr	−0.74
Zn^{2+}	+			$2e^-$	→	Zn	−0.76
Mn^{2+}	+			$2e^-$	→	Mn	−1.18
Al^{3+}	+			$3e^-$	→	Al	−1.66
Be^{2+}	+			$2e^-$	→	Be	−1.70
Mg^{2+}	+			$2e^-$	→	Mg	−2.37
Na^+	+			e^-	→	Na	−2.71
Ca^{2+}	+			$2e^-$	→	Ca	−2.87
Sr^{2+}	+			$2e^-$	→	Sr	−2.89
Ba^{2+}	+			$2e^-$	→	Ba	−2.90
Rb^+	+			e^-	→	Rb	−2.92
K^+	+			e^-	→	K	−2.92
Cs^+	+			e^-	→	Cs	−2.92
Li^+	+			e^-	→	Li	−3.05

EXERCISE

Write the balanced redox reaction for each of the following, determine E°_{cell}, and state whether the reaction is spontaneous:

(a) The single displacement of copper(II) by zinc metal

(b) The single displacement of H^+ by iron, forming Fe^{2+}

(c) The reduction of tin(IV) to tin(II) by iron(II) being oxidized to iron(III)

(d) The reduction of dichromate ions to chromium(III) ions by Mn^{2+} forming MnO_2

(e) The oxidation of AsO_3^{3-} to $H_2AsO_4^-$ by the reduction of iodine to iodide ions

Answers

(a) $Cu^{2+} + Zn \rightarrow Zn^{2+} + Cu$
$E^\circ_{cell} = +0.34\text{ V} - (-0.76\text{ V}) = +1.10\text{ V}$
The reaction is spontaneous.

(b) $Fe + 2H^+ \rightarrow Fe^{2+} + H_2$
$E^\circ_{cell} = 0.00\text{ V} - (-0.44\text{ V}) = +0.44\text{ V}$
The reaction is spontaneous.

(c) $Sn^{4+} + 2Fe^{2+} \rightarrow 2Fe^{3+} + Sn^{2+}$
$E^\circ_{cell} = +0.15\text{ V} - (+0.77\text{ V}) = -0.62\text{ V}$
This reaction is not spontaneous.

(d) $2H^+ + Cr_2O_7^{2-} + 3Mn^{2+} \rightarrow 3MnO_2 + 2Cr^{3+} + H_2O$
$E^\circ_{cell} = +1.33\text{ V} - (+1.23\text{ V}) = +0.10$
The reaction is spontaneous.

(e) $H_2O + AsO_3^{3-} + I_2 \rightarrow H_2AsO_4^- + 2I^-$
$E^\circ_{cell} = +0.54\text{ V} - (+0.58\text{ V}) = -0.04\text{ V}$
The reaction is not spontaneous.

Nernst Equation

Cell Voltages

On pages 209–210 standard reduction potentials are used to determine the standard cell voltage expected from a galvanic cell. Standard conditions specify the temperature, pressure, and concentrations necessary to obtain these standard voltages. The voltage measured will differ from the standard voltage if any of these variables are changed. Walter Nernst developed the **Nernst equation**:

$$E = E^\circ - \frac{RT}{n\mathcal{F}} \ln Q$$

Here E is the reduction potential under nonstandard conditions; $E°$ *is the standard reduction potential;* R = 0.0821 liter atmosphere per mole per Kelvin and is the universal gas law constant, T is the Kelvin temperature, n is the number of electrons in the half-reaction, F = 96,485 coulombs per mole and is Faraday's constant, and Q is the reaction quotient for the reduction half-reaction. At 298 K the Nernst equation is often written as follows:

$$E = E° - \frac{0.0591}{n} \log Q$$

The constant R and $\mathcal{F}$, along with the temperature, 298 K, and a conversion factor from natural logarithms to base-10 logarithms, are combined into the constant 0.0591.

Both of the above equations become $E = E°$ at standard state, when all concentrations in the reaction quotient are equal to 1.00 and therefore Q = 1.00. The logarithm of 1.00 is zero, and the log term disappears so that $E = E°$.

The balanced half reaction for the reduction of dichromate ions is:

$$Cr_2O_7^{2-} \quad + \quad 14H^+ \quad + \quad 6e^- \quad \rightarrow \quad 2Cr^{3+} \quad + \quad 7H_2O$$

The corresponding Nernst equation is

$$E = E° - \frac{0.0591}{6} \log\left(\frac{[Cr^{3+}]^2}{[Cr_2O_7^{2-}][H^+]^{14}}\right)$$

Q in the Nernst equation is always based on the form of the equilibrium law for the reduction half-reaction.

For the complete galvanic cell, two Nernst equations are needed, one for the cathode and one for the anode. These are combined to obtain the nonstandard cell voltage (E_{cell}):

$$E_{cell} = E_{cathode} - E_{anode}$$

Here the symbols E without the superscript zero indicate electrode potentials and cell voltages that are not at standard state.

EXERCISES

1. One cell of a galvanic cell involves the half-reaction

$$Fe^{3+} \quad + \quad e^- \quad \rightarrow \quad Fe^{3+}$$

   What is the electrode potential if the concentration of Fe^{2+} ions is 0.0245 M and the concentration of Fe^{3+} ions is 0.187 M? ($E°$ = +0.771 V)

2. The reduction of iron(III) with tin(II) proceeds by the reaction

$$2Fe^{3+} \quad + \quad Sn^2 \quad \rightarrow \quad 2Fe^{3+} \quad + \quad Sn^{4+}$$

   In a galvanic cell, all of the product concentrations are 0.0355 M and the reactant concentrations are 0.100 M. What is the voltage of the galvanic cell? $E°(Sn) = +0.154$ V and $E°(Fe) = +0.771$ V.

Solutions

1. The Nernst equation is used to solve this problem.

$$E = E° - \frac{0.0591}{n} \log Q$$

Q for this half reaction is:

$$Q = \frac{[Fe^{2+}]}{[Fe^{3+}]} = \frac{0.0245}{0.187} = 0.131$$

and the remaining steps are:

$$E = 0.771 - \frac{0.0591}{1} \log 0.131$$

$$= 0.771 - (0.0591)(-0.883)$$

$$= 0.771 - (-0.052)$$

$$= 0.823 \text{ V}$$

2. The Nernst equation for the entire reaction is used:

$$E_{cell} = E^{\circ}_{cell} - \frac{0.0591}{n} \log Q$$

where $n = 2$ and

$$Q = \frac{[Fe^{2+}]^2[Sn^{4+}]}{[Fe^{3+}]^2[Sn^{2+}]} = \frac{(0.0355)^2(0.0355)}{(0.100)^2(0.100)} = 0.0447$$

Then

$$E^{\circ}_{cell} = 0.771 - 0.154 = 0.617 \text{ V}$$

and

$$E_{cell} = 0.617 - \frac{0.0591}{2} \log 0.0447$$

$$= 0.617 - (-0.040)$$

$$= 0.657 \text{ V}$$

Determining Concentrations Using Galvanic Cells

The Nernst equation suggests a method whereby Galvanic cell measurements can be used to determine concentrations. A redox reaction that is appropriate for this purpose is

$$2Ag^+(aq) + Cu(s) \rightarrow 2Ag(s) + Cu^{2+}(aq)$$

The cell diagram (see page 208) for this reaction is

$$Cu \mid Cu^{2+} \parallel Ag^{+} \mid Ag$$

and the cell voltage is represented as

$$E_{cell} = E_{cathode} - E_{anode}$$

Inserting the Nernst equations for $E_{cathode}$ and E_{anode}, we obtain

$$E_{cell} = \left(E^{\circ}_{Ag/Ag^{+}} - \frac{0.0591}{1}\log\frac{1}{[Ag^{+}]}\right) - \left(E^{\circ}_{Cu/Cu^{2+}} - \frac{0.0591}{2}\log\frac{1}{[Cu^{2+}]}\right) \quad (5.17)$$

If $[Ag^{+}]$ can be kept at a constant, known value, then the value of E_{cell} determined in an experiment can be used to calculate $[Cu^{2+}]$. A saturated solution of AgCl will maintain the silver ion concentration at a constant 1.0×10^{-5} M.

Since

$$E^{\circ}_{Ag/Ag^{+}} = +0.80 \text{ V} \quad \text{and} \quad E^{\circ}_{Cu/Cu^{2+}} = +0.34 \text{ V}$$

equation 5.17 becomes

$$E_{cell} = 0.164 - \frac{0.0591}{2}\log\frac{1}{[Cu^{2+}]}$$

when all three constants are entered.

Now we will look at an actual situation. If an experiment is performed where $E_{cell} = 0.128$ volt for a solution that contains an unknown concentration of Cu^{2+}, we can calculate the Cu^{2+} concentration as follows:

$$0.128 = 0.164 - \frac{0.0591}{2}\log\frac{1}{[Cu^{2+}]}$$

$$-0.036 = -\frac{0.0591}{2}\log\frac{1}{[Cu^{2+}]}$$

$$+1.218 = \log\frac{1}{[Cu^{2+}]}$$

$$16.52 = \frac{1}{[Cu^{2+}]}$$

$$[Cu^{2+}] = 0.0605 \text{ M } Cu^{2+}$$

The key to using galvanic cell measurements for determining concentrations is that all but one concentration in the Nernst equation must be known and must remain constant during the experiment. In this analysis the silver ion concentration around the cathode was held constant. This constant electrode is also known as the **reference electrode**. The anode that monitored the Cu^{2+} ion concentration is called the **indicator electrode**.

Part 2

TECHNIQUES AND CALCULATIONS OF CHEMISTRY

6
MATHEMATICS OF CHEMISTRY

SCIENTIFIC MATHEMATICS

Calculations

Most calculations in chemistry involve only simple algebra. There are two basic approaches for solving these equations. The first is the use of the factor-label method (see pages 230–239) to convert information from one set of units into another. Second is the use of a memorized equation or law into which data for all variables except one are inserted. The one remaining variable is the unknown for the problem. Correct use of the second approach also requires that all units be shown to ensure that they cancel properly to yield the desired units for the numerical answer.

Scientific calculators simplify mathematical operations to the touch of a few keys. Understanding the principles and concepts of chemistry enables us to decide in which order to press those keys. Understanding numbers tells us how to properly interpret the answer that appears on the calculator screen.

Accuracy and Precision

The **accuracy** of a measurement refers to the closeness between the measurement obtained and the true value. Since scientists rarely know the true value, it is generally impossible to determine the accuracy completely. One approach to evaluating accuracy is to make a measurement by two completely independent methods. If the results from independent measurements agree, scientists have more confidence in the accuracy of their results.

Accuracy is affected by **determinate errors**, that is, errors due to poor technique or incorrectly calibrated instruments. Careful evaluation of an experiment may eliminate determinate errors.

Precision refers to the closeness of repeated measurements to each other. If the mass of an object is determined as 35.43 grams, 35.41 grams and 35.44 grams in three measurements, the results may be considered precise because they differ only in the last of the four digits. There is no guarantee that they are accurate, however, unless the balance was properly calibrated and the correct methods were used in weighing the object. With proper techniques, precise results infer, but do not guarantee, accurate results.

Precision is affected by **indeterminate errors**, that is, errors that arise in estimating the last, uncertain digit of a measurement. Indeterminate errors are random errors and cannot be eliminated. Statistical analysis deals with the theory of random errors.

Significant Figures

Every experimental measurement is made in such a way as to obtain the most information possible from whatever instrument is used. As a result, measurements involve numbers in which the last digit is uncertain. Scientists characterize a measured number based on the number of **significant figures** it contains.

The number of significant figures in a measurement includes all digits in that number from the first nonzero digit on the left to the last digit on the right. For exponential numbers, the number of significant figures is determined from the digits to the left of the multiplication sign. Thus, 3.456×10^8 has four significant figures.

Sometimes there are trailing zeros on the right side of a number. If the number contains a decimal point, trailing zeros are always significant. Trailing zeros that are used to complete a number, however, may or may not be significant. Scientists avoid writing a number such as 12,000 since it does not definitely indicate the number of significant figures. Scientific notation is used instead. Twelve thousand can be written as 1.2×10^4, 1.20×10^4, 1.200×10^4, or 1.2000×10^4, indicating two, three, four, and five significant figures, respectively. It is the responsibility of the experimenter to write numbers in such a way that there is no ambiguity.

EXERCISE

Determine the number of significant figures in each of the following measured values:

(a) 23.46 mL

(b) 0.0036 s

(c) 854.236 g

(d) 6.02×10^{23} molecules

(e) 0.98 mol

(f) 0023 m

(g) 1.00026×10^{-3} cm

(h) 2.0000 J

(i) 824 mg

Answers

The numbers are repeated with the significant figures in bold type:

(a) **23.46** mL

(b) 0.00**36** s

(c) **854.236** g

(d) **6.02** $\times 10^{23}$ molecules

(e) 0.**98** mol

(f) 00**23** m

(g) **1.00026** $\times 10^{-3}$ cm (i) **824** mg

(h) **2.0000** J

Some numbers are **exact numbers**, which involve no uncertainty. The number of plates set on a table for dinner can be determined exactly. If five plates are observed and counted, that measurement is exactly five. In chemistry many defined equalities are exact. For instance, there are exactly 4.184 joules in each calorie and exactly two hydrogen atoms in each water molecule. Other exact numbers are stoichiometric coefficients and subscripts in chemical formulas.

The reason for determining the number of significant figures in a measured value is that this number tells us how to write the answers to calculations based on that value. There are two basic rules.

SIGNIFICANT FIGURE RULES IN CALCULATIONS

1. The number with the fewest significant figures in a **multiplication or division** problem determines the number of significant figures in the answer. In these calculations an exact number is considered to have an infinite number of significant figures.
2. The number with the fewest decimal places in an **addition or subtraction** problem determines the number of decimal places in the answer. Numbers expressed in scientific notation must all be converted to the same power of 10 before determining which decimal places can be retained.
3. A problem involving both multiplication or division and addition or subtraction in the same operation must be solved stepwise, applying rules 1 and 2 according to the type of operation performed.

EXERCISE

Perform each of the following calculations, and report the answer with the correct number of significant figures:

(a) $23.456 + 16.0094 + 9.21$

(b) $14.98 \times 0.00234 \times 1.5$

(c) $1.46 \times 10^3 + 5.83 \times 10^4$

(d) $(8.236 \times 10^2)(5.55 \times 10^{-3})$

(e) $(23.45 - 16.12)/6.233$

(f) $(6.02 \times 10^{23})(1.00 \times 10^{-6})/(18.23)$

(g) $\dfrac{44.23}{2.33 \times 10^2 - 2.25 \times 10^2}$

Answers

(a) 48.68
(b) 0.053
(c) 5.68×10^4
(d) 4.57
(e) 1.18
(f) 3.30×10^{16}
(g) 6.

Uncertainty

There are two types of uncertainty, absolute uncertainty and relative uncertainty.

The **absolute uncertainty** is the uncertainty of the last digit of a measurement. For example, 45.47 milliliters is a measurement of volume, and the last digit is uncertain. The absolute uncertainty is ± 0.01 milliliter. The measurement should be regarded as somewhere between 45.46 and 45.48 milliliters.

The **relative uncertainty** of a number is found by dividing the absolute uncertainty of the measurement by the number itself. For the above example, the relative uncertainty

$$\frac{0.01 \text{ mL}}{45.47 \text{ mL}} = 2 \times 10^{-4}$$

The absolute uncertainty governs the principles used for addition and subtraction. The relative uncertainty governs the principles used for multiplication and division.

Rounding

Calculations, especially those done using an electronic calculator, often generate more, but sometimes fewer, significant figures or decimal places than are required by rules 1 and 2 in the box on the previous page. Each of these answers must be rounded to the proper number of significant figures or decimal places. To do this, the following steps are followed:

1. The number of digits to be kept in a calculation is determined using rules 1 and/or 2 in the box.
2. If the digit just after the kept digits is less than 5, the remaining digits are dropped. For example, rounding 6.23499 to three significant figures yields 6.23 because 4 is less than 5.
3. If the digit just after the kept digit is greater than 5, or if it is 5 with additional nonzero digits, 1 is added to the last digit of the kept digits. For example, rounding 5.5589 to three significant figures yields 5.56. Similarly, 6.345002 is rounded to 6.35 because 5 has a nonzero digit after it.
4. If the digit to be rounded is just 5 or is 5 with only zeros after it, the last digit in the kept digits is rounded to the nearest even number. For example, rounding 2.335 to two decimal places yields 2.34, and rounding 6.785000 to two decimal places yields 6.78.

Significant Figures in Atomic and Molar Masses

All of the atomic masses listed in the Periodic Table have four or more significant figures. These are measured values and must be included in the determination of the correct number of significant figures in any calculation that uses them. However, most calculations involve fewer than four significant figures, and the significant figures in atomic and molar masses usually do not affect the number of significant figures in the answers. As a result, atomic and molar masses are often rounded to the nearest whole number. One exception is chlorine, which usually has its atomic mass rounded to 35.5.

EXERCISE

What percentage error is expected when the atomic mass of each of the following elements is rounded to the nearest whole number?

(a) Na

(b) Ag

(c) Pb

(d) Cl

Solutions

The percentage error is calculated as

$$\text{Percentage error} = \frac{\text{measured value} - \text{true value}}{\text{true value}} \times 100$$

In this problem, the measured value is the rounded atomic mass and the true value is the atomic mass.

(a) $\frac{0.01}{22.99} \times 100 = +0.04\%$

(b) $\frac{0.132}{107.868} \times 100 = +0.1\%$

(c) $\frac{0.2}{207.2} \times 100 = -0.1\%$

(d) $\frac{0.453}{35.453} \times 100 = -1.3\%$

The error for chlorine is ten times that of the other elements. When rounded to 35.5, however, the error for chlorine is +0.1%, which is in line with the magnitude of the rounding errors for other elements.

Graphs

A **graph** is used to illustrate the relationship between two variables. Graphs are often a more effective method of communication than tables of data. A graph has two axes: a horizontal axis, called the *x*-axis (abscissa), and a vertical axis, called the *y*-axis (ordinate).

It is customary to use the *x*-axis for the independent variable, and the *y*-axis for the dependent variable, in an experiment. An **independent variable** is one that the experimenter selects. For instance, the concentrations of standard solutions that a chemist prepares are independent variables since any concentrations may be chosen. The **dependent variable** is a measured property of the independent variable. For instance, the dependent variable may be the amount of light that each of the standard solutions prepared by the chemist absorbs, since the absorbed light is dependent on the concentration. Each data point is an *x,y* pair representing the value of the independent variable and the value of the dependent variable determined in the experiment.

The first step in constructing a graph is to label the *x*- and *y*-axes to indicate the identity of the independent and dependent variables. Next, the axes are numbered, usually from zero to the largest value expected for each variable. Finally, each data point is plotted by drawing a horizontal line at the value of the dependent variable and a vertical line upward from the value of the independent variable. The intersection of these two lines determines where that data point belongs on the graph, as shown in Figure 6.1.

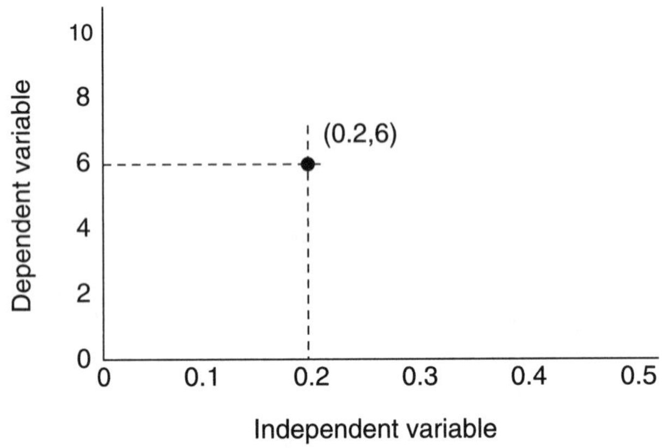

FIGURE 6.1. *The positions of the independent and dependent variables in a graph. Dashed lines show the placement of a point representing a value of 0.2 for the independent variable and 6 for the dependent variable.*

Most graphs show a linear relationship between two variables. Others, such as kinetic curves, have curved lines. In both cases, data points are plotted on the graph and then the best smooth line is drawn through the points. Lines are never drawn by connecting the data points with straight lines. In very accurate work a statistical analysis called the method of least squares is used to determine the best straight line for the data. In most cases, however, the line is drawn by eye, attempting to have all data points as close as possible to the line. The usual result is a line that has the same number of data points above and below it.

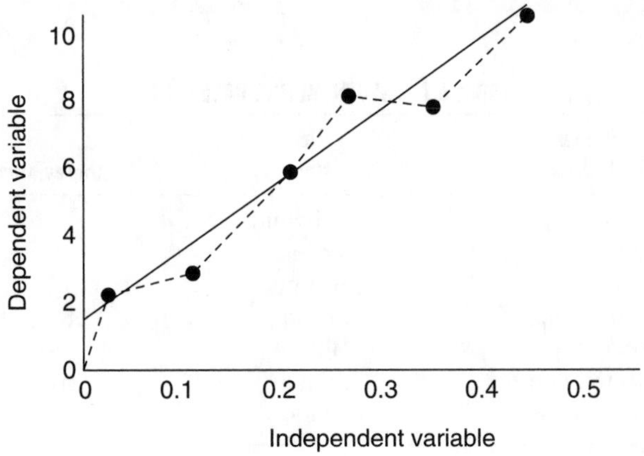

FIGURE 6.2. *The correct and incorrect ways to draw a line using data in a graph. The solid line is the correct line; the dashed line is incorrect.*

Figure 6.2 illustrates that it is incorrect to draw any line beyond the measured data points. The reason is that anything beyond the measured data is unknown. Extending the line implies information that has not been verified by experimental data.

The slope of a curve or line is often needed as an experimental result. To determine the slope of a line, two points on the line are chosen. The left-hand point has coordinates (x_1, y_1), and the right-hand point has coordinates (x_2, y_2). The values of x and y at these points are determined from the graph, and the following equation is used to determine the slope:

$$\text{Slope} = \frac{x_2 - x_1}{y_2 - y_1}$$

In a graph with a curved line the slope is determined by drawing a tangent to the curve and then determining the slope of the tangent, as is done for a straight line.

Metric Unit Conversions

Chemistry students in the United States are hampered by the fact that in this country chemistry is done using the metric system but everyday life is conducted with the English system of measurement. The modern metric system is called the Système International, or S.I. for short. The S.I. system and units are used for all chemistry problems in this book.

S.I. Units and Prefixes

The seven defined **base units** of the S.I. system are listed in Table 6.1.

TABLE 6.1. SEVEN METRIC BASE UNITS

Property Defined	Unit Name	Abbreviation
Mass	Kilogram	kg
Length	Meter	m
Time	Second	s
Temperature	Kelvin	K
Quantity	Mole	mol
Electric current	Ampere	A
Light intensity	Candela	cd

These seven base units can be combined in a variety of ways to give other common units. Area can be expressed as square meters (m^2). Volume may be written as cubic meters (m^3).

These base units, which are often too large or too small for practical use in a laboratory, can be modified by the addition of a metric prefix. Each **metric prefix** represents a number that multiplies the base unit. The most commonly used prefixes for metric units are given in Table 6.2.

TABLE 6.2. METRIC PREFIXES

Prefix Name	Prefix Symbol	Exponential Value
Mega-	M	10^6
Kilo-	k	10^3
Deci-	d	10^{-1}
Centi-	c	10^{-2}
Milli-	m	10^{-3}
Micro-	μ	10^{-6}
Nano-	n	10^{-9}
Pico-	p	10^{-12}

It is essential to remember the first five S.I. base units in Table 6.1 and all of the metric prefixes in Table 6.2. Factor labels between a unit with a prefix and the corresponding metric base unit may be quickly constructed by first writing an equality:

$$1 \text{ centimeter} = 1 \text{ centimeter}$$

and then replacing one of the prefixes with its corresponding exponent:

$$1 \text{ centimeter} = 10^{-2} \text{ meter}$$

This equality can then be used to write the two possible factor labels:

$$\left(\frac{\text{cm}}{10^{-2}\text{ m}}\right) \quad \text{and} \quad \left(\frac{10^{-2}\text{ m}}{\text{cm}}\right)$$

Conversion between a prefix and a base unit is a one-step calculation, but conversion between two different prefixes requires two steps.

Example Convert 2.38 centimeters to (a) meters and (b) millimeters.

Solutions:

(a) The problem is set up with the information desired and the given data:

$$? \text{ m} = 2.38 \text{ cm (setup)}$$

To convert centimeters to millimeters, one of the factor labels given above may be used:

$$? \text{ m} = 2.38 \text{ cm}\left(\frac{10^{-2}\text{ m}}{\text{cm}}\right)$$

$$= 0.0238 \text{ m} = 2.38 \times 10^{-2} \text{ m}$$

(b) Again, the problem is set up:

$$? \text{ mm} = 2.38 \text{ cm (setup)}$$

To convert centimeters to millimeters, two conversion factors are needed. The first converts the prefix to the base unit, meters, and the second converts the base unit to the new prefix:

$$? \text{ mm} = 2.38 \text{ cm}\left(\frac{10^{-2}\text{ m}}{\text{cm}}\right)\left(\frac{\text{mm}}{10^{-3}\text{ m}}\right)$$

$$= 23.8 \text{ mm}$$

$$= 2.38 \times 10^{1} \text{ mm}$$

The answers to both parts of this problem are written in exponential notation to show that the number 2.38 does not change in these conversions, but the exponent does. In (b), however, the answer 23.8 mm is preferred since it is simpler to write.

Conversion of Complex Units

Many times the data used in chemistry are expressed in complex units. There are area measurements that have squared units such as square kilometers (km^2), square meters (m^2), or square centimeters (cm^2). Volume measurements may have cubic units such as cubic centimeters (cm^3), cubic millimeters (mm^3), or cubic meters (m^3). For velocity the units may be meters per second ($m\ s^{-1}$ or m/s). Acceleration has units of meters per second squared ($m\ s^{-2}$), and for energy the units are kilogram-meters squared per second squared ($kg\ m^2\ s^{-2}$).

The same conversion technique can be used in a variety of situations. The following examples illustrate the method when the units have squared or cubed terms and when the units involve a ratio.

Example How many square centimeters are there in 180 square meters?

Solution: Setting up the problem as before, we have

$$? \text{ cm}^2 = 180 \text{ m}^2$$

The units are square centimeters and square meters. For clarity it is best to write the setup of this problem as

$$? \text{ cm} \times \text{cm} = 180 \text{ m} \times \text{m}$$

We can use the equality that says that 10^{-2} m is equal to 1 cm to write the needed factor label. Using the ratio that cancels out the meter units, we obtain

$$? \text{ cm} \times \text{cm} = 180 \text{ m} \times \text{m}\left(\frac{1 \text{ cm}}{10^{-2} \text{ m}}\right)$$

By using the conversion ratio only once, we have canceled only one of the meter units, leaving the mixed units of centimeters and meters. Applying the conversion ratio a second time cancels all of the meter units and leaves us with the desired square centimeters:

$$? \text{ cm} \times \text{cm} = 180 \text{ m} \times \text{m}\left(\frac{1 \text{ cm}}{10^{-2} \text{ m}}\right)\left(\frac{1 \text{ cm}}{10^{-2} \text{ m}}\right)$$

Once the units cancel properly, the answer can be calculated as 1.80×10^6 cm^2.

For volumes, which have cubic units, each factor label is repeated three times.

When the units involve ratios that must be converted, the units in the numerator are converted to the units required by the problem. Then the units in the denominator are converted.

Example A train is moving at a speed of 25 kilometers per hour. How fast is it moving in units of meters per minute?

Solution: We set up the problem as before with the desired units as the question and the given speed as the starting point in the conversion:

$$? \frac{\text{m}}{\text{min}} = \frac{25 \text{ km}}{\text{h}} \quad \text{(setup)}$$

The two equalities that must be used to construct factor labels for this problem are

$$1 \text{ kilometer} = 10^3 \text{ meters}$$

$$1 \text{ hour} = 60 \text{ minutes}$$

Applying the factor obtained from the first equality converts the kilometers to meters:

$$? \frac{\text{m}}{\text{min}} = \frac{25 \text{ km}}{\text{h}} \left(\frac{10^3 \text{ m}}{1 \text{ km}} \right)$$

Next, the second equality is used to convert the hour units in the denominator to the minutes required:

$$? \frac{\text{m}}{\text{min}} = \frac{25 \text{ km}}{\text{h}} \left(\frac{10^3 \text{ m}}{1 \text{ km}} \right) \left(\frac{1 \text{ h}}{60 \text{ min}} \right)$$

The units cancel properly, leaving the desired meters per minute units. The answer is then calculated to be 417 m min^{-1}.

The above examples illustrate three important principles about the factor-label method. First, it is necessary to know the equalities required to obtain the proper conversion factors. Second, there is often a proper sequence in which to use these conversion factors. Third, if the units for the given value are in the form of a ratio, they must be converted into other units that also represent a ratio.

EXERCISE

Convert each of the following:

(a) 8.89 nm to mm

(b) 3.89×10^5 cm^2 to μm^2

(c) 2.43×10^2 kg m^2s^{-2} to g cm^2 s^{-2}

Answers

(a) 8.89×10^{-6} mm (b) 3.89×10^{13} μm^2 (c) 2.43×10^9 g cm^2 s^{-2}

SCIENTIFIC CALCULATIONS

Factor-Label Method

Most chemistry problems can be solved by more than one specific method. Through the years, however, it has been found that the factor-label method has significant advantages. For instance, it forces us to perform the correct mathematical operations, dividing when we should divide and multiplying when we should multiply. Importantly, it also provides hints to guide us through each calculation. On the other hand, it requires that the calculations be performed in a careful, logical manner. Attention to detail is very important in chemistry calculations.

Defining Factor Labels

In conversion problems the units of a measurement, but not its magnitude, are changed. For instance, a distance of 2.0 yards can be converted into 72 inches by multiplying 2.0 by the factor 36. This does not indicate, however, how the units of yards became units of inches. Instead of just using the factor 36, multiplying by the factor with its units will show how the yards were converted into inches:

$$2.0 \text{ yards}\left(\frac{36 \text{ inches}}{1 \text{ yard}}\right) = 72 \text{ inches}$$

In this equation the units of yards cancel, and it is clear that the remaining units are inches. The ratio (36 inches/1 yard) is known as a **factor label**. Using labels along with the numerical factors enables us to keep track of the units and to produce an answer with the correct units. Proper use of factor labels also tells us when to multiply and when to divide.

A factor label is derived from a defined relationship between two sets of units. The factor label above was obtained from the definition of 1 yard:

$$1 \text{ yard} = 36 \text{ inches} \qquad (6.1)$$

This is the equality between yards and inches. Two factor labels can be made from every defined equality. For example, dividing both sides of equation 6.1 by 1 yard gives

$$\left(\frac{1 \text{ yard}}{1 \text{ yard}}\right) = \underbrace{\left(\frac{36 \text{ inches}}{1 \text{ yard}}\right)}_{\text{Factor Label}} = 1$$

Dividing both sides by 36 inches gives

$$\underbrace{\left(\frac{1 \text{ yard}}{36 \text{ inches}}\right)}_{\text{Factor Label}} = \left(\frac{36 \text{ inches}}{36 \text{ inches}}\right) = 1$$

These two possible factor labels are reciprocals of each other. In addition, both are equal to exactly 1.00. When a measurement is multiplied by a factor label, it is multiplied, in effect, by 1.00; its true value does not change, but the units do change.

Using Factor Labels

All conversion problems must start with two pieces of information: the number and the units that need to be converted, and the units that the answer requires. This is set up as

$$? \text{ answer units} = xxx \text{ given units} \tag{6.2}$$

The left side of the equal sign reminds us of the units needed for the solution to the problem. The *xxx* on the right side represents the number given in the problem and its units, which will be converted. The initial setup tells us where we start (right side) and where we want to end (left side) of equation 6.2. The setup is critical since we can never get to where we are going if we don't know where we are going!

After writing the setup, we need to find the proper equalities to use to produce the factor labels needed to solve the problem.

> In the following example and exercise the cancellation of units is not shown. It is suggested that you use a colored pencil to perform all of the cancellations to assure yourself that each factor label does indeed cancel properly.

Example Convert (a) 35 yards into feet and (b) 624 feet into yards.

Solutions: (a) The solution starts with the initial setup of the question, with the desired units as the unknown and the given data as the starting point in the equation:

$$? \text{ feet} = 35 \text{ yards (initial setup)}$$

We recall that in the English system of measurement 1 yard is defined as being 3 feet:

$$1 \text{ yard} = 3 \text{ feet}$$

From this equality two conversion factors, (3 feet/1 yard) and (1 yard/3 feet), can be written. The correct conversion factor is the one that allows us to cancel the yard units leaving units of feet. We use this factor label to multiply the given 35 yards:

$$? \text{ feet} = 35 \text{ yds}\left(\frac{3 \text{ ft}}{1 \text{ yd}}\right)$$

After canceling the yard units with a pencil, we calculate the answer as 105 ft.

The following equation illustrates what happens if the wrong form of the factor label is chosen:

$$? \text{ feet} = 35 \text{ yds}\left(\frac{1 \text{ yd}}{3 \text{ ft}}\right) = 11.7 \text{ yds}^2 \text{ ft}^{-1}$$

Wrong Factor Label

This conversion will not work since the units do not cancel; in fact, the final unit is the meaningless $\text{yd}^2/\text{ft}^{-1}$. As shown above, correct answer is

$$? \text{ feet} = 35 \text{ yds}\left(\frac{3 \text{ ft}}{1 \text{ yd}}\right) = 105 \text{ ft}$$

(b) Here, the reverse calculation, from feet to yards, is needed. Starting as before with the question and the data supplied, we write

$$? \text{ yards} = 624 \text{ feet}$$

Selecting the correct conversion factor gives

$$? \text{ yards} = 624 \text{ ft}\left(\frac{1 \text{ yd}}{3 \text{ ft}}\right)$$
$$= 208 \text{ yds}$$

For this conversion the same equality, but a different factor label, was used. One of these factor labels converts yards to feet, and the other converts feet to yards.

Many conversions require more than one step to reach the desired units. These problems can be solved stepwise, one factor label at a time, or the factor labels may be combined in one large equation. Both methods are illustrated in the following exercise.

EXERCISE

A typical school year includes 180. days of classes. How many minutes are there in those days?

Solution

We start with the question and the given information to obtain

$$? \text{ minutes} = 180.\text{ days}$$

Next, we find the conversion equalities that may be useful. These are as follows:

$$1 \text{ day} = 24 \text{ hours}$$
$$1 \text{ hour} = 60 \text{ minutes}$$

The given data have units of days, and the only equality that also has units of days is the first one. The ratio needed for the conversion must have the day units in the denominator so that these units will cancel. Multiplying by this ratio yields

$$? \text{ minutes} = 180.\text{ days}\left(\frac{24 \text{ hrs}}{1 \text{ day}}\right) = 4320 \text{ hrs}$$

This result still does not have the desired units, and another step is needed. Starting with the result obtained above, we write

$$? \text{ minutes} = 4320 \text{ hrs}$$

In our record conversion equality, there is a relationship between hours and minutes, 1 hour = 60 minutes. When the next factor label is inserted, the hour units cancel:

$$? \text{ minutes} = 4320 \text{ hrs}\left(\frac{60 \text{ min}}{1 \text{ hr}}\right) = 259 \times 10^5 \text{ min}$$

Since the units of this answer match the units that the question requires, the problem is solved. This answer is rounded off to 2.59×10^5 min since the initial question started with three significant figures (180.).

Solving this same problem with one large equation involves writing all of the factor labels needed until the units match:

$$? \text{ minutes} = 180.\text{ days}\left(\frac{24 \text{ hrs}}{1 \text{ day}}\right)\left(\frac{60 \text{ min}}{1 \text{ hr}}\right) = 259{,}200 \text{ min}$$

As before, 259,200 is rounded to three significant figures and the answer is reported as 2.59×10^5 min.

The step-by-step method and the combined method result in exactly the same answer, and either method is correct. The combined method has the advantage of allowing us to determine the correct number of significant figures directly, and it avoids the possibility of making rounding errors in intermediate steps.

Factor Labels from Balanced Chemical Equations

A balanced chemical equation gives us a group of relationships that can be used to generate a large number of factor labels. For example, the balanced equation for the combustion of benzene, C_6H_6, is as follows:

$$2C_6H_6 \quad + \quad 15O_2 \quad \rightarrow \quad 12CO_2 \quad + \quad 6H_2O$$

On a mole basis the following equality relationships can be obtained:

$$\begin{aligned}
2 \text{ moles } C_6H_6 &= 15 \text{ moles } O_2 \\
2 \text{ moles } C_6H_6 &= 12 \text{ moles } CO_2 \\
2 \text{ moles } C_6H_6 &= 6 \text{ moles } H_2O \\
15 \text{ moles } O_2 &= 12 \text{ moles } CO_2 \\
15 \text{ moles } O_2 &= 6 \text{ moles } H_2O \\
12 \text{ moles } CO_2 &= 6 \text{ moles } H_2O
\end{aligned}$$

Here the equal sign does not represent a mathematical equality, but should be read as "is equivalent to." These equalities apply only to the balanced equation from which they are derived.

As stated above, any BALANCED chemical equation is a source of a large number of factor-label equalities. Once the equation is balanced, we simply write "mol" (mole or moles) between the coefficient and the formula for each compound. Then we choose any desired pair for the equality. For example:

$$FeCl_3 \quad + \quad 3NaOH \quad \rightarrow \quad Fe(OH)_3 \quad + \quad 3NaCl$$

can be rewritten as

$$1 \text{ mol } FeCl_3 \quad + \quad 3 \text{ mol } NaOH \quad \rightarrow \quad 1 \text{ mol } Fe(OH)_3 \quad + \quad 3 \text{ mol } NaCl$$

Here we choose any two of the compounds in the equation to set up our equality and then our factor label.

If we are working at the molecular level, we can replace the moles (mol) by molecules to read

$$1 \text{ molecule } FeCl_3 + 3 \text{ molecules } NaOH \rightarrow 1 \text{ molecule } Fe(OH)_3 + 3 \text{ molecules } NaCl$$

As before, any pair can be equated to obtain an equality that will result in a factor label.

Factor Labels from Chemical Formulas

The formula for an ionic compound is an empirical formula representing the simplest ratio of atoms. A formula for a molecular compound gives the numbers and types of all the atoms that make up one molecule of that compound. Every chemical formula represents the ratio of atoms within the formula. This fact allows the chemist to write relationships between the formula as a whole and the individual atoms in that formula. For example, common table sugar, sucrose, has the formula $C_{12}H_{22}O_{11}$. On the atomic scale this formula gives the following relationships, where the equal sign is read as "is chemically equivalent to":

$$\begin{aligned}
&\text{1 molecule of } C_{12}H_{22}O_{11} &&= \text{12 atoms of carbon}\\
&\text{1 molecule of } C_{12}H_{22}O_{11} &&= \text{22 atoms of hydrogen}\\
&\text{1 molecule of } C_{12}H_{22}O_{11} &&= \text{11 atoms of oxygen}\\
&\text{12 atoms of carbon} &&= \text{22 atoms of hydrogen}\\
&\text{12 atoms of carbon} &&= \text{11 atoms of oxygen}\\
&\text{22 atoms of hydrogen} &&= \text{11 atoms of oxygen}
\end{aligned}$$

In short, one sucrose molecule gives the chemist six different relationships with which to construct factor labels. (It should be emphasized that these are not true equalities. As stated above, the equal sign in these equations should be read as "is chemically equivalent to"; then the first relationship actually says that one molecule of $C_{12}H_{22}O_{11}$ is chemically equivalent to 12 atoms of carbon.)

It is rare that the chemist thinks of these relationships in terms of molecules and atoms. Rather they are viewed as moles of molecules or moles of atoms. In essence, each side of the relationships given above is multiplied by Avogadro's number to obtain the chemical equivalences:

$$\begin{aligned}
&\text{1 mol } C_{12}H_{22}O_{11} &&= \text{12 mol C}\\
&\text{1 mol } C_{12}H_{22}O_{11} &&= \text{22 mol H}\\
&\text{1 mol } C_{12}H_{22}O_{11} &&= \text{11 mol O}\\
&\text{12 mol C} &&= \text{22 mol H}\\
&\text{12 mol C} &&= \text{11 mol O}
\end{aligned}$$

Finally, it is important to remember that these relationships are true only for the specified compound. They may be different for other compounds.

EXERCISE

How many different chemical equivalences can be written for each of the following formulas?

(a) NaCl

(b) $FeCl_3$

(c) $NiSO_4$

(d) $(NH_4)_3PO_4$

(e) O_3

Answers

(a) 3: 1 mol NaCl = 1 mol Na		(d) 10: 1 mol $(NH_4)_3PO_4$ = 3 mol N	
1 mol NaCl = 1 mol Cl		1 mol $(NH_4)_3PO_4$ = 12 mol H	
1 mol Na = 1 mol Cl		1 mol $(NH_4)_3PO_4$ = 1 mol P	
(b) 3: 1 mol $FeCl_3$ = 1 mol Fe		1 mol $(NH_4)_3PO_4$ = 4 mol O	
1 mol $FeCl_3$ = 3 mol Cl		3 mol N = 12 mol H	
1 mol Fe = 3 mol Cl		3 mol N = 1 mol P	
(c) 6: 1 mol $NiSO_4$ = 1 mol Ni		3 mol N = 4 mol O	
1 mol $NiSO_4$ = 1 mol S		12 mol H = 1 mol P	
1 mol $NiSo_4$ = 4 mol O		12 mol H = 4 mol O	
1 mol Ni = 1 mol S		1 mol P = 4 mol O	
1 mol Ni = 4 mol O		(e) 1: 1 mol O_3 = 3 mol O	
1 mol S = 4 mol O			

In addition to the equivalences discussed above and illustrated in the preceding exercise, the chemist also recognizes groupings of atoms, particularly the groups classified as polyatomic ions. For instance, phosphoric acid, H_3PO_4, in addition to the expected equivalences of the atoms, also has the following equivalences:

$$1 \text{ mol } H_3PO_4 = 1 \text{ mol } PO_4^{3-}$$
$$1 \text{ mol } PO_4^{3-} = 3 \text{ mol } H^+$$

These relationships often come in handy when considering ionic and net ionic equations. The chemical formula is a rich source of equivalences for the construction of factor labels used in problem solving.

Concentration and Density as Factor Labels

Two very common units in chemistry, molarity and density, provide important **factor labels** for problem solving.

Molarity

The concentration of a solution can be used as a factor label for conversion calculations. The most common concentration unit used in chemistry is molarity. **Molarity** is defined as the number of moles of a solute that are dissolved in 1 liter of solution. When a problem gives the concentration of a solution, for example, as 0.250 molar NaOH (often written as 0.250 M NaOH), this can be made into a factor label by specifying the basic units:

$$0.250 \text{ M NaOH} = \left(\frac{0.250 \text{ mol NaOH}}{1 \text{ L NaOH}}\right)$$

This ratio is a factor label for conversions between moles of NaOH and liters of the NaOH solution. As with all conversion factors, its inverse is also a factor label:

$$\left(\frac{1\text{ L NaOH}}{0.250\text{ mol NaOH}}\right)$$

In many problems the volume is expressed in milliliters (mL). The two factor labels shown above can be rewritten as

$$\left(\frac{0.250\text{ mol NaOH}}{1000\text{ mL NaOH}}\right) \quad \text{and} \quad \left(\frac{1000\text{ mL NaOH}}{0.250\text{ mol NaOH}}\right)$$

since there are 1000 milliliters in each liter (1000 mL = 1 L).

Example 1 How many milliliters of 0.0324 M $CaCl_2$ are needed to obtain 23.8 grams of $CaCl_2$?

Solution: The setup for the question is

$$?\text{ mL CaCl}_2 = 23.8\text{ g CaCl}_2$$

The first step is to convert grams of $CaCl_2$ to moles of $CaCl_2$:

$$?\text{ mL CaCl}_2 = 23.8\text{ g CaCl}_2\left(\frac{1\text{ mol CaCl}_2}{111\text{ g CaCl}_2}\right)$$

Now the molarity (here, the inverse of the molarity) can be used to convert moles to mL:

$$?\text{ mL CaCl}_2 = 23.8\text{ g CaCl}_2\left(\frac{1\text{ mol CaCl}_2}{111\text{ g CaCl}_2}\right)\left(\frac{1000\text{ mL CaCl}_2}{0.0324\text{ mol CaCl}_2}\right)$$

$$= 6620\text{ mL CaCl}_2 \text{ or } 6.62\text{ L CaCl}_2$$

Density

The **density** of a substance may also provide a factor label for conversions between volume and mass. Most densities in chemistry have units of grams per cubic centimeter (g cm^{-3}). If an organic liquid has a density of 0.741 gram per cubic centimeter, the density may be written as the factor label:

$$\text{Density} = \left(\frac{0.741\text{ g}}{1\text{ cm}^3}\right)$$

As with molarity, the inverse of this ratio is the other factor label obtained from the density:

$$\left(\frac{1\text{ cm}^3}{0.741\text{ g}}\right)$$

Since 1 cubic centimeter is the same as 1 milliliter, we can interchange the two terms as needed to obtain

$$\left(\frac{1\ \text{mL}}{0.741\ \text{g}}\right) \quad \text{and} \quad \left(\frac{0.741\ \text{g}}{1\ \text{mL}}\right)$$

Example 2 It is much easier to measure volumes than to weigh substances. How many milliliters of ethyl alcohol, C_2H_6O, density = 0.7893 gram per cubic centimeter, will be needed to have a 75.0 gram sample of ethyl alcohol?

Solution: Set up the question as follows:

$$?\ \text{mL}\ C_2H_6O = 75.0\ \text{g}\ C_2H_6O$$

Use the inverse of the density to have the factor label that will cancel properly:

$$?\ \text{mL}\ C_2H_6O = 75.0\ \text{g}\ C_2H_6O\left(\frac{1\ \text{cm}^3}{0.7893\ \text{g}}\right)$$

$$= 95.0\ \text{cm}^3$$

This is the same as 95.0 mL since 1 cm^3 = 1 mL.

In both of the conversions illustrated in this chapter we are able to change a tedious measurement (weighing mass) into a relatively easier measurement (determining volume). Chemists often take advantage of these techniques to simplify experiments.

Useful Conversion Factors

Another useful equality for constructing a factor label is the relationship between the number of moles of a gas and the volume of the gas at standard temperature and pressure. Standard temperature is 0°C, and standard pressure is 1 atmosphere. Under these conditions 1 mole of a gas occupies 22.4 liters. The equality and its two conversion factors are as follows:

$$1\ \text{mol gas} = 22.4\ \text{L gas}$$

$$1 = \left(\frac{22.4\ \text{L gas}}{1\ \text{mol gas}}\right)$$

and

$$1 = \left(\frac{1\ \text{mol gas}}{22.4\ \text{L gas}}\right)$$

Although this expression refers specifically to an ideal gas, either of these factor labels can be used in calculations for most real gases. When used for real gases, the error is usually less than 5 percent.

Finally, there are other equalities that are also very useful to remember:

$$1\ \text{cm}^3 = 1\ \text{mL}$$

$$1\ \text{L} = 1000\ \text{mL} = 1000\ \text{cm}^3$$

$$1\ \text{cm}^3\ H_2O = 1\ \text{g}\ H_2O$$

The last equality is also true for dilute solutions (generally less than 1 mol L^{-1}) and can be written as

$$1\ \text{cm}^3\ \text{dilute sol'n} = 1\ \text{g dilute sol'n}$$

MEASUREMENTS

Gas Measurements

The physical properties of gases depend on four variables: volume, temperature, number of moles of gas, and pressure. Each of these quantities may be measured independently. Often, however, it is easier to measure three of these four variables and then calculate the last from the ideal gas law equation. The measurement methods are reviewed briefly on the following page.

Volume is expressed in liters (sometimes milliliters) and may be determined in several ways, including:

1. Careful measurement of the dimensions of the container and the appropriate geometric calculations.
2. Measurement of the mass of a liquid of known density that fills the container to capacity.

Temperature is measured with a standard thermometer. Special instruments may be used, however, if high accuracy and precision are needed or if the temperature of the gas is extremely high or low. Since gases have very low densities, sufficient time must be allowed for a thermometer to reach the correct temperature. Often, the temperature of a gas is determined by measuring the temperature of its surroundings after sufficient time has elapsed for the container and the surroundings to reach the same temperature (come to thermal equilibrium). The container may be submersed in a liquid such as water to thermostat the system (i.e., keep the temperature constant) and also to speed the attainment of thermal equilibrium.

The **number of moles** of gas, symbolized by n, is determined by measuring the mass (in grams) of the gas and then using the molecular mass to convert it to moles (n = g/molar mass = mol). Measuring the mass of any gas involves evacuating a vessel to a very low pressure, using a vacuum pump. The evacuated vessel is weighed, and then the gas sample is introduced to the vessel, which is weighed again to determine the increase in mass due to the gas.

Pressure, which is the force exerted per unit area, may be recorded in SI units of kilopascals (kPa) or in experimental units such as millimeters of mercury (also known as torr) or atmospheres (1 atm = 760 mm Hg). Pressures may be determined either by direct reading from an instrument called a pressure gauge or by using a **manometer**. Two types of manometer, the open-end and closed-end, are used. Both manometers are U-shaped pieces of glass with one end always open so that it can be attached to a vessel containing a gas. The other end may be open or closed, as the names suggest. The two types are shown in Figure 6.3.

The closed-end manometer (also called a **eudiometer**) is completely filled with mercury. When set up, the difference in the mercury levels gives the pressure, P_{gas} of whatever system is attached to the apparatus. If this device is not attached to an experimental setup but is left open to the atmosphere, it measures atmospheric pressure and is called a **barometer**.

The open-end manometer is usually filled with mercury, and each end is connected to gases at different pressures. The difference in pressures (ΔP), is equal to the difference in the levels of the liquid on the two sides of the manometer. This method gives only the difference in pressure between the two sides of the manometer. If, however, one side of the manometer is left open to the atmosphere, the difference represents the difference in pressure between the gas and the atmospheric pressure.

If a fluid other than mercury is used in a manometer, the difference in the heights of this fluid represents the pressure. However, in order to compare this measurement with the measurement obtained with a normal mercury manometer, the readings must be corrected for the relative densities of the fluid used and of mercury by using this equation:

$$\text{mm of Hg} = \text{mm of fluid}\left(\frac{\text{density of fluid}}{\text{density of Hg}}\right)$$

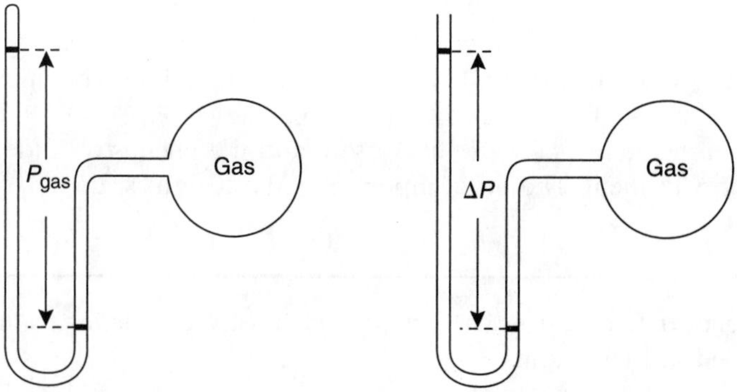

FIGURE 6.3. *Left: A closed-end manometer measures the gas pressure directly. Right: An open-end manometer measures the difference between the atmospheric pressure and the gas pressure.*

EXERCISE

Medical devices for anesthesiology must not restrict the air flow to the patient by more than 2.50 millimeters of mercury. Suggest a method for making such measurements accurately.

Solution

It is difficult to measure 2.50 mm Hg with accuracy in a mercury-filled manometer. If a water manometer is used, the difference in liquid levels will be much greater and hence will be easier to measure. For example, the difference in water levels equal to 2.50 mm Hg is calculated from the equation above as follows:

$$2.50 \text{ mm Hg} = \text{mm } H_2O \frac{1.00 \text{ g } H_2O/\text{mL } H_2O}{13.6 \text{ g Hg/mL Hg}}$$

$$\text{mm } H_2O = 34 \text{ mm } H_2O$$

It is much easier to measure 34 mm than 2.5 mm.

Temperature

Temperature Scales

In chemistry temperatures are measured by thermometers calibrated to read in degrees **Celsius**. On this scale, the melting point for water is 0.0°C, and the boiling point is 100°C. At 1.00 atmosphere of pressure, these points can be duplicated very easily. In fact, you can calibrate your thermometer by inserting it into an ice-water mixture for 0.00°C and by inserting it into boiling water for 100°C.

Other temperature scales are also used, and the chemistry student should be familiar with them. The most important is the **Kelvin** scale.

> Whenever temperature is part of a chemistry calculation, it must be expressed in Kelvin units.

The Kelvin temperature is easily calculated from the Celsius temperature by adding 273.16 to the Celsius temperature:

$$K = {}^\circ C + 273.16$$

(We often round this off and simply add 273 to the Celsius temperature.)

Another temperature scale is the **Fahrenheit** scale. Daniel Farenheit mixed salt and water to get a solution that had the lowest melting point possible and defined it as 0 on his scale. He also measured his body temperature and defined it as 100 on his scale. However, the salt and water solution crystallized at a temperature much below the freezing point of pure water. Also, Farenheit had a low-grade fever when determining his 100 degree point; the average normal body temperature is 98.6. It is clear that the Farenheit temperature scale was established on two factors that are not easily reproduced. Nevertheless, since many people use the Farenheit scale, the chemistry student must be able to convert between the Fahrenheit temperature, and the Celsius temperature, by using the equations

$$^\circ F = \frac{9}{5}\,^\circ C + 32^\circ$$

or to calculate the Celsius temperature, given the Farenheit temperature:

$$^\circ C = \frac{5}{9}(^\circ F - 32^\circ)$$

Temperature and Energy

The **temperature** of a substance defines the average kinetic energy of the substance. The specific heat of a substance is the conversion factor between temperature change and energy change for a system, so

Change in energy = (specific Heat) (grams of substance) (temperature change)

or

$$\Delta E = \text{sp. ht.} \times \text{g} \times \Delta T$$

The energy calculated in either of the preceding equations is called heat energy.

7
COMPOUNDS

NAMING COMPOUNDS

Ionic Compound Nomenclature

An **ionic compound** contains a metal and a nonmetal. (The ammonium ion, NH_4^+, is considered a metal, and polyatomic anions are considered nonmetals for these purposes.) Ionic compounds are also known as **salts.** An ionic compound is designated by giving the name of the cation first and then the name of the anion.

For a cation that has only one possible charge, the name is the same as the name of the element. When a cation may have more than one charge, as is true of most of the transition elements, the Stock system is used. In the **Stock system** the element name is followed by parentheses enclosing the charge written in roman numerals; examples are lead(II) and lead(IV). The Stock system is the preferred method for naming ionic compounds. An older system using Latin names and the suffixes *-ic* and *-ous* is being phased out. In this naming system, the higher charged cation is given the suffix *-ic* and the lower charged cation has the *-ous* ending.

Table 7.1 lists some ions with their Stock and old names. For reading and understanding the older chemical literature, this list of old names is useful.

TABLE 7.1. OLD AND STOCK NAMES FOR SOME COMMON IONS

Ion	Stock Name	Old Name
Fe^{2+}	Iron(II)	Ferrous
Fe^{3+}	Iron(III)	Ferric
Sn^{2+}	Tin(II)	Stannous
Sn^{4+}	Tin(IV)	Stannic
Pb^{2+}	Lead(II)	Plumbous
Pb^{4+}	Lead(IV)	Plumbic
Cu^{+}	Copper(I)	Cuprous
Cu^{2+}	Copper(II)	Cupric
Hg^{+} or Hg_2^{2+}	Mercury(I)	Mercurous
Hg^{2+}	Mercury(II)	Mercuric

The name given to an anion depends on whether the anion is a monatomic ion or a polyatomic anion. Monatomic anions are named by taking the root or first portion of the element name and then changing the ending to *-ide*, as shown in the list below. Polyatomic anions have unique names, given in Table 7.2, that must be memorized.

COMMON ANION NAMES

sulf**ur**	sulf**ide**
ox**ygen**	ox**ide**
chlor**ine**	chlor**ide**
nitr**ogen**	nitr**ide**
brom**ine**	brom**ide**
fluor**ine**	fluor**ide**

TABLE 7.2 COMMON POLYATOMIC IONS

Ion Formula	Ion Name
NH_4^+	ammonium ion
CO_3^{2-}	carbonate ion
HCO_3^-	bicarbonate ion
PO_4^{3-}	phosphate ion
ClO^-	hypochlorite ion
ClO_2^-	chlorite ion
ClO_3^-	chlorate ion
ClO_4^-	perchlorate ion
NO_2^-	nitrite ion
NO_3^-	nitrate ion
SO_3^{2-}	sulfite ion
SO_4^{2-}	sulfate ion
MnO_4^-	permanganate
$Cr_2O_7^{2-}$	dichromate ion
CrO_4^{2-}	chromate ion
$S_2O_3^{2-}$	thiosulfate ion

A compound is designated by writing the name of the cation followed by the name of the anion; the two names are separate words [e.g., MgF_2 is magnesium fluoride, and FeI_3 is iron(III) iodide]. Names of chemical compounds are not capitalized except at the beginning of a sentence.

EXERCISES

1. Name each of the following compounds:

(a) $MgCl_2$
(b) MnO_2
(c) Cr_2O_3
(d) Ca_3N_2
(e) Na_2CrO_4
(f) $Fe(ClO_2)_2$
(g) $Hg(NO_3)_2$
(h) $Al(NO_3)_3$
(i) $TiBr_4$
(j) Na_3PO_4
(k) $(NH_4)_2SO_3$
(l) $HgSO_4$
(m) $SnCl_4$
(n) BiF_3

2. Write the formulas for the following compounds:

(a) aluminum sulfate
(b) magnesium oxide
(c) vanadium(III) bromide
(d) barium nitrite
(e) cobalt(II) chloride
(f) gold(III) nitrate
(g) lithium sulfite
(h) ammonium phosphate
(i) strontium fluoride
(j) lead(IV) carbonate

Answers

1. (a) magnesium chloride
(b) manganese(IV) oxide
(c) chromium(III) oxide
(d) calcium nitride
(e) sodium chromate
(f) iron(II) chlorite
(g) mercury(II) nitrate
(h) aluminum nitrate
(i) titanium(IV) bromide
(j) sodium phosphate
(k) ammonium sulfite
(l) mercury(II) sulfate
(m) tin(IV) chloride
(n) bismuth(III) fluoride

To name a compound that contains a metal that may have more than one possible charge, we must know the charge on the ion. The charge is determined by "taking apart" the formula unit to learn the charges on the ions before they combined. This is possible because we will know the charge on the anion (it will be either a representative monatomic anion or one of the polyatomic anions in the list to be memorized). If the charge and the number of anions are known, the charge on the cation can be deduced since the formula must always have a net charge of zero. A review of how formulas are made will indicate how the process can be reversed is given on pages 103–104.

2. Writing formulas from names is often easier than writing names from formulas since the charges on the nonrepresentative elements are given in parentheses. Remember the requirement for electric neutrality: no net charge on any chemical compound.

(a) $Al_2(SO_4)_3$
(b) MgO
(c) VBr_3
(d) $Ba(NO_2)_2$
(e) $CoCl_2$
(f) $Au(NO_3)_3$
(g) Li_2SO_3
(h) $(NH_4)_3PO_4$
(i) SrF_2
(j) $Pb(CO_3)_2$

Nomenclature of Nonmetals

The naming of covalently bonded binary molecules (these are predominately molecules with two nonmetals) is quite different from the naming of ionic compounds, and the two methods should not be confused. In addition, many of these substances were discovered long before the modern

TABLE 7.3. TRIVIAL AND SYSTEMATIC NAMES OF SOME COMMON COVALENT MOLECULES

Molecule	Trivial Name	Systematic Name
H_2O	water	dihydrogen oxide*
CH_4	methane	carbon tetrahydride*
N_2O	nitrous oxide	dinitrogen oxide
NO	nitric oxide	nitrogen oxide
N_2O_3	nitrous anhydride	dinitrogen trioxide
N_2O_5	nitric anhydride	dinitrogen pentoxide
NH_3	ammonia	nitrogen trihydride*
AsH_3	arsine	arsenic trihydride
H_2O_2	hydrogen peroxide	dihydrogen dioxide
N_2H_4	hydrazine	dinitrogen tetrahydride

*These names are almost never used.

TABLE 7.4. PREFIXES USED IN NAMING NON-METAL COMPOUNDS

1	Mono-	6	Hexa-
2	Di-	7	Hepta-
3	Tri-	8	Octa-
4	Tetra-	9	Nona-
5	Penta-	10	Deca-

method of naming compounds was developed. Common covalent compounds often have trivial (older) and systematic (modern) names. Some important trivial names, along with the corresponding systematic names, are shown in Table 7.3.

In the systematic naming of covalent compounds, the prefixes listed in Table 7.4 indicate the number of atoms of each element present in a molecular formula. Some of these prefixes appear also in the trivial names in Table 7.3.

Thus, the formula for carbon dioxide is CO_2, and the name indicates one carbon and two (because of the prefix *di-*) oxygen atoms. Sulfur forms two compounds, SO_2 and SO_3, named sulfur dioxide and sulfur trioxide, respectively. Again, the *di-* and *tri-* indicate the number of oxygen atoms bound to the sulfur in each compound. The prefix *mono-* is rarely used as shown in the previous example. As another example, the compound N_2O_4 has two nitrogen and four oxygen atoms and is named dinitrogen tetroxide.

EXERCISES

1. Name each of the following covalent molecules:

(a) SiO_2 (d) SF_4 (g) $BrCl_3$

(b) P_4O_{10} (e) PBr_5 (h) S_4N_4

(c) N_2O_3 (f) XeF_4

2. Give the formula for each of the following compounds:
   (a) diboron tetrabromide
   (b) boron trifluoride
   (c) carbon tetrafluoride
   (d) carbon monoxide
   (e) diphosphorous pentoxide
   (f) carbon disulfide
   (g) sulfur trioxide
   (h) nitrogen triiodide

Answers

1. (a) sulfur dioxide
   (b) tetraphosphorus decoxide
   (c) dinitrogen trioxide
   (d) sulfur tetrafluoride
   (e) phosphorus pentabromide
   (f) xenon tetrafluoride
   (g) bromine trichloride
   (h) tetrasulfur tetranitride
2. (a) B_2Br_4
   (b) BF_3
   (c) CF_4
   (d) CO
   (e) P_2O_5
   (f) CS_2
   (g) SO_3
   (h) NI_{32}

Additional rules and names for compounds are given on pages 250–251 on acids and bases and in pages 251–256 on common organic compounds.

Acid-Base Nomenclature

Naming Acids

The first step in naming an acid is to identify the compound as an acid. In most introductory chemistry courses this task is simplified by writing the formulas of all acids with hydrogen as the first element in the formula. Once identified, acids can be divided into three groups:

Binary acids
Acids containing a polyatomic anion with a name ending in the suffix *-ate*
Acids containing a polyatomic anion with a name ending in the suffix *-ite*

Binary Acids: These acids contain hydrogen and one other atom. The names of all binary acids start with the prefix *hydro-* and end with the suffix *-ic* on the root of the second atom in the formula. Table 7.5 lists several binary acid names

TABLE 7.5. SOME BINARY ACID NAMES

Formula	Name
HF	hydrofluoric acid
HCl	hydrochloric acid
HI	hydroiodic acid
H_2S	hydrosulfuric acid

Acids with Polyatomic Anions Ending in *-ate*: When an acid has more than one other atom along with hydrogen, the other atoms are combined in a polyatomic anion. The common polyatomic anions are listed in Table 7.2 and should be memorized. To name an acid containing a polyatomic anion ending in *-ate*, the *-ate* ending of the anion is changed to *-ic* and the word *acid* is added, as shown in Table 7.6.

TABLE 7.6. SOME ACIDS OBTAINED FROM -*ATE* ENDING POLYATOMIC ANIONS

Polyatomic Anion	Anion Name	Acid Formula	Acid Name
SO_4^{2-}	sulfate	H_2SO_4	sulfuric acid
HSO_4^-	hydrogen sulfate	H_2SO_4	sulfuric acid
NO_3^-	nitrate	HNO_3	nitric acid
ClO_3^-	chlorate	$HClO_3$	chloric acid
ClO_4^-	perchlorate	$HClO_4$	perchloric acid
PO_4^{3-}	phosphate	H_3PO_4	phosphoric acid
HPO_4^{2-}	hydrogen phosphate	H_3PO_4	phosphoric acid
$H_2PO_4^-$	dihydrogen phosphate	H_3PO_4	phosphoric acid
$C_2H_3O_2^-$	acetate	$HC_2H_3O_2$	acetic acid
CO_3^{2-}	carbonate	H_2CO_3	carbonic acid
HCO_3^-	bicarbonate	H_2CO_3	carbonic acid
CrO_4^{2-}	chromate	H_2CrO_4	chromic acid

In this table we notice that the SO_4^{2-} and the HSO_4^{2-} anions both result in sulfuric acid, H_2SO_4.

Acids with Polyatomic Anions Ending in *-ite*: In a similar fashion, acids with polyatomic anions ending in *-ite* have names with *-ous* endings and the word *acid* added, as shown in Table 7.7.

TABLE 7.7. SOME ACIDS OBTAINED FROM -*ITE* ENDING POLYATOMIC ANIONS

Polyatomic Anion	Anion Name	Acid Formula	Acid Name
SO_3^{2-}	sulfite	H_2SO_3	sulfurous acid
HSO_3^-	hydrogen sulfite	H_2SO_3	sulfurous acid
NO_2^-	nitrite	HNO_2	nitrous acid
ClO_2^-	chlorite	$HClO_2$	chlorous acid
ClO^-	hypochlorite	$HClO$	hypochlorous acid

Here we notice that sulfurous acid is obtained from two related polyatomic anions, SO_3^{2-} and HSO_{3-}.

Organic Acids: These acids have both common and systematic names. The common name is based on the name of the polyatomic ion in the acid. For instance, the formate ion gives formic acid its common name. In the systematic name for an organic acid, the suffix *-oic* and the word *acid* are added to the name of the molecule.

The names and formulas of some organic acids are listed in Table 7.8.

TABLE 7.8. ORGANIC ACID NAMES

Systematic Name	Common Name	Formula
methanoic acid	formic acid	$HCOOH$
ethanoic acid	acetic acid	CH_3COOH
propanoic acid	propanoic acid	CH_3CH_2COOH
butanoic acid	butyric acid	$CH_3CH_2CH_2COOH$
pentanoic acid	valeric acid	$CH_3CH_2CH_2CH_2COOH$
benzoic acid	benzoic acid	C_6H_5COOH

Here we notice that the formulas for organic acids do not start with hydrogen. The formulas for all organic acids may be written in two forms, one with an initial hydrogen and the other with the organic functional group for acids, $-COOH$. For example, the formula for acetic acid can be written as $HC_2H_3O_2$ or as CH_3COOH.

Naming Bases

There are two types of bases, those that have hydroxide ions in their formulas and those that contain nitrogen. A hydroxide base is designated by using the name of the metal, with a roman numeral if necessary (see page 245), and then the word *hydroxide*. Thus $NaOH$, $Al(OH)_3$, and $Fe(OH)_3$ are named sodium hydroxide, aluminum hydroxide, and iron(III) hydroxide, respectively.

Nitrogen bases related to ammonia are amines. Replacing a hydrogen on ammonia with a methyl, $-CH_3$, group produces methylamine. If two methyl groups replace two hydrogen atoms, the compound is called dimethylamine.

The chloride salt of ammonia is called ammonium chloride. The name of the chloride salt of methylamine is methylammonium chloride or, alternatively, methylamine hydrochloride. Chloride salts of nitrogen bases with common names, such as hydrazine, usually use the hydrochloride ending, as in hydrazine hydrochloride.

Organic Compound Nomenclature

The original, or common, names of organic compounds were usually related to the source of the compound. Butyric acid is found in butter, caproic acid is isolated from goats, and vinegar is obtained from wine. When the number of known compounds caused this haphazard naming system to become unwieldy, the International Union of Pure and Applied Chemistry (IUPAC) developed the systematic nomenclature system. Consequently, many organic compounds have both common and systematic (IUPAC) names.

The systematic name for an organic compound consists of a prefix, a parent name, and a suffix. In most cases the parent name is based on the longest carbon chain in the molecule. For ring compounds, however, parent names are based on fundamental ring structures such as the benzene ring.

Steps in Naming an Organic Compound

Four steps are involved in constructing a name:

1. For the parent name, identify the longest continuous chain of carbon atoms in the structure. The longest chain is not always obvious, as Figure 7.1 indicates. Here only the carbon backbone is shown. The task is more difficult when all the hydrogen atoms are added to the structure, as in Figure 7.2.

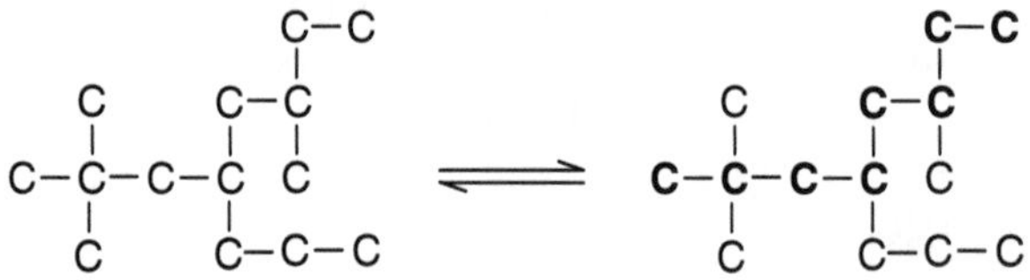

FIGURE 7.1. *The longest carbon chain for the structure on the left is eight carbon atoms, as shown in bold type on the right.*

2. Number the carbon atoms in the longest chain, starting at the end closest to a side chain or functional group. In Figure 7.2 the carbon atoms are numbered from left to right.

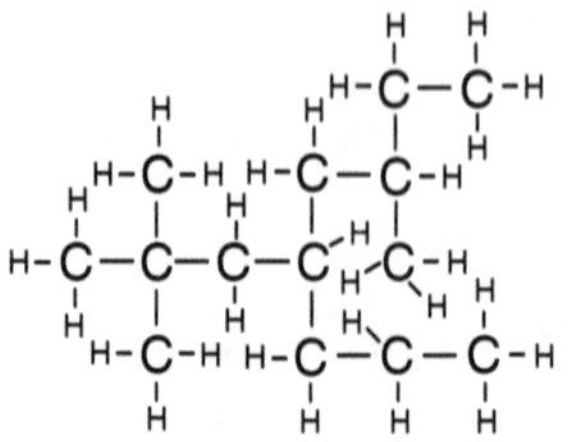

FIGURE 7.2. *Compound in Figure 7.1 with all hydrogen atoms attached.*

3. Identify the substituents and the number of the carbon atom to which each is attached. For the compounds in Figure 7.2:

$-CH_3$ on carbon 2	2-methyl
$-CH_3$ on carbon 2	2-methyl
$-CH_2CH_2CH_3$ on carbon 4	4-propyl
$-CH_3$ on carbon 6	6-methyl

4. The following suffixes are added for alcohol (-ol), aldehyde (-al), ketone (-one), and acid (-oic acid) functional groups. If identical functional groups are found, combine them and indicate the quantity by the prefix *di-*, *tri-*, *tetra-*, and so on.

The name of the compound in Figure 7.2 is 2,2,6-trimethyl-4-propyloctane.

EXERCISES

1. Name the following compound:

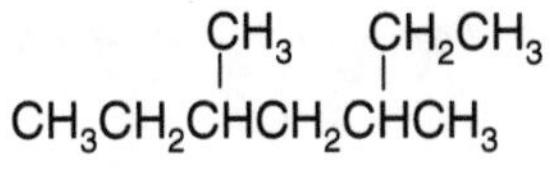

2. What is the name of the following molecule?

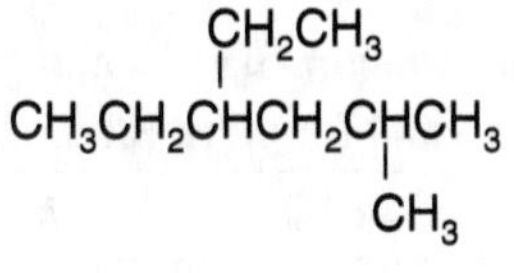

Solutions

1. We follow the four steps for constructing a name:
   1. The longest carbon chain has seven carbon atoms, so the parent name is heptane.
   2. This compound can be numbered from either the left or the right (the numbers for the side chains will be the same either way).
   3. The compound contains a 3-methyl and a 5-methyl group.
   4. Combining the methyl groups, we conclude that the name is 3,5-dimethylheptane.
2. Again, we follow the four steps:
   1. The longest chain contains six carbon atoms, so the parent name is hexane.

2. The carbon atoms are numbered from right to left (this gives the lowest numbers for the side chains).
3. The substituents are 2-methyl and 4-ethyl.
4. The name is 2-methyl-4-ethylhexane.

As stated in Step 4 above, systematic nomenclature for the alcohols, ketones, and acids uses suffixes. For alcohols the suffix is #-ol; for ketones, #-one; and for acids, #-oic acid. The # symbol represents the number of the carbon atom to which the functional group is attached. To illustrate this principle, Figure 7.3 shows the structures and systematic names for three different alcohols on a pentane carbon chain.

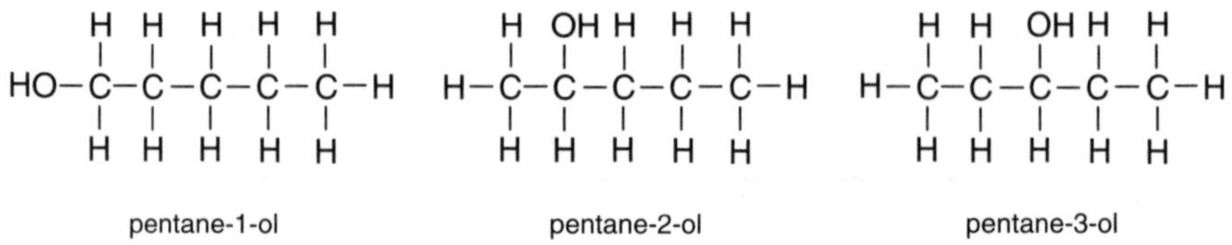

FIGURE 7.3. *Systematic names for the three alcohols made from* n*-pentane. Their common names are 1-pentanol, 2-pentanol, and 3-pentanol, respectively.*

The simpler organic compounds are often designated by common names. Some of these are summarized in Table 7.9.

TABLE 7.9. COMMON NAMES FOR STRAIGHT-CHAIN ORGANIC COMPOUNDS WITH ONE TO FIVE CARBON ATOMS AND THE GIVEN FUNCTIONAL GROUPS

Alkane	Acid	Alcohol	Aldehyde	Amine
methane	formic acid	methanol	formaldehyde	methylamine
ethane	acetic acid	ethanol	acetaldehyde	ethylamine
propane	propanoic acid	1-propanol	propionaldehyde	propylamine
n-butane	butanoic acid	1-butanol	butyraldehyde	butylamine
n-pentane	pentanoic acid	1-pentanol	pentanaldehyde	1-aminopentane

Ring Compounds

Benzene and cyclohexane are the simplest and most common parent structures for ring compounds. Because a ring is circular, the numbering of the carbon atoms follows a different procedure. If there is only one substituent, no number is necessary and the substituent is assumed to be attached to carbon 1. Chlorobenzene is a compound in which a chlorine atom replaces one of the hydrogen atoms of the benzene ring. When two or more substituents are present on a ring, the carbon atoms are numbered clockwise, so that one substituent is on carbon 1 and the remaining substituents are on the lowest numbered carbon atoms possible. In Figure 7.4 two dichlorobenzene molecules shows the correct and incorrect methods of numbering.

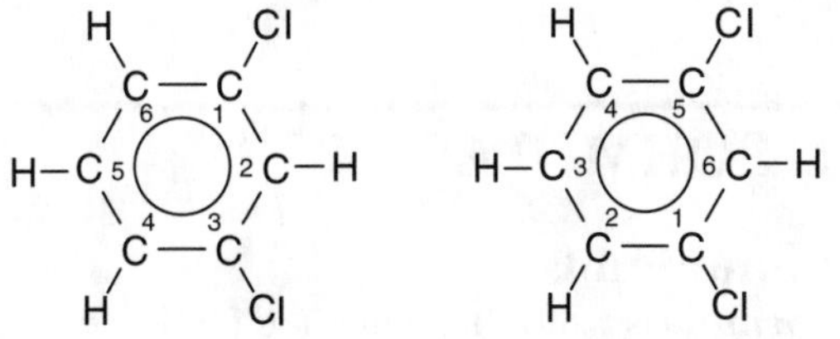

FIGURE 7.4. *The correct (left) and incorrect (right) methods for numbering a ring structure.*

Once the correct numbering sequence is determined, a ring compound is named in the same way as an alkane.

A frequently used system uses the prefix *ortho-*, *meta-*, or *para-* to indicate the position of the second substituent on a ring. Since the first substituent is always at carbon 1, a substituent in the ortho position is on carbon 2. The meta position is carbon 3, and the para position is carbon 4. 1,3-Dichlorobenzene, shown in Figure 7.4, is also known as meta-dichlorobenzene.

EXERCISE

Draw the structure for para-aminobenzoic acid, the most common sunscreen chemical sold commercially.

Solution

The para position is position 4 on the benzene ring. This structure should have the carboxyl group and the amino group on opposing carbons of the benzene ring, as shown in Figure 7.5.

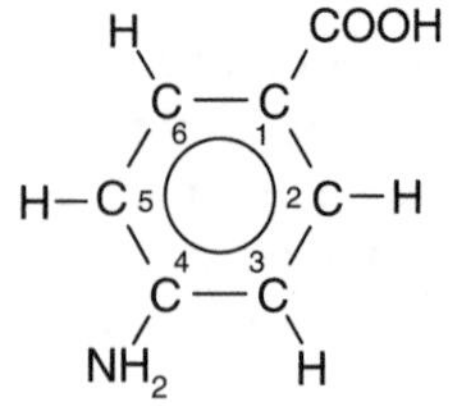

FIGURE 7.5. *Structure of para-aminobenzoic acid.*

CHEMICAL FORMULAS

Formulas of Compounds

The **chemical formula** is a shorthand method for describing the compounds formed in a chemical reaction. It uses the symbols in the Periodic Table to designate the elements that are in the compound. A subscript to the right of the symbol indicates how many atoms of each element are in the compound.

> The ability to count the number of atoms of each element in a chemical formula is absolutely essential to success in chemistry.

For example, potassium permanganate contains:

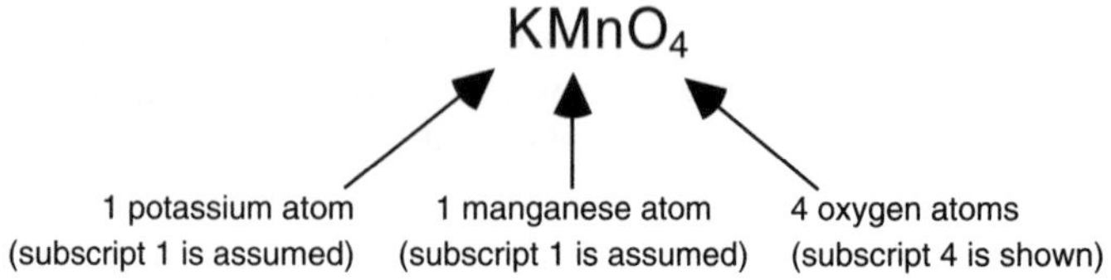

Parentheses in chemical formulas are used to clarify and to provide additional information. A subscript placed after a parentheses multiplies everything within the parentheses. For example, aluminum nitrate contains:

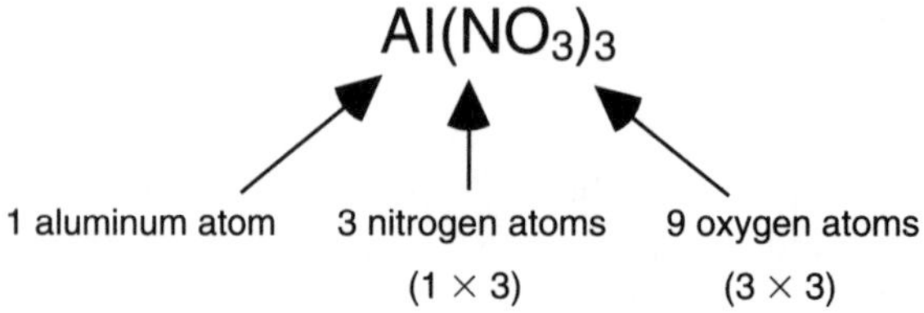

This formula could have been written as AlN_3O_9, which represents the same number of each atom as $Al(NO_3)_3$. However, the parentheses around the (NO_3) provide the added information that the nitrogen and oxygen atoms are in three groups of NO_3 units. (As was shown in the section on polyatomic ions, the NO_3 is the nitrate ion, NO_3^-.)

The formula for ammonium phosphate:

$$(NH_4)_3PO_4$$

shows that this compound contains 3 nitrogen, 12 hydrogen, 1 phosphorus, and 4 oxygen atoms. Here the parentheses show three ammonium groups (actually NH_4^+ ions).

Another type of formula is used for compounds called hydrates. These compounds have fixed numbers of water molecules, called the water of hydration, in their crystal lattices. To show the water of hydration clearly in the chemical formula, it is written after a dot that is placed in the middle of the line. The dot links two separate compounds into one formula unit.

For example, the hexahydrate of cobalt(II) chloride is written as and contains

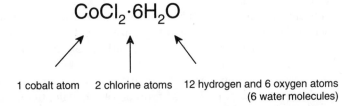

$$CoCl_2{\cdot}6H_2O$$

1 cobalt atom ↗ 2 chlorine atoms ↑ 12 hydrogen and 6 oxygen atoms (6 water molecules) ↖

The compound

$$Na_2SO_4{\cdot}10H_2O$$

contains 2 sodium, 1 sulfur, 14 oxygen, and 20 hydrogen atoms.

The names of these compounds are cobalt(II) chloride hexahydrate and sodium sulfate decahydrate. Hexahydrate indicates six water molecules in the formula; decahydrate, ten water molecules. Common prefixes for hydrates are listed in Table 7.10.

TABLE 7.10. PREFIXES USED WITH HYDRATES

1	mono-	6	hexa-
2	di-	7	hepta-
3	tri-	8	octa-
4	tetra-	9	nona-
5	penta-	10	deca-

The formulas for organic compounds can be written in a variety of ways. For example, *n*-decane has the formula $C_{10}H_{22}$, which can also be written as $CH_3(CH_2)_8CH_3$ or as $CH_3CH_2CH_2CH_2CH_2CH_2CH_2CH_2CH_2CH_3$. All of these formulas have the same numbers of carbon and hydrogen atoms, but the numbers are presented in different ways. No matter which way the formula for *n*-decane is written, however, it always contains 10 carbon atoms and 22 hydrogen atoms.

Chemical Formula Types

Empirical Formulas

A compound that is ionic in nature has a formula that represents the simplest ratio of the atoms in a crystal of that substance. This simplest ratio of atoms is called an **empirical formula**. For instance, potassium bromide is assigned the formula KBr. In a crystal, there are no distinct molecules containing one potassium ion and one bromide ion. In fact, each potassium ion is attracted to six oppositely charged bromide ions. At the same time, each bromide ion is attracted to six neighboring potassium ions. A feature of these ionic substances is that no distinct pairs of potassium and bromine atoms, or ions, are observed in the solid, liquid, or gas state or in solution. Instead, the individual ions K^+ and Br^- are observed.

Empirical formulas can also be written for covalently bonded molecules. For instance, the empirical formula of benzene, C_6H_6, is CH; the simplest ratio of the carbon and hydrogen atoms. Empirical formulas of compounds can be determined from analytical information about the percentage composition of the substance. Knowledge of the molar mass then allows the chemist to determine the actual molecular formula of a covalent compound.

Molecular Formulas

For compounds that have covalent bonds **molecular formulas** are used. Benzene has a molecular formula of C_6H_6, and acetic acid has a formula of $HC_2H_3O_2$. These are not empirical formulas; they represent the actual number of each atom present in a single molecule of each of these compounds. The covalent bonds are not easily broken, and the entire group of atoms is usually intact in the solid, liquid, and gas states. In solution, the molecule often remains intact.

Another form of molecular formula is the **structural formula**. The structural formula shows the chemist how the atoms are connected to each other and the covalent bonds between the atoms. The structural formulas for benzene and acetic acid are shown in Figure 7.6. Structural formulas are used when the chemist wishes to show more detail about a molecule than is possible with the simple molecular formula.

With the advent of desktop publishing and readily available computer software, the **three-dimensional structure** has become popular. This type of structure shows a molecule as it actually exists. For example, butane has the formula C_4H_{10}. Its expanded formula can be written as $CH_3CH_2CH_2CH_3$, and the structural formula can be shown in Figure 7.7.

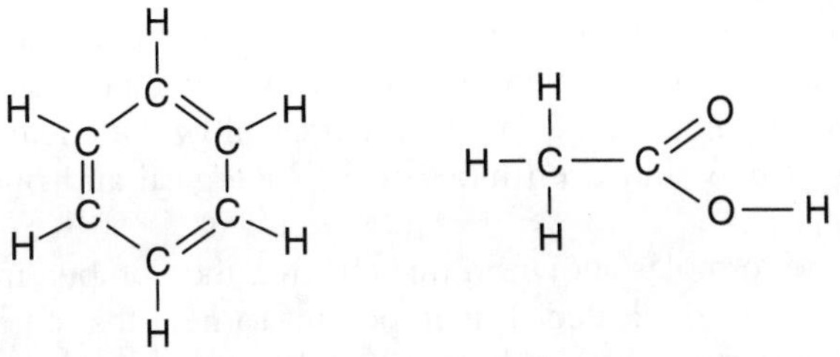

FIGURE 7.6. *Structural formulas for benzene (left) and acetic acid (right).*

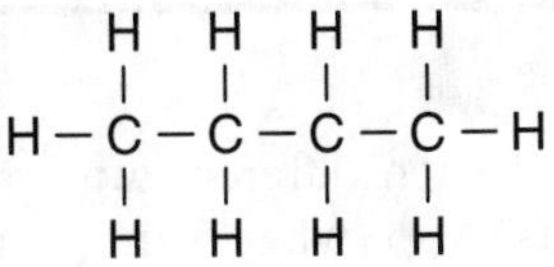

FIGURE 7.7. *Structural formula of butane, C_4H_{10}*

However, each carbon atom is bonded to four other atoms; therefore, their geometries are all tetrahedral (*sp^3* hybrid). When the four tetrahedral structures are placed together, the three-dimensional structure shown in Figure 7.8 can be drawn.

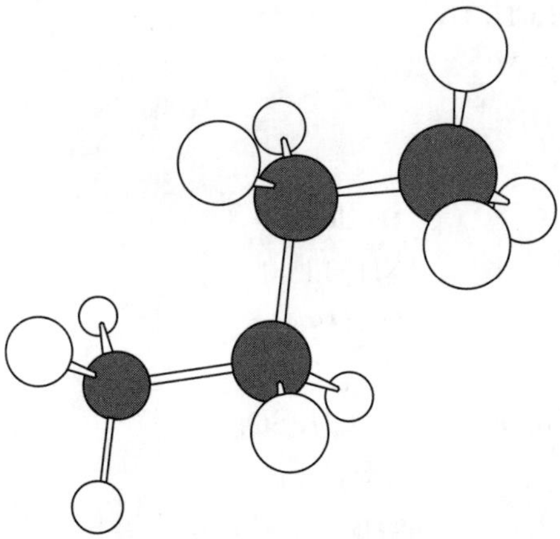

FIGURE 7.8. *Three-dimensional computer-generated structure for butane, showing the tetrahedral arrangement of atoms around each carbon (shaded circles).*

To understand the functions of many important biologically active molecules (proteins, DNA, etc.) and of many pharmaceutical compounds, three-dimensional structures are important. These structures allow the chemist to visualize the possibility of new and even more active biological and pharmacological molecules.

In each of the formulas shown in this chapter, the number and the type of each atom are clearly indicated. It is important to note that all the structures, whether molecular, structural, or three-dimensional, have the same atoms. The chemist chooses the structural representation in order to clearly illustrate the nature of the molecule.

EXERCISE

Determine the number of each different atom represented in each of the following chemical formulas. Also, what is the empirical formula for each of these substances?

(a) $NaClO_4$

(b) $NH_4C_2H_3O_2$

(c) LiH_2AsO_3

(d) $Ca(C_3H_5O_3)_2 \cdot 5H_2O$

(e) $Cu(NH_3)_4SO_4 \cdot H_2O$

(f) $C_{10}H_{18}$

(g) $CH_3(CH_2)_4CHCH_2$

(h) $CH_3OCH_2CH_3$

(i) $CHCCH(OH)CH_2CHCH_2$

(j) Butane in Figure 7.8

Answers

(a) 1Na, 1Cl, 4O	$NaClO_4$
(b) 1N, 7H, 2C, 2O	$NH_4C_2H_3O_2$
(c) 1Li, 2H, 1As, 3O	LiH_2AsO_3
(d) 1Ca, 6C, 20H, 11O	$CaC_6H_{20}O_{11}$
(e) 1Cu, 4N, 14H, 1S, 5O	$CuN_4H_{14}SO_5$
(f) 10C, 18H	C_5H_9
(g) 7C, 14H	CH_2
(h) 3C, 8H, 1O	C_3H_8O
(i) 6C, 8H, 1O	C_6H_8O
(j) 4C, 10H	C_2H_5

PERCENT COMPOSITION

Percentages of Elements in Compounds

The **percent composition** of a chemical substance tells the chemist how much of each element or each polyatomic ion is present in a compound on a percent basis. In other words, the percent composition is the mass of an element or polyatomic ion that is present in 100 grams of a chemical compound.

EXERCISE

What is the percentage of (a) iron, (b) oxygen, (c) hydrogen, and (d) the OH^- polyatomic ion in $Fe(OH)_3$?

Solution

Each calculation will start with 100 g of $Fe(OH)_3$, and we calculate the grams of each element and the ion in turn to obtain the percentages.

$$\text{(a) ? g Fe} = 100\text{ g Fe(OH)}_3\left(\frac{1\text{ mol Fe(OH)}_3}{106.8\text{ g Fe(OH)}_3}\right)\left(\frac{1\text{ mol Fe}}{1\text{ mol Fe(OH)}_3}\right)\left(\frac{55.8\text{ g Fe}}{1\text{ mol Fe}}\right)$$

$$= 52.2\text{ g Fe} = 52.2\%\text{ Fe}$$

$$\text{(b) ? g O} = 100\text{ g Fe(OH)}_3\left(\frac{1\text{ mol Fe(OH)}_3}{106.8\text{ g Fe(OH)}_3}\right)\left(\frac{3\text{ mol O}}{1\text{ mol Fe(OH)}_3}\right)\left(\frac{16.0\text{ g O}}{1\text{ mol O}}\right)$$

$$= 44.9\text{ g O} = 44.9\%\text{ O}$$

$$\text{(c) ? g H} = 100\text{ g Fe(OH)}_3\left(\frac{1\text{ mol Fe(OH)}_3}{106.8\text{ g Fe(OH)}_3}\right)\left(\frac{3\text{ mol H}}{1\text{ mol Fe(OH)}_3}\right)\left(\frac{1.0\text{ g H}}{1\text{ mol H}}\right)$$

$$= 2.8\text{ g H} = 2.8\%\text{ H}$$

$$\text{(d) ? g OH}^- = 100\text{ g Fe(OH)}_3\left(\frac{1\text{ mol Fe(OH)}_3}{106.8\text{ g Fe(OH)}_3}\right)\left(\frac{3\text{ mol OH}^-}{1\text{ mol Fe(OH)}_3}\right)\left(\frac{17.0\text{ g OH}^-}{1\text{ mol OH}^-}\right)$$

$$= 47.8\text{ g OH}^- = 47.8\%\text{ OH}^-$$

The percentages of the elements add up to 99.9%. The sum should equal 100%, but does not because the numbers were rounded to one decimal place.

Another method for calculating the percentage composition is to use this formula:

$$\text{Percent of element} = \frac{\text{atomic mass of number of atoms in formula}}{\text{molar mass of compound}} \times 100$$

A close look at the stoichiometry equations in the exercise above shows that the equation given above is just a summary of these stoichiometric calculations.

EXERCISES

1. What is the percentage of each element in (a) $Ca(NO_3)_2$ and (b) $CH_3CH_2NH_2$?
2. A 1.376-gram sample of a compound composed of only carbon, hydrogen, and oxygen is burned, and 3.411 grams of CO_2, along with 1.221 grams of H_2O, are produced. What is the percentage of (a) C, (b) H, and (c) O in the compound?

Answers

1. (a) Ca = 34.52%, N = 24.13%, O = 41.35%

   (b) C = 53.28%, H = 15.65%, N = 31.07%

   Finally, if the formula of a chemical compound is not known, it may be determined by chemical analysis. For example, controlled burning of an organic compound that contains only carbon, hydrogen, and oxygen and measurement of the CO_2 and H_2O produced will allow us to calculate the carbon and hydrogen percentages. Then the oxygen content is calculated by difference.

Solutions

2. (a) $? \text{ g C} = 3.411 \text{ g } CO_2 \left(\frac{1 \text{ mol } CO_2}{44 \text{ g } CO_2}\right)\left(\frac{12 \text{ g C}}{1 \text{ mol C}}\right)$

$$= 0.930 \text{ g C}$$

$$? \text{ \% C} = \frac{0.930 \text{ g C}}{1.376 \text{ g}} \times 100$$

$$= 67.6\% \text{ C}$$

$$(b)\ ?\,g\,H = 1.221\,g\,H_2O\left(\frac{1\,mol\,H_2O}{18\,g\,H_2O}\right)\left(\frac{2\,mol\,H}{1\,mol\,H_2O}\right)\left(\frac{1\,g\,H}{1\,mol\,H}\right)$$

$$= 0.136\,g\,H$$

$$?\,\%\,H = \frac{0.136\,g\,H}{1.376\,g} \times 100$$

$$= 9.86\%\,H$$

(c) The percentage of O is obtained by subtracting the percentages of C and H from 100:

$$?\,\%\,O = 100.0 - (67.6 + 9.86)$$

$$= 22.5\%\,O$$

EMPIRICAL FORMULAS

Empirical Formulas of Compounds

The empirical formula is the simplest ratio of atoms in a chemical compound. For example, butyric acid has the molecular formula $HC_4H_7O_2$; its empirical formula is C_2H_4O. Benzene, C_6H_6, has the empirical formula CH. For the sugar glucose, $C_6H_{12}O_6$ the empirical formula is CH_2O. The fact that all sugars have the same empirical formula is the reason why they are called carbohydrates (*carbo-* for the carbon and *-hydrate* for the water molecule).

Determining Empirical Formulas

Early chemists realized that the empirical formula of a compound could be determined from only the atomic masses and the percentage composition of the compound. From this information the simplest whole-number ratio of the moles of all the atoms in a compound is found by using the following steps:

1. Calculate the moles of each element in a compound from the given data.
2. Divide the number of moles of each element found in Step 1 by the smallest of these values, to obtain whole-number subscripts for the empirical formula.
3. If Step 2 does not give whole numbers (within ±0.1), the decimal portion of the number will be close to a rational fraction: 0.5 = ½, 0.33 = ⅓, 0.67 = ⅔, and so on. In this case, each item of data is multiplied

by the denominator of the rational fraction to remove the fractions and obtain whole-number subscripts.

EXERCISES

1. What is the empirical formula of a compound that contains 4.0 grams of calcium and 7.1 grams of chlorine?

2. A compound containing carbon, hydrogen, and oxygen is found to contain 9.1 percent hydrogen and 54.5 percent carbon. What is its empirical formula?

3. A compound is analyzed and found to contain 74.1 percent oxygen and 25.9 percent nitrogen. What is its empirical formula?

4. Determine the empirical formula of each of the following compounds, given its composition:

   (a) A compound composed of 17.72 grams of chlorine and 3.10 grams of phosphorus

   (b) A compound that is 24.74 percent potassium, 40.50 percent oxygen, and 34.76 percent manganese

   (c) A compound containing carbon, hydrogen, and oxygen that is 40.0 percent carbon and 6.66 percent hydrogen

Solutions

1. The first step involves converting the grams of calcium and chlorine to moles:

$$\begin{aligned} ?\ \text{mol Ca} &= 4.0\ \text{g Ca}\left(\frac{1\ \text{mol Ca}}{40.0\ \text{g Ca}}\right) \\ &= 0.10\ \text{mol Ca} \\ ?\ \text{mol Cl} &= 7.1\ \text{g Cl}\left(\frac{1\ \text{mol Cl}}{35.5\ \text{g Cl}}\right) \\ &= 0.20\ \text{mol Cl} \end{aligned}$$

The second step is to divide each of these results by the smaller value, 0.10:

$$\frac{0.10\ \text{mol Ca}}{0.10} = 1\ \text{mol Ca}$$

$$\frac{0.20\ \text{mol Cl}}{0.10} = 2\ \text{mol Cl}$$

These results tell us that the empirical formula contains 1 mol of Ca and 2 mol of Cl and is written as $CaCl_2$. This is also the formula for calcium chloride since $CaCl_2$ is expected to be an ionic compound and the empirical formula of an ionic compound is also its actual formula.

2. Since the percentage of an element is the number of grams per 100 g of compound, we may assume a 100-g sample of compound and convert the percents to grams directly. Since the percent hydrogen and percent carbon do not add up to 100%, we conclude that the missing 36.4% is oxygen. Step 1 is used to calculate the number of moles of each element:

$$? \text{ mol C} = 54.5 \text{ g C}\left(\frac{1 \text{ mol C}}{12 \text{ g C}}\right) = 4.54 \text{ mol C}$$

$$? \text{ mol H} = 9.1 \text{ g H}\left(\frac{1 \text{ mol H}}{1 \text{ g H}}\right) = 9.1 \text{ mol H}$$

$$? \text{ mol O} = 36.4 \text{ g O}\left(\frac{1 \text{ mol O}}{16 \text{ g O}}\right) = 2.28 \text{ mol O}$$

Using Step 2 and dividing each result by the smallest number, 2.28, we obtain

$$\frac{4.54 \text{ mol C}}{2.28} = 1.99 \text{ mol C}$$

$$\frac{9.1 \text{ mol H}}{2.28} = 3.99 \text{ mol H}$$

$$\frac{2.28 \text{ mol O}}{2.28} = 1.00 \text{ mol O}$$

Since these numbers are within ± 0.1 of a whole number, they are rounded to 2 mol C, 4 mol H, and 1 mol O, and the empirical formula is written as C_2H_4O.

3. Assuming a 100-g sample, we convert the percentages directly into gram units and the moles of N and O are calculated as

$$? \text{ mol N} = 25.9 \text{ g N}\left(\frac{1 \text{ mol N}}{14 \text{ g N}}\right) = 1.85 \text{ mol N}$$

$$? \text{ mol O} = 74.1 \text{ g N}\left(\frac{1 \text{ mol O}}{16 \text{ g O}}\right) = 4.63 \text{ mol O}$$

Dividing both answers by 1.85, Step 2, gives

$$\frac{1.85 \text{ mol N}}{1.85} = 1.00 \text{ mol N}$$

$$\frac{4.63 \text{ mol O}}{1.85} = 2.50 \text{ mol O}$$

The 2.50 mol O cannot be rounded to a whole number, but the decimal fraction 0.50 represents the rational fraction ½. Step 3 must be used; all of the data are multiplied by the denominator of the rational fraction, which in this case is 2:

$$1.00 \text{ mol N} \times 2 = 2.00 \text{ mol N}$$

$$2.50 \text{ mol O} \times 2 = 5.00 \text{ mol O}$$

These whole numbers are used to write the empirical formula, N_2O_5, of this compound.

Answers

4. (a) PCl_5 (b) $KMnO_4$ (c) CH_2O

Writing Empirical Formulas

In empirical formulas, the least electronegative elements are written first. For simplicity, the first element should be the one closest to the left edge of the Periodic Table, and the last element the one closest to the right edge of the Periodic Table. In empirical formulas for organic compounds, carbon is always first, hydrogen second, and oxygen third; other elements, if present, are then added in order of increasing electronegativity.

Laboratory Determination of Empirical Formulas

Chemists use a variety of methods for determining the compositions of compounds. These data may then be used to calculate the empirical formula. One method to determine composition is combustion analysis. An accurately weighed sample of an organic compound is burned in oxygen in the presence of a catalyst to convert it completely into CO_2 and H_2O. The carbon dioxide is absorbed onto asbestos saturated with sodium hydroxide, and the water is absorbed onto calcium carbonate. The increases in mass of the asbestos and the calcium carbonate indicate the masses of the two products. Once the masses of CO_2 and H_2O are known, the empirical formula can be determined.

EXERCISES

1. When a 0.200-gram sample of an organic compound is burned, 0.357 gram of CO_2 and 0.146 gram of H_2O are produced. What is the empirical formula of this compound?

2. A 0.150 gram sample of a compound containing only chromium and chlorine is titrated with 15.6 milliliters of 0.156 M $AgNO_3$. What is the empirical formula of this compound?

Solutions

1. We calculate the grams of carbon from the CO_2 and the grams of hydrogen from the H_2O. The grams of oxygen are determined by adding the grams

of H and the grams of C and subtracting this sum from the total weight of the sample:

$$? \text{ g C} = 0.357 \text{ g } CO_2 \left(\frac{1 \text{ mol } CO_2}{44 \text{ g } CO_2}\right)\left(\frac{1 \text{ mol C}}{1 \text{ mol } CO_2}\right)\left(\frac{12 \text{ g C}}{1 \text{ mol C}}\right)$$

$$= 0.0974 \text{ g C}$$

$$? \text{ g H} = 0.146 \text{ g } H_2O \left(\frac{1 \text{ mol } H_2O}{18 \text{ g } H_2O}\right)\left(\frac{2 \text{ mol H}}{1 \text{ mol } H_2O}\right)\left(\frac{1 \text{ g H}}{1 \text{ mol H}}\right)$$

$$= 0.0162 \text{ g H}$$

$$\text{g O} = 0.200 \text{ g total} - (0.0974 \text{ g C} + 0.0162 \text{ g H})$$

$$= 0.0864 \text{ g O}$$

Now the moles of C, H, and O are calculated:

$$? \text{ mol C} = 0.0974 \text{ g C}\left(\frac{1 \text{ mol C}}{12 \text{ g C}}\right) = 0.00812 \text{ mol C}$$

$$? \text{ mol H} = 0.0162 \text{ g H}\left(\frac{1 \text{ mol H}}{1 \text{ g H}}\right) = 0.0162 \text{ mol H}$$

$$? \text{ mol O} = 0.0864 \text{ g O}\left(\frac{1 \text{ mol O}}{16 \text{ g O}}\right) = 0.00540 \text{ mol O}$$

The smallest number of moles is 0.00540, and we divide each result by 0.00540:

$$\frac{0.00812 \text{ mol C}}{0.00540} = 1.503 \text{ mol C}$$

$$\frac{0.0162 \text{ mol H}}{0.00540} = 3.00 \text{ mol H}$$

$$\frac{0.00540 \text{ mol O}}{0.00540} = 1.00 \text{ mol O}$$

Finally, we multiply all of these results by a factor of 2 to eliminate the fraction and obtain the empirical formula $C_3H_6O_2$.

2. Chromium is a transition element that can have more than one oxidation state, and therefore the formula of its chloride cannot be readily predicted. The reaction of silver nitrate, $AgNO_3$, with chloride ions is:

$$AgNO_3(aq) \quad + \quad Cl^-(aq) \quad \rightarrow \quad AgCl(s) \quad + \quad NO_3^-(aq)$$

$$? \text{ g Cl} = 15.6 \text{ mL AgNO}_3 \left(\frac{0.156 \text{ mol AgNO}_3}{1000 \text{ mL AgNO}_3}\right)\left(\frac{1 \text{ mol Cl}^-}{1 \text{ mol AgNO}_3}\right)\left(\frac{35.45 \text{ g Cl}^-}{1 \text{ mol Cl}^-}\right)$$

$$= 0.08627 \text{ g Cl}$$

$$? \text{ g Cr} = 0.150 \text{ g total} - 0.08627 \text{ g Cl}$$

$$= 0.06373 \text{ g Cr}$$

The moles of Cr and Cl are then calculated as

$$? \text{ mol Cr} = 0.06373 \text{ g Cr}\left(\frac{1 \text{ mol Cr}}{51.99 \text{ g Cr}}\right) = 0.001226 \text{ mol Cr}$$

$$? \text{ mol Cl} = 0.08627 \text{ g Cl}\left(\frac{1 \text{ mol Cl}^-}{35.45 \text{ g Cl}^-}\right) = 0.002434 \text{ mol Cl}$$

Dividing both values by the smaller number, 0.001226, gives 1 mol Cr and 2 mol Cl for an empirical formula of $CrCl_2$. You may wish to review titration calculations starting on page 352.

Molecular Formulas

An **empirical formula** gives the simplest ratio of atoms in a molecule. For ionic compounds, the empirical formula is used to represent the compound. For molecular covalent compounds, the actual **molecular formula** may be the empirical formula or a whole-number multiple of the empirical formula. Once the empirical formula has been determined as shown on pages 263–267, the molar mass of the compound can be used to determine the molecular formula.

The number of empirical formula units in a molecular formula is determined by dividing the molar mass by the mass of the empirical formula. The result will be a small whole number:

$$\frac{\text{Molar mass}}{\text{Empirical formula mass}} = \text{small whole number} \tag{7.1}$$

Each subscript in the empirical formula is then multiplied by this small whole number to obtain the molecular formula.

EXERCISES

1. A compound has the empirical formula CH_2O, and its molar mass is determined in a separate experiment to be 180 grams per mole. What is the molecular formula of this compound.
2. For each of following empirical formulas, the molar mass is given in parentheses after the formula. Determine each corresponding molecular formula.

   (a) C_3H_7 (86)

(b) CH_2 (70)
(c) $C_4H_3O_2$ (166)
(d) BH_3 (27.7)
(e) CH_2ON (176)

3. What methods can be used to determine the molar masses used in problems similar to the preceding exercise?

Solutions

1. The number of CH_2O units in the molecule is determined from equation 7.1:

$$\frac{180 \text{ g mol}^{-1}}{30 \text{ g/empirical formula unit}} = 6 \text{ empirical formula units per mole}$$

Each subscript, including the unwritten 1's, in the empirical formula is multiplied by 6 to obtain $C_6H_{12}O_6$.

2. (a) C_6H_{14} (b) C_5H_{10} (c) $C_8H_6O_4$ (d) B_2H_6 (e) $C_4H_8O_4N_4$

3. There are several methods:
   a. The density of a gas at STP can be converted into the molar mass (see page 63).
   b. Osmotic pressure can be used to determine the molar masses of large molecules (see pages 293–295).
   c. Freezing point depression and boiling point elevation are often used for small, soluble, compounds (see pages 289–292).
   d. Vapor-pressure measurements of ideal mixtures can be used (see pages 295–299)
   e. Titrations can be used to determine equivalent weights (see pages 352–356).

LEWIS STRUCTURES

Lewis Electron Dot Structures

Covalent Bonds

The covalent bond represents the second way in which elements can attain a noble is gas electronic configuration; ionic bonding is the other method. The covalent bond involves sharing of electrons between two or more atoms. When electrons are shared, the atoms are firmly bonded together into individual **molecules**.

The sharing of electrons is usually represented by drawing Lewis electron-dot structures. In the Lewis representation the outermost *s* and *p* electrons (valence electrons) are shown as dots arranged around the atomic symbol. The Lewis structures for the elements most commonly used are shown in Figure 7.9.

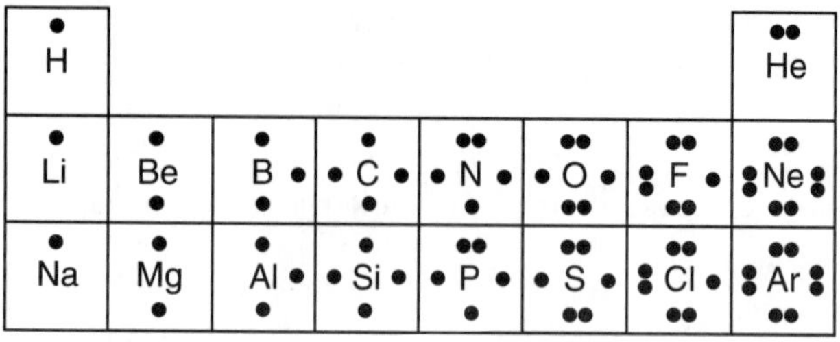

FIGURE 7.9. *Elements normally represented with Lewis electron-dot symbols.*

When electrons are shared between atoms, one atom donates one electron and the other atom donates the second electron. The shared pair of electrons represents a **covalent bond**. If two pairs of electrons are shared, a **double bond** exists; if three pairs are shared, a **triple bond** is formed.

The sharing of electrons follows the same basic principle that governs the formation of ions: the atoms are trying to attain a noble-gas electron configuration. There are two *s* electrons and six *p* electrons in each complete sublevel of a noble gas. These eight electrons represent the octet of the octet rule that governs covalent compounds. The **octet rule** states that the noble-gas configuration will be achieved if the Lewis structure shows eight electrons around each atom. Hydrogen is an exception; its "octet" consists of two electrons, similar to the two outermost electrons in the noble gas helium.

Below are the Lewis electron dot structures for some diatomic gases. In each, dots represent the electrons of one atom, and small circles represent the electrons of the other atom:

In these molecules, each atom considers the shared electrons as its own. Each hydrogen in H_2 "thinks" it has two electrons and an electron configuration like that of helium. Nitrogen molecules have triple bonds with three pairs of shared electrons. Fluorine and chlorine look very similar, as expected since both are halogens.

Lewis structures of other molecules and polyatomic ions may be drawn using the basic octet rule. For larger molecules it is first necessary to determine the general arrangement of the atoms, often called the "skeleton." In most molecules this skeleton consists of a central atom with surrounding atoms bonded to it. Five general rules apply in determining the skeleton:

1. Carbon is usually a central atom in the structure. In compounds with more than one carbon atom, the carbon atoms are joined in a chain to start the skeleton.
2. Hydrogen is never a central atom because it can form only one covalent bond.
3. A halogen forms only a single covalent bond when oxygen is not present and therefore will generally not be a central atom.
4. Oxygen forms only two covalent bonds and is rarely a central atom. However, it may link two carbon atoms in a carbon chain.
5. In the simpler molecules, the atom that appears only once in the formula will be the central atom.

Once the molecular skeleton is determined, the available valence electrons must be arranged in octets around each atom. This is done in steps as follows:

1. The valence electrons of all atoms are added together.
2. If the substance is an anion, the charge represents additional electrons that must be added to the total for the valence electrons. If the substance is a cation, the charge represents missing electrons that must be subtracted from the valence electrons.
3. Two electrons are placed between each outer atom and the central atom in the skeletal structure to represent a covalent bond. These are called the bonding pairs of electrons.
4. The remaining electrons are used to complete the octets of all the outer atoms in the skeleton. These are nonbonding pairs of electrons (also called lone pairs).
5. If there are still electrons left over, they are added in pairs to the central atom. These are also nonbonding pairs of electrons (lone pairs)
6. When all electrons have been placed, the outer atoms will all have octets. The central atom may have an octet, or it may have more or less than eight electrons.
   a. If the central atom has an octet, the structure is complete (see pages 276–278 on formal charges).
   b. If the central atom has fewer than eight electrons, (6 is permissible if the atom is boron) double bonds must be constructed to make the central atom have an octet. This is done by taking a nonbonding pair of electrons from an outer atom and placing these electrons as a bonding pair to make a double bond. Enough double bonds are constructed to give the central atom an octet.
   c. The central atom may have more than eight electrons only if it is in Period 3–7 of the Periodic Table. If the central atom is in Period 2, it cannot have more than an octet of electrons.

For example, we can construct the Lewis diagram for methane, CH_4, in the following manner. First, we draw the skeleton with carbon in the center and the hydrogens arranged symmetrically around it. Next we count the valence electrons; there are four valence electrons on carbon and one each on the four hydrogen atoms for a total of eight electrons. In the next step we add the bonding pairs of electrons to the structure. This step uses up all of the electrons, and we check to see whether each atom has an octet, bearing in mind that for hydrogen an "octet" is just one pair of electrons. Finally, to simplify the structure, we replace the bonding pairs of electrons by a line representing a bond. These steps are shown below:

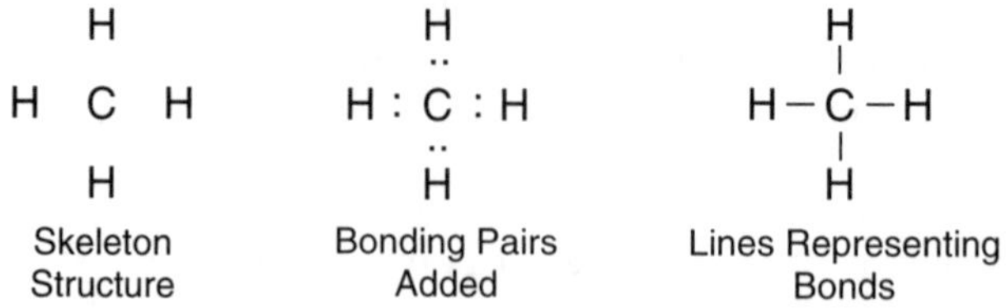

The most common electron-deficient molecules contain boron. The Lewis diagram for boron trifluoride, BF_3, is constructed as shown below:

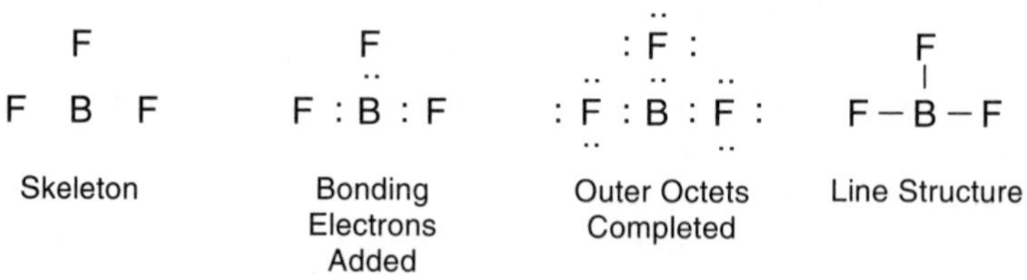

The skeleton is arranged with boron as the central atom. Adding up the valence electrons gives 3 + 7 + 7 + 7 = 24 electrons for the boron and the three fluorine atoms. Next, bonding pairs are added between each fluorine atom and the boron atom, using six electrons and leaving 18 to be placed. Six more electrons are placed around each fluorine atom to complete its octet. This step uses up all of the electrons. Since boron is commonly found with an electron-dificient structure (i.e., less than an octet of electrons on the B atom), the structure is complete. The line structure may be drawn for simplicity. The nonbonding pairs of the outer atoms are generally not shown in a line structure.

Phosphorus forms two compounds with chlorine, PCl_3 and PCl_5, and we may examine their Lewis structures. First, PCl_3 is constructed as follows:

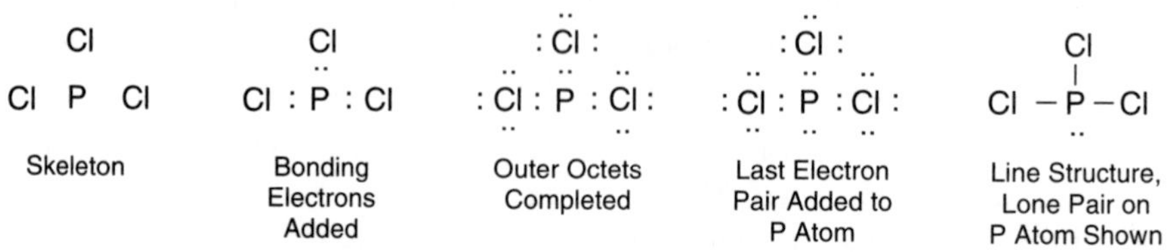

As before, the skeleton is drawn and the valence electrons counted; there are $5 + 7 + 7 + 7 = 26$ electrons. The bonding electron pairs and the outer octets are completed in the next two steps, leaving two electrons unused. These are placed on the central phosphorus atom as shown, completing its octet. All atoms are then checked to see that each has an octet of electrons. One way to do this is to draw a circle around all of the electrons adjacent to each atom, as shown below:

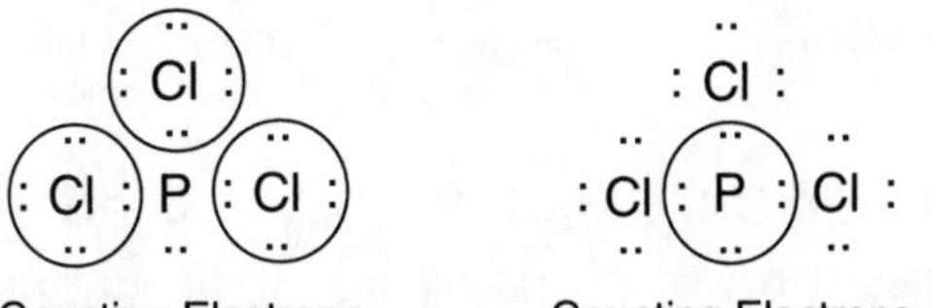

Counting Electrons Around Cl Atoms

Counting Electrons Around P Atom

Finally, the line structure may be used. Since nonbonding pairs on the central atom are often important in chemical reactions, they are shown in the line structure.

In constructing PCl_5, we observe some differences as shown in this diagram:

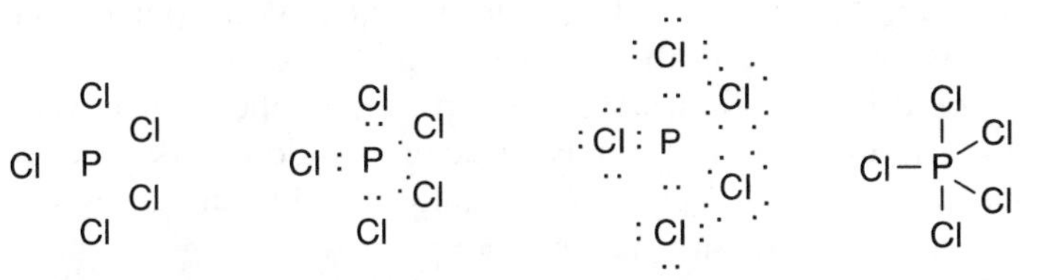

Skeleton

Bonding Electron Pairs Added

Complete Octets for Cl Atoms

Line Structure

In PCl_5 a total of 40 valence electrons must be placed. The skeleton is drawn, bonding pairs are added, and then the octets for chlorine are completed. These steps use up all 40 electrons, and the line structure may be drawn for clarity. Notice that the phosphorus atom has more than an octet (10) of electrons, as is reasonable since phosphorus is in Period 3 and may have an excess of electrons.

The maximum number of electrons that a central atom can have is 12 (six bonding pairs), as shown in the construction of SF_6:

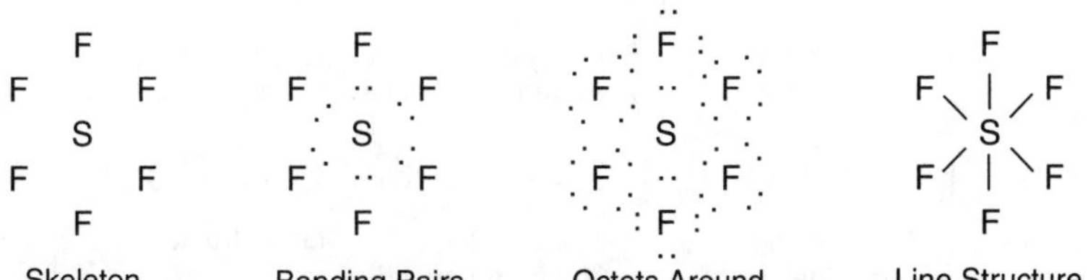

Skeleton

Bonding Pairs Added

Octets Around F Atoms Completed

Line Structure

Some compounds require the use of double bonds, which are represented by two pairs of electrons between atoms. Sulfur dioxide, SO_2, is one of those substances. Eighteen valence electrons must be distributed on the OSO skeleton to obtain octets on all the atoms. The steps are illustrated in the following diagram:

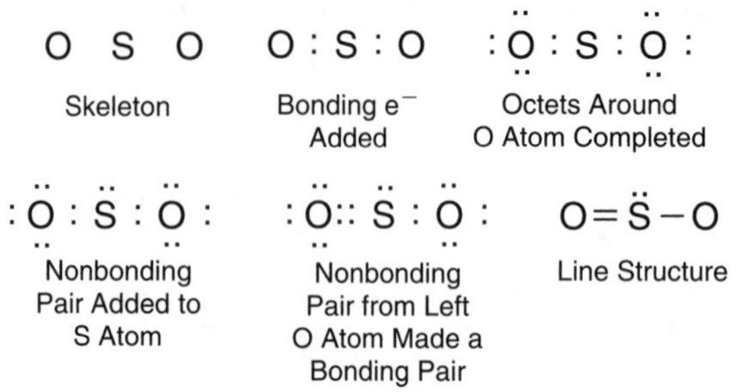

After the last two electrons are added to the sulfur atom as a nonbonding pair, the sulfur still has only six electrons. To obtain an octet, a nonbonding electron pair from the left oxygen atom is moved between the sulfur and oxygen, converting it into a bonding pair and creating a double bond. The oxygen still has an octet, and the sulfur now also has an octet, of electrons. Finally, we could have chosen to move a pair from the right oxygen atom instead of the left; in that case, a similar structure would be obtained, with the double bond on the right. These two, equally probable structures are called **resonance structures** and are shown below:

In addition to the Lewis structures for molecules, we can draw Lewis structures for covalently bonded polyatomic ions. The methods are the same as for molecules except that we must account for the electrons that give an ion its charge. One electron is added for each negative charge on an ion, and one electron is subtracted for each positive charge. The structure of the nitrite ion, NO_2^-, is an example:

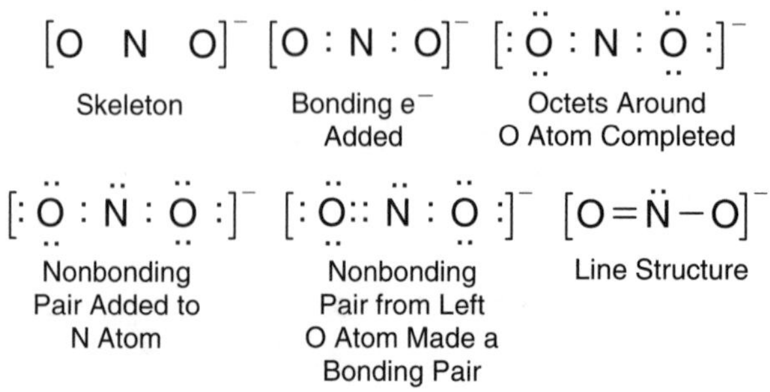

Using the SO_2 molecule as a reference, can you construct the resonance structures for the NO_2^- ion?

EXERCISES

1. Construct the Lewis structure for each of the following compounds:
   (a) CH_3Cl
   (b) CS_2
   (c) PH_3
   (d) SiF_4
   (e) H_2S
2. Construct the Lewis structure for each of the following ions:
   (a) NO_3^-
   (b) CO_3^{2-}
   (c) PO_4^{3-}
   (d) SO_3^{2-}
   (e) ClO_4^-

Answers

1.

(a) 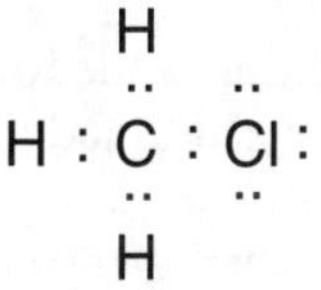

(c) 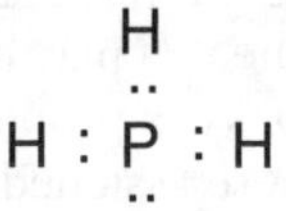

(e) 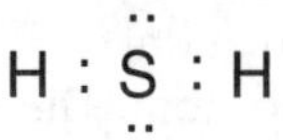

(b) 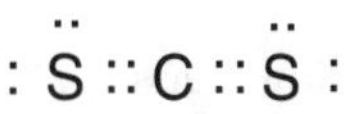

(d) 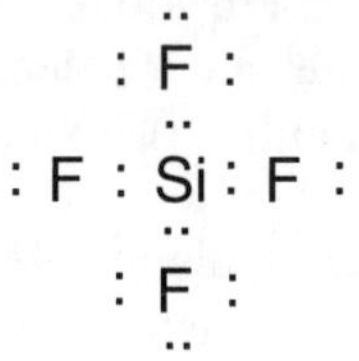

2.

(a) 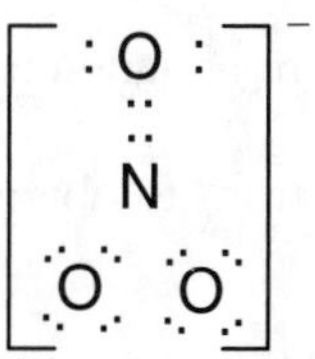

(b) 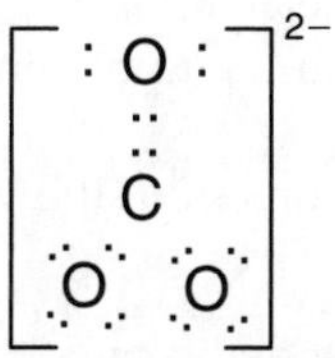

(c) 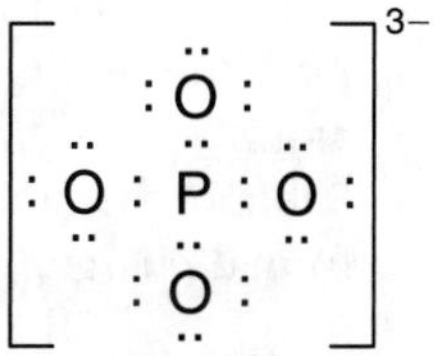

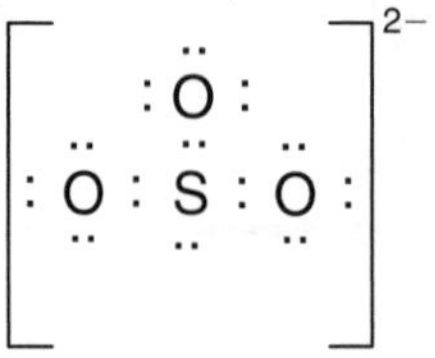
(d)

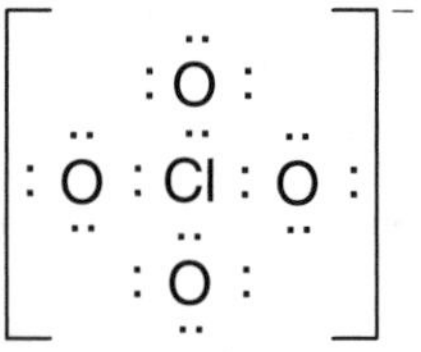
(e)

Lewis Structures of Odd Electron Compounds; Free Radicals

A compound may have a formula in which the total number of valence electrons is an odd number. In this case it will be impossible to construct a Lewis structure with an octet around each atom. Nitrogen dioxide, NO_2, is one such compound. One possible Lewis structure is shown below:

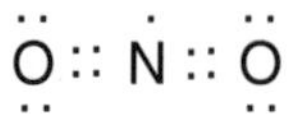

Molecules that have Lewis structures with an unpaired electron are often called **free radicals**. The unpaired electron makes the molecule unusually reactive. Free radicals have been implicated in such biological processes as aging and cancer. Free radicals, in an effort to pair up the single electron, may also form **dimers** or pairs of molecules. Thus the NO_2 molecule dimerizes to produce the N_2O_4 molecule in this reaction:

$$2\ NO_2 \rightleftharpoons N_2O_4$$

Formal Charges

How do we know if a Lewis structure is reasonable? Calculation of the **formal charge** on each atom is one technique that may be used to make this judgment. In this calculation, each of the electrons in a proposed Lewis structure is assigned to a specific atom. The number of these assigned electrons is then compared to the number of valence electrons of that atom. If the number of assigned electrons and the number of valence electrons are equal, the formal charge is zero. If more electrons are assigned than there are valence electrons, the formal charge will be a negative value equal to the number of extra electrons. Similarly, if fewer electrons are assigned than there are valence electrons, the atom has a positive formal charge equal to the number of missing electrons.

To calculate the formal charge on each atom the following steps are taken:

1. For each atom, count all electrons not used for bonding by the atom.
2. Count half of the bonding electrons.
3. Add the results of Steps 1 and 2 to obtain the electrons assigned to that atom.
4. Subtract the assigned electrons from the valence electrons to obtain the formal charge.

In equation form the formal charge is calculated as follows:

$$\text{Formal charge} = \text{valence electrons} - \text{number of nonbonding electrons} - \frac{1}{2}(\text{bonding electrons})$$

The formal charges on the atoms in a molecule must add up to zero, by the law of electroneutrality. For polyatomic ions the formal charges must add up to the charge on the ion. A molecule or polyatomic ion with the lowest possible formal charges on the individual atoms is judged to be a more probable structure than any others.

As an example consider the sulfate ion, SO_4^{2-}. Its Lewis structure may be drawn as

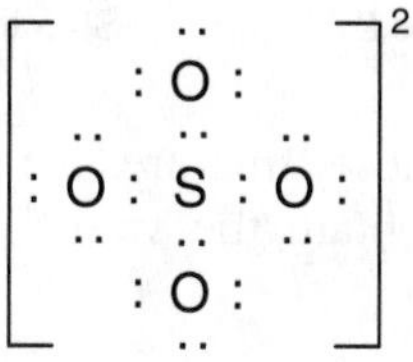

From this structure we calculate the formal charges on the sulfur and oxygen as follows:

$$\text{Formal charge on sulfur} = 6 - 0 - \frac{1}{2}(8) = +2$$

$$\text{Formal charge on oxygen} = 6 - 6 - \frac{1}{2}(2) = -1$$

All of the oxygens are identical, and each has a formal charge of -1; the formal charge on sulfur is $+2$. The sum of these adds up to the charge of the ion, as it should.

Can a structure be designed with a lower formal charge? The structure below:

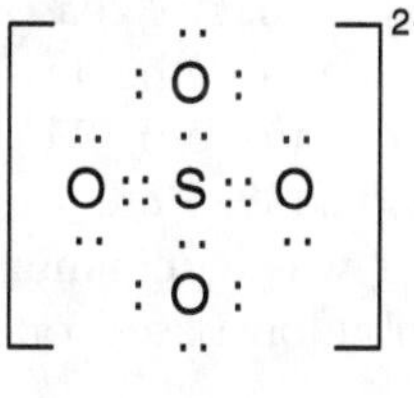

is one possible variation. There are now two types of oxygen-sulfur bonds (two O-S single bonds and two O-S double bonds). When the formal charges are calculated:

$$\text{Formal charge on sulfur} = 6 - 0 - \frac{1}{2}(12) = 0$$

$$\text{Formal charge on oxygen with double bond} = 6 - 4 - \frac{1}{2}(4) = 0$$

$$\text{Formal charge on oxygen with single bond} = 6 - 6 - \frac{1}{2}(2) = -1$$

This seems to be the preferred structure since it has the minimum number of formal charges, which happen to be equal to the overall charge on the sulfate ion itself.

Formal charges may also be used to deduce the appropriate structure for a compound that has many possible Lewis structures. For instance, the NOCl molecule can be drawn with the following four structures:

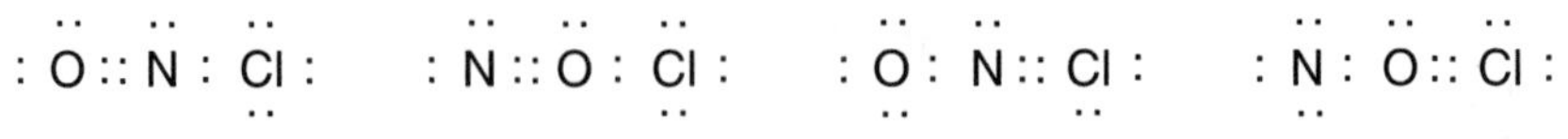

Structure 1 Structure 2 Structure 3 Structure 4

The question is, Which of these structures is the most reasonable? Using the rules for formal charge, we obtain the values in Table 7.11 for these four structures.

TABLE 7.11. FORMAL CHARGES ON NOCl STRUCTURES

Element	Structure 1	Structure 2	Structure 3	Structure 4
Oxygen	0	+1	−1	+1
Nitrogen	0	−1	0	−2
Chlorine	0	0	+1	+1

From the table we see that NOCl structure 1 has the lowest formal charges on all atoms, and it is preferred. Structure 2 has larger formal charges, and also has a negative charge on nitrogen, even though nitrogen is less electronegative than oxygen. This is not a reasonable situation. Structures 3 and 4 also have negative charges on the less electronegative elements as well as formal charges greater than zero. We therefore conclude that only structure 1 is possible and reasonable. Here we used two criteria for judging a reasonable structure: the first was a minimum formal charge, and second was a reasonable charge distribution based on the electronegativities of the elements.

Resonance Structures

Often several different Lewis structures can be drawn for a particular substance. Usually, the most reasonable one can be selected using the concept of formal charges. However, at times we can construct several Lewis structures that are totally equivalent, even down to the formal charges on the atoms. In such cases we call these structures **resonance structures**.

Most resonance structures are very similar for a given substance, usually differing in the geometry of the molecule or ion. In addition, it is also found experimentally that none of the resonance Lewis structures properly describes the substance, whose true properties are determined by blending all of the resonance structures together. The resonance of the SO_3 molecule is shown below with the three possible Lewis structures:

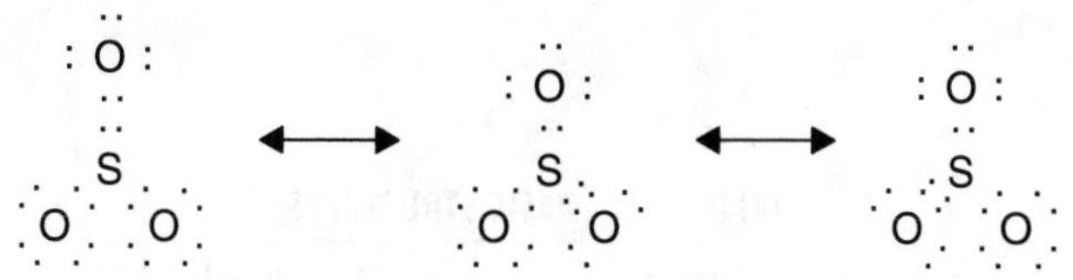

It is important to understand the nature of resonance. In the above diagrams, each SO_3 has two single bonds and one double bond. Experiments show that all of the sulfur-oxygen bonds are identical, and the experimental properties of these bonds indicate that they lie between those of a single bond and a double bond. To properly visualize the SO_3 molecule, therefore, we must think of three identical S-O bonds that are two-thirds single bond and one-third double bond in character. A similar procedure must be used for other resonance structures.

MOLECULAR SHAPES

Molecular Geometry

Once a valid Lewis structure has been determined, the overall geometry of the molecule can be determined. This can be done readily for a simple molecule with one central atom. The process can be extended to very large macromolecules such as proteins and DNA by determining the geometries around individual atoms and then combining these geometries to obtain the entire structure. **Molecular geometry** is extremely important in understanding the properties of chemical compounds. The key to the discovery of DNA's double helix was the geometric structures of the four bases that must hydrogen-bond to each other to hold the structure together.

The **valence shell electron-pair repulsion (VSEPR) theory** allows us to determine the three-dimensional shapes of covalently bonded molecules with a minimum of information. This theory states that the geometry around each atom will depend on the repulsion of the valence-shell electrons (bonding electrons and nonbonding pairs) away from each other. As a result of this repulsion, due to the negative charges on the electrons, the electron pairs are aligned as far away from each other as possible.

Basic Structures

To determine the three-dimensional geometry around a central atom, A, all we need to know is how many other atoms, each represented as X, are covalently bonded to it. The one restriction is that central atom A must have no nonbonding pairs of electrons.

Table 7.12 lists the **basic structures** for the six possible geometries. In this table the AX_n notation is used to indicate the number of atoms X attached to the central atom, A.

TABLE 7.12. BASIC STRUCTURES

Notation	Shape	Example	Angle(s)
AX	Linear	HBr	180°
AX_2	Linear	CS_2	180°
AX_3	Planar triangle	BCl_3	120°
AX_4	Tetrahedron	CCl_4	109.5°
AX_5	Trigonal bipyramid	PCl_5	120°, 90°
AX_6	Octahedron	XeF_6	90°

The angles listed in this table are the angles between the bonds, assuming that the central atom is the vertex of the angle. For structures AX to AX_4, all bonds are equidistant from all others. For the AX_5 and AX_6 structures, bond angles are measured between the nearest neighbors. Thus, for the AX_5 structure, the 120° angle is for the three equatorial atoms and the 90° angle is the angle between the axial atoms and the equatorial atoms. Figure 7.10 shows these shapes in diagram format. (The AX structure is omitted from many textbooks as being trivial since any molecule that contains only two atoms must be linear.)

The geometry around a central atom that does not have any nonbonding electron pairs is determined by counting the atoms bonded to it. For instance, three atoms bound to a central atom that has no nonbonding pairs must have a planar triangle geometry. Five atoms bound to a central atom must have the shape of a trigonal bipyramid.

Derived Structures

Nonbonding electron pairs on the central atom take up space, just as an atom does. In fact, a nonbonding electron pair takes up slightly more space than an atom. As a result, we count the number of nonbonding pairs as well as the atoms attached to the central atom to determine the basic structure as we did above. Since the nonbonding pairs of electrons are not "seen" when a structure is drawn, the actual geometry of the atoms we do see will be only part of this basic structure. The part that we see is called the **derived structure**. The possible derived structures are listed in Table 7.13. The symbol E in the notation represents a nonbonding pair of electrons on the central atom.

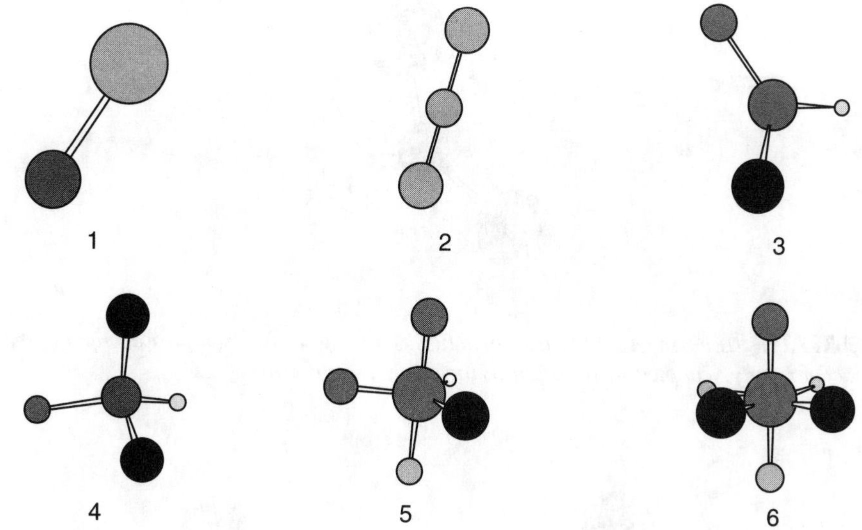

FIGURE 7.10. *Perspective diagrams of the six basic geometric structures. Darker atoms are closest to the viewer, structures are tilted to show all atoms. (1) Linear diatomic; (2) linear triatomic; (3) planar triangle; (4) tetrahedron; (5) trigonal bipyramid; (6) octahedron. In diagrams (5) and (6) the axial atoms are at the top and bottom of the figures, while the equatorial atoms are in the center.*

Since the first two entries in Table 7.13 represent a single atom and a diatomic substance, respectively, these structures need not be drawn. The geometries that correspond to the remaining derived structures are shown in Figures 7.11 to 7.14.

TABLE 7.13. DERIVED STRUCTURES WITH NONBONDING ELECTRON PAIRS ON THE CENTRAL ATOM

Basic Structure Notation	Derived Structure Notation	Shape	Example	Angle(s)
AX	AE	single atom	—	—
AX_2	AXE	linear diatomic	CN^-	180°
AX_3	AX_2E	bent	$SnCl_2$	120°
AX_4	AX_3E	triangular pyramid	NH_3	109.5°
AX_4	AX_2E_2	bent	H_2O	109.5°
AX_5	AX_4E	distorted tetrahedron	SF_4	120°, 90°
AX_5	AX_3E_2	t-shape	ICl_3	90°
AX_5	AX_2E_3	linear	I_3^-	180°
AX_6	AX_5E	square pyramid	IF_5	90°
AX_6	AX_4E_2	square planar	XeF_4	90°

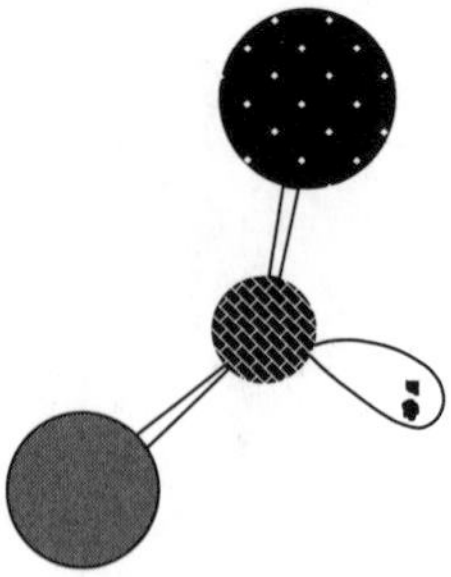

FIGURE 7.11. *The bent AX_2E derived structure, showing an electron pair occupying the space formerly occupied by an atom in the basic AX_3 structure.*

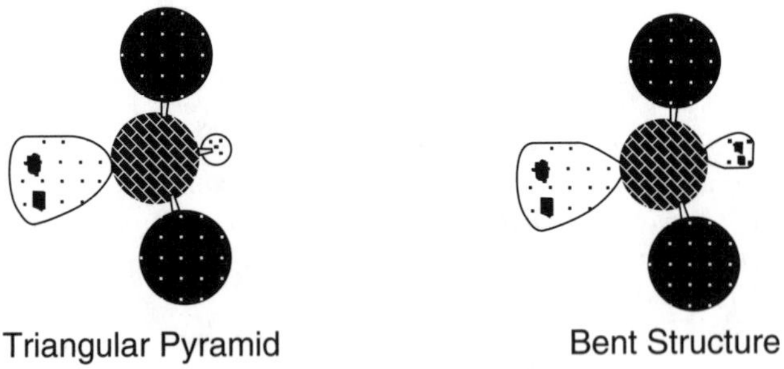

FIGURE 7.12. *The AX_3E (triangular pyramid) and AX_2E_2 (bent) derived structures that are derived from the AX_4 basic structure.*

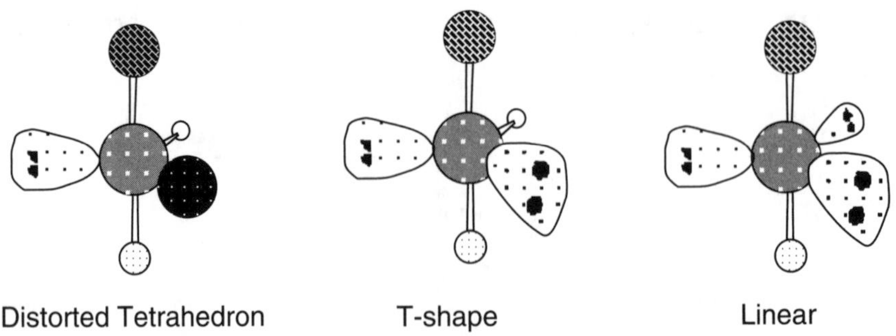

FIGURE 7.13. *The three possible derived structures—the AX_4E (distorted tetrahedron), AX_3E_2 (T-shape), and AX_2E_3 (linear)—obtained from the AX_5 basic structure. The equatorial atoms are replaced by electron pairs, shown by pairs of dots.*

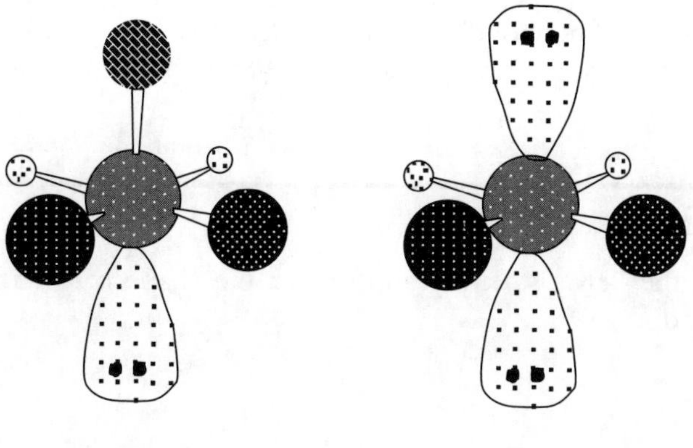

FIGURE 7.14. *The AX_5E (square pyramid) and AX_4E_2 (square planar) derived structures obtained from the AX_6 basic structure. The second atom replaced by an electron pair is on the opposite side of the molecule so that the electron clouds have the extra space they need.*

In Table 7.13 and Figures 7.11 and 7.12 there are two bent structures, AX_2E and AX_2E_2. The bond angles for these two structures are very different, and we can distinguish between the two based on their angles. Since the AX_2E structure is derived from the trigonal planar AX_3, we expect its angle to be approximately 120°. The AX_2E_2 bent structure is derived from the tetrahedral AX_4 basic structure; therefore, we expect bond angles of approximately 109.5°.

In forming the derived structures from the basic structures, we are, in effect, replacing one or more atoms with pairs of nonbonding electrons. Up to the AX_4 structure, it does not matter which atom is replaced by an electron pair; we get the same derived structure. However, the AX_5 and AX_6 structures will have different shapes, when we make the derived structures, depending on which atom is replaced by a nonbonding electron pair. The actual shapes of the molecules can be explained by the fact that a nonbonding pair of electrons takes up relatively more space than a bonded atom. In the AX_5 structure, the nonbonding electron pairs will replace the equatorial atoms since more room is available (120° between the atoms, compared to 90° for the axial atoms) for the electron cloud. In the AX_6 structure, it does not matter which atom is replaced by the first nonbonding electron pair; the result is always a square pyramid. The second nonbonding electron pair replaces the atom opposite the first nonbonding electron pair. This positioning allows the nonbonding pairs the most room possible on the molecule.

When determining the geometry around a central atom, it is necessary to count the number of atoms bound to the central atom and also the number of nonbonding pairs of electrons, if any. We now understand why it is so important to show nonbonding pairs of electrons in a Lewis structure. Of particular importance are the nonbonding pairs on the central atom.

EXERCISE

Construct the Lewis structure and predict the shape of each of following molecules and ions:

(a) CH_3Cl
(b) CS_2
(c) PH_3
(d) SiF_4
(e) H_2S
(f) NO_3^-
(g) CO_3^{2-}
(h) PO_4^{3-}
(i) SO_3^{2-}
(j) ClO_4^-

Answers

The shapes are the same whether we use the simple Lewis structures or the structures optimized for the best formal charges.

(a) tetrahedral
(b) linear
(c) triangular pyramid
(d) tetrahedron
(e) bent
(f) triangular planar
(g) triangular planar
(h) tetrahedral
(i) tetrahedral
(j) tetrahedral

POLARITY OF BONDS AND MOLECULES

Covalent-Bond Polarity and Electronegativity

Bond Polarity

Electrons are shared equally only in a covalent bond between two identical atoms (for example, H_2, F_2, and N_2). If the electrons are not shared equally by the two atoms, they will spend more time localized near one atom or the other. As a result, the atom that attracts the electrons will be relatively more negative than the atom that does not. In this case, we say that the bond is **polar** with a positive end and a negative end. An understanding of bond polarities allows the chemist to explain many physical properties of chemical compounds. We will use this concept many times in later chapters.

To understand polarities we need a method to determine how effectively different atoms attract electrons. Linus Pauling developed the concept of **electronegativity** in order to represent numerically the ability of an atom to attract electrons. Figure 7.15 shows the Periodic Table along with the electronegativity value for each element. In this figure, we see that electronegativity increases from left to right in each period (row) of the Periodic Table. In addition, the electronegativity within any group (column) increases from the bottom of the group to the top. In general, the electronegativity increases from the lower left corner of the Periodic Table to the upper right corner. This is one of the important **diagonal relationships** in the Periodic Table.

H																
2.1																
Li	Be											B	C	N	O	F
1.0	1.5											2.0	2.5	3.1	3.5	4.0
Na	Mg											Al	Si	P	S	Cl
1.0	1.3											1.5	1.8	2.1	2.4	2.9
K	Ca	Sc	Ti	V	Cr	Mn	Fe	Co	Ni	Cu	Zn	Ga	Ge	As	Se	Br
0.8	1.1	1.2	1.3	1.5	1.6	1.6	1.7	1.7	1.8	1.8	1.7	1.8	2.0	2.2	2.5	2.8
Rb	Sr	Y	Zr	Nb	Mo	Tc	Ru	Rh	Pd	Ag	Cd	In	Sn	Sb	Te	I
0.8	1.0	1.1	1.2	1.3	1.3	1.4	1.4	1.5	1.4	1.4	1.5	1.5	1.7	1.8	2.0	2.5
Cs	Ba	La	Hf	Ta	W	Re	Os	Ir	Pt	Au	Hg	Tl	Pb	Bi	Po	At
0.7	0.9	1.1	1.2	1.4	1.4	1.5	1.5	1.6	1.5	1.4	1.5	1.5	1.6	1.7	1.8	2.2
Fr	Ra	Ac														
0.7	0.9	1.0														

FIGURE 7.15. *Periodic Table showing the electronegativities of the elements.*

The diagonal relationship of the electronegativities allows chemists to determine quickly which end of a bond is negative and which end is positive. In a bond, the element closest to fluorine will be relatively negative and the element furthest from fluorine will be relatively positive. Polarities are indicated with the symbols $\delta+$ and $\delta-$ to indicate partially positive and partially negative atoms, respectively. Full positive and negative charges are used only for ions. For example, the B$-$O bond is written as $^{\delta+}B-O^{\delta-}$ to show that the boron atom is more positive than the oxygen atom.

EXERCISE

1. Without using Figure 7.15, show the positive and negative ends of each of the following bonds, using the symbols δ+ and δ− :
   (a) S−O
   (b) C−N
   (c) S−P
   (d) C−F
   (e) Si−O
   (f) H−Br
   (g) H−O
2. For each of the bonds in the previous exercise, determine ΔEN value and predict which bond is the least polar and which is the most polar.
3. Using a Periodic Table, but not a table of electronegativities, estimate, in each of the following cases, which bond pair is more polar:
   (a) C−N or C−O
   (b) H−Cl or H−Br
   (c) S−O or S−Br
   (d) H−S or H−O
   (e) P−Br or S−Br

Answers

1. Using the diagonal relationships in the Periodic Table, we can write the following:
   (a) $^{\delta+}S-O^{\delta-}$ (d) $^{\delta+}C-F^{\delta-}$ (f) $^{\delta+}H-Br^{\delta-}$
   (b) $^{\delta+}C-N^{\delta-}$ (e) $^{\delta+}Si-O^{\delta-}$ (g) $^{\delta+}H-O^{\delta-}$
   (c) $^{\delta-}S-P^{\delta+}$
2. Figure 7.15 gives numerical data that may be used to evaluate the magnitude of the bond polarity. This is done by taking the absolute value of the difference in the electronegativities of the two atoms participating in a bond. We may call this value delta EN or ΔEN.

$$\Delta EN = \underset{\text{electronegativity}}{\text{atom with largest}} - \underset{\text{electronegativity}}{\text{atom with smallest}}$$

The larger the value of ΔEN, the greater is the polarity of the bond. If ΔEN is zero, the bond is considered to be nonpolar.

2. (cont.) (a) 1.7 (d) 1.5 (f) 0.7
(b) 0.6 (e) 1.7 (g) 1.4
(c) 0.3

The S—O and Si—O bonds are the most polar, and the S—P bond is the least polar.

3. Even without an electronegativity table, it is possible to determine which of two bonds is more polar if they contain one atom in common. The more polar bond will be the one where the second atom is furthest in the Periodic Table from the common atom. For instance, the N-F bond is more polar than the O-F bond since nitrogen is further from fluorine than is oxygen.

(a) C—O (c) S—O (e) P—Br
(b) H—Cl (d) H—O

Electronegativity and Ionic Character

At one end of the polarity scale are the completely **nonpolar bonds** between diatomic elements. We may visualize ionic compounds as being at the other end of the polarity scale since the electrons are actually transferred from one atom to another. From the table of electronegativities we can infer that the largest ΔEN is 3.3 for the ionic compound FrF. The well-known ionic compounds NaCl and $CaBr_2$ have ΔEN values of 3.0 and 1.7, respectively.

The ΔEN of a bond has been used to estimate the percentage **ionic character** of a bond. Chemists say that a ΔEN of 1.7 represents a bond that is 50 percent ionic and 50 percent covalent in character. A bond with a ΔEN of 1.7 or greater is considered ionic. Bonds with ΔEN values less than 1.7 are polar covalent, and those whose ΔEN values are 0 are nonpolar.

Molecular Polarity

The polarity of a bond depends on the electronegativities of the two elements bonded together. Very few molecules are diatomic, meaning that for most molecules more than one bond must be considered in determining molecular polarity. In compounds the bonds are arranged geometrically in space as described on pages 279–284. The result is that, even if a bond is polar, the molecule as a whole may or may not be polar.

Four general rules apply to the determination of whether a molecule is polar:

1. A molecule that is totally symmetrical will be nonpolar. It does not matter how polar the individual bonds are.
2. A nonsymmetrical molecule will be polar if the bonds are polar.
3. A molecule with more than one type of atom attached to the central atom will often be nonsymmetrical and therefore polar.
4. Central atoms with nonbonding electron pairs are often nonsymmetrical and polar.

When determining the polarity of a molecule, we must remember that there is only one positive end and one negative end, which will be directly opposite each other. The CH_3Cl and CBr_4 molecules are recognized as tetrahedral structures. CH_3Cl is polar because it is not symmetrical, and CBr_4 is symmetrical and nonpolar, as shown in Figure 7.16.

FIGURE 7.16. *Line structures of CH_3Cl, showing the polarity of the molecule, and CBr_4, showing its symmetry and nonpolarity.*

From our knowledge of electronegativities we can predict that the negative end of the CH_3Cl molecule is located near the chlorine atom and that the positive end is opposite the chlorine atom in a region of space between the three hydrogen atoms.

For the water molecule we may draw the structure shown in Figure 7.17.

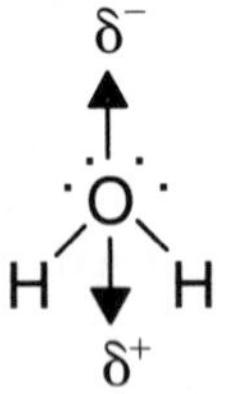

FIGURE 7.17. *Structure of water and its polarity.*

The electronegativity of the oxygen indicates that it is the negative end of the molecule, and the region of space between the two hydrogen atoms is the positive end.

EXERCISE

Construct the Lewis structure and predict the polarity of each of the following:

(a) CH_3Cl

(b) CS_2

(c) PH_3

(d) SiF_4

(e) H_2S

(f) NO_3^-

(g) CO_3^{2-}

(h) PO_4^{3-}

(i) SO_3^{2-}

(j) ClO_4^-

Answers

(a) Polar with Cl the negative end

(b) Nonpolar

(c) Polar with P the negative end

(d) Nonpolar

(e) Polar with S the negative end

(f–j) Nonpolar due to resonance; charge on ion is distributed evenly over the ion. These ions are charged but not polar.

COLLIGATIVE PROPERTIES

Boiling-Point Elevation and Freezing-Point Depression of Nonvolatile Nonelectrolyte Solutions

Calculating Boiling-Point Elevation and Freezing-Point Depression

A solute decreases the vapor pressure of a solvent according to Raoult's law (see pages 295–299). The phase diagram (see pages 96–98 for a complete description of phase diagrams) for a solution is similar to the phase diagram for the pure solvent except that the decrease in vapor pressure lowers the liquid-gas equilibrium line.

Figure 7.18 illustrates the change in the position of the liquid-gas equilibrium line, which establishes a new triple point and also shifts the solid-liquid equilibrium line for the solution compared to the pure solvent.

From Figure 7.18 it is apparent that a decrease in the vapor pressure must also cause the boiling point to increase from b.p. to *B* on the graph. At the same time the melting point will decrease from m.p. to *A*.

The increase in the boiling point is directly related to the decrease in vapor pressure caused by the presence of a solute. As a result, higher temperatures are needed to make the vapor pressure of the solution equal to atmospheric pressure. The decrease in the melting point occurs because the attractive forces between the solute and solvent interfere with the crystallization process.

The changes in the boiling and freezing temperatures of a solution depend on the solvent and the molal concentration (molality) of the solute. For the increase in the boiling point (ΔT):

$$\Delta T = k_b \text{ m}$$

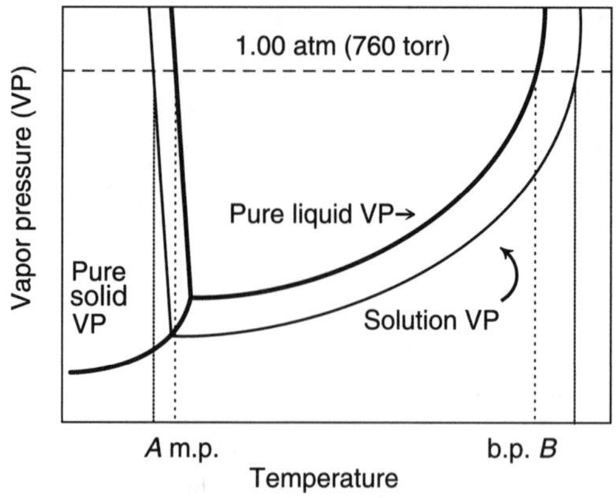

FIGURE 7.18. *Change in the phase diagram of a solvent when a nonvolatile solute is dissolved in it. Heavy lines represent the phase diagram for the pure solvent. Lighter lines represent changes to the solid-liquid and liquid-gas equilibrium lines. The abbreviations m.p. and b.p. indicate the melting and boiling points of the pure solvent, while* A *and* B *are the new melting and boiling points.*

where m is the molality of the solution and k_b is the boiling-point-elevation constant for the solvent. For the decrease in the freezing point (ΔT) the equation is similar except that it has a negative sign to indicate the decrease in temperature and a different constant (k_f), called the freezing-point-depression constant:

$$\Delta T = k_f \text{ m}$$

For either boiling or freezing, the calculated ΔT must be added to the boiling point or freezing point of the pure solvent to determine the actual boiling and freezing points. Table 7.14 lists some common solvents with their boiling-point-elevation and freezing-point-depression constants.

TABLE 7.14. FREEZING-POINT-DEPRESSION AND BOILING-POINT-ELEVATION CONSTANTS FOR SOME COMMON SOLVENTS

Solvent	Freezing Point(°C)	k_f (°C/m)	Boiling Point(°C)	k_b (°C/m)
Acetic acid	16.66	3.90	117.90	2.53
Benzene	5.50	5.10	80.10	2.53
Cyclohexane	6.50	20.2	80.72	2.75
Camphor	178.40	40.0	207.42	5.61
p-Dichlorobenzene	53.10	7.1	174.1	6.2
Naphthalene	80.29	6.94	217.96	6.2
Water	0.00	1.86	100.0	0.52

EXERCISE

What are the freezing and boiling points of naphthalene when 10.0 grams of nicotine, $C_{10}H_{14}N_2$, is dissolved in 50.0 grams of naphthalene?

Solution

The constants k_f and k_b for naphthalene are given in Table 7.14. The molality of nicotine in this solution must be calculated to obtain ΔT:

$$? \frac{\text{mol } C_{10}H_{14}N_2}{\text{kg naphthalene}} = \frac{10.0 \text{ g } C_{10}H_{14}N_2 / 162 \text{ g mol}^{-1}}{0.050 \text{ kg naphthalene}}$$

$$= 1.23 \text{ m } C_{10}H_{14}N_2$$

Using the proper k_f and k_b values, along with this molality, gives the temperature changes:

$$\Delta T_f = -(6.94°\text{C/m})\ (1.23 \text{ m})$$

$$= -8.54°\text{C}$$

and

$$\Delta T_b = (6.2°\text{C/m})\ (1.23 \text{ m})$$

$$= 7.63°\text{C}$$

These changes are then added to the freezing and boiling temperatures of the pure solvent to obtain the freezing point and boiling point of the solution:

$$\text{Freezing point} = 80.29°\text{C} - 8.54°\text{C} = 71.75°\text{C}$$

$$\text{Boiling point} = 217.96°\text{C} + 7.63°\text{C} = 225.59°\text{C}$$

Uses of Boiling-Point Elevation and Freezing-Point Depression

Chemists take advantage of freezing point depression to make cooling baths below 0°C. This is done by dissolving large quantities of salt in water and then adding ice. Fahrenheit used this method to make the coldest possible solution for zero on his temperature scale. Automobile owners take advantage of antifreeze in their car engines. Antifreeze works by lowering the freezing point of water with a high concentration of ethylene glycol. Antifreeze solutions also increase the boiling point of water, allowing engines to run at temperatures above 100°C without boiling over.

The most important use of the freezing-point-depression and boiling-point-elevation phenomenon involves the determination of molar masses. The molality used to calculate ΔT_f and ΔT_b is defined as moles of solute/per kilogram of solvent. The moles of solute are equal to the mass of solute divided by its molar mass (g of solute/molar mass of solute). The equations for ΔT_f and ΔT_b (respectively the freezing and boiling points) may be expanded to read

$$\Delta T_f = -k_f\left(\frac{\text{g solute/molar mass solute}}{\text{kg solvent}}\right)$$

and

$$\Delta T_b = k_b\left(\frac{\text{g solute/molar mass solute}}{\text{kg solvent}}\right)$$

To obtain the molar mass of a substance, we need to know the number of grams of solute dissolved in each kilogram of solvent, the change in either the boiling point or the freezing point, and the appropriate constant, k_f or k_b, for the solvent.

EXERCISE

When 10.0 grams of elemental sulfur is dissolved in 100.0 grams of cyclohexane, the freezing point of this solution is found to be −1.40°C. What is the molar mass of sulfur?

Solution

The expanded equation for ΔT_f is used:

$$\Delta T_f = -k_f\left(\frac{\text{g solute/molar mass solute}}{\text{kg solvent}}\right)$$

Since the freezing point of the solution is −1.40°C and the normal freezing point of cyclohexane is 6.50°C,

$$\Delta T_f = -7.90°\text{C}.$$

From Table 7.13 the freezing-point-depression constant of cyclohexane is 20.2°C m^{-1}. Entering the data in the equation yields

$$-7.90°\text{C} = -20.2°\text{C } m^{-1}\left(\frac{10.0\text{ g sulfur/molar mass sulfur}}{0.100\text{ kg cyclohexane}}\right)$$

$$= \frac{-2020°\text{C g S/mol S}}{\text{molar mass sulfur}}$$

$$\text{Molar mass sulfur} = \frac{-2020°\text{C g S/mol S}}{-7.90°\text{C}}$$

$$= 256\text{ g S/mol S}$$

This result tells us that one molecule of sulfur has a mass of 256. Since the atomic mass of sulfur is 32 in the Periodic Table, there must be eight sulfur atoms in one molecule of sulfur (256/32 = 8). Therefore elemental sulfur exists as S_8 molecules.

Osmotic Pressure

Freezing-point-depression and boiling-point-elevation measurements have constants in the range of a few °C per molal. Therefore, to attain any reasonable measurements, solutions must be fairly concentrated since small temperature changes are often difficult to measure.

Modern chemistry is often interested in extremely large molecules such as polymers, proteins, peptides, and DNA. These molecules, which have molar masses exceeding 100,000 grams per mole, are impossible to dissolve in sufficient quantities to obtain the large temperature changes needed for accurate results. However, dilute solutions produce an osmotic pressure that is easily measured.

Osmosis is the selective flow of solvent, but not solute, molecules through a semipermeable membrane. It is the basic mechanism that transports water from the roots to the upper branches of trees that may be over 100 feet tall.

In explaining osmosis, we start by observing that two solutions of different concentrations will diffuse into each other. There is a natural tendency for solutes and solvents to diffuse from regions of high concentration to regions of lower concentration until the concentration is uniform throughout the solution.

A semipermeable membrane placed between two solutions with different concentrations will allow only the solvent to diffuse. Solute molecules, which are often larger than solvent molecules, cannot pass through the semipermeable membrane. Diffusion of the solvent through the semipermeable membrane will stop when the concentrations have been equalized or when an opposing force is applied to stop the diffusion. This opposing force is known as **osmotic pressure**.

The osmotic pressure experiment is diagrammed in Figure 7.19. At the start of the experiment two compartments, one holding a solution containing the solute and the other holding a pure solvent, are separated by a semipermeable membrane. The heights of the two liquids in the measuring tubes are equal at the start, indicating that the same atmospheric pressure is acting on both compartments.

Since the solution is more concentrated than the solvent, solvent immediately starts diffusing from the solvent side to the solution side in an attempt to equalize the concentrations. This dilutes the solution by increasing its volume, and the liquid level in the measuring tube for the solution increases. Conversely, the loss of solvent from the solvent compartment causes the liquid level in its measuring tube to decrease. The difference in liquid levels between the two compartments represents the opposing force, that is, the osmotic pressure. The liquid levels will stop changing when the osmotic pressure is large enough to oppose the force of the solvent diffusing through the membrane.

The equation that governs this process is the ideal gas law equation from pages 59–62:

$$PV = nRT$$

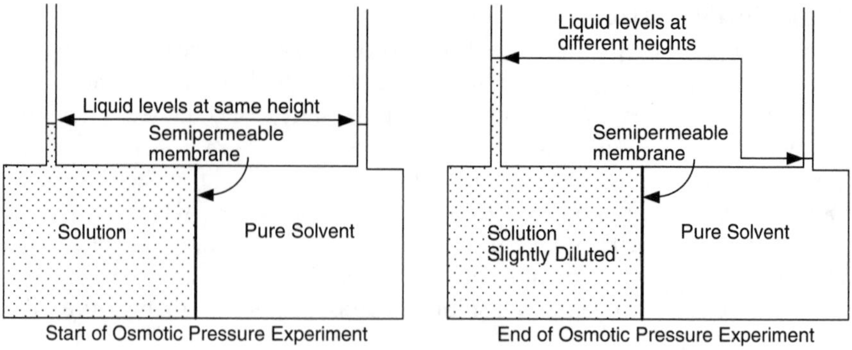

FIGURE 7.19. *The start (left) of an osmotic pressure experiment with equal pressures acting on both compartments. At the end (right) of the experiment the pressure developed is the osmotic pressure.*

For an osmotic pressure experiment the symbol Π replaces P:

$$\Pi V = nRT$$

EXERCISE

Calculate the osmotic pressure that would result from dissolving 2 teaspoons of sucrose in a cup of coffee at 80°C. (R = 0.0821 L atm mol^{-1} K^{-1})

Solution

One cup of coffee is approximately 0.250 L, and 2 teaspoons (about 18 g each) of sucrose (molar mass = 342) is equivalent to 0.105 mole of sucrose. Since the temperature is given in degrees, Celsius, it must be converted to kelvins: 353 K. Substituting these data into the osmotic pressure equation yields

$$\Pi\ (0.250\ \text{L}) = (0.105\ \text{mol})(0.0821\ \text{L atm mol}^{-1}\ \text{K}^{-1})(353\ \text{K})$$

Solving, we obtain an osmotic pressure of 12.17 atm or 9251 mm Hg (9.3 m of mercury or 126 m of water!).

This exercise shows that the osmotic pressure is very large for a 0.1 molar solution. A 10 micromolar solution will result in more than a centimeter difference in pressure for aqueous solutions. This fact makes osmotic pressure measurements ideal for very dilute solutions. Of particular interest is the use of osmotic pressure to determine molar mass, as shown in the next exercise.

EXERCISE

A large polymer is synthesized, and 100 milliliters of a solution containing 1.00 gram of this polymer is tested in an osmotic pressure experiment. The resulting pressure is measured as 55.0 millimeters of water at 25°C. From this information calculate the molar mass of the polymer.

Solution

First, the pressure must be converted to millimeters of mercury and then to atmospheres since we intend to use a value of R that is 0.0821 L atm mol^{-1} K^{-1}. This is done by using the ratio of the densities of water and mercury:

$$? \text{ atm} = 55 \text{ mm } H_2O \left(\frac{1.00 \text{ g } H_2O \text{ mL}^{-1}}{13.5 \text{ g Hg mL}^{-1}}\right) = 4.07 \text{ mm Hg}$$

$$= 4.07 \text{ mm Hg}\left(\frac{1 \text{ atm}}{760 \text{ mm Hg}}\right) = 5.36 \times 10^{-3} \text{ atm}$$

Entering the data into the osmotic pressure equation yields the number of moles of polymer:

$$(5.36 \times 10^{-3} \text{ atm})(0.100 \text{ L}) = n(0.0821 \text{ L atm mol}^{-1} \text{ K}^{-1})(353 \text{ K})$$

$$n = 1.85 \times 10^{-5} \text{ mol}$$

Since the number of moles of any substance is the mass divided by the molar mass, we can write

$$n = \frac{\text{g compound}}{\text{molar mass compound}}$$

or

$$\text{Molar mass compound} = \frac{\text{g compound}}{n}$$

Entering the data, we obtain

$$\text{Molar mass compound} = \frac{1.00 \text{ g}}{1.85 \times 10^{-5} \text{ mol}}$$

$$= 54{,}054 \text{ g mol}^{-1} \text{ or } 54.1 \text{ kg mol}^{-1}$$

Since this solution is very dilute, the molarity and molality can be considered to be the same. In aqueous solution the freezing-point depression would have been 4×10^{-4} °C. This temperature change is difficult to measure precisely, and it indicates the utility of osmotic pressure measurement.

Raoult's Law

If a nonvolatile nonelectrolyte is dissolved in a liquid, it causes the vapor pressure to decrease in proportion to its concentration. This effect is explained by the fact that some of the nonvolatile solute molecules take up space at the surface of the liquid, thereby decreasing the effective surface area available for evaporation. The result is to decrease the evaporation rate and the vapor pressure.

In the 1880s F.M. Raoult was the first to observe this effect and to show that vapor pressure is proportional to the mole fraction of the solvent in a solution. This is entirely reasonable since the mole fraction of the solvent represents the proportion of solvent molecules at the surface of the liquid. **Raoult's law** states that the actual vapor pressure, P_{actual}, is equal to the vapor pressure of the pure solvent, (P^0) multiplied by the mole fraction of the solvent:

$$P_{actual} = P^0 \, X_{solvent} \qquad (7.2)$$

EXERCISE

Some people use two teaspoons (18 g each) of sugar in each cup (250 mL) of coffee. The vapor pressure of water at 80°C is 355 millimeters of mercury. What is the vapor pressure of sugared coffee at the same temperature?

Solution

Sucrose, $C_{12}H_{22}O_{11}$, has a molar mass of 342 g mol^{-1}, and 2 teaspoons represents 0.105 mol. The cup of coffee has a volume of 250 mL, and since the density of water is 1.00 g mL^{-1}, the mass of the solvent is 250 g, or 13.89 mol, of water. The mole fraction of solvent is

$$X = \frac{\text{mol } H_2O}{\text{mol sucrose} + \text{mol } H_2O}$$

$$= \frac{13.89 \text{ mol } H_2O}{0.108 \text{ mol sucrose} + 13.89 \text{ mol } H_2O}$$

$$= 0.992$$

The vapor pressure of the sugared coffee is then

$$P = (355 \text{ mm Hg})(0.992)$$

$$= 352 \text{ mm Hg}$$

We see from this result that the small mole fraction of sucrose, X = 0.008, has only a minor effect on the vapor pressure. Larger amounts will of course have larger effects.

Vapor Pressures of Volatile Liquid Mixtures

When one liquid is dissolved in another, the vapor pressure of the entire mixture depends on the vapor pressures of the individual liquids and the proportions in which these liquids are mixed. We may use the same logic as for nonvolatile solutes to understand what is happening. In this case the solute liquid occupies space at the surface of the solution, inhibiting the vaporization of the solvent and decreasing its vapor pressure. Now, however, the solute itself is volatile, and the space it occupies at the surface allows it to vaporize. By using Raoult's law for both volatile components of the solution, we obtain

$$P_{total} = X_1P_1^0 + X_2P_2^0 + \cdots \qquad (7.3)$$

Subscripts 1 and 2 represent the two different volatile liquids in the mixture, and the ellipsis indicates that for mixtures of more than two volatile liquids similar terms are added. At this point we notice that, if one solute is not volatile ($P_2^0 = 0$), equation 7.3 becomes the same as equation 7.2.

EXERCISE

Calculate the vapor pressure of a mixture consisting of equal weights of water and ethyl alcohol, C_2H_5OH, at 35°C, where the vapor pressures of these liquids are 42.0 and 100 millimeters of mercury, respectively.

Solution

If there are equal masses of the two liquids, let us assume that 100 g of each liquid is mixed with 100 g of the other. This gives us 5.56 mol of water and 2.17 mol of ethyl alcohol. The mole fraction of water is

$$X_1 = \frac{5.56 \text{ mol } H_2O}{5.56 \text{ mol } H_2O + 2.17 \text{ mol } C_2H_5OH} = 0.719$$

and the mole fraction of ethyl alcohol is

$$X_2 = \frac{2.17 \text{ mol } C_2H_5OH}{5.56 \text{ mol } H_2O + 2.17 \text{ mol } C_2H_5OH} = 0.281$$

Entering this information, along with the corresponding vapor pressures of the pure liquids, into the Raoult's law equation yields

$$P_{total} = (42 \text{ mm Hg})\ (0.719) + (100 \text{ mm Hg})\ (0.281)$$
$$= 58.3 \text{ mm Hg}$$

From Raoult's law for mixtures of liquids, we can make the generalization that the total vapor pressure of a liquid-liquid solution will always lie somewhere between the lowest and highest vapor pressures of the two liquids.

Figure 7.20 is a pictoral representation of Raoult's law for a binary mixture of water and ethyl alcohol.

Ideal Solutions and Raoult's Law

The preceding discussion of Raoult's law assumes that the solutions are ideal solutions. An ideal solution is one in which the energy used to disrupt the attractive forces in the solvent and solute is exactly balanced by the energy

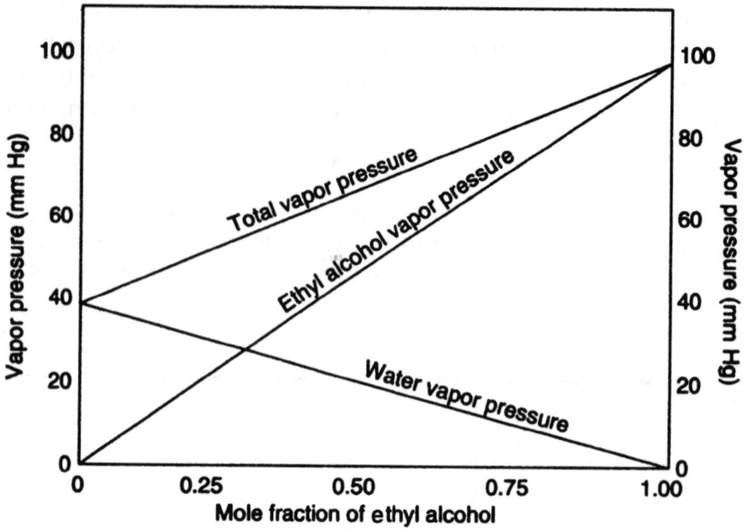

FIGURE 7.20. *Graphical representation of Raoult's law for a binary liquid mixture of water and ethyl alcohol at 35°C.*

released when the solution forms. In other words, the attractive forces in the pure solute and solvent are the same as the attractive forces in the solution. If the solution is not ideal, however, deviations from Raoult's law occur. These deviations may be summarized as follows:

1. *Positive deviation*: The vapor pressure of the solution is greater than expected from Raoult's law because the attractive forces in the solution

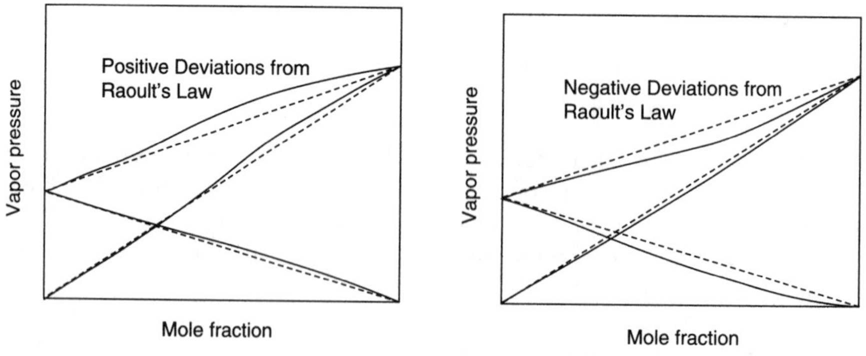

FIGURE 7.21. *Positive (left) and negative (right) deviations from Raoult's law. Dashed lines indicate Raoult's law for ideal solutions.*

are weaker than those in the pure solute and solvent. A solution with positive deviation cools as the mixture is prepared since the extra energy needed to break the attractions is absorbed from the surroundings.

2. *Negative deviation*: The vapor pressure is lower than expected from Raoult's law because the attractive forces in the solution are greater than those in the pure solute and solvent. These solutions produce extra energy that is given off as heat. A mixture of this type becomes warmer when the solution is prepared.

Figure 7.21 illustrates the effects of positive and negative deviations from Raoult's law.

8
REACTIONS

BALANCING EQUATIONS

Balancing Equations by Inspection

All chemical reactions involve the conversion of **reactants** into **products**. Chemical equations are written to describe these reaction processes. The formulas of the reactants are written on the left and the products on the right side of an arrow that indicates that the reactants are converted into products:

$$\text{Reactants} \quad \rightarrow \quad \text{products}$$

The order of the reactants and products in an equation does not matter.

$$Zn \quad + \quad H_2SO_4 \quad \rightarrow \quad H_2 \quad + \quad ZnSO_4$$

is the same as

$$H_2SO_4 \quad + \quad Zn \quad \rightarrow \quad ZnSO_4 \quad + \quad H_2$$

Chemical equations must be **balanced** so that there is the same number of each atom on both sides of the arrow. (The arrow is similar to an equal sign in an ordinary mathematical equation.) A balanced chemical equation satisfies the law of conservation of matter.

Equations are balanced by inserting **coefficients** for the reactants and products in order to equalize the atoms on the two sides of the arrow. Here are two very important rules of balancing:

1. A coefficient multiplies all of the atoms in the compound immediately to the right of the coefficient.
2. Subscripts or formulas are *never altered* to balance an equation.

One common reaction is the combustion of propane fuel, $CH_3CH_2CH_3$, in a barbecue grill:

$$CH_3CH_2CH_3 \quad + \quad O_2 \quad \rightarrow \quad CO_2 + H_2O \quad \text{(unbalanced)}$$

As written, this equation simply gives the reactants ($CH_3CH_2CH_3$ + O_2) and the products (CO_2 + H_2O). It can be balanced by using the appropriate coefficients for the two reactants and two products. There are two methods for determining these coefficients. One is the **inspection method**, and the other is the **ion-electron method**, which is used for complex oxidation-reduction equations and is described on pages 305–310.

The Inspection Method of Balancing

The first step in the inspection method of balancing an equation involves counting the number of each atom in the equation on the reactant side and then on the product side. This step requires care and attention to detail since the smallest mistake will ruin the entire effort. The next step is to balance one atom at a time by adding a coefficient and then recounting all of the atoms. Adding coefficients and recounting continue until the same number of atoms is present on each side of the arrow.

Chemists find that the balancing process works most smoothly if they start balancing the atoms in the most complex molecule first, leaving the simple compounds and elements until last. Also, atoms that appear in more than one compound on either the reactant or product side are left to the end. Finally, it is faster to balance groups of atoms, such as the polyatomic sulfate or nitrate ions, as listed in Table 7.2 on page 246, as if they were individual atoms.

Returning to the combustion of propane, we count 3 carbon, 8 hydrogen, and 2 oxygen atoms on the reactant side. On the product side we have 1 carbon, 2 hydrogen, and 3 oxygen atoms. Since propane, $CH_3CH_2CH_3$, is the most complex molecule in the reaction, it is used as the starting point. Its 3 carbon atoms can be balanced by adding the coefficient 3 to the CO_2:

$$CH_3CH_2CH_3 \quad + \quad O_2 \quad \rightarrow \quad \mathbf{3}CO_2 \quad + \quad H_2O \qquad \text{(still unbalanced)}$$

There are now 3 carbon atoms on both sides, but the numbers of hydrogen and oxygen atoms are still not equal.

Next, the 8 hydrogen atoms may be balanced by adding the coefficient 4 to the water molecules:

$$CH_3CH_2CH_3 \quad + \quad O_2 \quad \rightarrow \quad 3CO_2 \quad + \quad \mathbf{4}H_2O \qquad \text{(still unbalanced)}$$

Recounting the atoms, we find that there are now 3 carbon, 8 hydrogen, and 2 oxygen atoms on the reactant side and 3 carbon, 8 hydrogen, and 10 oxygen atoms on the product side. Only the oxygen atoms remain to be balanced. This can be done by using the coefficient 5 for the O_2 molecule:

$$CH_3CH_2CH_3 \quad + \quad \mathbf{5}O_2 \quad \rightarrow \quad 3CO_2 \quad + \quad 4\,H_2O \qquad \text{(balanced)}$$

The equation now has 3 carbon, 8 hydrogen, and 10 oxygen atoms on both the reactant and the product side. The coefficient for $CH_3CH_2CH_3$ is 1, but it is not written.

The procedure for balancing combustion reactions usually follows the sequence shown above: balance the CO_2 and H_2O, starting with the organic compound; balance the oxygen atoms last.

Another type of reaction is the double-replacement reaction. One reaction of this type is as follows:

$$Cr_2(SO_4)_3 \quad + \quad KOH \quad \rightarrow \quad Cr(OH)_3 \quad + \quad K_2SO_4 \qquad \text{(unbalanced)}$$

In this unbalanced form there are 2 chromium, 3 sulfur, 13 oxygen, 1 hydrogen, and 1 potassium atom on the reactant side and 1 chromium, 1 sulfur, 7 oxygen, 3 hydrogen, and 2 potassium atoms on the product side. The chemist would see 2 chromium atoms, 3 sulfate ions (the SO_4 unit), 1 potassium atom, and 1 hydroxide ion (the OH unit) on the reactant side and 1 chromium atom, 1 sulfate ion, 2 potassium atoms, and 3 hydroxide ions on the product side. Focusing on the $Cr_2(SO_4)_3$, we can balance the chromium atoms by placing the coefficient 2 in front of $Cr(OH)_3$:

$$Cr_2(SO_4)_3 \quad + \quad KOH \quad \rightarrow \quad \mathbf{2}Cr(OH)_3 \quad + \quad K_2SO_4 \quad \text{(unbalanced)}$$

Next, the 6 hydroxide ions in $2Cr(OH)_3$ can be balanced by using the coefficient 6 in front of KOH:

$$Cr_2(SO_4)_3 \quad + \quad \mathbf{6}KOH \quad \rightarrow \quad 2Cr(OH)_3 \quad + \quad K_2SO_4 \quad \text{(unbalanced)}$$

Then the 6 potassium atoms in 6KOH can be balanced by using 3 for the coefficient of K_2SO_4; this also balances the sulfate ion, and the equation is balanced:

$$Cr_2(SO_4)_3 \quad + \quad 6KOH \quad \rightarrow \quad 2Cr(OH)_3 \quad + \quad \mathbf{3}K_2SO_4 \quad \text{(balanced)}$$

The balancing of this equation followed a ping-pong sequence from one side of the arrow to the other and back again. Double-replacement equations are often balanced in this manner, starting with the most complex compound.

Properly Balanced Equations

There is a difference between an equation that is balanced and one that is properly balanced. In a **properly balanced equation** the coefficients are the smallest possible whole numbers. For example, the reaction of hydrogen with oxygen to form water is properly balanced as

$$2H_2 \quad + \quad O_2 \quad \rightarrow \quad 2H_2O$$

It is also balanced if written as

$$4H_2 \quad + \quad 2O_2 \quad \rightarrow \quad 4H_2O$$
$$16H_2 \quad + \quad 8O_2 \quad \rightarrow \quad 16H_2O$$
$$H_2 \quad + \quad \frac{1}{2}O_2 \quad \rightarrow \quad H_2O$$

Although the last three equations are technically balanced since each has the same number of hydrogen atoms and oxygen atoms on either side of the arrow, they are not in the best form.

> Properly balanced equations have the smallest whole-number coefficients possible.

To be properly balanced, the first of the three equations above should be divided by 2, the second should be divided by 8 and the third should be multiplied by 2. An equation may be multiplied or divided as necessary, but it must be remembered that all coefficients in the equation must be multiplied or divided by the same factor.

Balancing equations requires practice to develop skill and speed. The following exercise provides equations to balance by inspection.

EXERCISE

Balance each of the following reactions by inspection:

(a) $C_6H_6 + O_2 \rightarrow CO_2 + H_2O$

(b) $MgCl_2 + AgNO_3 \rightarrow AgCl + Mg(NO_3)_2$

(c) $Al + O_2 \rightarrow Al_2O_3$

(d) $CaO + H_2SO_4 \rightarrow H_2O + CaSO_4$

(e) $Al + Fe_3O_4 \rightarrow Fe + Al_2O_3$

(f) $NO_2 + O_2 \rightarrow N_2O_5$

(g) $HCl + CaCO_3 \rightarrow CaCl_2 + H_2O + CO_2$

(h) $O_2 + C_4H_9NH_2 \rightarrow CO_2 + H_2O + N_2$

(i) $Mg + HCl \rightarrow H_2 + MgCl_2$

(j) $Zn + Cu(NO_3)_2 \rightarrow Cu + Zn(NO_3)_2$

(k) $CoCl_3 + Ba(OH)_2 \rightarrow BaCl_2 + Co(OH)_3$

(l) $Ba(OH)_2 + H_3PO_4 \rightarrow H_2O + Ba_3(PO_4)_2$

(m) $C_6H_8 + H_2 \rightarrow C_6H_{12}$

(n) $SO_2 + O_2 \rightarrow SO_3$

(o) $C_4H_{10} + O_2 \rightarrow CO_2 + H_2O$

(p) $H_2S + AuCl_3 \rightarrow Au_2S_3 + HCl$

Answers

(a) $2C_6H_6 + 15O_2 \rightarrow 12CO_2 + 6H_2O$

(b) $MgCl_2 + 2AgNO_3 \rightarrow 2AgCl + Mg(NO_3)_2$

(c) $4Al + 3O_2 \rightarrow 2Al_2O_3$

(d) $CaO + H_2SO_4 \rightarrow H_2O + CaSO_4$

(e) $8Al + 3Fe_3O_4 \rightarrow 9Fe + 4Al_2O_3$

(f) $4NO_2 + O_2 \rightarrow 2N_2O_5$

(g) $2HCl + CaCO_3 \rightarrow CaCl_2 + H_2O + CO_2$

(h) $27O_2 + 4C_4H_9NH_2 \rightarrow 16CO_2 + 22H_2O + 2N_2$

(i) $Mg + 2HCl \rightarrow H_2 + MgCl_2$

(j) $Zn + Cu(NO_3)_2 \rightarrow Cu + Zn(NO_3)_2$

(k) $2CoCl_3 + 3Ba(OH)_2 \rightarrow 3BaCl_2 + 2Co(OH)_3$

(l) $3Ba(OH)_2 + 2H_3PO_4 \rightarrow 6H_2O + Ba_3(PO_4)_2$

(m) C_6H_8	+	$2H_2$	→	C_6H_{12}		
(n) $2SO_2$	+	O_2	→	$2SO_3$		
(o) $2C_4H_{10}$	+	$13O_2$	→	$8CO_2$	+	$10H_2O$
(p) $3H_2S$	+	$2AuCl_3$	→	Au_2S_3	+	$6HCl$

BALANCING REDOX EQUATIONS

Redox Reaction Balancing Methods

Chemical equations for redox reactions tend to be complex, and attempts to balance them by inspection often fail to achieve results within a reasonable length of time. Over the years, however, chemists have devised several other methods for balancing redox reactions. The simplest and most versatile is the **ion-electron method**.

To use the ion-electron method properly, six steps must be performed in the order given below. These six steps are used for reactions that occur in acid (H^+) solution. A seventh step must be added if the reaction occurs in basic (OH^-) solution, or if H^+ appears on one side and OH^- appears on the other side of the equation.

ION-ELECTRON METHOD

1. Write two half-reactions, one to represent the oxidation, and the other the reduction, that occur in the reaction. It is not necessary to know which is which at this point.
2. In each half-reaction, balance all atoms except hydrogen and oxygen.
3. Balance the oxygen atoms in each half-reaction by adding one H_2O molecule for each oxygen atom needed. Never use O_2, OH^-, or any other form of oxygen.
4. Balance the hydrogen atoms by adding hydrogen ions, H^+. Never use H_2, OH^-, H_2O, or any other form of hydrogen.
5. Balance the charges by adding electrons. If Steps 1–4 have been done properly, electrons will be added to the left side of one half-reaction and to the right side of the other.
6. Multiply each half-reaction by the appropriate number so that both half-reactions have the same number of electrons. Add the half-reactions. Cancel the electrons (they must cancel) and all common ions and molecules. Write the equation and simplify the coefficients if possible.
7. Add one OH^- ion for each H^+ ion to each side of the equation written in Step 6. Combine the H^+ and OH^- ions on one side of the reaction into H_2O molecules. Cancel H_2O molecules that appear on both sides of the equation, and simplify if possible.

Before using the ion-electron method, it is worthwhile to try balancing any equation by inspection. For example, the unbalanced equation for the reaction of magnesium metal with acid may be written as

$$Mg + H^+ \rightarrow Mg^{2+} + H_2$$

By inspection we see that a coefficient 2 in front of the H^+ will balance the hydrogen atoms and the charge in this equation:

$$Mg + 2H^+ \rightarrow Mg^{2+} + H_2$$

There is no need to use the longer ion-electron method.

The next example illustrates a situation where the ion-electron method must be used. Standard solutions of iodine are prepared by reacting iodide ions with the iodate ion in acid solution. The skeleton reaction is as follows:

$$I^- + IO_3^- \rightarrow I_2$$

Step 1 of the ion-electron method requires two half-reactions, which are written as

$$I^- \rightarrow I_2$$

and

$$IO_3^- \rightarrow I_2$$

Even though the skeleton reaction has only one product, there is no reason why both reactants cannot yield the same product, one by oxidation and the other by reduction. Usually two pairs of reactants can be identified to obtain the half reactions. In addition, H^+, OH^-, or H_2O may be included in the skeleton reaction. These are generally ignored, however, since they are added in later steps.

Step 2 requires the balancing of the iodine atoms, using the coefficient 2 in both half-reactions:

$$2I^- \rightarrow I_2$$
$$2IO_3^- \rightarrow I_2$$

Step 3 applies only to the second half-reaction since the first one contains no oxygen. Six water molecules are needed to balance the 6 oxygen atoms in two iodate ions:

$$2I^- \rightarrow I_2$$
$$2IO_3^- \rightarrow I_2 + 6H_2O$$

Step 4 also involves only the second half-reaction, where $12H^+$ ions are required:

$$2I^- \rightarrow I_2$$
$$12H^+ + 2IO_3^- \rightarrow I_2 + 6H_2O$$

At this point all atoms are balanced, and we use Step 5 to balance the charges with electrons. To balance the charges, we first count the charge on each side of the half-reactions. The first half-reaction has a total charge of −2 on the left and 0 on the right. The second half-reaction has a total charge of +10 on the left and 0 on the right. Two electrons are added on the right side of the first half-reaction to equalize the total charge at −2. Ten electrons are added to the left side of the second half-reaction, resulting in a total charge of 0 on both sides:

$$2I^- \rightarrow I_2 + 2e^-$$
$$10e^- + 12H^+ + 2IO_3^- \rightarrow I_2 + 6H_2O$$

Now we can equalize the electrons in the two half-reactions by multiplying the entire first half-reaction by 5:

$$10I^- \rightarrow 5I_2 + 10e^-$$
$$10e^- + 12H^+ + 2IO_3^- \rightarrow I_2 + 6H_2O$$

In Step 6, adding the equations yields

$$10I^- + 10e^- + 12H^+ + 2IO_3^- \rightarrow 5I_2 + 10e^- + I_2 + 6H_2O$$

We then cancel 10 electrons from each side, add the $5I_2$ to the I_2 to obtain $6I_2$, and divide all of the coefficients by 2 to obtain the final balanced equation:

$$5I^- + 6H^+ + IO_3^- \rightarrow 3I_2 + 3H_2O$$

All atoms must be balanced in the equation. In addition, in ionic reactions the charges must also balance. In this equation we count a total charge of 0 on both sides of the arrow.

As a third example, the reaction between the permanganate ion and iodide ions occurs in basic solution. The unbalanced skeleton reaction is as follows:

$$MnO_4^- + I^- \rightarrow I_2 + MnO_2$$

Step 1: The obvious pairs for the half-reactions are

$$MnO_4^- \rightarrow MnO_2$$

and

$$I^- \rightarrow I_2$$

Step 2: In the second half-reaction the iodine atoms must be balanced:

$$MnO_4^- \rightarrow MnO_2$$
$$2I^- \rightarrow I_2$$

Step 3: In the first half-reaction 2 water molecules are needed to balance the oxygen atoms:

$$MnO_4^- \rightarrow MnO_2 + 2H_2O$$
$$2I^- \rightarrow I_2$$

Step 4: Four hydrogen ions are needed to balance the hydrogen atoms:

$$4H^+ + MnO_4^- \rightarrow MnO_2 + 2H_2O$$
$$2I^- \rightarrow I_2$$

Step 5: The charges must be balanced. The first half-reaction has a total charge of +3 on the left and 0 on the right, while the second half-reaction has a total charge of −2 on the left and 0 on the right. Three electrons are added to the first half-reaction on the left, and 2 electrons to the second half-reaction on the right:

$$3e^- + 4H^+ + MnO_4^- \rightarrow MnO_2 + 2H_2O$$
$$2I^- \rightarrow I_2 + 2e^-$$

To equalize the electrons, the first half-reaction is multiplied by 2 and the second half-reaction is multiplied by 3:

$$6e^- + 8H^+ + 2MnO_4^- \rightarrow 2MnO_2 + 4H_2O$$
$$6I^- \rightarrow 3I_2 + 6e^-$$

Step 6: The equations are added:

$$6e^- + 6I^- + 8H^+ + 2MnO_4^- \rightarrow 2MnO_2 + 4H_2O + 3I_2 + 6e^-$$

and the 6 electrons are canceled:

$$6I^- + 8H^+ + 2MnO_4^- \rightarrow 2MnO_2 + 4H_2O + 3I_2$$

This equation is balanced for a reaction in acid solution.

To convert to a basic solution, Step 7 is used and 8 OH^- ions are added to each side:

$$8OH^- + 6I^- + 8H^+ + 2MnO_4^- \rightarrow 2MnO_2 + 4H_2O + 3I_2 + 8OH^-$$

The $8H^+$ and 8 OH^- ions on the left are combined to make 8 H_2O molecules:

$$8H_2O + 6I^- + 2MnO_4^- \rightarrow 2MnO_2 + 4H_2O + 3I_2 + 8OH^-$$

We then cancel 4 H_2O molecules from each side to simplify the final equation:

$$4H_2O + 6I^- + 2MnO_4^- \rightarrow 2MnO_2 + 3I_2 + 8OH^-$$

EXERCISES

1. Balance each of the following half-reactions in acid solution:
   (a) $Fe(s) \rightarrow Fe^{3+}$
   (b) $Cl_2 \rightarrow Cl^-$
   (c) $Cr^{3+} \rightarrow CrO_4^{2-}$
   (d) $NO_3^- \rightarrow NO_2$
   (e) $S_2O_3^{2-} \rightarrow SO_4^{2-}$

2. Balance each of the following skeleton redox reactions in the solution indicated:

(a) $Cr_2O_7^{2-} + CH_3CH_2OH \rightarrow Cr^{3+} + CH_3COOH$ (acid solution)
(b) $Cu + NO_3^- \rightarrow NO_2 + Cu^{2+}$ (acid solution)
(c) $MnO_2 + ClO_3^- \rightarrow MnO_4^- + Cl^-$ (basic solution)
(d) $Al + H_2O \rightarrow Al(OH)_4^- + H_2$ (basic solution)

Answers

1. (a) $Fe(s) \rightarrow Fe^{3+} + 3e^-$
(b) $2e^- + Cl_2 \rightarrow 2Cl^-$
(c) $4H_2O + Cr^{3+} \rightarrow CrO_4^{2-} + 8H^+ + 3e^-$
(d) $e^- + 2H^+ + NO_3^- \rightarrow NO_2 + H_2O$
(e) $5H_2O + S_2O_3^{2-} \rightarrow 2SO_4^{2-} + 10H^+ + 8e^-$

2. (a) The balanced half-reactions are

$$6e^- + 14H^+ + Cr_2O_7^{2-} \rightarrow 2Cr^{3+} + 7H_2O$$
$$H_2O + CH_3CH_2OH \rightarrow CH_3COOH + 4H^+ + 4e^-$$

and the balanced equation is

$$16H^+ + 2Cr_2O_7^{2-} + 3CH_3CH_2OH \rightarrow 4Cr^{3+} + 3CH_3COOH + 11H_2O$$

(b) The balanced half-reactions are

$$Cu \rightarrow Cu^{2+} + 2e^-$$
$$e^- + 2H^+ + NO_3^- \rightarrow NO_2 + H_2O$$

and the balanced equation is

$$Cu + 2NO_3^- + 4H^+ \rightarrow 2NO_2 + Cu^{2+}$$

(c) The balanced half-reactions are

$$6e^- + 6H^+ + ClO_3^- \rightarrow Cl^- + 3H_2O$$
$$2H_2O + MnO_2 \rightarrow MnO_4^- + 4H^+ + 3e^-$$

and the balanced equation is

$$2MnO_2 + 2OH^- + ClO_3^- \rightarrow 2MnO_4^- + Cl^- + H_2O$$

(d) The balanced half-reactions are

$$4H_2O + Al \rightarrow Al(OH)_4^- + 4H^+ + 3e^-$$
$$2e^- + 2H^+ \rightarrow H_2$$

and the balanced equation is

$$2Al + 2OH^- + 6H_2O \rightarrow 2Al(OH)_4^- + 3H_2$$

OXIDATION NUMBERS

Determining Oxidation Numbers

In complex reactions it is not always obvious that electrons are transferred. To determine whether a substance has lost or gained electrons, chemists calculate the **oxidation number**, also called the **oxidation state**, of each element. If the oxidation number changes during a reaction, electrons have been transferred.

To determine the oxidation numbers of the elements in a compound or polyatomic ion, we follow the six rules shown below.

OXIDATION NUMBER RULES

1. The oxidation numbers of all atoms add up to the charge on the molecule or ion.
2. For an alkali metal the oxidation number is +1, for an alkaline earth element it is +2, and for the metal in Group IIIA it is +3.
3. The oxidation number of hydrogen is +1, and the oxidation number of fluorine is −1.
4. The oxidation number of oxygen is −2.
5. The oxidation number of a halogen is −1.
6. The oxidation number for a nonmetal in Group VIA is −2.

The oxidation number rules constitute a hierarchy, meaning that rule 1 is the most important and rule 6 the least important. If two rules conflict, the rule closest to the top of the list is obeyed and the other is ignored.

Many simple compounds follow the above rules directly. Since the charge of any element is zero, the oxidation number of any element must also be zero according to rule 1. The charge of any monatomic ion is also its oxidation number, again according to rule 1. For example, $Al^{3+}(aq)$ has an oxidation number of +3.

For a compound such as $CaCl_2$, the calcium is +2 according to rule 2 and each chlorine is −1 by rule 5; the most important rule, rule 1, is obeyed since +2 and $2 \times (-1)$ add up to zero, the charge on $CaCl_2$. In potassium hydroxide, KOH, the potassium is +1 according to rule 2, hydrogen is +1 according to rule 3, and oxygen is −2 according to rule 4. Again, rule 1 is obeyed since $+1 +1 - 2 = 0$, which is the charge on KOH.

Some compounds, however, do not obey one or more of the rules. Since all substances must obey rule 1, we can write rule 1 as an equation where the total charge is equal to the sum of the elements' oxidation numbers multiplied in each case, by the number of times the element appears in the formula:

$$\text{Total charge} = (\text{Ox. No. } 1)(\text{subscript } 1) + (\text{Ox. No. } 2)(\text{subscript } 2) + \ldots$$

For example, the oxidation numbers for the elements in perchloric acid, $HClO_4$, would be +1 for hydrogen, −1 for chlorine, and −2 for each oxygen if rules 3, 4, and 5 were strictly followed. These assignments violate rule 1, however, since the oxidation numbers add up to −8 and the $HClO_4$ molecule must have a charge of zero. We therefore ignore rule 5 in favor of rule 1 and write the equation as

$$\begin{aligned} 0 \text{ charge} &= (\text{Ox. No. H})(1) + (\text{Ox. No. Cl})(1) + (\text{Ox. No. O})(4) \\ 0 &= (+1)(1) + (\text{Ox. No. Cl})(1) + (-2)(4) \\ 0 &= \text{Ox. No. Cl} - 7 \\ \text{Ox. No. Cl} &= +7 \end{aligned}$$

For perchloric acid the oxidation number of the chlorine atom is +7.

Oxidation numbers for polyatomic ions are determined using the same method, being sure that the numbers add up to the charge on the ion. For example, we might assign all the oxygen and sulfur atoms in the sulfate ion, SO_4^{2-}, oxidation numbers of −2 according to rules 4 and 6. The result disobeys rule 1, however, since the sum of the oxidation numbers is −2 − 8 = −10, whereas the actual charge on the sulfate ion is −2. Consequently, we abandon rule 6 and work with rules 1 and 4 to write this equation:

$$\begin{aligned} -2 \text{ charge} &= (\text{Ox. No. S})(1) + (\text{Ox. No. O})(4) \\ -2 &= \text{Ox. No. S} + (-2)(4) \\ &= \text{Ox. No. S} - 8 \\ \text{Ox. No. S} &= +6 \end{aligned}$$

For the sulfate ion, SO_4^{2}, the oxidation number of the sulfur atom is +6. Finally, the rules specify oxidation numbers for only 25 of the 111 known elements. Using the oxidation number rules and the formulas of the polyatomic ions, we can determine the numbers of most of the remaining elements. For example, the oxidation numbers for chromium and oxygen in the dichromate ion, $Cr_2O_7^{2-}$, can be determined. We use rules 1 and 4 to calculate the oxidation number of chromium:

$$\begin{aligned} -2 \text{ charge} &= (\text{Ox. No. Cr})(2) + (\text{Ox. No. O})(7) \\ -2 &= (\text{Ox. No. Cr})(2) + (-2)(7) \\ &= (\text{Ox. No. Cr})(2) - 14 \\ (\text{Ox. No. Cr})(2) &= +12 \\ \text{Ox. No. Cr} &= +\frac{12}{2} = +6 \end{aligned}$$

Some compounds may have two elements that are not covered by the oxidation number rules. One such compound is nickel(II) carbonate, $NiCO_3$. In this situation we take advantage of our knowledge that the carbonate ion is CO_3^{2-} and the nickel must be a +2 ion, Ni^{2+}. For Ni^{2+}, rule 1 applies directly, and its oxidation number is +2. For the carbonate ion we write

$$-2 \text{ charge} = (\text{Ox. No. C})(1) + (\text{Ox. No. O})(3)$$
$$-2 = (\text{Ox. No. C})(1) + (-2)(3)$$
$$= \text{Ox. No. C} - 6$$
$$\text{Ox. No. C} = +4$$

Using this procedure, we have determined that the oxidation numbers for $NiCO_3$ are Ni = +2, C = +4, and each oxygen = −2.

EXERCISE

Determine the oxidation number of each element in each of the following formulas:

(a) H_2O_2
(b) $HClO_3$
(c) K_3PO_4
(d) Na_2CrO_4
(e) $Fe_2(C_2O_4)_3$
(f) MnO_4^-
(g) SO_3^{2-}
(h) N_2O_5
(i) NH_4^+
(j) $FeNH_4(SO_4)_2$

Answers
(a) H = +1; O = −1
(b) H = +1; Cl = +5; O = −2
(c) K = +1; P = +5; O = −2
(d) Na = +1; Cr = +6; O = −2
(e) Fe = +3; C = +3; O = −2
(f) Mn = +7; O = −2
(g) S = +4; O = −2
(h) N = +5; O = −2
(i) N = −3; H = +1
(j) Fe = +2, N = −3; H = +1; S = +6; O = −2

Using Oxidation Numbers

We use oxidation numbers to determine whether a substance has been oxidized or reduced and, if so, which reaction has occurred. When the permanganate ion, MnO_4^-, reacts in acid solution to form Mn^{2+}, we find that the oxidation number of the manganese, which is +7 in the permanganate ion, is now +2 Mn^{2+} ion. This change in oxidation number indicates that an oxidation-reduction process has taken place. In addition, the fact that the oxidation number was reduced from +7 to +2 tells us that a reduction of manganese has occurred. The fact that the oxidation number changes by 5 tells us that five electrons were gained in the reduction process.

EXERCISE

Possible reactants and products for oxidation-reduction reactions are given below. In each case, determine the oxidation numbers to tell whether an oxidation or a reduction has occurred and, if so, how many electrons the starting substance lost or gained.

(a) $Cr_2O_7^{2-}$ to CrO_4^{2-}
(b) $C_2O_4^{2-}$ to CO_2
(c) NO_3^- to NO_2
(d) I_3^- to IO_3^-
(e) $BaCl_2$ to $BaSO_4$
(f) Fe to Fe_2O_3
(g) C_2H_4 to C_2H_6

Solutions

(a) No change in any oxidation number, not a redox process.
(b) C changes from +3 to +4. Two electrons per $C_2O_4^{2-}$ are gained in this oxidation.
(c) N changes from +5 to +4. One electron per NO_3^- is used in this reduction.
(d) I changes from $-1/3$ to +5. Five electrons per I (10 electrons per I_2) are used in this oxidation.
(e) Ba does not change; no redox process occurs (chloride and sulfate are ignored).
(f) Fe changes from 0 to +3. Three electrons per Fe are lost in this oxidation.
(g) C changes from −2 to −3. Two electrons per C_2H_4 are gained in this reduction.

In a redox equation we can now identify what is occurring. For example, Fe^{2+} is often titrated with permanganate ions. The unbalanced reaction is as follows:

$$Fe^{2+} + MnO_4^- \rightarrow Mn^{2+} + Fe^{3+}$$

From this equation we see that Fe^{2+} is oxidized to Fe^{3+}. The manganese in the permanganate ion has an oxidation number of +7 as a reactant and of +2 as a product; therefore the manganese is reduced. We may also say that the permanganate ion is reduced. Because the substance oxidized causes the reduction of the other reactant, it is called the **reducing agent**. In the above reaction, Fe^{2+} is oxidized, and it is also the reducing agent. Similarly, the MnO_4^- ion is reduced and is therefore the **oxidizing agent**. Although the manganese in the permanganate ion is the atom actually reduced, chemists never say that the manganese atom is the reducing agent. The terms *oxidizing agent* and *reducing agent* are used with the entire formula unit that contains the elements that are reduced or oxidized; these terms are not applied to the elements themselves.

EXERCISE

For each of the following unbalanced equations, identify the element that is oxidized, the element that is reduced, and the changes in oxidation number. Also identify the oxidizing and reducing agents.

(a) $O_2 + N_2H_4 \rightarrow H_2O_2 + N_2$

(b) $XeO_3 + I^- \rightarrow Xe + I_2$

(c) $I_2 + OCl^- \rightarrow IO_3^- + Cl^-$

(d) $NO_3^- + Cu \rightarrow NO_2 + Cu^{2+}$

(e) $PbO_2 + Cl^- \rightarrow PbCl_2 + Cl_2$

Answers

The table below lists the requested information; the changes in oxidation number are indicated after the elements in the first two columns.

	Element Oxidized	Element Reduced	Oxidizing Agent	Reducing Agent
(a)	N (+2)	O (−1)	O_2	N_2H_4
(b)	I (+1)	Xe (−6)	XeO_3	I^-
(c)	I (+5)	Cl (−2)	OCl^-	I_2
(d)	Cu (+2)	N (−1)	NO_3^-	Cu
(e)	Cl (+1)	Pb (−2)	PbO_2	Cl^-

NET IONIC EQUATIONS

Molecular, Ionic, and Net-Ionic Equations

Ionic Reactions; Ions in Solution

Most ionic compounds dissolve in water, and in the process the compound separates into the cations and anions. This solution process may be written as follows:

$$NaBr(s) \rightarrow Na^+(aq) + Br^-(aq)$$

The symbol in parentheses designates the state of each substance in the reaction: (s) means that the substance is a solid; (aq), that the substance is in an aqueous solution. Other symbols used are (ℓ) for liquid and (g) for gas. Chromium(III) nitrate dissolves according to this equation:

$$Cr(NO_3)_3(s) \rightarrow Cr^{3+}(aq) + 3NO_3^-(aq)$$

One chromium(III) ion and three nitrate ions are obtained from one formula unit of chromium(III) nitrate.

Three general principles apply to the dissolution of ionic compounds:

1. Only one cation and one anion are formed. Compounds containing three or more different atoms will break apart into the appropriate polyatomic ion(s). Exceptions are discussed in higher level chemistry courses.
2. The charges on all the ions must add up to zero for a compound.
3. The subscripts of monatomic ions become coefficients for the ions. For polyatomic ions, only the subscripts after parentheses become coefficients.

The next exercise illustrates these principles.

EXERCISE

Write the ions expected when each of the following compounds is dissolved in water:

(a) K_2S
(b) $FeCl_3$
(c) $MgBr_2$
(d) $AlCl_3$
(e) $Na_2Cr_2O_7$
(f) $(NH_4)_2S$
(g) K_3PO_4
(h) $Ti(NO_3)_4$

Answers

(a) $K_2S(s) \rightarrow 2K^+(aq) + S^{2-}(aq)$
(b) $FeCl_3(s) \rightarrow Fe^{3+}(aq) + 3Cl^-(aq)$
(c) $MgBr_2(s) \rightarrow Mg^{2+}(aq) + 2Br^-(aq)$
(d) $AlCl_3(s) \rightarrow Al^{3+}(aq) + 3Cl^-(aq)$
(e) $Na_2Cr_2O_7(s) \rightarrow Na^+(aq) + Cr_2O_7^{2-}(aq)$
(f) $(NH_4)_2S(s) \rightarrow 2NH_4^+(aq) + S^{2-}(aq)$
(g) $K_3PO_4(s) \rightarrow 3K^+(aq) + PO_4^{3-}(aq)$
(h) $Ti(NO_3)_4(s) \rightarrow Ti^{4+}(aq) + 4NO_3^-(aq)$

Notice that the charges on Fe^{3+} and Ti^{4+} must be calculated from the known negative charge on the anion and the fact that the total charge must add up to zero. Also observe which subscripts have become coefficients and which have remained as part of a polyatomic ion. Finally, be aware that this entire process is just the reverse of the method used to determine the formulas of ionic compounds.

Any ionic compound can be broken apart into its cations and anions in this manner. Whether or not an ionic compound will dissolve to an appreciable extent in water depends on which cations and anions make up the compound. Six general guidelines for predicting solubility should be remembered.

SOLUBILITY RULES

1. All compounds containing alkali metal cations or the ammonium ion are **soluble**.
2. All compounds containing the NO_3^-, ClO_4^-, ClO_3^-, or $C_2H_3O_2^-$ anion are **soluble**.
3. All chlorides, bromides, and iodides are **soluble except** those containing Ag^+, Pb^{2+}, or Hg_2^{2+}.
4. All sulfates are **soluble except** those containing Hg_2^{2+}, Pb^{2+}, Sr^{2+}, Ca^{2+}, or Ba^{2+}.
5. All hydroxides are **insoluble except** compounds of the alkali metals Ca^{2+}, Sr^{2+}, and Ba^{2+}.
6. All compounds containing the PO_4^{3-}, S^{2-}, CO_3^{2-}, or SO_3^{2-} ions are **insoluble except** those that also contain alkali metals or NH_4^+.

Double-Replacement Reactions; Predicting Products

If we know how to determine which ions make up an ionic compound, we can then mix two ionic compounds together and predict the possible products. These predictions are based on the principles involved in **double-replacement reactions**. As its name suggests, two replacements occur in any of these reactions:

1. The cation of the first salt replaces the cation in the second salt.
2. The cation of the second salt replaces the cation in the first salt.

For example, if we mix together solutions containing $AgNO_3$ and Na_2CrO_4, what products should we predict? The first step is to determine the ions that make up the two reacting compounds; they are Ag^+, NO_3^-, $2Na^+$, and CrO_4^{2-}. The next step is to pair up these four ions in various ways to try to make two new ionic compounds that will be the predicted products. It is worthwhile to look at all of the possible pairs to see how we arrive at our conclusions.

$Ag^+ + NO_3^- \rightarrow AgNO_3$	We already have this as a reactant.
$Ag^+ + Na^+ \rightarrow$	We cannot form an ionic compound from two positive ions.
$2Ag^+ + CrO_4^{2-} \rightarrow Ag_2CrO_4$	This is a possibility.
$Ag^+ + Ag^+ \rightarrow$	We cannot form an ionic compound from two identical ions.
$NO_3^- + Na^+ \rightarrow NaNO_3$	This is a possibility.
$NO_3^- + CrO_4^{2-} \rightarrow$	We cannot form an ionic compound from two negative ions.
$NO_3^- + NO_3^- \rightarrow$	We cannot form an ionic compound from two identical ions.
$2Na^+ + CrO_4^{2-} \rightarrow Na_2CrO_4$	We already have this as a reactant.
$Na^+ + Na^+ \rightarrow$	We cannot form an ionic compound from two identical ions.
$CrO_4^{2-} + CrO_4^{2-} \rightarrow$	We cannot form an ionic compound from two identical ions.

All of the possible combinations of the four ions are given above, along with the reason why each is a good or bad choice. Only two of the combinations give reasonable new compounds; all of the others lead either to impossible situations or back to the original compounds.

Using the only reasonable results as the products, we begin to construct a chemical equation as follows:

$$AgNO_3 + Na_2CrO_4 \rightarrow Ag_2CrO_4 + NaNO_3$$

The next step is to balance the chemical equation so that it has the same number of each atom on both sides of the arrow:

$$2AgNO_3 + Na_2CrO_4 \rightarrow Ag_2CrO_4 + 2NaNO_3$$

Reviewing what was done to predict this equation, we see that only the positions of the silver and sodium ions have been switched on the reactant and product sides. In switching the positions of the metal atoms, we were careful to write the new formulas based on the charges on the ions.

EXERCISE

1. Predict the products obtained from each of the following pairs of ionic compounds. In each case, write a balanced chemical equation for the reaction.
   (a) KCl and $Pb(NO_3)_2$
   (b) $CaCl_2$ and $MgSO_4$
   (c) $NaOH$ and $Fe_2(SO_4)_3$

Solutions

(a) 1. The ions involved are K^+, Cl^-, Pb^{2+}, and $2NO_3^-$. The new products are KNO_3 and $PbCl_2$.

$$2KCl \quad + \quad Pb(NO_3)_2 \quad \rightarrow \quad 2KNO_3 \quad + \quad PbCl_2$$

(b) 1. The ions involved are Ca^{2+}, 2 Cl^-, Mg^{2+}, and SO_4^{2-}. The new products are $CaSO_4$ and $MgCl_2$.

$$CaCl_2 \quad + \quad MgSO_4 \quad \rightarrow \quad CaSO_4 \quad + \quad MgCl_2$$

(c) 1. The ions involved are Na^+, OH^-, 2 Fe^{3+}, and 3 SO_4^{2-}. Note that the charge of the iron is determined by calculation. The new products are Na_2SO_4 and $Fe(OH)_3$.

$$6NaOH \quad + \quad Fe_2(SO_4)_3 \quad \rightarrow \quad 3Na_2SO_4 \quad + \quad 2Fe(OH)_3.$$

We can write these chemical reactions, but the major question is whether a chemical reaction will actually occur if the compounds are mixed together in the laboratory. In the next sections some ways in which the chemist can predict whether an actual reaction will occur are considered.

Chemical Driving Forces

So far, we can predict the products of any mixture of two ionic compounds, but not all such mixtures react. Chemists rely on three fundamental principles to make an educated guess about the possibility that a reaction will occur in a double-replacement reaction. These principles are sometimes described as **driving forces**.

1. The formation of water is perhaps the strongest driving force. In ionic reactions in which water is a product, a reaction will almost certainly occur.
2. Formation of a precipitate (insoluble compound) is another indicator of a strong driving force.
3. Formation of a nonionic (covalent) compound from ionic reactants is also a driving force. Many of these compounds are organic acids (acetic, formic, benzoic acids) or gases such as NH_3, SO_2, and CO_2.

Here are some common examples of these principles:

$HCl + NaOH \rightarrow H_2O + NaCl$ (water formed)

$Na_2SO_4 + Ba(NO_3)_2 \rightarrow BaSO_4(s) + 2NaNO_3$ (precipitate formed)

$KC_2H_3O_2 + HCl \rightarrow HC_2H_3O_2 + KCl$ (covalent compound formed)

$K_2SO_3 + 2HNO_3 \rightarrow 2KNO_3 + H_2O + SO_2$ (gas formed)

In some reactions two of these driving forces may be present. The driving force to form water is the strongest of all and will overcome another force that may be driving the reaction in the opposite direction. In the following equation, the formation of water overcomes the fact that calcium oxide, CaO, is a solid. Since CaO is on the reactant side, it is driving the equation toward the reactants. The production of water is a stronger driving force, however, and the net result is that this reaction actually occurs:

$$CaO(s) \quad + \quad 2HCl \quad \rightarrow \quad CaCl_2 \quad + \quad H_2O$$

Net Ionic Equations

In the process of determining the products of a reaction between ionic compounds, the ions for each substance are determined. In fact, in equations for aqueous solutions the soluble compounds appear only as ions while the insoluble compounds (precipitates), gases, and covalent compounds are written as molecules. It is possible to take a balanced double-replacement reaction and convert it into a net ionic equation that shows the actual reactants, if any, for the given reaction.

For example, here is a simple neutralization reaction:

$$HCl \quad + \quad NaOH \quad \rightarrow \quad NaCl \quad + \quad H_2O$$

The ionic reaction is obtained by writing all of the soluble ionic compounds as ions:

$$H^+ + Cl^- + Na^+ + OH^- \rightarrow Na^+ + Cl^- + H_2O$$

Since the Na^+ and the Cl^- are identical on both sides of the equation, they can be canceled to give the net ionic equation:

$$H^+ \quad + \quad OH^- \quad \rightarrow \quad H_2O$$

This allows the chemist to show that H^+ and OH^- are the active components of the reaction.

A reaction does not occur if potassium chloride and sodium nitrate solutions are mixed. We can demonstrate that no reaction occurs by deducing the reaction products as sodium chloride and potassium nitrate and then writing the net ionic equation. First, we write the molecular equation:

$$KCl \quad + \quad NaNO_3 \quad \rightarrow \quad NaCl \quad + \quad KNO_3 \quad \text{(molecular equation)}$$

Next, we obtain the ionic equation by separating each compound into its ions:

$$K^+ + Cl^- + Na^+ + NO_3^- \rightarrow Na^+ + Cl^- + K^+ + NO_3^- \quad \text{(ionic equation)}$$

Finally, when we cancel identical ions from both sides of this equation, nothing remains. That means that no reaction occurs and

no net ionic equation is possible.

Taking a close look at this reaction, we see also that no driving force is present. There is no water, no precipitate, no covalent molecule, and no gas being formed.

A common laboratory experiment involves the determination of sulfate ions by precipitation with barium ions. The precipitate is carefully collected, dried, and weighed. The reaction between potassium sulfate and barium nitrate may be predicted to produce barium sulfate and sodium nitrate. The balanced equation is as follows:

$$K_2SO_4 \quad + \quad Ba(NO_3)_2 \quad \rightarrow \quad BaSO_4(s) \quad + \quad 2KNO_3$$

and the ionic equation is

$$2K^+ + SO_4^{2-} + Ba^{2+} + 2NO_3^- \rightarrow BaSO_4 + 2K^+ + 2NO_3^-$$

From the ionic equation the two potassium and two nitrate ions can be canceled with this result:

$$SO_4^{2-} \quad + \quad Ba^{2+} \quad \rightarrow \quad BaSO_4$$

This balanced net ionic equation represents the reaction implied above when we mentioned the "determination of sulfate ions by precipitation with barium ions." In chemical analysis, the chemist is usually interested in a specific ion, such as the sulfate ion in this example. The net ionic equation shows how to isolate the sulfate ion from all other ions by precipitation with barium ions.

Reactions that evolve gases are a bit more complex. Archaeologists typically carry a small bottle of hydrochloric acid with them on field trips. Carbonate rocks can be quickly identified since they give off CO_2 (evidenced by bubbling and fizzing) when a few drops of HCl are placed on them. Most of these rocks are made of calcium carbonate. Using the techniques for a double-replacement reaction, we can predict the products as calcium chloride and carbonic acid:

$$2HCl \quad + \quad CaCO_3(s) \quad \rightarrow \quad CaCl_2 \quad + \quad H_2CO_3$$

The ionic equation is as follows:

$$2H^+ + 2Cl^- + CaCO_3(s) \rightarrow Ca^{2+} + 2Cl^- + H_2CO_3$$

In this reaction only the Cl^- ions will cancel:

$$2H^+ \quad + \quad CaCO_3(s) \quad \rightarrow \quad Ca^{2+} \quad + \quad H_2CO_3$$

This equation does not show any carbon dioxide gas. However, H_2CO_3 may also be written as $CO_2 + H_2O$. Substituting this, we obtain the final reaction:

$$2H^+ \quad + \quad CaCO_3(s) \quad \rightarrow \quad Ca^{2+} \quad + \quad H_2O \quad + \quad CO_2(g)$$

Table 8.1 lists the common gases and their equivalents when dissolved in water. The gas and aqueous forms are interchangeable in reactions as needed.

TABLE 8.1. COMMON GASES AND THEIR EQUIVALENTS IN AQUEOUS SOLUTION

Gas Name	Aqueous Form	Gas Form
Carbon dioxide	H_2CO_3	$CO_2 + H_2O$
Sulfur dioxide	H_2SO_3	$SO_2 + H_2O$
Hydrogen sulfide	H_2S	H_2S
Ammonia	NH_4OH*	$NH_3 + H_2O$

*NH_4OH does not actually exist and should always be written as the gas form.

EXERCISE

1. Write the balanced equation for each of the following pairs of ionic substances. Write the ionic and net ionic equations for each reaction.
   (a) $AgNO_3$ and $CaCl_2$
   (b) Na_2CO_3 and $Fe(NO_3)_3$
   (c) CaF_2 and HCl
   (d) NH_4Cl and KOH

Answers

(a) 1. $2AgNO_3 + CaCl_2 \rightarrow 2AgCl + Ca(NO_3)_3$

$2Ag^+ + 2NO_3^- + Ca^{2+} + 2Cl^- \rightarrow 2AgCl + Ca^{2+} + 2NO_3^-$

$Ag^+ + Cl^- \rightarrow AgCl$

(b) 1. $3Na_2CO_3 + 2Fe(NO_3)_3 \rightarrow Fe_2(CO_3)_3 + 6NaNO_3$

$6Na^+ + 3CO_3^{2-} + 2Fe^{3+} + 6NO_3^- \rightarrow Fe_2(CO_3)_3 + 6Na^+ + 6NO_3^-$

$3CO_3^{2-} + 2Fe^{3+} \rightarrow Fe_2(CO_3)_3$

(c) 1. $CaF_2 + 2HCl \rightarrow 2HF + CaCl_2$

$Ca^{2+} + 2F^- + 2H^+ + 2Cl^- \rightarrow 2HF + Ca^{2+} + 2Cl^-$

$F^- + H^+ \rightarrow HF$

(d) 1. $NH_4Cl + KOH \rightarrow NH_4OH + KCl$

$NH_4^+ + Cl^- + K^+ + OH^- \rightarrow NH_3 + H_2O + K^+ + Cl^-$

$NH_4^+ + OH^- \rightarrow NH_3 + H_2O$

Note that the NH_4OH obtained from the double-replacement technique was replaced by NH_3 and H_2O in the ionic equations since NH_4OH does not exist. It should not appear in the first equation either.

Stoichiometry of Redox Reactions

In Chapter 9 stoichiometric calculations that may be used for any chemical reaction are discussed. Stoichiometric calculations for redox reactions are no different from those for other reactions once the balanced equation has been obtained.

Converting Ionic Equations into Molecular Equations

Use of the ion-electron method for balancing a redox equation results in a net ionic equation. The chemicals on the stockroom shelf are complete compounds, not individual ions. It is, necessary, therefore, to translate the ions in a net ionic equation into compounds that can be measured in the laboratory. To do this, we convert a net ionic equation back into a molecular equation by adding the spectator ions that were previously eliminated. The general principles are as follows:

1. A cation is converted into molecular form by adding negatively charged spectator ions to form a soluble compound. There are many possible choices; but unless the problem specifies otherwise, the most convenient spectator ion to add is the chloride ion. If the chloride compound is insoluble, the next most convenient choice is the nitrate ion.
2. An anion is converted into molecular form by adding positively charged spectator ions to form a soluble compound. Again, there are many possible choices; but unless otherwise specified in the problem, the most convenient spectator ion to add is the sodium ion. Hydrogen ions are added to anions to make acids if needed.
3. The spectator ions chosen to convert anions and cations into molecular forms must be added in equal numbers to both sides of the equation to maintain balance.

To illustrate the process, let us determine the molecular equation for this net ionic equation:

$$Ba^{2+} + SO_4^{2-} \rightarrow BaSO_4$$

Adding two chloride ions and two sodium ions to each side yields the molecular equation, with NaCl on the product side:

$$BaCl_2 + Na_2SO_4 \rightarrow BaSO_4 + 2NaCl$$

As another example, in the reaction used to determine ethyl alcohol, CH_3CH_2OH, the net ionic equation is

$$8H^+ + Cr_2O_7^{2-} + 3CH_3CH_2OH \rightarrow 2Cr^{3+} + 3CH_3CHO + 7H_2O$$

Sometimes, however, a question is phrased in terms of potassium dichromate, $K_2Cr_2O_7$. For example, how many grams of potassium dichromate will react with 5.00 grams of ethyl alcohol? It is not necessary to write a complete molecular equation to solve the problem. We simply add on the left the necessary number of K^+ ions to form potassium dichromate and on the right an equal number of potassium ions:

$$8H^+ + K_2Cr_2O_7 + 3CH_3CH_2OH \rightarrow 2Cr^{3+} + 3CH_3CHO + 7H_2O + 2K^+$$

This form of the equation now has the information necessary to produce the factor-label:

$$\left(\frac{1 \text{ mol } K_2Cr_2O_7}{3 \text{ mol } CH_3CH_2OH}\right),$$

required to solve the problem.

9
STOICHIOMETRY

THE MOLE

The Mole and Avogadro's Number

The **mole** (mol) is the central unit of measurement in chemistry. It allows the chemist to relate a chemical equation to masses measured on a balance and to a variety of other units. The modern definition is that exactly 1 mole represents the number of carbon atoms in exactly 12 grams of the carbon-12 isotope. Numerically this represents 6.02×10^{23} carbon atoms, and 6.02×10^{23} is called **Avogadro's number** in honor of his pioneering work in stoichiometry and his recognition that such a number would be very important in chemistry. Avogadro never actually determined the number named as a tribute to him.

For an element 1 mole represents 6.02×10^{23} atoms. For a molecular compound such as CH_4 or CO_2, 1 mole represents 6.02×10^{23} molecules. For an ionic compound such as NaCl or $MgBr_2$, 1 mole represents 6.02×10^{23} empirical formula units.

In mathematical equations the number of moles is represented by the letter n. This symbol is also used for many other purposes, and a thorough understanding of an equation is needed to determine whether n represents moles or some other quantity.

In general for substance X:

$$1 \text{ mole of X} = 6.02 \times 10^{23} \text{ units of X}$$

Here are some specific examples:

$$\begin{aligned} &\text{1 mole of argon atoms} &&= 6.02 \times 10^{23} \text{ Ar atoms} \\ &\text{1 mole of } CH_4 \text{ molecules} &&= 6.02 \times 10^{23}\ CH_4 \text{ molecules} \\ &\text{1 mole of } Mg^{2+} \text{ ions} &&= 6.02 \times 10^{23}\ Mg^{2+} \text{ ions} \\ &\text{1 mole NaCl formula units} &&= 6.02 \times 10^{23} \text{ NaCl formula units} \end{aligned}$$

On the left side of all four equations the chemical symbol specifically shows the type of unit involved. Therefore chemists avoid "1 mole of argon atoms" but write instead "1 mol Ar." Since argon always exists as atoms, it is not necessary to specify that fact. The four equations above are commonly written by chemists as follows:

$$1 \text{ mol Ar} = 6.02 \times 10^{23} \text{ argon atoms}$$

$$1 \text{ mol } CH_4 = 6.02 \times 10^{23}\ CH_4 \text{ molecules}$$

$$1 \text{ mol } Mg^{2+} = 6.02 \times 10^{23}\ Mg^{2+} \text{ ions}$$

$$1 \text{ mol NaCl} = 6.02 \times 10^{23} \text{ NaCl formula units}$$

The definition of the mole also allows the chemist to determine how many moles of atoms, or molecules, are in a sample just by weighing it. We may write:

1 mole X = gram-formula mass X

The **gram-formula mass**, often called the **gram-molar mass**, is the sum of all the masses of all the elements in the formula of the substance.

The importance of the mole is that it is larger by a factor of 6.02×10^{23} times than the fundamental atomic units of atoms, molecules, and ions, and therefore is readily measured by simple laboratory balances. In addition, the elements and compounds in chemical reactions must obey the rules of the balanced chemical equation. The coefficients in a balanced equation tell the chemist the number of atoms that react. When all of the coefficients in a balanced chemical equation are multiplied by 6.02×10^{23}, those coefficients then represent the number of moles of the various substances that react.

These relationships define the equalities between different units in chemistry. We can use them to construct factor labels so that conversions between different substances can be made in stoichiometric calculations.

Molar mass (Alternative names: formula mass, molecular weight, formula weight)

The Periodic Table gives the **relative atomic mass** of each element directly underneath the chemical symbol. Relative atomic mass has no units and simply indicates the mass of one element as compared to that of another. In chemistry it is customary to add gram units to the atomic masses listed in the Periodic Table, calling them **gram-atomic masses**. One mole of an element is equal to the gram-atomic mass of that element:

1 mole of an element = gram-atomic mass of an element

For a chemical compound the gram-molar mass is equal to the sum of the gram-atomic masses of all of the atoms in the chemical formula of that compound:

Gram-molar mass of a compound = Σ (gram-atomic masses in formula)

Similarly to the situation for elements, the **gram-molar mass** of a compound is equal to 1 mole of that compound. Gram-molar masses of ions are determined by the atom(s) present in the ion. A gain or loss of electrons has virtually no effect on the total mass of a molecule.

1 mole of any compound = gram-molar mass of that compound

Some specific examples of molar-mass relationships are as follows:

1 mole of argon = 39.948 grams of argon

1 mole of uranium = 238.029 grams of uranium

1 mole of CH_4 = 16.043 grams of CH_4

1 mole of NaCl = 58.4424 grams of NaCl

1 mole of Mg^{2+} = 24.3050 grams of Mg^{2+}

For most calculations, masses may be rounded off to whole numbers. For simplicity we write the equalities using standard abbreviations. Thus the five examples above are more commonly written as follows:

1 mol Ar = 40 g Ar

1 mol U = 238 g U

1 mol CH_4 = 16 g CH_4

1 mol NaCl = 58 g NaCl

1 mol Mg^{2+} = 24 g Mg^{2+}

Although it is correct to refer to the gram-atomic mass of an element and the gram-molar mass of a compound, it is common usage to refer to these quantities as simply the atomic mass (A), and the molar mass, respectively.

The equalities defined in this section can be used to produce the appropriate factor labels for conversion calculations. Currently more than 10 million compounds are registered, and these definitions will give twice as many factor labels.

The **molar mass** of a substance is the sum of all of the atomic masses in the formula for that substance. A formula may be written as a complete chemical compound with no charge or as an ion. To determine the molar mass the following steps are used:

1. Make a list of how many atoms of each element are present in the formula.
2. Use the Periodic Table to find the atomic mass of each element in the formula.
3. Multiply each atomic mass by the corresponding number of atoms in the formula.
4. Add the results of Step 3 to obtain the total molar mass.

To obtain the correct result, each of the above steps must be done carefully and accurately.

Example Determine the molar mass of dichloroethane, $C_2H_4Cl_2$.

Solution: Step 1: There are two carbon, four hydrogen, and two chlorine atoms in this formula.

Step 2: The atomic masses are as follows:
C = 12.011, H = 1.00794, and Cl = 35.4527.
Step 3: $2 \times 12.011 = 24.022$ mass of C in the compound
$4 \times 1.00794 = 4.03176$ mass of H in the compound
$2 \times 35.4527 = 70.9045$ mass of Cl in the compound
Step 4: Adding these results gives a molar mass of 98.958 (rounded to the correct number of decimal places).

It is rare that the exact molar mass will be needed in calculations. In all but the most exact calculations, atomic masses (except for chlorine) may be rounded to whole numbers to calculate molar masses. The mass of chlorine is rounded to 35.5. In the above example, using masses of C = 12, H = 1, and Cl = 35.5 results in a molar mass of 99. The slight error, 0.05 percent, is compensated by the ease of calculation. It should be recognized, however, that the true molar mass has five significant figures even if they are abbreviated to two for convenience.

EXERCISES

1. Determine the molar mass of each of the following compounds. First, use the exact atomic masses; then round the atomic masses to whole numbers.
   (a) $CaCl_2$
   (b) SO_3
   (c) $Cr(NO_3)_3$
   (d) NH_4Br

(e) $C_4H_{10}O_2$
(f) $C_{22}H_{44}N_2O_6$
(g) $Fe_2O_3 \cdot FeO$
(h) $CH_3(CH_2)_{12}CH_3$
(i) $Ce_2(CO_3)_3 \cdot 5H_2O$
(j) $Sc_2(SO_4)_3$
(k) $SnCl_3Br$
(l) $BrCH_2CO_2C_2H_5$
(m) $H_2NCO_2C_2H_5$
(n) C_9H_7N
(o) $PtO_2 \cdot 3H_2O$
(p) NH_4^+
(q) $Cr_2O_7^{2-}$
(r) $C_2H_3O_2^-$

2. Using the Periodic Table determine the molar mass of each of the following compounds to two decimal places:
(a) $Cd(NO_3)_2$
(b) $CH_3(CH_2)_4Br$
(c) $(NH_4)_2SO_4$
(d) $(CH_3CH_2CH_2)_2O$
(e) $CuSO_4^- \cdot 5H_2O$

Solutions

1.	(a)	1(Ca) × 40.078	=	40.078	or	40
		2(Cl) × 35.4527	=	70.9054	or	71
		Molar mass	=	110.983	or	111
	(b)	1(S) × 32.066	=	32.066	or	32
		3(O) × 15.9994	=	47.9982	or	48
		Molar mass	=	80.064	or	80
	(c)	1(Cr) × 51.9961	=	51.9961	or	52
		3(N) × 14.0067	=	42.0201	or	42
		9(O) × 15.9994	=	143.995	or	144
		Molar mass	=	238.011	or	238
	(d)	1(N) × 14.0067	=	14.0067	or	14
		4(H) × 1.00794	=	4.03176	or	4
		1(Br) × 79.904	=	79.904	or	80
		Molar mass	=	97.943	or	98

(e)	4(C) × 12.011	=	48.044	or	48
	10(H) × 1.00794	=	10.0794	or	10
	2(O) × 15.9994	=	31.9988	or	32
	Molar mass	=	90.122	or	90
(f)	22(C) × 12.011	=	264.24	or	264
	44(H) × 1.00794	=	44.3494	or	44
	2(N) × 14.0067	=	28.0134	or	28
	6(O) × 15.9994	=	95.9964	or	96
	Molar mass	=	432.60	or	432
(g)	3(Fe) × 55.847	=	167.54	or	168
	4(O) × 15.9994	=	63.9976	or	64
	Molar mass	=	231.54	or	232
(h)	14(C) × 12.011	=	168.15	or	168
	30(H) × 1.00794	=	30.2382	or	30
	Molar mass	=	198.39	or	198
(i)	2(Ce) × 140.115	=	280.230	or	280
	3(C) × 12.011	=	36.033	or	36
	14(O) × 15.9994	=	223.992	or	224
	10(H) × 1.00794	=	10.0794	or	10
	Molar mass	=	550.334	or	550
(j)	2(Sc) × 44.9559	=	89.9118	or	90
	3(S) × 32.066	=	96.198	or	96
	12(O) × 15.9994	=	191.993	or	192
	Molar mass	=	378.103	or	378
(k)	1(Sn) × 118.710	=	118.710	or	119
	3(Cl) × 35.4527	=	106.358	or	106.5
	1(Br) × 79.904	=	79.904	or	80
	Molar mass	=	304.972	or	305.5
(l)	1(Br) × 79.904	=	79.904	or	80
	4(C) × 12.011	=	48.044	or	48
	7(H) × 1.00794	=	7.05558	or	7
	2(O) × 15.9994	=	31.9988	or	32
	Molar mass	=	167.003	or	167
(m)	7(H) × 1.00794	=	7.05558	or	7
	1(N) × 14.0067	=	14.0067	or	14
	3(C) × 12.011	=	36.033	or	36

	2(O) × 15.9994	=	31.9988	or	32
	Molar mass	=	89.095	or	89
(n)	9(C) × 12.011	=	108.10	or	108
	7(H) × 1.00794	=	7.05558	or	7
	1(N) × 14.0067	=	14.0067	or	14
	Molar mass	=	129.16	or	129
(o)	1(Pt) × 195.08	=	195.08	or	195
	5(O) × 15.9994	=	79.9970	or	80
	6(H) × 1.00794	=	6.04764	or	6
	Molar mass	=	281.13	or	281
(p)	1(N) × 14.0067	=	14.0067	or	14
	4(H) × 1.00794	=	4.03176	or	4
	Molar mass	=	18.0385	or	18
(q)	2(Cr) × 51.9961	=	103.992	or	104
	7(O) × 15.9994	=	111.996	or	112
	Molar mass	=	215.988	or	216
(r)	2(C) × 12.011	=	24.022	or	24
	2(O) × 15.9994	=	31.9988	or	32
	3(H) × 1.00794	=	3.02382	or	3
	Molar mass	=	59.045	or	59

2. Use all of the digits listed in the Periodic Table, and round to two decimal places at the end:

(a) 236.42 (c) 132.14 (e) 249.69

(b) 151.04 (d) 102.18

NOTE: Different Periodic Tables may give different values. Your answers may differ slightly from the ones given above if the Periodic Table in this book is not used. These slight differences are inconsequential in most calculations.

MOLE CALCULATIONS

Stoichiometric Calculations

Stoichiometric calculations allow the chemist to predict the result of a chemical reaction in terms of the mass or moles of products produced and reactants consumed. All of these calculations are conversion calculations in which a factor label is used to convert data from one set of units to another.

Therefore, familiarity with the source of conversion factor labels, as discussed on pages 230–239, is essential. Also, it is important to understand the sequence of operations in order to apply these factor labels successfully and efficiently to any conversion. Figure 9.1 illustrates how the conversions in chemistry are related to each other. This figure shows the sequence of conversions from any given item of chemical information to any other that may be desired. The notations along the arrows indicate the type of factor label needed to perform each conversion. At most, a conversion will require three steps, not including any conversions of metric prefixes.

To start analyzing this diagram, we note that all of the boxes on the left side all refer to "SUBSTANCE A" and the boxes on the right side to "SUBSTANCE B." The exercises that follow are divided into two groups. The first group involves the conversion of units for one substance and uses only the SUBSTANCE A side of the diagram. A typical question might be, "How many atoms of iron are in a 2.00-gram sample of iron?" The second group involves problems in which SUBSTANCE A is converted into an equivalent amount of SUBSTANCE B. These conversions start on the SUBSTANCE A side of the diagram and end on the SUBSTANCE B side. Here a typical question might be, "How many grams of carbon are there in 10.0 grams of $Fe_2(CO_3)_3$?" In all problems we will find the given information as one of the boxes on the SUBSTANCE A side of the diagram. The box representing the desired units is then located, and conversions are made step by step, using the indicated factor labels.

Calculations Involving One Substance

Some stoichiometric questions involve converting from one set of units to another set for the same chemical substance. In this case we focus entirely on the left side of Figure 9.1 Some sample exercises, with their solutions, follow.

EXERCISES

1. How many grams of $FeCl_3$ (molar mass = 162.3) need to be weighed to have 0.456 mole of $FeCl_3$?

2. A sample contains 24.6 grams of CaO. How many moles of CaO (molar mass = 56.0) are in this sample?

3. A solution has a molarity of 0.658 mol $MgBr_2$ per liter. How many moles of $MgBr_2$ are in 0.400 liter of this solution?

4. The same $MgBr_2$ solution as in the last problem must be used to obtain 0.500 mole of $MgBr_2$. How many milliliters of this solution are needed?

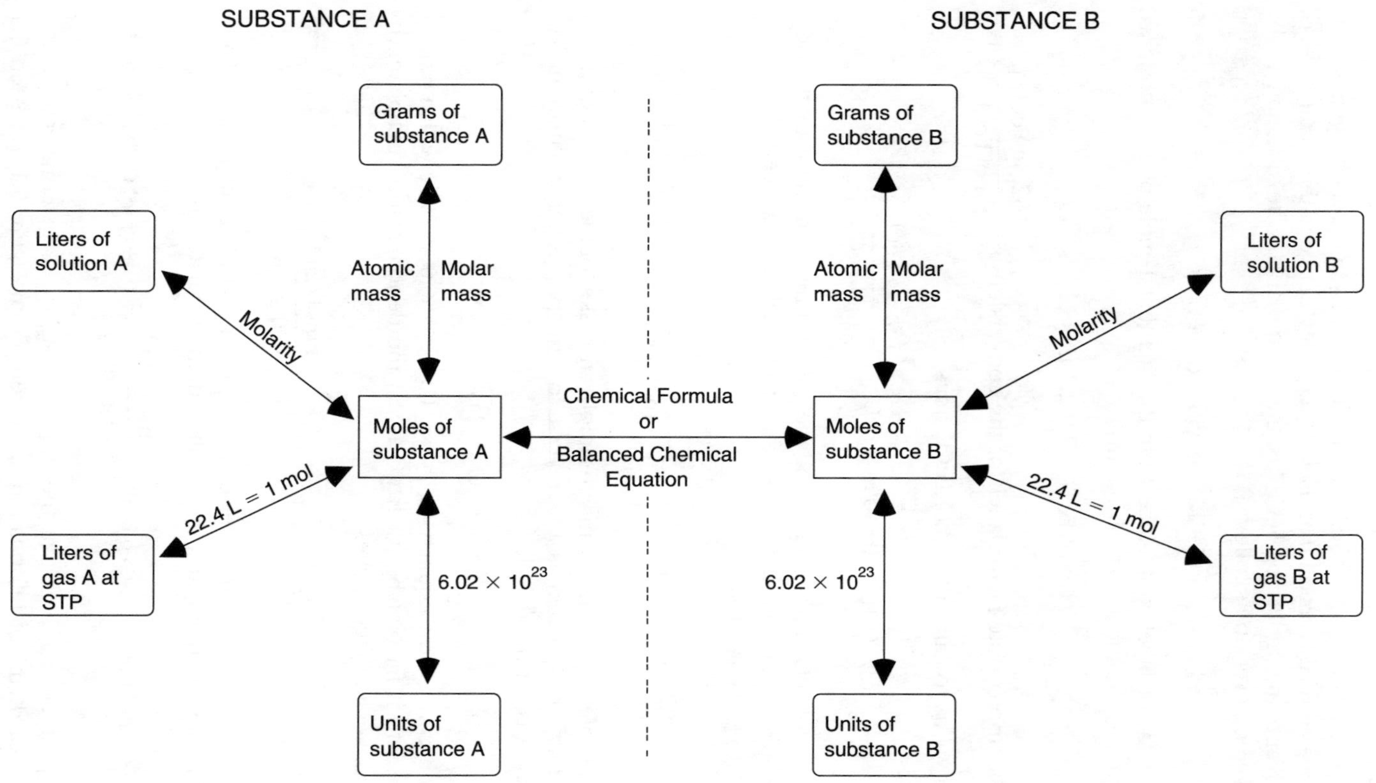

FIGURE 9.1. *The sequence of steps used in stoichiometry calculations. The box corresponding to the given data is on the SUBSTANCE A side of the diagram. The box corresponding to the desired data is then located. Arrows between boxes indicate where factor labels are derived from.*

Solutions

1. This problem gives the number of moles of $FeCl_3$ and asks for grams. This is a one-step conversion that uses the equality between the molar mass of $FeCl_3$ and the moles of $FeCl_3$ as the factor label. The question to be answered is set up as follows:

$$? \text{ g } FeCl_3 = 0.456 \text{ mol } FeCl_3$$

The factor label is obtained from the fact that 1 mole of any substance is equal to the molar mass in grams:

$$1 \text{ mol } FeCl_3 = 162.3 \text{ g } FeCl_3$$

The appropriate factor label for the conversion is $\left(\frac{162.3\text{g } FeCl_3}{1 \text{ mol } FeCl_3}\right)$ since it allows us to cancel the mol $FeCl_3$ units:

$$? \text{ g } FeCl_3 = 0.456 \text{ mol } FeCl_3 \left(\frac{162.3 \text{ g } FeCl_3}{1 \text{ mol } FeCl_3}\right)$$

The mol $FeCl_3$ units cancel, and the g $FeCl_3$ units remaining are the ones requested in the question. No other factor labels are needed, and the answer is calculated as

$$? \text{ g } FeCl_3 = 0.456 \text{ mol } FeCl_3 \left(\frac{162.3 \text{ g } FeCl_3}{1 \text{ mol } FeCl_3}\right) = 74.0 \text{ g } FeCl_3$$

2. This problem gives the grams of sample and asks for the number of moles. In effect, the process is the reverse of the preceding calculation. The question is set up as follows:

$$? \text{ mol CaO} = 24.6 \text{ g CaO}$$

The conversion equality is 1 mol CaO = 56.0 g CaO, which can be made into a factor label with g CaO in the denominator. Multiplying by the conversion factor gives

$$? \text{ mol CaO} = 24.6 \text{ g CaO} \left(\frac{1 \text{ mol CaO}}{56.0 \text{ g CaO}}\right)$$

When the g CaO units are canceled, the mol CaO units remain. These are the desired units, and the result is then calculated as

$$? \text{ mol CaO} = 24.6 \text{ g CaO} \left(\frac{1 \text{ mol CaO}}{56.0 \text{ g CaO}}\right) = 0.439 \text{ mol CaO}$$

3. This problem requires that the moles of $MgBr_2$ be determined, but two numerical items of information are given. From Figure 9.1 it is seen that the molarity is used as a factor label, and therefore the starting point is the liters of solution given. The question is set up as follows:

$$? \text{ mol MgBr}_2 = 0.400 \text{ L MgBr}_2$$

The molarity is already a ratio:

$$0.658 \text{ M MgBr}_2 = \frac{0.658 \text{ mol MgBr}_2}{1 \text{ L MgBr}_2}$$

and may be used as the conversion factor:

$$? \text{ mol MgBr}_2 = 0.400 \text{ L MgBr}_2 \left(\frac{0.658 \text{ mol MgBr}_2}{1 \text{ L MgBr}_2}\right)$$

Canceling the units and solving, we obtain the answer:

$$? \text{ mol MgBr}_2 = 0.263 \text{ mol MgBr}_2$$

4. Now we must calculate the volume from the number of moles given, and the setup starts as follows:

$$? \text{ mL MgBr}_2 = 0.500 \text{ mol MgBr}_2$$

Once again, the molarity is the conversion factor. However, it cannot be used directly since the units will not cancel. The molarity ratio is inverted and then used in the equation as follows:

$$? \text{ mL MgBr}_2 = 0.500 \text{ mol MgBr}_2 \left(\frac{1 \text{ L MgBr}_2}{0.658 \text{ mol MgBr}_2}\right)$$

Although the mol units cancel properly, the answer will be calculated in liters, not the milliliters requested. We must use an additional factor label to change the prefix of the liter units. The appropriate factor label is $\left(\frac{1 \text{ mL}}{10^{-3} \text{ L}}\right)$. Multiplying by this factor label and canceling the L units, we obtain the desired mL units:

$$? \text{ mL MgBr}_2 = 0.500 \text{ mol MgBr}_2 \left(\frac{1 \text{ L MgBr}_2}{0.658 \text{ mol MgBr}_2}\right)\left(\frac{1 \text{ mL}}{10^{-3} \text{ L}}\right)$$

The answer is 760 mL $MgBr_2$ solution.

The previous exercises demonstrated the one-step conversion of data to and from mole units. Many common calculations, however, involve two steps, as shown below.

EXERCISES

1. How many grams of KCl (molar mass = 74.6) are there in 0.250 liter of a 0.300 molar solution of KCl?

2. What is the mass of one molecule of CH_4 (molar mass = 16)?

3. How many molecules of CO_2 are contained in a 3.00-liter flask at standard temperature and pressure? (Assume that CO_2 behaves as an ideal gas.)

4. Perform each of the following conversions:
   (a) 26.5 g of $MgCl_2$ into moles of $MgCl_2$
   (b) 3.456 mol of CH_4 into grams of CH_4
   (c) 6.57×10^{18} atoms of Fe into moles of Fe
   (d) 2.22 mol of O_2 into molecules of O_2
   (e) 1.45 mol of KCl into liters of KCl with a molarity of 0.135
   (f) 23.5 mL of 0.766 M HF into moles of HF
   (g) 1.46 L of CO_2 at STP into moles of CO_2
   (h) 0.025 mol of N_2 into liters of N_2
   (i) 26.5 g of $MgCl_2$ into liters of 0.200 M $MgCl_2$ solution
   (j) 3.456 g of CH_4 into liters of CH_4 at STP
   (k) 6.57×10^{18} atoms of Fe into grams of Fe
   (l) 2.22 g of O_2 into molecules of O_2
   (m) 1.45 L of HCl at STP into liters of HCl with a molarity of 0.135
   (n) 23.5 mL of 0.766 M HF into liters of HF gas at STP
   (o) 0.025 g of N_2 into liters of N_2
   (p) 1.46 L of CO_2 at STP into molecules of CO_2

Solutions

1. From Figure 9.1 we see that to get from the given volume of the solution to the grams required for the answer involves two steps. The first step uses the molarity as a factor label to convert to moles, and then the second step uses the molar-mass factor label to convert to grams. The setup for the problem starts with the volume of the solution:

$$? \text{ g KCl} = 0.250 \text{ L KCl}$$

Next the molarity is used to convert to moles:

$$? \text{ g KCl} = 0.250 \text{ L KCl}\left(\frac{0.300 \text{ mol KCl}}{1 \text{ L KCl}}\right)$$

Then the conversion factor for the molar mass is used:

$$? \text{ g KCl} = 0.250 \text{ L KCl}\left(\frac{0.300 \text{ mol KCl}}{1 \text{ L KCl}}\right)\left(\frac{74.6 \text{ g KCl}}{1 \text{ mol KCl}}\right)$$

After canceling the L KCl and mol KCl units, we have the desired g KCl units, and the calculation can be made:

$$? \text{ g KCl} = 0.250 \text{ L KCl}\left(\frac{0.300 \text{ mol KCl}}{1 \text{ L KCl}}\right)\left(\frac{74.6 \text{ g KCl}}{1 \text{ mol KCl}}\right)$$
$$= 5.60 \text{ g KCl}$$

2. This problem involves another two-step calculation, starting with one molecule of CH_4 and ending with the number of grams of CH_4. The conversion involves the use of Avogadro's number and the molar mass of CH_4 as the factor labels. The question is set up as follows:

$$? \text{ g } CH_4 = 1 \text{ molecule } CH_4$$

Figure 9.1 shows that the first conversion uses Avogadro's number, 6.02×10^{23}, as a factor label. The units for Avogadro's number in this problem are molecules of CH_4, and the equality used is:

$$1 \text{ mol } CH_4 = 6.02 \times 10^{23} \text{ molecules } CH_4.$$

Setting up the factor label properly, so that the units of molecules CH_4 cancel, gives

$$? \text{ g } CH_4 = 1 \text{ molecule } CH_4 \left(\frac{1 \text{ mol } CH_4}{6.02 \times 10^{23} \text{ molecules } CH_4} \right)$$

The next step is to use the molar mass to convert from moles to grams:

$$? \text{ g } CH_4 = 1 \text{ molecule } CH_4 \left(\frac{1 \text{ mol } CH_4}{6.02 \times 10^{23} \text{ molecules } CH_4} \right) \left(\frac{16 \text{ g } CH_4}{1 \text{ mol } CH_4} \right)$$

Since all units cancel properly, the calculation can now be performed to obtain

$$? \text{ g } CH_4 = 2.66 \times 10^{-23} \text{ g } CH_4$$

as the mass of one molecule of CH_4.

3. This problem starts with the volume of a gas and ends with the number of molecules of CO_2. In the first step the volume of gas is converted to moles, using the fact that 1 mol of a gas occupies 22.4 L at standard temperature and pressure; then the moles are converted to molecules using Avogadro's number. The setup of the question is as follows:

$$? \text{ molecules } CO_2 = 3.00 \text{ L } CO_2$$

This is multiplied by the ratio $\left(\frac{1 \text{ mol } CO_2}{22.4 \text{ L } Co_2} \right)$:

$$? \text{ molecules } CO_2 = 3.00 \text{ L } CO_2 \left(\frac{1 \text{ mol } CO_2}{22.4 \text{ L } CO_2} \right)$$

The next step is to convert moles to molecules, being sure that the units of the ratio cancel properly:

$$? \text{ molecules } CO_2 = 3.00 \text{ L } CO_2 \left(\frac{1 \text{ mol } CO_2}{22.4 \text{ L } CO_2} \right) \left(\frac{6.02 \times 10^{23} \text{ molecules } CO_2}{1 \text{ mol } CO_2} \right)$$

The units do cancel properly, and the answer is calculated as

$$? \text{ molecules } CO_2 = 8.06 \times 10^{22} \text{ molecules } CO_2$$

Answers

4. (a) 0.278 mol $MgCl_2$
(b) 55.44 g CH_4
(c) 1.09×10^{-5} mol Fe
(d) 1.34×10^{24} molecules O_2
(e) 10.7 L KCl
(f) 0.0180 mol HF
(g) 0.0652 mol CO_2
(h) 0.56 L N_2
(i) 1.39 L $MgCl_2$
(j) 4.83 L CH_4
(k) 6.10×10^{-4} g Fe
(l) 4.18×10^{22} molecules O_2
(m) 0.479 L HCl(aq)
(n) 0.403 L HF(g)
(o) 0.0200 L N_2
(p) 3.92×10^{22} molecules CO_2

Calculations Involving Two Substances

To this point all the sample calculations have involved the same substance. In Figure 9.1, all of the conversions took place between boxes in the part labeled "SUBSTANCE A." When we start with one substance and end up with a different one, however, the conversion **must always** include the central conversion from moles of SUBSTANCE A to moles of SUBSTANCE B. There is simply no other possible way to perform the conversions. These conversions must use information obtained from a given chemical formula or from a balanced chemical reaction.

The simplest two-substance conversions are mole-to-mole conversions, as shown in the following exercises.

EXERCISES

1. How many moles of nitrogen are there in 6.50 moles of ammonium phosphate, $(NH_4)_3PO_4$?

2. How many moles of water will be formed in the complete combustion of 2.50 moles of methane, CH_4?

Solutions

1. The chemical formula gives the information for the factor label; this compound has three nitrogen atoms for each unit of ammonium phosphate. The setup states the question as follows:

$$? \text{ mol N} = 6.50 \text{ mol } (NH_4)_3PO_4$$

This is then multiplied by the factor label $\left(\frac{3 \text{ mol N}}{1 \text{ mol } (NH_4)_3PO_4}\right)$:

$$? \text{ mol N} = 6.50 \text{ mol } (NH_4)_3PO_4 \left(\frac{3 \text{ mol N}}{1 \text{ mol } (NH_4)_3PO_4} \right)$$

The mol $(NH_4)_3PO_4$ units cancel, leaving the mol N units that the problem requests. The answer is calculated as

$$? \text{ mol N} = 19.5 \text{ mol N}$$

2. This problem starts with one substance, methane, and asks a question about a second, very different substance, water. A chemical reaction will be needed to solve the problem. Every combustion reaction has oxygen as a reactant and carbon dioxide and water as products:

$$CH_4 \quad + \quad 2O_2 \quad \rightarrow \quad CO_2 \quad + \quad 2H_2O$$

This equation tells us that 1 mol of methane will form 2 mol of water, and this information is used to construct the factor label, $\left(\frac{2 \text{ mol } H_2O}{1 \text{ mol } CH_4} \right)$.

The question is written as follows:

$$? \text{ mol } H_2O = 2.50 \text{ mol } CH_4$$

Multiplying this by the factor label yields

$$? \text{ mol } H_2O = 2.50 \text{ mol } CH_4 \left(\frac{2 \text{ mol } H_2O}{1 \text{ mol } CH_4} \right)$$

After canceling the mol CH_4 units and verifying that the proper mol H_2O units have been obtained to satisfy the question, the answer is calculated:

$$? \text{ mol } H_2O = 2.50 \text{ mol } CH_4 \left(\frac{2 \text{ mol } H_2O}{1 \text{ mol } CH_4} \right) = 5.00 \text{ mol } H_2O$$

More complex calculations involve adding a step before the mole-to-mole calculation and a step afterwards. In all of these calculations the units for the given information in the problem are found in one of the five boxes on the left, or SUBSTANCE A side Figure 9.1. Then the units requested in the problem are found on the SUBSTANCE B side of the diagram. Proceeding from the given units to the requested units defines the sequence of conversions, and the factor labels, that are needed to obtain the correct answer.

One of the more common calculations involves calculating the mass of a compound, given the mass of another compound and the balanced chemical reaction. Figure 9.1 shows that this procedure involves three steps

1. Convert the starting mass into moles, using the formula mass of the given compound.
2. Convert the moles of compound A to moles of compound B, using the equivalencies derived from the balanced chemical reaction.

3. Convert from moles back to grams, using the formula mass of the requested compound.

EXERCISE

Propane, C_3H_8, is a common heating and cooking fuel in rural areas of the United States. Propane is also the fuel used in outdoor grills and in small hand-held torches. If 100 grams of propane is burned in excess oxygen, how many grams of oxygen will be needed, and how many grams of carbon dioxide and water will be formed, in the reaction?

Solution

This problem asks for the calculation of three quantities: O_2, CO_2, and H_2O. These will be calculated as three separate problems. First, a balanced chemical equation is needed to determine the relationships among all of the compounds. The reactants and products are listed in equation (a):

$$C_3H_8 \quad + \quad O_2 \quad \rightarrow \quad CO_2 \quad + \quad H_2O \tag{a}$$

This is then balanced to obtain

$$C_3H_8 \quad + \quad 5O_2 \quad \rightarrow \quad 3CO_2 \quad + \quad 4H_2O \tag{b}$$

To calculate the oxygen needed, the question is set up as

$$?\text{ g } O_2 = 100\text{ g } C_3H_8$$

The first conversion uses the molar mass of the C_3H_8 molecule, which is $3(12) + 8(1) = 44$:

$$?\text{ g } O_2 = 100\text{ g } C_3H_8 \left(\frac{1\text{ mol } C_3H_8}{44\text{ g } C_3H_8}\right)$$

The next conversion factor comes from the chemical equation (b), which says that 1 mol of C_3H_8 is equivalent to 5 mol of O_2:

$$?\text{ g } O_2 = 100\text{ g } C_3H_8 \left(\frac{1\text{ mol } C_3H_8}{44\text{ g } C_3H_8}\right)\left(\frac{5\text{ mol } O_2}{1\text{ mol } C_3H_8}\right)$$

The final factor label uses the molar mass of the O_2 molecule, $16 + 16 = 32$:

$$?\text{ g } O_2 = 100\text{ g } C_3H_8 \left(\frac{1\text{ mol } C_3H_8}{44\text{ g } C_3H_8}\right)\left(\frac{5\text{ mol } O_2}{1\text{ mol } C_3H_8}\right)\left(\frac{32\text{ g } O_2}{1\text{ mol } O_2}\right)$$

All the units cancel except the g O_2 units. The answer is calculated as

$$?\text{ g } O_2 = 364\text{ g } O_2$$

The other two quantities, CO_2 and H_2O, are calculated using the following setups:

$$? \text{ g } CO_2 = 100 \text{ g } C_3H_8 \left(\frac{1 \text{ mol } C_3H_8}{44 \text{ g } C_3H_8}\right)\left(\frac{3 \text{ mol } CO_2}{1 \text{ mol } C_3H_8}\right)\left(\frac{44 \text{ g } CO_2}{1 \text{ mol } CO_2}\right)$$

$$= 300 \text{ g } CO_2$$

$$? \text{ g } H_2O = 100 \text{ g } C_3H_8 \left(\frac{1 \text{ mol } C_3H_8}{44 \text{ g } C_3H_8}\right)\left(\frac{4 \text{ mol } H_2O}{1 \text{ mol } C_3H_8}\right)\left(\frac{18 \text{ g } H_2O}{1 \text{ mol } H_2O}\right)$$

$$= 164 \text{ g } H_2O$$

Reviewing this problem we see that it started with 100 grams of C_3H_8, which was found to require 364 grams of O_2 for complete combustion. The masses of the products were calculated as 300 grams of CO_2 and 164 grams of H_2O.

The total mass of the reactants, 100 grams of C_3H_8 and 364 grams of O_2, is 464 grams. At the end of the reaction, the total mass of the products, 300 grams of CO_2 and 164 grams of H_2O, is 464 grams. The law of conservation of mass states that matter cannot be created or destroyed in a chemical reaction. Since this law cannot be violated, we expect that the results of our calculations will obey it. The fact that the same total mass is obtained for the reactants and the products indicates that the law of conservation of mass was not violated.

It is always an advantage to be able to estimate an answer to a problem to ensure that no errors were made in calculations. In stoichiometry problems, only mass-to-mass conversions allow us to do this. In 95 percent of these problems, the calculated mass is between one-fifth and five times the given mass. In the problem above this means that the answers should be between 20 and 500 grams. All of our results fall in that range, giving added confidence that the conversions were correctly done. If the results did not fit the estimates, however, we would be well advised to check our calculations carefully. The 5 percent of reactions that do not follow this general rule are those in which the molar masses of the compounds are very different, for example, those of H_2 and zinc.

In some instances the information given is in the form of the volume and molarity of a reactant that produces a precipitate. These calculations also involve a three-step conversion:

1. Start with the given volume and convert to moles, using the molarity of the given solution.
2. Use the balanced chemical reaction to calculate the number of moles of product.
3. Convert the moles of product into grams using the molar mass.

EXERCISES

1. A 45.0-milliliter sample of 0.300 molar $FeCl_3$ is reacted with enough NaOH solution to precipitate $Fe(OH)_3$. How many grams of $Fe(OH)_3$ will be precipitated?

2. How many milliliters of a 0.250 M NaOH solution are needed to completely neutralize 65.0 milliliters of a 0.400 M solution of sulfuric acid?

3. Metallic copper reacts with concentrated nitric acid, HNO_3, to produce nitrogen dioxide, NO_2. Calculate the volume of NO_2 at STP that will form when 1.25 grams of copper is completely reacted according to the equation

$$Cu(s) + 4HNO_3(aq) \rightarrow Cu(NO_3)_2(aq) + 2NO_2(g) + 2H_2O(\ell)$$

4. Blackboard chalk is almost 100 percent calcium carbonate, $CaCO_3$. What volume of carbon dioxide, CO_2, will be evolved at STP if an excess of chalk is reacted with 35.0 milliliters of 0.888 molar hydrochloric acid, HCl?

5. (a) How many grams of water are obtained when 35.6 grams of benzene, C_6H_6, is burned in excess oxygen? (b) How many liters of CO_2 at STP will be produced in the same reaction?

6. (a) How many milliliters of 0.248 M HCl are needed to react with 1.36 grams of zinc to produce hydrogen gas? (b) How many milliliters of hydrogen gas are expected at STP?

Solutions

1. First a balanced chemical reaction must be written:

$$FeCl_3 \quad + \quad 3NaOH \quad \rightarrow \quad Fe(OH)_3 \quad + \quad 3NaCl$$

The question is set up as follows:

$$? \text{ g } Fe(OH)_3 = 45.0 \text{ mL } FeCl_3$$

The first factor label is the molarity, written as

$$0.300 \text{ M } FeCl_3 = \left(\frac{0.300 \text{ mol } FeCl_3}{1000 \text{ mL } FeCl_3} \right)$$

The denominator of this factor label includes the conversion from liters to milliliters without using another factor label:

$$? \text{ g } Fe(OH)_3 = 45.0 \text{ mL } FeCl_3 \left(\frac{0.300 \text{ mol } FeCl_3}{1000 \text{ mL } FeCl_3} \right)$$

The next factor label is obtained from the relationships found in the chemical reaction, where 1 mol of $FeCl_3$ is equivalent to 1 mol of $Fe(OH)_3$:

$$? \text{ g } Fe(OH)_3 = 45.0 \text{ mL } FeCl_3 \left(\frac{0.300 \text{ mol } FeCl_3}{1000 \text{ mL } FeCl_3} \right) \left(\frac{1 \text{ mol } Fe(OH)_3}{1 \text{ mol } FeCl_3} \right)$$

Finally, the moles of $Fe(OH)_3$ are converted to grams by using the molar mass, 107, for $Fe(OH)_3$:

$$? \text{ g Fe(OH)}_3 = 45.0 \text{ mL FeCl}_3 \left(\frac{0.300 \text{ mol FeCl}_3}{1000 \text{ mL FeCl}_3}\right)\left(\frac{1 \text{ mol Fe(OH)}_3}{1 \text{ mol FeCl}_3}\right)\left(\frac{107 \text{ g Fe(OH)}_3}{1 \text{ mol Fe(OH)}_3}\right)$$

Since the units cancel, the answer is calculated as

$$? \text{ g Fe(OH)}_3 = 1.44 \text{ g Fe(OH)}_3$$

2. In many reactions it is important to know the volume of one reactant that will react with a given volume of a second reactant. The molarities of both reactants must be given if this type of problem is to be solved.

   A balanced equation is required. Since the problem states that the sulfuric acid, H_2SO_4, is completely neutralized, both protons on the sulfuric acid react with the NaOH:

$$H_2SO_4 \quad + \quad 2NaOH \quad \rightarrow \quad Na_2SO_4 \quad + \quad 2\,H_2O$$

   Figure 9.1 indicates another three-step calculation:

   1. Convert milliliters of H_2SO_4 to moles of H_2SO_4, using the molarity of the acid.
   2. Convert moles of H_2SO_4 to moles of NaOH.
   3. Convert moles of NaOH to milliliters, using the molarity of the base.

   The initial setup of the question is

$$? \text{ mL NaOH} = 65.0 \text{ mL H}_2\text{SO}_4$$

   Using the molarity of H_2SO_4 as a conversion factor gives

$$? \text{ mL NaOH} = 65.0 \text{ mL H}_2\text{SO}_4 \left(\frac{0.400 \text{ mol H}_2\text{SO}_4}{1000 \text{ mL H}_2\text{SO}_4}\right)$$

   Then the balanced chemical equation written above is used to convert to moles of NaOH:

$$? \text{ mL NaOH} = 65.0 \text{ mL H}_2\text{SO}_4 \left(\frac{0.400 \text{ mol H}_2\text{SO}_4}{1000 \text{ mL H}_2\text{SO}_4}\right)\left(\frac{2 \text{ mol NaOH}}{1 \text{ mol H}_2\text{SO}_4}\right)$$

   Finally, the molarity of NaOH is used to convert moles of NaOH to milliliters of NaOH:

$$? \text{ mL NaOH} = 65.0 \text{ mL H}_2\text{SO}_4 \left(\frac{0.400 \text{ mol H}_2\text{SO}_4}{1000 \text{ mL H}_2\text{SO}_4}\right)\left(\frac{2 \text{ mol NaOH}}{1 \text{ mol H}_2\text{SO}_4}\right)\left(\frac{1000 \text{ mL NaOH}}{0.250 \text{ mol NaOH}}\right)$$

   Since the units cancel properly, the answer may be calculated as 208 mL.

3. In some chemical reactions a gas is evolved as one of the products. The most common cases are the reactions of active metals with mineral acids and the reactions of carbonates with acids. It is possible to calculate the volume of gas from a chemical reaction at standard temperature and pressure (STP = 0°C and 1 atm). (If the final conditions are not at STP, pages 59–62 should be consulted for further calculations using the ideal

gas law.) The volume of gas evolved from a given mass or volume of reactant is calculated in the following exercises.

We have the chemical reaction and need to follow three steps to convert the grams of copper to the volume of $NO_2(g)$ formed:

1. Convert grams of Cu to moles of Cu, using the atomic mass.
2. Convert moles of Cu to moles of NO_2, using the balanced equation.
3. Convert moles of NO_2 to volume, using the molar volume of an ideal gas.

We start with the setup of the question:

$$? \text{ L } NO_2 = 1.25 \text{ g Cu}$$

Using the atomic mass as a factor label, we get

$$? \text{ L } NO_2 = 1.25 \text{ g Cu}\left(\frac{1 \text{ mol Cu}}{63.55 \text{ g Cu}}\right)$$

The next factor label involves the balanced chemical equation given in the statement of the problem, which tells us that 1 mole of Cu will form two moles of NO_2. The equation becomes

$$? \text{ L } NO_2 = 1.25 \text{ g Cu}\left(\frac{1 \text{ mol Cu}}{63.55 \text{ g Cu}}\right)\left(\frac{2 \text{ mol } NO_2}{1 \text{ mol Cu}}\right)$$

Finally the fact that 1 mol of an ideal gas at STP occupies 22.4 L is used as a conversion factor to obtain

$$? \text{ L } NO_2 = 1.25 \text{ g Cu}\left(\frac{1 \text{ mol Cu}}{63.55 \text{ g Cu}}\right)\left(\frac{2 \text{ mol } NO_2}{1 \text{ mol Cu}}\right)\left(\frac{22.4 \text{ L } NO_2}{1 \text{ mol } NO_2}\right)$$

Since all of the units cancel properly, the answer may be calculated as

$$? \text{ L } NO_2 = 0.881 \text{ L } NO_2 \text{ at STP}$$

4. The balanced chemical reaction may be obtained from the facts given in the problem. Knowing that the reactants are HCl and $CaCO_3$ and that one of the products is CO_2, we can readily deduce the other products, H_2O and $CaCl_2$:

$$2HCl(aq) + CaCO_3(s) \rightarrow CO_2(g) + H_2O(\ell) + 2CaCl_2(aq)$$

The starting point for the calculation is

$$? \text{ L } CO_2 = 35.0 \text{ mL HCl}$$

The molarity of the HCl is used to convert to moles of HCl:

$$? \text{ L } CO_2 = 35.0 \text{ mL HCl}\left(\frac{0.888 \text{ mol HCl}}{1000 \text{ mL HCl}}\right)$$

Next, the balanced chemical reaction written above is used to obtain a factor label for the conversion from moles of HCl to moles of CO_2:

$$? \text{ L } CO_2 = 35.0 \text{ mL HCl}\left(\frac{0.888 \text{ mol HCl}}{1000 \text{ mL HCl}}\right)\left(\frac{1 \text{ mol } CO_2}{2 \text{ mol HCl}}\right)$$

Finally, the molar volume of a gas at STP is used to convert to the volume of CO_2 formed:

$$? \text{ L } CO_2 = 35.0 \text{ mL HCl}\left(\frac{0.888 \text{ mol HCl}}{1000 \text{ mL HCl}}\right)\left(\frac{1 \text{ mol } CO_2}{2 \text{ mol HCl}}\right)\left(\frac{22.4 \text{ L } CO_2}{1 \text{ mol } CO_2}\right)$$

The units cancel properly, and the answer is calculated as

$$? \text{ L } CO_2 = 0.348 \text{ L } CO_2 \text{ at STP}$$

5. The balanced reaction is

$$2C_6H_6 \quad + \quad 15O_2 \quad \rightarrow \quad 12CO_2 \quad + \quad 6H_2O$$

$$\text{(a) } ? \text{ g } H_2O = 35.6 \text{ g } C_6H_6\left(\frac{1 \text{ mol } C_6H_6}{78.0 \text{ g } C_6H_6}\right)\left(\frac{6 \text{ mol } H_2O}{2 \text{ mol } C_6H_6}\right)\left(\frac{18 \text{ g } H_2O}{1 \text{ mol } H_2O}\right)$$

$$= 16.4 \text{ g } H_2O$$

$$\text{(b) } ? \text{ L } CO_2 = 35.6 \text{ g } C_6H_6\left(\frac{1 \text{ mol } C_6H_6}{78.0 \text{ g } C_6H_6}\right)\left(\frac{12 \text{ mol } CO_2}{2 \text{ mol } C_6H_6}\right)\left(\frac{22.4 \text{ L } CO_2}{1 \text{ mol } CO_2}\right)$$

$$= 61.3 \text{ L } CO_2$$

6. The balanced reaction is

$$2HCl \quad + \quad Zn \quad \rightarrow \quad ZnCl_2 \quad + \quad H_2$$

$$\text{(a) } ? \text{ mL HCl} = 1.36 \text{ g Zn}\left(\frac{1 \text{ mol Zn}}{65.38 \text{ g Zn}}\right)\left(\frac{2 \text{ mol HCl}}{1 \text{ mol Zn}}\right)\left(\frac{1000 \text{ mL HCl}}{0.248 \text{ mol HCl}}\right)$$

$$\text{HCl} = 168 \text{ mL HCl}$$

$$\text{(b) } ? \text{ mL } H_2 = 1.36 \text{ g Zn}\left(\frac{1 \text{ mol Zn}}{65.38 \text{ g Zn}}\right)\left(\frac{1 \text{ mol } H_2}{1 \text{ mol Zn}}\right)\left(\frac{22400 \text{ mL } H_2}{1 \text{ mol } H_2}\right)$$

$$= 466 \text{ mL } H_2$$

Limiting Reactant Calculations

When chemicals are mixed together under the appropriate conditions, a chemical reaction is started. The reaction will stop when one of the reactants

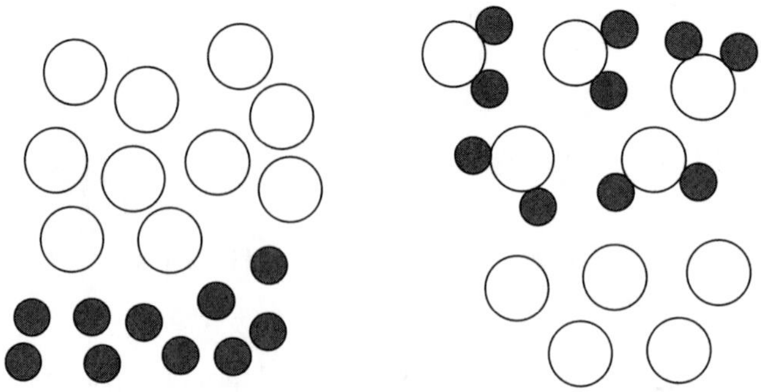

FIGURE 9.2. *The vending-machine example of a limiting reactant. On the left are the ten quarters (large circles) and ten dimes (small circles). On the right are five groups of one quarter and two dimes each used for purchases. Also on the right are five left-over quarters. The dimes are the limiting reactant.*

is completely used up. The reactant that is totally consumed, thereby stopping the reaction, is called the **limiting reactant** or **limiting reagent**. The other reactant(s) are the excess reactant(s). The amount of the limiting reactant will determine how much of the other reactant(s) react and how much of each product is formed. Up to this point, only the amount of one reactant has been given in a problem, and this reactant was assumed to be the limiting reactant. When the amounts of two or more reactants are given, special procedures for limiting reactant calculations must be used.

To understand the concept of a limiting reactant more fully, we consider a vending machine that accepts only quarters and dimes and does not give any change. If an item in that machine costs 45 cents, the only way it can be purchased is with two dimes and one quarter. If we have ten dimes and ten quarters, only five items can be obtained from this machine since the dimes will run out before the quarters do. Figure 9.2 illustrates this example. A variety of problems can be solved in the context of the limiting reactant concept. These include determinations of which reactant is the limiting reactant, how much product is formed, and how much of the excess reactant does not react. Examples of these calculations are shown below.

Procedures for solving limiting reactant problems must always be used when the amounts of two or more reactants are given in the statement of the problem.

Identifying the Limiting Reactant

Example For the reaction below, determine the limiting reactant if 100 grams of $FeCl_3$ is reacted with 50.0 grams of H_2S:

$$2FeCl_3(aq) \quad + \quad 3H_2S(g) \quad \rightarrow \quad Fe_2S_3(s) \quad + \quad 6HCl(aq)$$

Solution: The first thing to remember is that the reactant present in the smallest amount is not necessarily the limiting reactant. To determine the limiting reactant, we convert the amount given for one reactant into the amount of the other reactant that is needed to react with it. We can convert 50 g of H_2S into the grams of $FeCl_3$ needed to react with it as follows:

$$? \text{ g } FeCl_3 = 50.0 \text{ g } H_2S\left(\frac{1 \text{ mol } H_2S}{34 \text{ g } H_2S}\right)\left(\frac{2 \text{ mol } FeCl_3}{3 \text{ mol } H_2S}\right)\left(\frac{162 \text{ g } FeCl_3}{1 \text{ mol } FeCl_3}\right)$$

$$= 159 \text{ g } FeCl_3$$

From this result we see that 159 g of $FeCl_3$ is needed to react with all 50.0 g of H_2S. However, the amount of $FeCl_3$ given in the problem is only 100 g. The conclusion must be that we will run out of $FeCl_3$ before all of the H_2S can be reacted. Since the $FeCl_3$ is used up first, it is the limiting reactant.

(If the problem had stated that 200 g of $FeCl_3$, instead of 100 grams, was available, it would be apparent that we had more than enough $FeCl_3$ to react with all of the H_2S. Then we would have concluded that H_2S was the limiting reactant)

Once a limiting reactant is identified, all further calculations are based on the amount of the limiting reactant given in the statement of the problem.

Determining the Theoretical Yield

The term **theoretical yield** refers to the maximum amount of product formed in a reaction, with the determination based on the amounts of reactants used. This amount is a theoretical yield since no laboratory work is done. Many events can occur in the laboratory that may result in less than the theoretical amount of product. One is that the reactants do not combine completely. Another is the possibility of side reactions that produce different products. Products can also be lost by poor lab techniques or in the purification process. In most cases, the term *theoretical yield* refers to the maximum mass of product that can be produced.

Example

What is the theoretical yield of Fe_2S_3 that can be made from 100 grams of $FeCl_3$ and 50.0 grams of H_2S?

Solution: These are the same data as in the previous example but the question has changed. To solve this problem, we need to know the limiting reactant. In this case it has already been determined as the $FeCl_3$. We now use the 100 g of $FeCl_3$ given in the problem to calculate the mass of Fe_2S_3:

$$? \text{ g } Fe_2S_3 = 100 \text{ g } FeCl_3 \left(\frac{1 \text{ mol } FeCl_3}{162.5 \text{ g } FeCl_3}\right)\left(\frac{1 \text{ mol } Fe_2S_3}{2 \text{ mol } FeCl_3}\right)\left(\frac{208 \text{ g } Fe_2S_3}{1 \text{ mol } Fe_2S_3}\right)$$

$$= 64.0 \text{ g } Fe_2S_3$$

(It is essential that the 100 g of $FeCl_3$ given in the statement of the problem be used for this calculation. If we had used the 159 g calculated in the previous example a totally incorrect answer would have been obtained.)

Finding the Amount of Excess (Unused) Reactant

Another question that can be asked in a limiting reactant question is how much of the excess reactant is left over when the reaction stops. For the vending machine example in Figure 9.2 this can be done by counting the number of quarters that combine with the ten dimes (the limiting coin) and then subtracting this number from the number of quarters we started with. In the chemical calculation the amount of the excess reactant that reacts with the limiting reactant is calculated. This is then subtracted from the starting amount of the excess reactant given in the problem. The next exercise illustrates the procedure.

Example

When 100 grams of $FeCl_3$ is reacted with 50.0 grams of H_2S, how many grams of which reactant will be left over when the reaction is complete?

Solution: To determine which reactant is left over, the limiting reactant is identified and then all other reactants become the excess reactants. Since the data are the same for this problem as for the previous two examples, we know that $FeCl_3$ is the limiting reactant and, therefore, H_2S is the excess reactant. To determine how much H_2S is left over, we first calculate the grams of H_2S that react:

$$? \text{ g } H_2S = 100 \text{ g } FeCl_3 \left(\frac{1 \text{ mol } FeCl_3}{162.5 \text{ g } FeCl_3}\right)\left(\frac{3 \text{ mol } H_2S}{2 \text{ mol } FeCl_3}\right)\left(\frac{34 \text{ g } H_2S}{1 \text{ mol } H_2S}\right)$$

$$= 31.4 \text{ g } H_2S$$

Since the problem started with 50.0 g of H_2S and 31.4 g reacted, the remaining amount of H_2S must be

$$? \text{ g } H_2S \text{ (left)} = 50.0 \text{ g } H_2S \text{ (start)} - 31.4 \text{ g } H_2S \text{ (reacted)}$$
$$= 18.6 \text{ g } H_2S$$

To complete all of the information about this reaction, we can calculate the theoretical yield of HCl in the same way that the theoretical yield of Fe_2S_3 was calculated.

Example When 100 grams of $FeCl_3$ is reacted with 50.0 grams of H_2S, how many grams of HCl are formed?

Solution: As before, the limiting reactant must be identified; we already know that it is $FeCl_3$. The 100 grams of $FeCl_3$ given in the problem serves as the starting point for the calculation:

$$? \text{ g HCl} = 100 \text{ g } FeCl_3$$
$$= 100 \text{ g } FeCl_3 \left(\frac{1 \text{ mol } FeCl_3}{162.5 \text{ g } FeCl_3}\right)\left(\frac{6 \text{ mol HCl}}{2 \text{ mol } FeCl_3}\right)\left(\frac{36.5 \text{ g HCl}}{1 \text{ mol HCl}}\right)$$
$$= 67.4 \text{ g HCl}$$

This calculation follows the same principles as the calculation of the amount of Fe_2S_3 in the example on page 348.

With this calculation we have determined the masses of all reactants and products in the chemical equation. Listing the masses of the substances before and after reaction demonstrates once again that the law of conservation of matter is obeyed.

Reaction:	$2FeCl_3$	+	$3H_2S$	→	Fe_2S_3	+	$6HCl$
Start	100 g		50.0 g		0 g		0 g
End	0 g		18.6 g		64.0 g		67.4 g

In this table we can add up the masses of all of the substances at the start of the reaction to get a total of 150 grams. At the end of the reaction the masses again add up to 150 grams. These results show that the law of conservation of mass has been obeyed since the total mass at the start and at the end of the reaction is the same.

EXERCISES

1. Silver tarnishes in air because of a complex reaction with the oxygen and hydrogen sulfide, H_2S, in the air, which that may be written as

$$4Ag \quad + \quad O_2 \quad + \quad 2H_2S \quad \rightarrow \quad 2Ag_2S \quad + \quad 2H_2O$$

What is the theoretical yield of silver sulfide, Ag_2S, that can be produced from a mixture of 0.200 gram of silver, 1.50 liters of oxygen at STP, and 65.0 milliliters of 0.0350 molar hydrogen sulfide solution?

2. Silver nitrate reacts with sodium chromate to form silver chromate and sodium nitrate.
   a. If 45.5 milliliters of 0.200 M silver nitrate is mixed with 35.8 milliliters of 0.436 M sodium chromate, what is the theoretical yield of the precipitate silver chromate?
   b. How many grams of which reactant are left over?

Solutions

1. This is a limiting reactant problem since the amounts of three different reactants are given. An added complexity is that each of these amounts has a different unit. To begin, the identity of the limiting reactant must be determined. Because there are three reactants, Ag, O_2, and H_2S, a process of elimination is used. First, one pair of reactants is selected to determine which *might be* the limiting reactant, while eliminating the excess reactant. Next, the third reactant and the possible limiting reactant from the first step are used to determine the actual limiting reactant.

   Choosing silver and oxygen as the first pair, we have

$$? \text{ L O}_2 = 0.200 \text{ g Ag}\left(\frac{1 \text{ mol Ag}}{108 \text{ g Ag}}\right)\left(\frac{1 \text{ mol O}_2}{4 \text{ mol Ag}}\right)\left(\frac{22.4 \text{ L O}_2}{1 \text{ mol O}_2}\right)$$
$$= 0.0104 \text{ L O}_2$$

Since the problem gives the amount of oxygen as 1.5 L, there is plenty of oxygen and it cannot be the limiting reactant, but silver may be. The next step determines the amount of hydrogen sulfide needed to react with the silver:

$$? \text{ mL H}_2\text{S} = 0.200 \text{ g Ag}\left(\frac{1 \text{ mol Ag}}{108 \text{ g Ag}}\right)\left(\frac{2 \text{ mol H}_2\text{S}}{4 \text{ mol Ag}}\right)\left(\frac{1000 \text{ mL H}_2\text{S}}{0.0350 \text{ mol H}_2\text{S}}\right)$$
$$= 26.5 \text{ mL H}_2\text{S}$$

Since the question states that 65.0 mL of H_2S solution is available and all that is needed is 26.5 mL, H_2S cannot be the limiting reactant.

Because both H_2S and O_2 have been ruled out, the limiting reactant must be the silver. Now the mass of Ag_2S can be calculated based on the 0.200 g of silver given in the statement of the problem:

$$? \text{ g Ag}_2\text{S} = 0.200 \text{ g Ag}\left(\frac{1 \text{ mol Ag}}{108 \text{ g Ag}}\right)\left(\frac{2 \text{ mol Ag}_2\text{S}}{4 \text{ mol Ag}}\right)\left(\frac{248 \text{ g Ag}_2\text{S}}{1 \text{ mol Ag}_2\text{S}}\right)$$
$$= 0.230 \text{ g Ag}_2\text{S}$$

2a. The reaction is

$$2AgNO_3(aq) \quad + \quad Na_2CrO_4(aq) \quad \rightarrow \quad Ag_2CrO_4(s) \; + \quad 2NaNO_3(aq)$$

To determine the limiting reactant, the calculation is

$$? \text{ mL } Na_2CrO_4 = 45.5 \text{ mL } AgNO_3 \left(\frac{0.200 \text{ mol } AgNO_3}{1000 \text{ mL } AgNO_3}\right)\left(\frac{1 \text{ mol } Na_2CrO_4}{2 \text{ mol } AgNO_3}\right)\left(\frac{1000 \text{ mL } Na_2CrO_4}{0.436 \text{ mol } Na_2CrO_4}\right)$$
$$= 10.4 \text{ mL } Na_2CrO_4$$

Since we were given 35.8 mL of Na_2CrO_4, the conclusion is that $AgNO_3$ is the limiting reactant, and all further calculations are based on the given amount of $AgNO_3$:

$$? \text{ g } Ag_2CrO_4 = 45.5 \text{ mL } AgNO_3 \left(\frac{0.200 \text{ mol } AgNO_3}{1000 \text{ mL } AgNO_3}\right)\left(\frac{1 \text{ mol } Ag_2CrO_4}{2 \text{ mol } AgNO_3}\right)\left(\frac{332 \text{ g } Ag_2CrO_4}{1 \text{ mol } Ag_2CrO_4}\right)$$
$$= 1.51 \text{ g } Ag_2CrO_4$$

2b. Since $AgNO_3$ is the limiting reactant, Na_2CrO_4 is the excess reactant. The amount of Na_2CrO_4 that reacts is calculated and subtracted from the given amount:

$$? \text{ g } Na_2CrO_4 = 45.5 \text{ mL } AgNO_3 \left(\frac{0.200 \text{ mol } AgNO_3}{1000 \text{ mL } AgNO_3}\right)\left(\frac{1 \text{ mol } Na_2CrO_4}{2 \text{ mol } AgNO_3}\right)\left(\frac{162 \text{ g } Na_2CrO_4}{1 \text{ mol } Na_2CrO_4}\right)$$
$$= 0.737 \text{ g } Na_2CrO_4 \text{ reacts}$$

The original amount of Na_2CrO_4 is calculated as

$$? \text{ g } Na_2CrO_4 = 35.8 \text{ mL } Na_2CrO_4 \left(\frac{0.436 \text{ mol } Na_2CrO_4}{1000 \text{ mL } Na_2CrO_4}\right)\left(\frac{162 \text{ g } Na_2CrO_4}{1 \text{ mol } Na_2CrO_4}\right)$$
$$= 2.529 \text{ g } Na_2CrO_4 \text{ initially present}$$

The amount of Na_2CrO_4 left over is calculated by subtraction:

$$? \text{ g } Na_2CrO_4 \text{ left over} = 2.529 \text{ g } Na_2CrO_4 - 0.737 \text{ g } Na_2CrO_4$$
$$= 1.792 \text{ g } Na_2CrO_4$$

TITRATION

Titrations and Titration Calculations

The titration technique used for chemical analysis utilizes the reaction of two solutions. One reactant solution is placed in a beaker, and the other in a buret, which is a long, graduated tube with a stopcock. The stopcock is a valve that allows the chemist to add controlled amounts of solution from the buret to the beaker. An indicator, that is, a compound that changes color when the reaction is complete, is added to the solution in the beaker. The chemist reads the volume of solution in the buret at the start of the experiment and again, at the point where the indicator changes color. The difference in these volumes represents the volume of reactant delivered from the buret. Figure 9.3 illustrates the experimental setup.

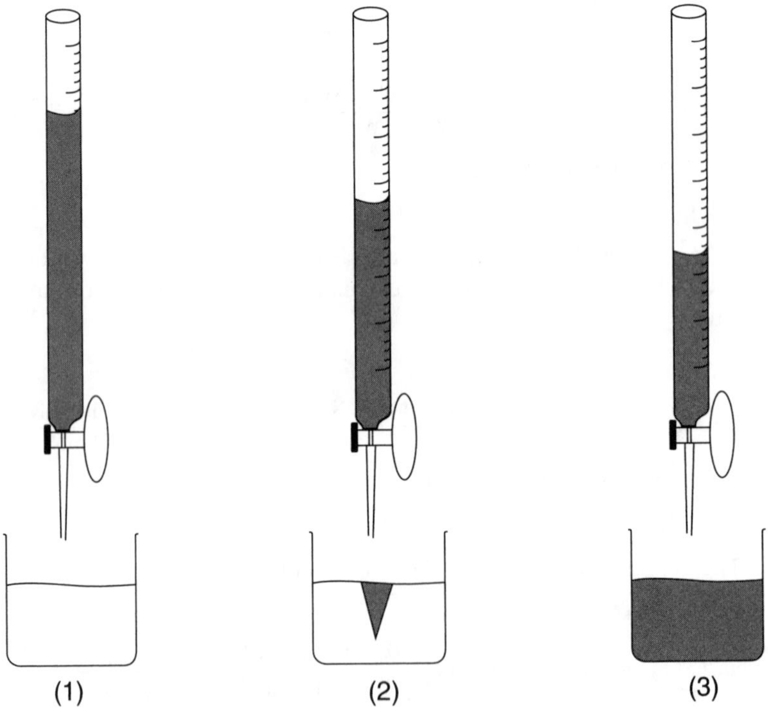

FIGURE 9.3. *Titration experiment showing (1) the initial setup, (2) the slight color change of the indicator before the end point, and (3) the colored solution at the end point.*

The crucial point about the titration experiment is that the indicator is designed to change color when the amount of reactant delivered from the buret is exactly the amount needed to react with the solution in the beaker. From this experiment a variety of calculations may be made as shown below.

A classic chemical reaction is the one between Fe^{2+} and the permanganate ion MnO_4^-:

$$5Fe^{2+} + MnO_4^- + 8H \rightarrow Mn^{2+} + 5Fe^{3+} + 4H_2O \qquad (9.1)$$

In titrations, the purple permanganate ion is the indicator of the point where the correct amount has been added to completely react all of the Fe^{2+} ions in the sample. Up to the end point the solution remains virtually colorless. At the end point, however, excess permanganate colors the solution purple and the titration is stopped. The calculations in exercises 1–3 refer to Equation 9.1.

EXERCISES

1. It takes 34.35 milliliters of a 0.240 M solution of $KMnO_4$ (0.240 M MnO_4^-) to titrate an unknown sample of Fe^{2+} to its end point. How many grams of Fe^{2+} are in the sample?

2. How many milliliters of 0.240 M MnO_4^- solution will be needed to titrate a 1.56-gram sample of pure $Fe(NO_3)_2$?

3. What is the molarity of an Fe^{2+} solution if 4.53 milliliters of a 0.687 M MnO_4^- solution is required to titrate 30.00 milliliters of the Fe^{2+}containing solution to the end point?

4. A 0.235 M solution of hydrochloric acid, HCl, is titrated to the end point with 23.4 milliliters of 0.216 M NaOH.
    1. How many grams of HCl were in the titrated sample?
    2. If the volume of the HCl sample is 50.0 milliliters, what is the molarity of the HCl solution?

Solutions

1. This problem gives the amount of MnO_4^-, and the grams of Fe^{2+} are to be calculated. Figure 9.1 shows the sequence of steps, and the required factor labels can be determined. The question is set up as follows:

$$? \text{ g } Fe^{2+} = 34.35 \text{ mL } MnO_4^-$$

Then the necessary factor labels are entered:

$$? \text{ g } Fe^{2+} = 34.35 \text{ mL } MnO_4^- \left(\frac{0.240 \text{ mol } MnO_4^-}{1000 \text{ mL } MnO_4^-}\right)\left(\frac{5 \text{ mol } Fe^{2+}}{1 \text{ mol } MnO_4^-}\right)\left(\frac{55.85 \text{ g } Fe^{2+}}{1 \text{ mol } Fe^{2+}}\right)$$

$$= 2.30 \text{ g } Fe^{2+}$$

2. The question is set up as follows:

$$? \text{ mL } MnO_4^- = 1.56 \text{ g } Fe(NO_3)_2$$

The result is converted into moles by using the molar mass of $Fe(NO_3)_2$:

$$? \text{ mL } MnO_4^- = 1.56 \text{ g } Fe(NO_3)_2 \left(\frac{1 \text{ mol } Fe(NO_3)_2}{180 \text{ g } Fe(NO_3)_2}\right)$$

Using the ionization reaction

$$Fe(NO_3)_2 \rightarrow Fe^{2+} + 2NO_3^-$$

we can apply the factor label $\left(\frac{1 \text{ mol } Fe^{2+}}{1 \text{ mol } Fe(NO_3)_2}\right)$ to convert to the Fe^{2+} ion:

$$? \text{ mL } MnO_4^- = 1.56 \text{ g } Fe(NO_3)_2 \left(\frac{1 \text{ mol } Fe(NO_3)_2}{180 \text{ g } Fe(NO_3)_2}\right)\left(\frac{1 \text{ mol } Fe^{2+}}{1 \text{ mol } Fe(NO_3)_2}\right)$$

Equation 9.1 is used to convert to moles of MnO_4^-, and then the molarity is used to convert to milliliters of MnO_4^-:

$$? \text{ mL } MnO_4^- = 1.56 \text{ g } Fe(NO_3)_2 \left(\frac{1 \text{ mol } Fe(NO_3)_2}{180 \text{ g } Fe(NO_3)_2}\right)\left(\frac{1 \text{ mol } Fe^{2+}}{1 \text{ mol } Fe(NO_3)_2}\right)$$
$$\left(\frac{1 \text{ mol } MnO_4^-}{5 \text{ mol } Fe^{2+}}\right)\left(\frac{1000 \text{ mL } MnO_4^-}{0.240 \text{ mol } MnO_4^-}\right)$$
$$= 7.22 \text{ mL } MnO_4^-$$

This exercise demonstrates that the stoichiometry calculations described above can be used also for calculations in titration experiments. There is one calculation, however, that stoichiometry calculations do not address—conversion of the molarity of one solution into the molarity of another. Since the answer desired is molarity, and molarity is not one of the possible starting points in Figure 9.1, a new method, done totally with factor labels, must be developed.

3. The desired information is the molarity of the Fe^{2+} solution. The units of molarity are a ratio, and we look for a ratio of units with which start the problem. The appropriate term is the molarity of the MnO_4^- solution. The setup of the question is as follows:

$$? \left(\frac{\text{mol } Fe^{2+}}{\text{L } Fe^{2+}}\right) = \frac{0.687 \text{ mol } MnO_4^-}{1 \text{ L } MnO_4^-}$$

We need factor labels to convert the numerator from mol MnO_4^- into mol Fe^{2+} and the denominator from L MnO_4^- into L Fe^{2+}. The factor label for the numerator comes from the balanced chemical equation 9.1. We obtain the factor label for the denominator from the ratio of the volumes of the two solutions at the end point of the titration:

$$? \left(\frac{\text{mol Fe}^{2+}}{\text{L Fe}^{2+}}\right) = \left(\frac{0.687 \text{ mol MnO}_4^-}{1 \text{ L MnO}_4^-}\right)\left(\frac{5 \text{ mol Fe}^{2+}}{1 \text{ mol MnO}_4^-}\right)\left(\frac{4.53 \text{ mL MnO}_4^-}{30.00 \text{ mL Fe}^{2+}}\right)$$

$$= \frac{0.519 \text{ mol Fe}^{2+}}{\text{L Fe}^{2+}} = 0.519 \text{ M Fe}^{2+}$$

4. The reaction is

$$\text{HCl} + \text{NaOH} \rightarrow \text{NaCl} + \text{H}_2\text{O}$$

1. To calculate the grams of HCl, we use this equation:

$$? \text{ g HCl} = 23.4 \text{ mL NaOH}\left(\frac{0.216 \text{ mol NaOH}}{1000 \text{ mL NaOH}}\right)\left(\frac{1 \text{ mol HCl}}{1 \text{ mol NaOH}}\right)\left(\frac{36.461 \text{ g HCl}}{1 \text{ mol HCl}}\right)$$

$$= 0.184 \text{ g HCl}$$

2. Calculating the molarity of HCl involves units that are a ratio. The starting point is the molarity of the NaOH, which is also a ratio of units:

$$? \frac{\text{mol HCl}}{\text{L HCl}} = \left(\frac{0.216 \text{ mol NaOH}}{1 \text{ L NaOH}}\right)\left(\frac{1 \text{ mol HCl}}{1 \text{ mol NaOH}}\right)\left(\frac{23.4 \text{ mL NaOH}}{50.0 \text{ mL HCl}}\right)$$

$$= \frac{0.101 \text{ mol HCl}}{\text{L HCl}} = 0.101 \text{ M HCl}$$

pH Indicators

An **indicator** consists of a conjugate acid-base pair in which the conjugate acid has a different color than the conjugate base. The reason for the difference in colors is that the loss or gain of a proton changes the energy of the electrons within their structures. This, in turn, changes the energy of the light absorbed, and this change is observed as a difference in color.

If, for example, a conjugate acid is yellow, and its conjugate base is blue, the eye will see these colors clearly only if there is at least ten times more of one color than of the other. Thus we will see yellow if the conjugate acid is at least ten times more concentrated than the conjugate base, and blue if the conjugate base is at least ten times more concentrated than the conjugate acid. Mathematically we can write:

$$\frac{[\text{conjugate base}]}{[\text{conjugate acid}]} < 0.1 \qquad \text{yellow observed}$$

$$\frac{[\text{conjugate base}]}{[\text{conjugate acid}]} > 10 \qquad \text{blue observed}$$

Between these two ratios various shades of green will be seen. Since the indicators are weak acids and bases, the significance of this, in terms of pH, is best understood using the Henderson-Hasselbach equation (see Chapter 10, pages 410–411). Substituting 0.1 into the log term yields

$$\text{pH} = \text{p}K_a + \log 0.1 = \text{p}K_a - 1.0$$

so that the yellow color is observed when the pH of the solution is at least 1 pH unit below the pK_a of the indicator. To see the blue color

$$\text{pH} = \text{p}K_a + \log 10 = \text{p}K_a + 1.0$$

the pH of the solution must be 1 pH unit above the pK of the indicator.

This example illustrates two important properties of the indicators used in titrations. First, the pH at the end point of a titration curve must change by at least 2 units very rapidly in order to see a distinct color change. Second, the pK of the indicator must be close to the end-point pH of the titration. Proper selection of an indicator requires that the pH at the titration end point and the pK of the indicator be close to each other.

EQUIVALENT WEIGHTS AND NORMALITY

Equivalent Weight, Equivalents, and Normality

The equivalent weight system, along with equivalents and normality, is an old system of measurement that is being phased out. However, this system is often encountered in the older literature. For completeness we discuss equivalent weights in relation to neutralization and redox reactions.

Equivalent Weights (masses)

Equivalent Weights in Neutralization Reactions: The **equivalent weight** of a compound in a neutralization reaction is equal to its molar mass divided by the number of protons that the substance provides or reacts with in the chemical reaction:

$$\text{Equivalent weight} = \frac{\text{molar mass}}{\text{number of H}^+ \text{ per mole}}$$

For example, arsenic acid, H_3AsO_4, can be neutralized with potassium hydroxide in the following molecular equation:

$$H_3AsO_4 \quad + \quad 2KOH \quad \rightarrow \quad K_2HAsO_4 \quad + \quad 2H_2O \qquad (9.2)$$

provides two hydrogen ions in this reaction since H_3AsO_4 starts with three hydrogens, and K_2HAsO_4 has only one hydrogen. The equivalent weight of H_3AsO_4 is therefore half of its molar mass. An acid with only one ionizable proton in its formula always has an equivalent weight identical to its molar mass.

Strong bases, that is, the hydroxy bases (with distinct OH units), always ionize completely when dissolved. As a result the number of hydrogen ions that a strong base will react with is equal to the number of OH units in the formula. The equivalent weight of $Al(OH)_3$ is always the molar mass divided by 3 or 26 grams per equivalent.

Equivalent Weights in Redox Reactions: The equivalent weight of a compound in a redox reaction is equal to the molar mass of the compound divided by the number of electrons per mole in its balanced half-reaction:

$$\text{Equivalent weight} = \frac{\text{molar mass}}{\text{number of electrons per mole}}$$

Example

The half-reaction for the reduction of the iodate ion to the iodide ion is as follows:

$$10e^- \quad + \quad 2IO_3^- \quad + \quad 12H^+ \quad \rightarrow \quad I_2 \quad + \quad 6H_2O$$

What is the equivalent weight (a) F_2 and (b) KIO_3 according to this half reaction?

Solution: (a) There are 10 electrons for each mole of I_2 in this equation, and the equivalent weight will be one-tenth of the molar mass of I_2. (b) KIO_3 does not appear in the balanced half-reaction, but the IO_3^- ion does. We may add two K^+ ions to each side of the half-reaction to obtain KIO_3:

$$10e^- + 2IO_3^- + 12H^+ + 2K^+ \rightarrow I_2 + 6H_2O + 2K^+$$

This may be rewritten as

$$10e^- + 2KIO_3 + 12H^+ \rightarrow I_2 + 6H_2O + 2K^+$$

We now see that 10 electrons are required for the reduction of 2 mol of KIO_3, or five electrons per mole. Therefore the equivalent weight of KIO_3 is one-fifth of its molar mass.

Using Equivalent Weights in Calculations: The equivalent weight is similar to the molar mass and is used to convert grams into equivalents of a given compound, and also to convert equivalents into grams.

Example The equivalent weight of potassium dichromate, $K_2Cr_2O_7$, is 47.9 grams per equivalent. How many equivalents are there in 3.68 grams of $K_2Cr_2O_7$?

Solution: We can use the equivalent weight as a factor label to perform this conversion. We start with the mathematical setup of the question:

$$? \text{ eq } K_2Cr_2O_7 = 3.68 \text{ g } K_2Cr_2O_7$$

Inserting the factor label, we obtain

$$\begin{aligned} ? \text{ eq } K_2Cr_2O_7 &= 3.68 \text{ g } K_2Cr_2O_7 \left(\frac{1 \text{ eq } K_2Cr_2O_7}{47.9 \text{ g } K_2Cr_2O_7} \right) \\ &= 0.0768 \text{ equivalent } K_2Cr_2O_7 \end{aligned}$$

It is important to remember that a correctly balanced chemical equation is needed to determine the equivalent weight of any compound.

Equivalents

The **equivalent** is a quantity similar to the mole. It is determined by dividing the mass of a compound by the equivalent weight or, as shown above, by using the equivalent weight as a factor-label conversion factor:

$$\begin{aligned} \text{Equivalents} &= \frac{\text{mass of compound}}{\text{equivalent weight}} \\ &= (\text{mass of compound}) \left(\frac{1 \text{ equivalent of X}}{\text{equivalent weight of X}} \right) \end{aligned}$$

We can also convert between moles and equivalents by using the following relationship:

$$\boxed{\text{Equivalents} = (\text{moles of X}) \times n}$$

In this equation n is the number of protons exchanged in a neutralization reaction or the number of electrons per mole in a redox half-reaction.

Example Using the preceding equations, determine the number of moles or equivalents for each of the following moles or equivalents:

(a) 0.234 mole of I_2
(b) 2.86 equivalents of H_3AsO_4

(c) 0.832 mol of KOH
(d) 0.0348 equivalent of KIO_3

Solution: In each case we need to convert moles to equivalents or equivalents to moles. We will use the relationship between moles and equivalents given above and the preceding balanced equations.

$$\text{(a) equivalents } I_2 = (0.234 \text{ mol } I_2) \times n \quad (n = 10 \text{ eq mol}^{-1})$$
$$= (0.234 \text{ mol } I_2) \times 10 \text{ eq mol}^{-1}$$
$$= 2.34 \text{ eq } I_2$$
$$\text{(b) } 2.86 \text{ eq } H_3AsO_4 = (\text{mol } H_3AsO_4) \times n \quad (n = 2 \text{ eq/mol})$$
$$\text{mol } H_3AsO_4 = \frac{2.86 \text{ eq } H_3AsO_4}{2 \text{ eq/mol}}$$
$$= 1.43 \text{ mol } H_3AsO_4$$
$$\text{(c) equivalents KOH} = (0.832 \text{ mol KOH}) \times n \quad (n = 1 \text{ eq/mol})$$
$$= 0.832 \text{ eq KOH}$$
$$\text{(d) } 0.348 \text{ eq } KIO_3 = (\text{mol } KIO_3) \times n \quad (n = 5 \text{ eq/mol})$$
$$\text{mol } KIO_3 = \frac{0.348 \text{ eq } KIO_3}{5 \text{ eq/mol}}$$
$$= 0.0696 \text{ eq } KIO_3$$

For a given chemical reaction 1 equivalent of one substance will always react with 1 equivalent of another substance in that reaction.

Example How many grams of KOH will react with 0.739 equivalent of H_3AsO_4 if the reaction in equation 9.2 is carried out?

Solution: Set up the question as

$$? \text{ g KOH} = 0.739 \text{ eq } H_3AsO_4$$

Use the factor label that shows that 1 eq of one substance reacts with 1 eq of another, in this case KOH and H_3AsO_4:

$$? \text{ g KOH} = 0.739 \text{ eq } H_3AsO_4 \left(\frac{1 \text{ eq KOH}}{1 \text{ eq } H_3AsO_4}\right)$$

Use the equivalent weight of KOH to obtain the grams of KOH:

$$? \text{ g KOH} = 0.739 \text{ eq } H_3AsO_4 \left(\frac{1 \text{ eq KOH}}{1 \text{ eq } H_3AsO_4}\right)\left(\frac{56.1 \text{ g KOH}}{1 \text{ eq KOH}}\right)$$
$$= 41.5 \text{ g KOH}$$

Normality (N)

Normality (N) is a concentration unit used in the equivalent weight system. The normality of a solution is defined as the number of equivalents of a solute, X, dissolved in 1 liter of solution:

$$\text{Normality of X} = N_x = \frac{\text{equivalents of X}}{\text{liters of solution}}$$

The conventional concentration unit is Molarity (M). We can quickly convert from Molarity to Normality by using the definitions given above for the equivalent weight and the determination of N, the number of equivalents per mole.

$$\text{Normality} = \text{molarity} \times n$$

Example What is the Normality of a 0.0375 M solution of H_3AsO_4 if it participates in the reaction of equation 9.2?

Solution: In reaction equation 9.2, n is 2 eq mol^{-1}, and we calculate

$$\text{Normality of } H_3AsO_4 = 0.0375 \text{ M} \times 2 \text{ eq mol}^{-1}$$

$$= \left(\frac{0.0375 \text{ mol } H_3AsO_4}{\text{L } H_3AsO_4}\right)\left(\frac{2 \text{ eq } H_3AsO_4}{1 \text{ mol } H_3AsO_4}\right)$$

$$= \frac{0.0750 \text{ eq } H_3AsO_4}{\text{L } H_3AsO_4}$$

$$= 0.0750 \text{ N } H_3AsO_4$$

Normality is used as a factor label in stoichiometry calculations in exactly the same way that molarity units are used. It is important to remember that a normality of 0.236 N KOH should be rewritten as

$$\left(\frac{0.236 \text{ eq KOH}}{\text{L KOH}}\right) \quad \text{or} \quad \left(\frac{0.236 \text{ eq KOH}}{1000 \text{ mL KOH}}\right)$$

before using it in any calculations.

Example Using the half-reaction in equation 9.2, calculate how many grams of KIO_3 are needed to produce 0.173 gram of I_2.

Solution: We set up the problem as

$$? \text{ g } KIO_3 = 0.173 \text{ g } I_2$$

The next step is to calculate the equivalents of I_2:

$$? \text{ g KIO}_3 = 0.173 \text{ g I}_2 \left(\frac{1 \text{ eq I}_2}{25.4 \text{ g I}_2}\right)$$

One equivalent of I_2 is formed from 1 eq of KIO_3:

$$? \text{ g KIO}_3 = 0.173 \text{ g I}_2 \left(\frac{1 \text{ eq I}_2}{25.4 \text{ g I}_2}\right)\left(\frac{1 \text{ eq KIO}_3}{1 \text{ eq I}_2}\right)$$

The equivalent weight of KIO_3 is one-fifth of its molar mass or 42.8 g eq^{-1}:

$$\begin{aligned} ? \text{ g KIO}_3 &= 0.173 \text{ g I}_2 \left(\frac{1 \text{ eq I}_2}{25.4 \text{ g I}_2}\right)\left(\frac{1 \text{ eq KIO}_3}{1 \text{ eq I}_2}\right)\left(\frac{42.8 \text{ g KIO}_3}{1 \text{ eq KIO}_3}\right) \\ &= 0.292 \text{ g KIO}_3 \end{aligned}$$

One advantage of the equivalent weight system is that titration calculations are very simple. For any titration experiment the equation

$$N_a V_a = N_b V_b$$

will always be true.

10
EQUILIBRIUM

CONCENTRATION UNITS

Concentration Unit Definitions

In chemistry the concentration of a solution defines the amount of substance in a given volume or mass of the mixture. The chemist uses the concentration unit that is appropriate to the experiment being performed. The most common concentration units are defined below.

Molarity (M)

The most common unit of concentration used by chemists is **molarity**. It is defined as the number of moles of solute dissolved in 1 liter of solution and is abbreviated as the upper-case letter M.

$$\text{Molarity (M)} = \frac{\text{number of moles of solute}}{\text{1 liter of solution}}$$

Preparation of solutions with molarity units involves measuring the solute in an appropriate manner and quantitatively transferring it to a volumetric flask of the desired size. Solvent is added until the flask is about half full, and the mixture is then shaken to dissolve all of the solute. Once the solute is dissolved, solvent is added exactly to the mark on the volumetric flask and mixed again.

Molality (m)

Molality is defined as the number of moles of solute dissolved in 1 kilogram of solvent and is abbreviated as the lower-case letter m:

$$\text{Molality (m)} = \frac{\text{number of moles of solute}}{\text{number of kilograms of solvent}}$$

Preparation of solutions with molality units involves measuring the moles of solute and then mixing the solute with the required mass of solvent.

Care must be used in distinguishing the terms *molarity* and *molality*, which have very similar spellings and abbreviations. Chemists use upper-case M only for molarity and lower-case m only for molality. In addition, the denominator for molarity is the total volume of the solvent and solute (i.e., of the solution) after mixing, while the denominator for molality is kilograms of the solvent only, as we saw previously.

Mole Fraction (X)

The **mole fraction** (X) is defined as the number of moles of one component, A, of a mixture divided by the sum of all of the moles in the solution:

$$X_A = \frac{mol_A}{mol_A + mol_B + \cdots}$$

The three dots (an ellipsis) in the denominator means that we must also add in the moles for any additional components of the solution. The mole fraction does not distinguish the solute from the solvent. For any solution, we can calculate the mole fraction of each of the compounds in the mixture. The mole fraction always has a value from 0.00 to 1.00, and the sum of the mole fractions of all the compounds in a solution must add up to 1.00. At times, chemists may use a variation of this procedure and report the mole percent, which is the mole fraction multiplied by 100.

A solution with mole fraction units is prepared by carefully measuring the desired number of moles of each of the components and then mixing.

Mass (Weight) Fraction (Wt/Wt)

The **mass fraction** is also known as the **weight fraction**. It may be used in situations in which it is inconvenient or impossible to use mole units. One such situation occurs when the molar mass of a compound has not been accurately determined. The mass fraction is the ratio of the mass of the solute to the mass of the entire solution and is usually symbolized as wt/wt:

$$\text{Mass fraction}_A \left(\frac{wt}{wt}\right) = \frac{mass_A}{mass_A + mass_B + \cdots}$$

Like the mole fraction, the mass fraction must be a value between 0.00 and 1.00. The sum of all mass fractions must be 1.00. The mass percent is obtained by multiplying the mass fraction by 100.

Mass-fraction solutions are prepared by weighing each component and then mixing the components together.

Weight-Volume Fraction (Wt/Vol)

The mass of substance dissolved per liter of solution, the **weight-volume fraction**, is another common concentration unit. It is also known as the mass-volume fraction. It may be used to specify the concentration as

$$\frac{\text{wt}}{\text{vol}} = \frac{\text{number of grams of solute}}{\text{number of milliliters of solution}}$$

Since 1 milliliter of water weighs very close to 1 gram, the wt/vol unit is considered as identical to the (wt/wt) unit in dilute aqueous solutions. The wt/vol unit is often expressed as a percentage by multiplying the above equation by 100. Dilute weight/volume solutions are prepared in the same manner as solutions with molarity units.

Volume Fraction (Vol/Vol)

When working with a solution of one liquid dissolved in another liquid, it is often more convenient to measure the volumes of the liquids rather than their masses:

$$\frac{\text{vol}}{\text{vol}} = \left(\frac{\text{volume of liquid}_A}{\text{volume of liquid}_A + \text{volume of liquid}_B + \cdots}\right)$$

and to use the **volume fraction** as the unit of concentration.

Preparation of volume/volume solutions involves measuring the appropriate volume of each liquid separately and then mixing the liquids thoroughly. It should be noted, however, that for many liquids the total volume of a mixture is not equal to the sum of the combined volumes. For example, 50 milliliters of water and 50 milliliters of ethyl alcohol yield a mixture with a volume of about 93 milliliters.

Parts per Million and Parts per Billion

A value expressed in any of the units above that has the word *fraction* in its name (mole fraction, mass fraction, mass-volume fraction, and volume fraction) can be converted to a percent (literally parts per hundred) by multiplying it by 100. In addition, the mass fraction and the volume fraction are often multiplied by 1 million (10^6) to obtain **parts per million** or by 1 billion (10^9) to obtain **parts per billion**. This is particularly true for solutions that contain trace quantities of solutes.

Before performing any calculations, however, percent, parts per million, and parts per billion units must be converted back to their fractional forms by dividing them by 100, 10^6, and 10^9, respectively. The fractional form must always have a value between 0.00 and 1.00.

EXERCISE

1. Calculate the molarity and weight-volume fraction of a solution prepared by dissolving 25.0 grams of $MgCl_2$ in enough water to make 450 milliliters of solution.
2. Calculate the molality, mole fraction of decane, and mass fraction of hexane in a solution prepared by mixing 85.0 grams of hexane, C_6H_{14}, and 45.0 grams of decane, $C_{10}H_{22}$.

Solutions

1. The number of moles of $MgCl_2$ is calculated:

$$? \text{ mol } MgCl_2 = 25.0 \text{ g } MgCl_2 \left(\frac{1 \text{ mol } MgCl_2}{95.21 \text{ g } MgCl_2}\right)$$
$$= 0.263 \text{ mol } MgCl_2$$

Then, from the definitions:

$$\text{Molarity} = \frac{0.263 \text{ mol } MgCl_2}{0.450 \text{ L sol'n}}$$
$$= 0.584 \text{ M } MgCl_2$$
$$\frac{\text{wt}}{\text{vol}} = \frac{25.0 \text{ g } MgCl_2}{450 \text{ mL}}$$
$$= 0.0556 \text{ g ml}^{-1} \text{ } MgCl_2$$

2. The number of moles of each compound is calculated:

$$? \text{ mol } C_6H_{14} = 85.0 \text{ g } C_6H_{14} \left(\frac{1 \text{ mol } C_6H_{14}}{86.0 \text{ g } C_6H_{14}}\right)$$
$$= 0.988 \text{ mol } C_6H_{14}$$
$$? \text{ mol } C_{10}H_{22} = 45 \text{ g } C_{10}H_{22} \left(\frac{1 \text{ mol } C_{10}H_{22}}{142 \text{ g } C_{10}H_{22}}\right)$$
$$= 0.317 \text{ mol } C_{10}H_{22}$$

Then the molality is calculated by assuming that hexane is the solvent since it is the major component in the mixture:

$$\text{Molality} = \frac{0.317 \text{ mol } C_{10}H_{22}}{0.085 \text{ kg } C_6H_{14}}$$
$$= 3.73 \text{ molal } C_{10}H_{22}$$

Finally, the mole fraction and mass fraction are determined:

$$X_{decane} = \frac{0.317 \text{ mol } C_{10}H_{22}}{0.317 \text{ mol } C_{10}H_{22} + 0.988 \text{ mol } C_6H_{14}}$$

$$= 0.243$$

$$\text{Mass fraction}_{hexane} = \frac{45.0 \text{ g } C_6H_{14}}{45.0 \text{ g } C_6H_{14} + 85.0 \text{ g } C_{10}H_{22}}$$

$$= 0.346$$

Concentration Conversion Calculations

Concentration units may be classified as temperature dependent or temperature independent. The temperature-dependent units are molarity, volume fraction, and mass-volume fraction. A liquid's volume changes slightly with temperature, and temperature-dependent concentrations are identified as those that have volume units in their definitions. Molality, mole fraction, and mass fraction units are not affected by temperature changes. These units are defined on the basis of mass or moles, which do not change with temperature.

There are three methods for making conversions between concentration units. They may be summarized as follows:

1. Conversion from one temperature-independent unit to another temperature-independent unit.
2. Conversion from one-temperature dependent unit to another temperature-dependent unit.
3. Conversions between temperature-dependent units and temperature-independent units.

Each of these has its own techniques and logic.

Conversions between Temperature-Independent Concentration Units

When converting one temperature-independent concentration into another, we use the given concentration to construct a table listing the masses and moles of all of the components of the solution. This table of information is then used to calculate the desired concentration from its definition.

EXERCISE

Given an aqueous solution that is 0.500 molal in NaCl, what are the corresponding mass and mole fractions of NaCl?

Solution

To solve the problem, we construct a table as shown:

	Grams	Moles
NaCl		
H_2O		

A 0.500 molal solution is defined to mean that 0.500 mol of solute is dissolved in 1.00 kg of solvent, which is water. We enter these values directly in the table:

	Grams	Moles
NaCl		0.500
H_2O	1000	

To complete the table, we convert the 1000 g of water into moles (55.5 mol of water) and the molar mass of NaCl can be used to convert the 0.500 mol of NaCl to (29.22 g of NaCl). (Review pages 325–338 on stoichiometry calculations like these if necessary.)

	Grams	Moles
NaCl	29.22	0.500
H_2O	1000	55.5

With the information in this last table, we use the mass fraction and mole fraction (X) definitions to calculate their values:

$$\text{Mass fraction} = \frac{\text{g NaCl}}{\text{g NaCl} + \text{g } H_2O}$$

$$= \frac{29.22\text{ g}}{29.22\text{ g} + 1000\text{ g}}$$

$$= 0.0284$$

$$X = \frac{\text{mol NaCl}}{\text{mol NaCl} + \text{mol } H_2O}$$

$$= \frac{0.500 \text{ mol}}{0.500 \text{ mol} + 55.5 \text{ mol}}$$

$$= 0.00893$$

EXERCISE

An aqueous solution of glucose ($C_6H_{12}O_6$, MM = 180) has a mole fraction of 0.400 for glucose. What are the molality and the mass fraction of this solution?

Solution

We construct a table for the solution components:

	Grams	Moles
Glucose		
Water		

The definition of the mole fraction:

$$X = \frac{\text{mol glucose}}{\text{mol glucose} + \text{mol } H_2O} = 0.400$$

shows that, if the denominator is assumed to equal 1.00, then the numerator, that is, the moles of glucose, must equal 0.400. At the same time, since the denominator is assumed to equal 1.00

$$1.00 = \text{mol glucose} + \text{mol } H_2O$$

Therefore, there must be 0.600 mol of water. We enter these values in the table:

	Grams	Moles
Glucose		0.400
Water		0.600

Using the molar masses of water and glucose, we calculate the grams of each and enter them in the table:

	Grams	Moles
Glucose	72.0	0.400
Water	10.8	0.600

Finally, we calculate the mass fraction and molality from the tabulated data:

$$\text{Mass fraction X} = \frac{\text{g glucose}}{\text{g glucose} + \text{g H}_2\text{O}}$$

$$= \frac{72.0\text{ g}}{72.0\text{ g} + 10.8\text{ g}}$$

$$= 0.870$$

$$\text{Molality (m)} = \frac{\text{mol glucose}}{\text{kg water}}$$

$$= \frac{0.400\text{ mol}}{0.0108\text{ kg}}$$

$$= 37.0\text{ molal}$$

Conversions between Temperature-Dependent Concentration Units

These conversions involve the molarity (M) and the mass/volume concentration units. The direct conversion of units that is required can be done in a manner similar to the stoichiometry calculations on pages 335–340. The following exercise illustrates the technique.

EXERCISE

Suppose that a solution is 3.50×10^{-4} M in $Pb(NO_3)_2$. What is the mass/volume concentration?

Solution

We start by writing the requested units and the given data:

$$? \text{ mass/volume} = 3.50 \times 10^{-4}\text{ M Pb(NO}_3)_2$$

Although M represents molarity, when solving problems it is better to write the actual ratio of units (mol/L) that the molarity represents. Therefore we rewrite the expression above as

$$?\left(\frac{\text{g}}{\text{mL}}\right) = \frac{3.50 \times 10^{-4}\ \text{mol Pb(NO}_3)_2}{1\ \text{L Pb(NO}_3)_2}$$

This setup indicates that conversion from moles of $Pb(NO_3)_2$ to grams of $Pb(NO_3)_2$ in the numerator, and from liters of $Pb(NO_3)_2$ to milliliters of $Pb(NO_3)_2$ in the denominator, is needed. These conversions may be done as follows:

$$?\left(\frac{\text{g}}{\text{mL}}\right) = \left(\frac{3.50 \times 10^{-4}\ \text{mol Pb(NO}_3)_2}{1\ \text{L Pb(NO}_3)_2}\right)\left(\frac{331\ \text{g Pb(NO}_3)_2}{1\ \text{mol Pb(NO}_3)_2}\right)\left(\frac{1\ \text{L}}{1000\ \text{mL}}\right)$$

$$= \frac{0.116\ \text{g Pb(NO}_3)_2}{1000\ \text{mL Pb(NO}_3)_2}$$

$$= \frac{0.000116\ \text{g Pb(NO}_3)_2}{\text{mL Pb(NO}_3)_2}$$

Conversions Between Temperature-Dependent and Temperature-Independent Concentration Units

Temperature-dependent units include volume units in their definitions and temperature-independent units have masses or moles. At some time during the conversion process a conversion between mass and volume will be needed. *Density is the factor that must be used for such a conversion.* Aside from the need for density to convert between mass and volume, the calculations for this type of problem are a combination of the two types discussed above.

EXERCISES

1. The molarities of concentrated reagents such as ammonia, hydrochloric acid, sulfuric acid, and nitric acid are not specified when purchased from a supplier. Instead, these reagents have labels that list their weight percents (% wt/wt) and densities. For instance, commercial hydrochloric acid, HCl, is 36 percent (wt/wt) hydrogen chloride and has a density of 1.18 grams per milliliter. Determine the molarity of commercial HCl.
2. The molarity of a solution of potassium fluoride, KF, is 0.748.
   a. What is the molality of this solution if the density is 1.035 grams per mL?
   b. What is the mole fraction of KF?

Solutions

1. The 36 percent (wt/wt) indicates that every 100 g of commercial HCl contains 36 g of HCl. Setting up a stoichiometry type conversion for this problem yields

$$? \text{ M HCl} = \frac{36 \text{ g HCl}}{100 \text{ g sol'n}}$$

Converting the symbol M to its actual units gives

$$? \frac{\text{mol HCl}}{\text{L sol'n}} = \frac{36 \text{ g HCl}}{100 \text{ g sol'n}}$$

The molar mass of HCl is used to convert the grams of HCl into moles of HCl, while density is the conversion factor needed to convert grams of solution into liters:

$$? \frac{\text{mol HCl}}{\text{L sol'n}} = \left(\frac{36 \text{ g HCl}}{100 \text{ g sol'n}}\right)\left(\frac{1 \text{ mol HCl}}{36.46 \text{ g HCl}}\right)\left(\frac{1.18 \text{ g sol'n}}{1 \text{ mL sol'n}}\right)\left(\frac{1000 \text{ mL sol'n}}{1 \text{ L sol'n}}\right)$$
$$= 11.7 \text{ M HCl}$$

2a. The given solution contains 0.748 mol in 1.00 L of solution. From this the mass of KF is calculated to be 31.4 g. The mass of the 1.00 L of solution is obtained by multiplying by the density to get 1035 g, of which 31.4 g is KF and the remaining 1003.6 g is water. If needed, a table may be constructed to summarize this information:

	Grams	Moles
KF	31.4	0.748
H_2O	1003.6	

This table contains enough information to calculate the molality of KF from the definition:

$$\text{Molality KF} = \frac{\text{mol KF}}{\text{kg } H_2O} = \frac{0.748 \text{ mol KF}}{1.0036 \text{ kg } H_2O}$$
$$= 0.745 \text{ m}$$

2b. To calculate the mole fraction of KF, the 1003.6 g of water is converted into 55.8 mol of water. The data are then substituted into the definition of mole fraction:

$$X = \frac{\text{mol KF}}{\text{mol KF} + \text{mol } H_2O}$$

$$= \frac{0.748 \text{ mol KF}}{0.748 \text{ mol glucose} + 55.8 \text{ mol } H_2O}$$

$$= 0.0132$$

In this last exercise an aqueous solution that had a molarity of 0.748 was calculated to have a molality of 0.745. This is less than 0.5 percent difference. We may generalize that dilute aqueous solutions (less than 0.5 M or 0.5 m) have the same molality and molarity.

In dilute aqueous solutions, Molarity = molality

In addition to the molarity-molality simplification, the same concepts are used to conclude that in a dilute solution the mass fraction is equal to the mass-volume fraction. This is particularly true for solutions where the solute is present in trace quantities and the concentrations are expressed as parts per million or parts per billion.

Dilution

Definition

Dilution is a laboratory process whereby pure solvent is added to a solution in order to decrease the concentration of the solute. The mathematics developed to describe dilution can be used in different problem-solving situations.

Dilution with Pure Solvent

If the pure solvent added is the same as the solvent of the solution, the volumes will be additive:

$$V_{\text{total}} = V_{\text{sol'n}} + V_{\text{solvent}} \tag{10.1}$$

If, however, the diluting solvent is different from the solution solvent, the volumes are usually not additive. For instance, if 50 milliliters of water is mixed with 50 milliliters of ethyl alcohol, the total volume will be 93, not 100, milliliters.

When a dilution is performed in the laboratory, a measured volume of pure solvent is added to a measured volume of solution. The beaker contains a certain number of moles of solute before dilution and the SAME number of moles after dilution. We can write this equation:

$$\text{moles}_{\text{before}} = \text{moles}_{\text{after}}$$

Since the molarity multiplied by the volume, in liters, will result in moles, we can substitute M × L for moles in the equation above to obtain

$$M_{\text{before}}\, L_{\text{before}} = M_{\text{after}}\, L_{\text{after}} \qquad (10.2)$$

All dilution problems provide three of the four variables in Equation 10.2. Careful reading is needed to understand exactly what the problem requires, as shown in the following two examples.

Example 1 To what volume must 500 milliliters of a 2.50 M NaCl solution be diluted in order to have a 1.00 M solution?

Solution: We enter the data ($M_{\text{before}} = 2.50$; $L_{\text{before}} = 0.500$ L; $M_{\text{after}} = 1.00$) into the dilution equation:

$$(2.50\text{ M})(0.500\text{ L}) = (1.00\text{ M})(x\text{ L})$$

Solving yields

$$x\text{ L} = \frac{(2.50\text{ M})(0.500\text{ L})}{1.00\text{ M}}$$

The M units cancel, and

$$x\text{ L} = 1.25\text{ L}$$

Example What volume of distilled water must be added to 500 milliliters of a 2.50 M NaCl solution in order to have a 1.00 M solution?

Solution: Notice the slight difference in wording between this problem and Example 1. As in the preceding problem we find that the total volume should be 1.25 L. However, this question asks for the added water, not the total volume. We use Equation 10.1:

$$\begin{aligned} V_{\text{total}} &= V_{\text{sol'n}} + V_{\text{solvent}} \\ 1.25\text{ L} &= 0.500\text{ L} + V_{\text{solvent}} \\ V_{\text{solvent}} &= 1.25\text{ L} - 0.50\text{ L} \\ &= 0.75\text{ L} \end{aligned}$$

The answer is that 0.75 L or 750 ml *must be added* to the original solution.

In the previous examples the answers are distinctly different although the basic calculations are identical. The first example tells us we must end up with a total volume of 1.25 liters; the second tells us we must add 0.75 liter (750 mL) to the original solution. In both cases we end up with a 1.00 M solution.

We developed an equation (eq. 10.2) based on molarity and liters of solution. In fact, we may generalize this equation to read

$$C_{before}V_{before} = C_{after}V_{after}$$

where the concentration (C) may be in any units desired, but the units must be the SAME for C_{before} and C_{after}. Similarly, the volume units can be any units as long as V_{before} has the same units as V_{after}.

Mixing Two or More Solutions

As an advanced topic, we can consider mixing two or more solutions (with the same solvent). In this situation, the sum of the moles mixed must be equal to the total moles in the final mixture (law of conservation of matter). We can write the equation

$$\text{moles}_1 + \text{moles}_2 + \ldots = \text{moles}_{total}$$

As before, we expand this by recognizing that M × L = moles:

$$M_1L_1 + M_2L_2 + \ldots = M_{total}L_{total}$$

Continuing the derivation, we may use any concentration units, as long as the units are the same throughout the equation. Similarly, we may use any volume units, as long as the units are the same throughout the equation.

We can generalize the equation for mixing solutions as

$$C_1V_1 + C_2V_2 + \ldots = C_{total}V_{total} \tag{10.3}$$

Example

What is the concentration (C) of the aqueous solution produced by mixing 25 milliliters of 0.250 M HNO_3 with 85 milliliters of 0.600 M HNO_3?

Solution: We enter the given data (C_1 = 0.250 M; C_2 = 0.600 M; V_1 = 25 mL; V_2 = 85 mL; V_{total} = 25 mL + 85 mL = 110 mL) into equation 10.3:

$$(0.250\ \text{M})(25\ \text{mL}) + (0.600\ \text{M})(85\ \text{mL}) = C_{total}(110\ \text{mL})$$

Rearranging this equation results in

$$C_{\text{total}} = \frac{(0.250 \text{ M})(25 \text{ mL}) + (0.600 \text{ M})(85 \text{ mL})}{110 \text{ mL}}$$

$$= \frac{57.25 \text{ M mL}}{110 \text{ mL}}$$

$$= 0.520 \text{ M } HNO_3$$

When mixing two solutions, it is important to remember that the final concentration must be somewhere between the lowest and highest concentrations given for the two solutions. In this problem the 0.520 M calculated is reasonable since it is between the 0.250 M and 0.600 M of the solutions being mixed.

If more than two solutions are mixed, equation 10.3 still applies. Again, the final solution must have a concentration somewhere between the highest and lowest concentrations of the solutions being mixed.

Example The aqueous discharge from a long-unused coal mine was sampled by collecting 200 milliliters of water from the mine. Then 25.0 milliliters of this sample was diluted to 1.00 liter for analysis of the iron content by atomic absorption spectrometry. The sample analyzed was found to have an iron concentration of 47.3 parts per million. What was the concentration of iron discharged from this mine?

Solution: We use equation 10.2 to determine the concentration of iron in the undiluted sample:

$$C_{\text{before}} V_{\text{before}} = C_{\text{after}} V_{\text{after}}$$

Substituting the information from the problem ($V_{\text{before}} = 25$ mL; $V_{\text{after}} = 1000$ mL; $C_{\text{after}} = 47.3$ ppm Fe) gives

$$C_{\text{before}} (25 \text{ mL}) = (47.3 \text{ ppm Fe})(1000 \text{ mL})$$

$$= \frac{(47.3 \text{ ppm Fe})(1000 \text{ mL})}{25 \text{ mL}}$$

$$= 1892 \text{ ppm Fe}$$

Determinations of this type are very common in modern chemical analysis when natural concentrations are much larger than instruments can handle. The sample is diluted before analysis, and then the original concentration is calculated as shown in the example.

pH, The Measurement of Acidity

The acidity of a solution is expressed either as the hydrogen ion concentration, $[H^+]$, or the **pH** of the solution. In 1909 Soren Sorensen developed the pH system to present acidity data in a simplified form. The pH and hydrogen ion concentration are related by the equation:

$$\boxed{pH = -\log [H^+]}$$

For aqueous solutions of acids and bases the pH ranges from 0 to 14. For very concentrated solutions the pH may be negative (extremely acidic) or greater than 14 (extremely basic).

In a similar fashion, the hydroxide ion concentration is related to the **pOH** as follows:

$$\boxed{pOH = -\log [OH^-]}$$

The pH, pOH, $[H^+]$, and $[OH^-]$ are all related to each other.

Water dissociates into hydrogen ions and hydroxide ions in this reaction:

$$H_2O \rightleftharpoons H^+ + OH^-$$

The equilibrium law for the dissociation of water is

$$K = [H^+]\,[OH^-] = 1.0 \times 10^{-14} \tag{10.4}$$

The equilibrium constant in this case is designated as K_w, the **autopyrolysis constant of water**. Taking the negative logarithm of equation 10.4 gives

$$\boxed{pK_w = pH + pOH = 14.00}$$

The three boxed equations given above show the relationships among pH, pOH, $[H^+]$, and $[OH^-]$. Given any one of these four values, the chemist can determine the other three by calculation.

EXERCISE

Determine pH, pOH, $[H^+]$, and $[OH^-]$ given the following data:

(a) $[H^+] = 2.3 \times 10^{-4}$ M

(b) $[OH^-] = 6.3 \times 10^{-2}$ M

(c) pH = 10.67
(d) pOH = 2.34

Answers

Starting with the given information, we use the equations given in this section to convert:

	$[H^+]$	$[OH^-]$	pH	pOH
(a)	2.3×10^{-4}	4.3×10^{-11}	3.64	10.36
(b)	1.6×10^{-13}	6.3×10^{-2}	12.80	1.20
(c)	2.1×10^{-11}	4.7×10^{-4}	10.67	3.33
(d)	2.2×10^{-12}	4.6×10^{-3}	11.66	2.34

When $[H^+]$ is larger than $[OH^-]$, the solution is considered to be acidic. A pH less than 7.0 also represents an acid solution. When $[H^+]$ is smaller than $[OH^-]$, the solution is basic and the pH is greater than 7.0. When $[H^+]$ is equal to $[OH^-]$, the solution is neutral and the pH is 7.0.

Many of the common substances with which we come into contact are either acidic or basic. Table 10.1 lists some of these substances and their approximate pH values.

TABLE 10.1. APPROXIMATE pH VALUES OF COMMON SUBSTANCES

Substance	Approximate pH
0.10 M HCl	1.00
Lemon juice	2.3
Vinegar	3.0
Wine	2.8–3.8
Pickles	3.3
Orange juice	3.5
Soft drinks	3.0–4.0
Beer	4.0–5.0
Pure rainwater	5.6
Milk	6.3–6.6
Blood	7.4
Seawater	8.3
Milk of magnesia	10.5
Ammonia	11.2
Lime water	12.4
0.1 M NaOH	13.00

EQUILIBRIUM CONCENTRATION CALCULATIONS

Equilibrium Calculations

In this section it is shown how the equilibrium constant may be calculated by measuring the concentrations of all the molecules in a chemical reaction. Chemists take advantage of stoichiometric relationships to make experiments easier and more reliable. With experience, these relationships become more obvious. A tabular system of analysis allows us to summarize our knowledge and to perform calculations of considerable sophistication without many years of experience.

The Equilibrium Table

In solving equilibrium problems, it is convenient to organize the data into a logical format so that the proper conclusions can be drawn. To do this, we construct a table of information called the **equilibrium table**. The basic table has five lines: the first line is used for the balanced chemical reaction; the second line is for the initial conditions stated in the problem; the third line represents the stoichiometric relationships and shows how the concentrations on line 2 will change; the fourth line is the sum of the second and third lines and represents the concentrations at equilibrium once the change is complete; and the fifth line is used for the actual equilibrium amounts, calculated by solving the expression on the EQUILIBRIUM line or determined by direct chemical analysis.

To illustrate the equilibrium table, we start with a chemical reaction such as this:

$$NO_2(g) \quad + \quad SO_2(g) \quad \rightleftharpoons \quad NO(g) \quad + \quad SO_3(g)$$

The table to be constructed will look as follows:

REACTION:	$NO_2(g)$	+	$SO_2(g)$	$\rightleftharpoons$	$NO(g)$	+	$SO_3(g)$
INITIAL CONDITIONS							
CHANGE							
EQUILIBRIUM							
ANSWER							

In this table, the first line is always the balanced chemical reaction, as shown. The second and last lines will contain numerical data with units of molarity or partial pressure. All entries on these lines must have identical units. The CHANGE line represents the stoichiometric relationships inherent in the reaction. In this line the change is represented as the unknown, x. The coefficient for each x is the same as the coefficient for the substance in the chemical reaction at the top of its column. The algebraic signs of all reactants are the opposites of the signs of the products of the reaction on the third line. (Mathematically it does not matter which x's are positive and which are negative as long as all reactants have the same sign and all products have the opposite sign.) The EQUILIBRIUM line is the sum of the INITIAL CONDITIONS and CHANGE lines. Finally, the ANSWER and EQUILIBRIUM lines are mathematically equal to each other.

To illustrate a table filled out with the data given in a problem, let us assume that a 4.00-liter flask is filled with 1 mole of each of the four compounds in the given reaction. The molarity of each compound is 1.00 mole/4.00 liters = 0.250 M. The table will look like this:

REACTION:	$NO_2(g)$	+	$SO_2(g)$	$\rightleftharpoons$	$NO(g)$	+	$SO_3(g)$
INITIAL CONDITIONS	0.250 M		0.250 M		0.250 M		0.250 M
CHANGE	$+x$		$+x$		$-x$		$-x$
EQUILIBRIUM	$0.250 + x$		$0.250 + x$		$0.250 - x$		$0.250 - x$
ANSWER							

Using this table, we can then perform appropriate algebraic calculations.

Calculation of Equilibrium Constants

On pages 194–196 the **equilibrium constant** was determined by measuring all of the concentrations of a mixture when equilibrium was established. Using stoichiometric relationships provides an easier way to do the same thing. To determine the equilibrium constant for the reaction

$$NO_2(g) \quad + \quad SO_2(g) \quad \rightleftharpoons \quad NO(g) \quad + \quad SO_3(g)$$

an experiment can be set up where the initial amount of every compound is 0.250 M. The equilibrium table is the same as before:

REACTION:	$NO_2(g)$	+	$SO_2(g)$	$\rightleftharpoons$	$NO(g)$	+	$SO_3(g)$
INITIAL CONDITIONS	0.250 M		0.250 M		0.250 M		0.250 M
CHANGE	$+x$		$+x$		$-x$		$-x$
EQUILIBRIUM	$0.250 + x$		$0.250 + x$		$0.250 - x$		$0.250 - x$
ANSWER							

When equilibrium is reached, the concentration of NO_2 is measured as 0.261 M. This value may be entered in the ANSWER line of the table under the $NO_2(g)$ column as shown below:

REACTION:	$NO_2(g)$	+	$SO_2(g)$	$\rightleftharpoons$	$NO(g)$	+	$SO_3(g)$
INITIAL CONDITIONS	0.250 M		0.250 M		0.250 M		0.250 M
CHANGE	$+x$		$+x$		$-x$		$-x$
EQUILIBRIUM	$0.250 + x$		$0.250 + x$		$0.250 - x$		$0.250 - x$
ANSWER	**0.261 M**						

Since the values in the last two lines of the $NO_2(g)$ column are mathematically equal, we write

$$0.261 \text{ M} = 0.250 \text{ M} + x$$

and

$$x = 0.011 \text{ M}$$

Since the value of x is now known, the ANSWER line can be completed for all of the other compounds in the reaction:

REACTION:	$NO_2(g)$	+	$SO_2(g)$	$\rightleftharpoons$	$NO(g)$	+	$SO_3(g)$
INITIAL CONDITIONS	0.250 M		0.250 M		0.250 M		0.250 M
CHANGE	$+x$		$+x$		$-x$		$-x$
EQUILIBRIUM	$0.250 + x$		$0.250 + x$		$0.250 - x$		$0.250 - x$
ANSWER	**0.261 M**		**0.261 M**		**0.239 M**		**0.239 M**

These values are then entered into the equilibrium law to determine the value of the equilibrium constant (K_c):

$$K_c = \frac{[NO][SO_3]}{[NO_2][SO_2]}$$

$$= \frac{(0.239)(0.239)}{(0.261)(0.261)}$$

$$= 0.839$$

Only one measurement, the concentration of NO_2, was needed to determine the value of K_c, instead of four measurements.

Determination of Equilibrium Concentrations by Direct Analysis

When a reaction is at equilibrium, it obeys the equilibrium law. The **equilibrium concentration** of any compound in a reaction can be determined by measuring the concentrations of all the other compounds involved in the reaction. For instance, the reaction between H_2 and I_2 to form HI has an equilibrium constant equal to 49. If at equilibrium $[I_2] = 0.200$ M and [HI] $= 0.050$ M, we can calculate the concentration of H_2 as follows:

$$H_2 \quad + \quad I_2 \quad \rightleftharpoons \quad 2HI$$

$$K_c = \frac{[HI]^2}{[H_2][I_2]}$$

$$49 = \frac{(0.050)^2}{[H_2](0.200)}$$

$$[H_2] = \frac{(0.050)^2}{(49)(0.200)}$$

$$= 2.6 \times 10^{-4}$$

Determination of Equilibrium Concentrations from Initial Concentrations and Stoichiometric Relationships

Given initial concentrations and a known value for the equilibrium constant, the equilibrium concentrations of all compounds can be determined. For example, the reaction

$$Br_2 \quad + \quad Cl_2 \quad \rightleftharpoons \quad 2BrCl$$

has an equilibrium constant of 6.90. If 0.100 mol of bromine chloride, BrCl, is introduced into a 500-milliliter flask, the equilibrium concentrations of Br_2, Cl_2, and BrCl can be calculated.

The solution starts with setting up the equilibrium table. The initial concentration of BrCl is 0.100 mole/0.500 liter = 0.200 M; and since no Br_2 or Cl_2 is added to the flask, the concentrations of these molecules are entered as zero. On the CHANGE line a positive x for Br_2 and for Cl_2 is chosen because their concentrations cannot be less than zero. The change in BrCl must then be entered as $-2x$. (If the signs of all the x's were reversed, the same answer would be obtained. Writing the table as suggested, however, indicates a better understanding of the equilibrium process.)

REACTION:	$Br_2(g)$	+	$Cl_2(g)$	$\rightleftharpoons$	$2BrCl(g)$
INITIAL CONDITIONS	**0 M**		**0 M**		**0.200 M**
CHANGE	$+x$		$+x$		$-2x$
EQUILIBRIUM	$+x$		$+x$		$0.200 - 2x$
ANSWER					

The algebraic expressions on the EQUILIBRIUM line are entered into the equilibrium law:

$$K_c = \frac{[BrCl]^2}{[Br_2][Cl_2]}$$

$$6.90 = \frac{(0.200 - 2x)^2}{(x)(x)}$$

Taking the square root of each side of this equation yields

$$2.63 = \frac{0.200 - 2x}{x}$$

$$2.63x = 0.200 - 2x$$

$$4.63x = 0.200$$

$$x = 0.0432$$

Once a value for x is determined, we return to the table and calculate the values for the ANSWER line:

REACTION:	$Br_2(g)$	+	$Cl_2(g)$	$\rightleftharpoons$	$2BrCl(g)$
INITIAL CONDITIONS	0 M		0 M		0.200 M
CHANGE	$+x$		$+x$		$-2x$
EQUILIBRIUM	$+x$		$+x$		$0.200 - 2x$
ANSWER	**0.0432 M**		**0.0432 M**		**0.114 M**

In another example using the same reaction, we mix together 0.200 M Br_2 and 0.300 M Cl_2 and determine the concentrations at equilibrium. We set up the equilibrium table as shown on the following page:

REACTION:	$Br_2(g)$	+	$Cl_2(g)$	$\rightleftharpoons$	$2BrCl(g)$
INITIAL CONDITIONS	**0.200 M**		**0.300 M**		**0 M**
CHANGE	$-x$		$-x$		$+2x$
EQUILIBRIUM	$0.200 - x$		$0.300 - x$		$+2x$
ANSWER					

Since BrCl cannot have a concentration less than zero, the change for it must be $+2x$, and Br_2 and Cl_2 must have $-x$ for their changes since the reactants must decrease if the product increases. Adding the INITIAL CONDITIONS to the CHANGE, we obtain the expressions for the EQUILIBRIUM line and enter them into the equilibrium law:

$$K_c = \frac{[BrCl]^2}{[Br_2][Cl_2]}$$

$$6.90 = \frac{(2x)^2}{(0.200 - x)(0.300 - x)}$$

The square root cannot be taken to simplify this equation since the denominator is not an exact square. Multiplying out the denominator of the ratio yields

$$6.90 = \frac{(2x)^2}{x^2 - 0.5\,x + 0.06}$$

$$6.90x^2 - 3.45x + 0.414 = 4x^2$$

$$2.90x^2 - 3.45x + 0.414 = 0$$

This is a quadratic equation of the form $ax^2 + bx + c = 0$, which is solved for x by using the quadratic formula:

$$x = \frac{-b \pm \sqrt{b^2 - 4ac}}{2a}$$

Making the substitutions into the quadratic equation yields

$$x = \frac{-(-3.45) \pm \sqrt{(-3.45)^2 - 4(2.90)(0.414)}}{2(2.90)}$$

$$= \frac{3.45 \pm 2.66}{5.8}$$

There are two roots to this equation:

$$x = 1.05 \quad \text{or} \quad 0.136$$

If root 1.05 is chosen, we calculate a negative value for the Br_2 and Cl_2 concentrations, which is clearly impossible. Root 0.136 gives the reasonable results shown in the complete table:

REACTION:	$Br_2(g)$	+	$Cl_2(g)$	$\rightleftharpoons$	$2BrCl(g)$
INITIAL CONDITIONS	0.200 M		0.300 M		0 M
CHANGE	$-x$		$-x$		$+2x$
EQUILIBRIUM	$0.200 - x$		$0.300 - x$		$+2x$
ANSWER	**0.064 M**		**0.164 M**		**0.272 M**

In the examples above, the answers were obtained by standard mathematical methods. Such solutions are called **analytical solutions**. For other problems the mathematics becomes very time consuming, if not impossible, to solve analytically. Sometimes, however, as shown below, the chemist can make simplifying assumptions that result in answers that are accurate to better than ±10 percent. Other problems, which will not be described further, can be solved by making logical estimates of the answer. One way of making logical estimates is called the **method of successive approximations**. Other methods use sophisticated computer programs to make repeated estimates.

When the equilibrium constant (K_c) is very small and the initial concentrations of the reactants are given, quadratic equations, or even higher order equations, may be avoided by making some assumptions based on our knowledge of the meaning of K_c. For example, for the reaction

$$2SO_3 \rightleftharpoons 2SO_2 + O_2$$

$K_c = 2.4 \times 10^{-25}$. An equilibrium constant this small indicates that very little SO_3 will dissociate into the products. If 2.00 moles of SO_3 is placed in a 1.00-liter flask, we can calculate the concentrations of all three molecules when the system comes to equilibrium. First the equilibrium table is constructed:

REACTION:	$2SO_3$	$\rightleftharpoons$	$2SO_2$	+	O_2
INITIAL CONDITIONS	**2.00 M**		**0 M**		**0 M**
CHANGE	$-2x$		$+2x$		$+x$
EQUILIBRIUM	$2.00 - 2x$		$+2x$		$+x$
ANSWER					

Entering the expressions from the EQUILIBRIUM line into the equilibrium law gives

$$K_c = \frac{[SO_2]^2[O_2]}{[SO_3]^2}$$

$$2.4 \times 10^{-25} = \frac{[2x]^2[x]}{(2.00 - 2x)^2}$$

This is a cubic equation and is difficult to solve analytically. Because K_c is very small, it is possible to simplify this equation with an assumption based on our knowledge that this reaction produces very little product. The assumption is that $2x$ will be very small and that, when $2x$ is subtracted from 2.00 M, we will still have 2.00 M SO_3 left.

ASSUME: $2x << 2.00$ M so that $2.00 \text{ M} - 2x = 2.00$ M

The assumption allows us to simplify the denominator in the equation above:

$$2.4 \times 10^{-25} = \frac{[2x]^2[x]}{[2.00]^2}$$

A solution can now be obtained with much simpler mathematical operations. Multiplying both sides by $(2.00)^2$ and placing the unknown x on the left, we obtain

$$4x^3 = 9.6 \times 10^{-25}$$

$$x^3 = 2.4 \times 10^{-25}$$

$$x = 6.2 \times 10^{-9} \text{ M}$$

Before using the calculated x to fill in the ANSWER line in the table, we check the assumption to be sure it is valid. The assumption is true since 2.00 − 6.2×10^{-9} is equal to 2.00. (The actual value is 1.9999999938, which rounds off to 2.00.) Once x is shown to be reasonable, we can complete the table as follows:

REACTION:	$2SO_3$	$\rightleftharpoons$	$2SO_2$	+	O_2
INITIAL CONDITIONS	2.00 M		0 M		0 M
CHANGE	$-2x$		$+2x$		$+x$
EQUILIBRIUM	$2.00 - 2x$		$+2x$		$+x$
ANSWER	**2.00 M**		**1.24×10^{-9} M**		**6.2×10^{-9} M**

Simplifying assumptions are used only for items on the EQUILIBRIUM line of the table. In addition, these assumptions can be used only with terms that are themselves sums or differences. In the preceding table it is impossible to make any simplifying assumptions regarding the $+2x$ for SO_2 or the $+x$ for O_2.

In another example, the reverse of the reaction just considered is

$$O_2 + 2SO_2 \rightleftharpoons 2SO_3 \qquad (10.5)$$

and its equilibrium constant will be

$$\frac{1}{2.4 \times 10^{-25}} = 4.2 \times 10^{24}$$

If the initial concentration of SO_3 is 2.00 M, the concentrations of O_2 and SO_2 can be determined using the same method as in the preceding example to obtain exactly the same results.

A general principle about equilibrium calculations and the assumptions used can be deduced from the two examples considered above. If the initial concentrations of the reactants are given for a reaction with a very small equilibrium constant, we may assume that these concentrations will not change significantly. Similarly, if the initial concentrations are given for the products of a reaction with a large equilibrium constant, we may assume that the concentrations of the products will not change significantly.

The above principle indicates also that, if the equilibrium constant is large and the initial concentrations of the reactants are given, there will be a significant change in concentration that cannot be ignored. For example, we will use the reaction in equation 10.5 with its equilibrium constant $K_c = 4.2 \times 10^{24}$, and determine the equilibrium concentrations if 2.00 moles of O_2 and 2.00 moles of SO_2 are placed in a 1.00-liter flask.

The equilibrium table can be set up as follows:

REACTION:	$O_2(g)$	+	$2SO_2(g)$	$\rightleftharpoons$	$2SO_3(g)$
INITIAL CONDITIONS	**2.00 M**		**2.00 M**		**0 M**
CHANGE	$-x$		$-2x$		$+2x$
EQUILIBRIUM	$2.00 - x$		$2.00 - 2x$		$+2x$
ANSWER					

The equilibrium law for this is written as

$$4.2 \times 10^{24} = \frac{(2x)^2}{(2.00 - x)(2.00 - 2x)^2}$$

Because the equilibrium constant is so large, we cannot ignore the x in the $2.00 - x$ and $2.00 - 2x$ terms on the EQUILIBRIUM line. This equation is too complex to solve analytically, and estimation methods are too time consuming. The method used to solve this problem requires us to assume that the reaction goes to completion. Calculating the concentrations that would be found if the reaction went to completion provides a new set of initial concentrations that allow us to solve the problem easily.

This is a limiting reactant problem since the concentrations of two reactants are given. Using the equilibrium table, we can simplify the calculation by remembering that the limiting reactant is completely consumed in the reaction. If a reactant is completely used up, its concentration will be zero on the ANSWER line of the table. We can determine if O_2 is the limiting reactant by entering zero for it on the ANSWER line:

REACTION:	$O_2(g)$	+	$2SO_2(g)$	$\rightleftharpoons$	$2SO_3(g)$
INITIAL CONDITIONS	2.00 M		2.00 M		0 M
CHANGE	$-x$		$-2x$		$+2x$
EQUILIBRIUM	$2.00 - x$		$2.00 - 2x$		$+2x$
ANSWER	**0 M**				

With zero on the ANSWER line for O_2 the other concentrations can be calculated since $0.00\ M = 2.00 - x$ and therefore $x = 2.00$.

REACTION:	$O_2(g)$	+	$2SO_2(g)$	$\rightleftharpoons$	$2SO_3(g)$
INITIAL CONDITIONS	2.00 M		2.00 M		0 M
CHANGE	$-x$		$-2x$		$+2x$
EQUILIBRIUM	$2.00 - x$		$2.00 - 2x$		$+2x$
ANSWER	0 M		**−2.00 M**		4.00 M

It is impossible for O_2 to be the limiting reactant since the concentration of SO_2 would then have to be negative, as shown in the table above. Next we try SO_2 as the limiting reactant:

REACTION:	$O_2(g)$	+	$2SO_2(g)$	$\rightleftharpoons$	$2SO_3(g)$
INITIAL CONDITIONS	2.00 M		2.00 M		0 M
CHANGE	$-x$		$-2x$		$+2x$
EQUILIBRIUM	$2.00 - x$		$2.00 - 2x$		$+2x$
ANSWER			**0 M**		

Since 0.00 M = 2.00 − 2x, then x = 1.00 and the remaining items on the ANSWER line can be calculated:

REACTION:	O_2(g)	+	$2SO_2$(g)	⇌	$2SO_3$(g)
INITIAL CONDITIONS	2.00 M		2.00 M		0 M
CHANGE	$-x$		$-2x$		$+2x$
EQUILIBRIUM	$2.00 - x$		$2.00 - 2x$		$+2x$
ANSWER	**1.00 M**		**0 M**		**2.00 M**

These results are reasonable. To continue with the solution of the problem, the concentrations on the ANSWER line of the table above are entered in a new table as the initial concentrations, and new CHANGE and EQUILIBRIUM lines are written:

REACTION:	O_2(g)	+	$2SO_2$(g)	⇌	$2SO_3$(g)
INITIAL CONDITIONS	1.00 M		0 M		2.00 M
CHANGE	$+x$		$+2x$		$-2x$
EQUILIBRIUM	$1.00 + x$		$+2x$		$2.00 - 2x$
ANSWER					

The terms on the EQUILIBRIUM line are entered into the equilibrium law:

$$4.2 \times 10^{24} = \frac{(2.00 - 2x)^2}{(1.00 + x)\,(2x)^2}$$

This equation is simplified by using two assumptions based on the fact that SO_3 is not expected to dissociate to a large extent:

ASSUME: $1.00 >> x$ so that $1.00 + x = 1.00$

ASSUME: $2.00 >> 2x$ so that $2.00 - 2x = 2.00$

Then the equilibrium law becomes

$$4.2 \times 10^{24} = \frac{(2.00)^2}{(1.00)\,(2x)^2}$$

Solving this equation yields

$$4x^2 = \frac{(2.00)^2}{(1.00)(4.2 \times 10^{24})}$$
$$= 9.5 \times 10^{-25}$$
$$x^2 = 2.4 \times 10^{-25}$$
$$x = 4.9 \times 10^{-13}$$

The assumptions above are valid, and the table with answers entered becomes as follows:

REACTION:	$O_2(g)$	+	2 $SO_2(g)$	⇌	2 $SO_3(g)$
INITIAL CONDITIONS	1.00 M		0 M		2.00 M
CHANGE	$+x$		$+2x$		$-2x$
EQUILIBRIUM	$1.00 + x$		$+2x$		$2.00 - 2x$
ANSWER	**1.00 M**		**4.9 × 10^{-13} M**		**2.00 M**

This problem illustrates another principle of chemical equilibrium. As long as the same number of moles of each *element* is present initially in different reacting mixtures, and these mixtures have identical volumes, the final composition of the equilibrium mixtures will always be the same.

There are a variety of methods for solving problems involving initial concentrations. If the equilibrium constant is very large or very small, approximations and stoichiometric calculations can be used to simplify the mathematics. In general, a very large equilibrium constant is one where the value of K_c is at least 100 times greater than the initial concentrations. The value of K_c is considered to be very small when the equilibrium constant is less than one one-hundredth (<0.01) of the initial concentrations. If K_c is neither very large nor very small, assumptions cannot be used and analytical solutions are then sought.

Solubility Product Calculations

Problems involving the **solubility product constant** (K_{sp}) are solved in the same way as other equilibrium problems. For example, K_{sp} has a value of 1.6 × 10^{-39} for $Fe(OH)_3$ (from Appendix 6). The molar solubility is calculated by setting up an equilibrium table:

REACTION:	$Fe(OH)_3$	$\rightleftharpoons$	Fe^{3+}	+	$3OH^-$
INITIAL CONDITIONS	**Solid**		**0 M**		**0 M**
CHANGE	$-x$		$+x$		$+3x$
EQUILIBRIUM	Not used		$+x$		$+3x$
ANSWER					

Entering the information from the EQUILIBRIUM line into the K_{sp} equilibrium law gives

$$\begin{aligned} K_{sp} &= [Fe^{3+}][OH^-]^3 \\ 1.6 \times 10^{-39} &= (x)(3x)^3 \\ &= (x)(27x^3) \\ &= 27x^4 \\ x^4 &= 5.9 \times 10^{-41} \\ x &= 8.8 \times 10^{-11} \end{aligned}$$

The value of x is used to calculate the molar concentrations of Fe^{3+} and OH^- in the solution.

We enter these data in the table:

REACTION:	$Fe(OH)_3$	$\rightleftharpoons$	Fe^{3+}	+	$3OH^-$
INITIAL CONDITIONS	Solid		0 M		0 M
CHANGE	$-x$		$+x$		$+3x$
EQUILIBRIUM	Not used		$+x$		$+3x$
ANSWER			**8.8×10^{-11} M**		**2.6×10^{-10} M**

The solubility of $Fe(OH)_3$ can be deduced from this table also. The CHANGE line indicates that $-x$ of the compound dissolves; therefore x represents the molar solubility of $Fe(OH)_3$, or 8.8×10^{-11} M.

If $Fe(OH)_3$ is dissolved in a solution that already contains the Fe^{+3} ion, the **common ion effect** is observed. The common ion effect is a decrease in the solubility of a compound when it is dissolved in a solution that already contains an ion in common with the salt being dissolved. As an example we can calculate the solubility of $Fe(OH)_3$ in a solution that already has a 6.5×10^{-5} M concentration of Fe^{3+}. The equilibrium table is constructed as follows:

REACTION:	$Fe(OH)_3$	$\rightleftharpoons$	Fe^{3+}	+	$3OH^-$
INITIAL CONDITIONS	**Solid**		**6.5×10^{-5} M**		**0 M**
CHANGE	$-x$		$+x$		$+3x$
EQUILIBRIUM	Not used		$6.5 \times 10^{-5} + x$		$+3x$
ANSWER					

Entering the information into the K_{sp} equation yields

$$1.6 \times 10^{-39} = (6.5 \times 10^{-5} + x)(3x)^3$$

To avoid a very complex fourth-order equation, we can use the results of the preceding example to assume that x will be very small compared to 6.5×10^{-5}.

ASSUME: $x << 6.5 \times 10^{-5}$ so that $6.5 \times 10^{-5} + x = 6.5 \times 10^{-5}$

Entering this assumption into the equation gives

$$1.6 \times 10^{-39} = (6.5 \times 10^{-5})(3x)^3$$

$$27x^3 = \frac{1.6 \times 10^{-39}}{6.5 \times 10^{-5}}$$

$$x^3 = 9.1 \times 10^{-37}$$

$$x = 9.7 \times 10^{-13}$$

This result satisfies the simplifying assumption and may be entered in the table:

REACTION:	$Fe(OH)_3$	$\rightleftharpoons$	Fe^{3+}	+	$3OH^-$
INITIAL CONDITIONS	Solid		6.5×10^{-5} M		0 M
CHANGE	$-x$		$+x$		$+3x$
EQUILIBRIUM	Not used		$6.5 \times 10^{-5} + x$		$+3x$
ANSWER			**6.5×10^{-5} M**		**2.9×10^{-12} M**

In addition, we can conclude that the molar solubility of $Fe(OH)_3$ is 9.7×10^{-13} M.

To determine the value of the solubility product, a variety of experiments may be performed. In one of the simplest, the molar solubility of a compound is determined by measuring the amount that dissolves in 1 liter of water. For example, if the molar solubility of MgF_2 is determined to be 0.00118 mole per liter, what is its K_{sp}? When solid MgF_2 dissolves, the reaction is

$$MgF_2(s) \rightleftharpoons Mg^{2+}(aq) + 2F^-(aq)$$

and the K_{sp} expression is

$$K_{sp} = [Mg^{2+}][F^-]^2$$

The equilibrium table is set up as follows:

REACTION:	MgF_2	$\rightleftharpoons$	Mg^{2+}	+	$2F^-$
INITIAL CONDITIONS	**Solid**		**0 M**		**0 M**
CHANGE	$-x$		$+x$		$+2x$
EQUILIBRIUM	Not used		$+x$		$+2x$
ANSWER					

The solubility given in the problem is the amount dissolved, which is also x. Knowing x, we can immediately complete the ANSWER line:

REACTION:	MgF_2	$\rightleftharpoons$	Mg^{2+}	+	$2F^-$
INITIAL CONDITIONS	Solid		0 M		0 M
CHANGE	$-x$		$+x$		$+2x$
EQUILIBRIUM	Not used		$+x$		$+2x$
ANSWER			**0.00118 M**		**0.00236 M**

Substituting the values from the ANSWER line into the K_{sp} expression yields

$$K_{sp} = (0.00118)(0.00236)^2$$

$$= 6.6 \times 10^{-9}$$

pH CALCULATIONS

pH, pOH, [OH], and [H$^+$] of Strong Acids and Bases

Except for sulfuric acid, all strong acids are monoprotic and dissociate completely when dissolved in water. Consequently, the hydrogen ion concentration of a strong acid solution is equal to the molar concentration of the acid itself:

$$[H^+] = M_{strong\ acid}$$

Sulfuric acid, H_2SO_4, is the only diprotic strong acid. When this acid is dissolved in water, only one hydrogen ion dissociates and it acts as a monoprotic acid:

$$H_2SO_4 \rightleftharpoons H^+(aq) + HSO_4^-(aq)$$

Also, for sulfuric acid the hydrogen ion concentration is equal to the molar concentration.

The pH of a strong acid solution can be written as

$$pH = -\log [H^+] = -\log M_{strong\ acid}$$

$[OH^-]$, pOH, and pH of Strong Bases

Strong bases dissociate completely when dissolved in water; however, they may have more than one hydroxide ion per formula unit. The concentration of hydroxide ions is equal to the molarity of the base multiplied by the number of hydroxide ions in its chemical formula:

$$[OH^-] = M_{strong\ base} \times \text{number of } OH^- \text{ ions per mole}$$

For example, a 0.200 M solution of $Ba(OH)_2$ has a hydroxide concentration of 0.400 M. Once the $[OH^-]$ value is known, the pOH is calculated as follows:

$$pOH = -\log [OH^-]$$

From the pOH, the pH is calculated using the relationship

$$pH = 14.00 - pOH$$

EXERCISE

Determine the pH of each of the following solutions:

(a) 0.020 M HNO_3

(b) 0.00043 M $Ba(OH)_2$

(c) 3.0 g of NaOH dissolved in 250 mL H_2O

(d) 0.00032 mol SO_3 dissolved in 3.4 L of H_2O

(e) 0.00098 mol $Ca(OH)_2$ dissolved in 1.3 L H_2O

Answers

(For each question the essential preliminary calculations are shown. For acids we need $[H^+]$ before calculating pH. For bases we need $[OH^-]$, the pOH and then pH is calculated.)

(a) $[H^+] = 0.020$ M; pH $= 1.70$

(b) $[OH^-] = (0.00043\ M)(2) = 0.00086$ M; pOH $= 3.07$ and pH $= 10.93$

(c) $M_{NaOH} = 0.30$; $[OH^-] = 0.30$; pOH $= 0.52$ and pH $= 13.48$

(d) $SO_3 + H_2O \rightarrow H_2SO_4$; $M_{H_2SO_4} = 9.4 \times 10^{-5}$;$[H^+] = 9.4 \times 10^{-5}$; pH $= 4.03$

(e) $M_{Ca(OH)_2} = 7.5 \times 10^{-4}$; $[OH^-] = (7.5 \times 10^{-4})(2) = 1.5 \times 10^{-3}$; pOH $=$ 2.82 and pH $= 11.18$

Calculations of pH can always be checked for reasonableness since acids must have pH values less than 7 and bases must have pH values greater than 7.

pH Values of Weak Acids and Weak Bases

The determination of the pH of weak acid or weak base solutions is one very important application of equilibrium calculations. In this chapter all of the calculations are very similar.

Weak Acids

Hydrofluoric acid is a weak acid that dissociates in water according to this equation:

$$HF \quad + \quad H_2O \quad \rightleftharpoons \quad H_3O^+ \quad + \quad F^-$$

which may also be written in a shorter, more convenient form by eliminating water:

$$HF \quad \rightleftharpoons \quad H^+ \quad + \quad F^-$$

Either of the above equations can be used to write the equilibrium law:

$$K_a = \frac{[H^+][F^-]}{[HF]} \quad \text{or} \quad \frac{[H_3O^+][F^-]}{[HF]}$$

The two forms are identical, and either may be used in the calculations that follow. The constant of K_a is the **acid dissociation constant**; values for selected weak acids are tabulated in Appendix 3. If we know the initial concentration of HF and the value of K_a, the hydrogen ion concentration can be calculated. The technique used is similar to those developed earlier in this chapter.

EXERCISE

If 0.100 mole of HF is diluted with distilled water to a volume of 500 milliliters, what is the pH of the final solution?

Solution

To solve this problem, we set up an equilibrium table, as shown on pages 379–380, with the reaction on the first line, and we enter the initial concentration on the INIT. CONC. line. The initial concentration of HF is calculated as

$$\frac{0.100 \text{ mol HF}}{0.500 \text{ L sol'n}} = 0.200 \text{ M HF}$$

REACTION:	HF	⇌	H^+	+	F^-
INIT. CONC.	**0.200 M**		**0.00***		**0.00**
CHANGE					
EQUILIBRIUM					
SOLUTION					

*The hydrogen ion concentration in pure water is 1.0×10^{-7}, but in most instances it is considered to be zero.

On the CHANGE line we enter x's to represent the stoichiometric relationships between the reactants and products. In this table we assign a positive value to the x's under the products since they start at zero and cannot possibly decrease. As a consequence any x's under the reactant(s) must be negative.

REACTION:	HF	⇌	H^+	+	F^-
INIT. CONC.	0.100/0.500		0.00		0.00
CHANGE	$\mathbf{-x}$		$\mathbf{+x}$		$\mathbf{+x}$
EQUILIBRIUM					
SOLUTION					

The EQUILIBRIUM line is then the sum of the INIT. CONC. and CHANGE lines of the table:

REACTION:	HF	⇌	H^+	+	F^-
INIT. CONC.	0.100/0.500		0.00		0.00
CHANGE	$-x$		$+x$		$+x$
EQUILIBRIUM	$\mathbf{0.200 - x}$		$\mathbf{+x}$		$\mathbf{+x}$
SOLUTION					

At this point we enter the terms in the EQUILIBRIUM line into the equilibrium law, along with the value of K_a:

$$K_a = \frac{[H^+][F^-]}{[HF]}$$

$$6.6 \times 10^{-4} = \frac{(x)(x)}{0.200 - x}$$

If this equation is solved for x by ordinary means, a quadratic equation results. As before, we can try a simplifying assumption to solve the problem more quickly. The only term that can be simplified is $0.200 - x$. We use the assumption that $x << 0.200$ and conclude that $0.200 - x = 0.200$. (Remember that, when we finally calculate x, we must verify that the assumption is true.) Returning to the equation, we find that the assumption yields

$$6.6 \times 10^{-4} = \frac{(x)(x)}{0.200}$$

$$(6.6 \times 10^{-4})(0.200) = x^2$$

$$x = 0.0115$$

Checking the assumption, we find that 0.0115 is less than one-tenth of 0.200 and therefore qualifies as being negligible for this type of calculation. Now we substitute 0.0115 for x in the EQUILIBRIUM line and calculate the SOLUTION line:

REACTION:	HF	$\rightleftharpoons$	H^+	+	F^-
INIT. CONC.	0.100/0.500		0.00		0.00
CHANGE	$-x$		$+x$		$+x$
EQUILIBRIUM	$0.200 - x$		$+x$		$+x$
SOLUTION	**0.189**		**0.0115**		**0.0115**

With the value $[H^+] = 0.0115$ M, the pH can be calculated as $-\log(0.0115) = 1.94$.

Looking at this process a little more closely, we find that, if the assumption is true, then

$$[H^+] = \sqrt{K_a C_a}$$

where C_a is the initial concentration of the weak acid. Evaluating the conditions under which the assumption holds, we find that it will always be true as long as $C_a > 100K_a$.

Weak Bases

Weak bases such as ammonia and methylamine, CH_3NH_2, behave similarly to weak acids. The dissociation reaction for methylamine is as follows:

$$CH_3NH_2 \quad + \quad H_2O \quad \rightleftharpoons \quad CH_3NH_3^+ \quad + \quad OH^-$$

Although water cannot be eliminated to simplify the reaction, it is not included in the equilibrium law.

$$K_b = \frac{[CH_3NH_3^+][OH^-]}{[CH_3NH_2]}$$

The constant K_b is the **base dissociation constant**. This equilibrium law is similar to the one for HF in the preceding section in that it has two concentration terms in the numerator and one in the denominator. The two laws are mathematically equivalent.

EXERCISE

What is the pH of a 0.500 M solution of methylamine, $K_b = 4.2 \times 10^{-4}$ (from Appendix 5)?

Solution

First we set up the equilibrium table in the same fashion as for the weak acid example in the previous exercise.

REACTION:	CH_3NH_2	+	H_2O	$\rightleftharpoons$	$CH_3NH_3^+$	+	OH^-
INIT. CONC.	0.500		55.5		0.00		0.00
CHANGE	$-x$		$-x$		$+x$		$+x$
EQUILIBRIUM	$0.500 - x$		$55.5 - x$		$+x$		$+x$
SOLUTION							

We entered the concentration of water;

$$\frac{1000 \text{ g } H_2O}{18 \text{ g } H_2O \text{ mol}^{-1}} = 55.5 \text{ M}$$

for completeness although it does not appear in the equilibrium law. The terms from the EQUILIBRIUM line are entered into the equilibrium law as follows:

$$4.2 \times 10^{-4} = \frac{(x)(x)}{0.500 - x}$$

To simplify, we assume that x is very small compared to 0.500, with the result that the $0.500 - x$ term in the denominator becomes 0.500:

$$4.2 \times 10^{-4} = \frac{(x)(x)}{0.500}$$

Solving for x, we obtain

$$(4.2 \times 10^{-4})(0.500) = x^2$$

$$x = 0.0145$$

Once x is determined, we check the assumption and find that it is valid since 0.500 is more than ten times larger than 0.0145. The equilibrium table is completed using the calculated value of x.

REACTION:	CH_3NH_2	+	H_2O	$\rightleftharpoons$	$CH_3NH_3^+$	+	OH^-
INIT. CONC.	0.500		55.5		0.00		0.00
CHANGE	$-x$		$-x$		$+x$		$+x$
EQUILIBRIUM	$0.500 - x$		$55.5 - x$		$+x$		$+x$
SOLUTION	**0.486**		**55.5**		**0.0145**		**0.0145**

Using the listed $[OH^-]$, we calculate the pOH as 1.84, and the pH is 12.16. As with the weak acid, this entire process can be summarized in a simple equation:

$$[OH^-] = \sqrt{K_b C_b}$$

as long as $C_b > 100K_b$.

In the previous two exercises there are two ways to check the accuracy of the results. First the answers must be reasonable; a weak acid should have an acidic pH (below 7), and a weak base should have a basic pH (above 7). Second, when the calculated values are entered into the equilibrium law, the result should be close to the given K_a or K_b.

Hydrolysis Reactions and the pH of Salt Solutions

Definition of Hydrolysis

When an acid is neutralized with a base, a salt is formed. If the anion of a salt is the conjugate base of a weak acid, it will react with water in a **hydrolysis reaction**. For the fluoride ion the hydrolysis reaction is as follows:

$$F^-(aq) + H_2O(\ell) \rightleftharpoons HF(aq) + OH^-(aq) \qquad (10.6)$$

The conjugate base, F^-, of the weak acid, HF, will produce a basic solution, as shown by the OH^- in equation 10.6.

If the cation of a salt is the conjugate acid of a weak base, the hydrolysis reaction will result in an acid solution, as shown by the reaction of the ammonium ion:

$$NH_4^+(aq) \quad + \quad H_2O(\ell) \quad \rightleftharpoons \quad NH_3(aq) \quad + \quad H_3O^+(aq)$$

The conjugate acid of a strong base and the conjugate base of a strong acid are extremely weak acids and bases. In fact, they are so weak that they do not hydrolyze in water. For example, Cl^- from the strong acid HCl and Na^+ from the strong base NaOH do not hydrolyze in water.

Acidity of Salt Solutions and Classification of Salts

The cation of a salt will be the conjugate acid of either a strong base or a weak base, and the anion of a salt will be the conjugate base of either a strong acid or a weak acid. Table 10.2 describes the different types of salts possible.

TABLE 10.2. CLASSIFICATION OF SALTS BASED ON ACID AND BASE USED TO PREPARE THEM

Acid Used	Base Used	Salt Type	Salt Solution Will Be
Strong	Strong	Neutral	Close to pH 7
Strong	Weak	Acidic	Acidic
Weak	Strong	Basic	Basic
Weak	Weak	Mixed	Depends on K_a and K_b

The first step in determining the pH of a salt solution is to determine the type of salt. This is done by adding H^+ to the anion and OH^- to the cation in the salt to determine the type of acid and base used to prepare the salt. For instance, ammonium chloride is composed of NH_4^+ and Cl^- ions. By adding OH^- to the ammonium cation and H^+ to the chloride anion, NH_4OH (NH_3 + H_2O) and HCl are obtained. From this result we conclude that ammonium chloride is an acid salt prepared from a weak base and a strong acid. We may quickly conclude that solutions of ammonium chloride will be acidic with pH values below 7.

A mixed salt has a cation that is the conjugate acid of a weak base, and an anion that is the conjugate base of a weak acid. The acidity or alkalinity of a mixed salt will depend on which of these conjugates is stronger. Considering the K_a of the weak acid and the K_b of the weak base that form the salt, we conclude that the solution will be acidic if $K_b > K_a$ and basic if $K_a > K_b$.

EXERCISE

Determine whether the aqueous solution of each of the following salts will be acid, neutral, or basic:

(a) $CaCl_2$
(b) $NaNO_3$
(c) Na_2SO_3
(d) $KC_2H_3O_2$
(e) SrF_2
(f) NH_4Br
(g) KNO_2
(h) Li_2CO_3

Answers

(a) Neutral
(b) Neutral
(c) Basic
(d) Basic
(e) Basic
(f) Acid
(g) Basic
(h) Basic

pH of Salt Solutions

The pH of a neutral salt is 7.00. Although we can tell whether a mixed salt solution will be acidic or basic, calculation of the pH is left for higher level courses. We concern ourselves here with calculating the pH values of acidic and basic salts only.

In the preceding section ammonium chloride was determined to be an acid salt. Since the chloride ion does not hydrolyze, we focus on the hydrolysis of the ammonium ion in order to develop the equilibrium law needed to solve problems.

NH_4^+ is the conjugate acid of ammonia, and the hydrolysis reaction is written as either

$$NH_4^+ \quad + \quad H_2O \quad \rightleftharpoons \quad NH_3 \quad + \quad H_3O^+$$

or

$$NH_4^+ \quad \rightleftharpoons \quad NH_3 \quad + \quad H^+$$

The equilibrium law for this reaction is either of the two equivalent expressions. In the equation

$$K = \frac{[H^+][NH_3]}{[NH_4^+]} = \frac{[H_3O^+][NH_3]}{[NH_4^+]} \qquad (10.7)$$

the equilibrium constant K is equal to K_w/K_b, where K_b is the base dissociation constant for ammonia. We can demonstrate this relationship by multiplying the equilibrium law in equation 10.7 by $\frac{[OH^-]}{[OH^-]}$ to obtain

$$K = \left(\frac{[NH_3][H^+]}{[NH_4^+]}\right) \times \left(\frac{[OH^-]}{[OH^-]}\right) \tag{10.8}$$

In the numerator the product $[H^+]\,[OH^-] = K_w$, which we substitute into equation 10.8:

$$K = K_w\left(\frac{[NH_3]}{[NH_4^+][OH^-]}\right)$$

The term in parentheses is the reciprocal of the equilibrium law for the dissociation of ammonia, and its value is K_b. Performing the substitution yields

$$K = \frac{K_w}{K_b}$$

This value is sometimes called the hydrolysis constant for a salt (K_h), but a more accurate name is the acid dissociation constant of the Brönsted-Lowry conjugate acid, symbolized as K_a.

This derivation leads to a universally useful equation and concept. For a conjugate acid-base pair the K_a of the conjugate acid multiplied by the K_b of the conjugate base will always be equal to K_w:

$$\boxed{K_w = K_a K_b}$$

K_w is the ion product for the dissociation of water:

$$H_2O \rightleftharpoons H^+ + OH^-,$$

and

$$K_w = [H^+]\,[OH^-] = 1.00 \times 10^{-14}$$

From this derivation we see also that the salt of a weak base acts as a Brönsted-Lowry acid, and we expect its solutions to be acidic.

Once the equilibrium law for acid and basic salts is derived, we can start solving problems. Let us calculate the pH of a 0.100 M solution of ammonium chloride. Since the equilibrium law has already been described, the next step is to set up the equilibrium table:

REACTION:	NH_4^+	$\rightleftharpoons$	NH_3	+	H^+
INIT. CONC.	**0.100**		**0.00**		**0.00**
CHANGE	$-x$		$+x$		$+x$
EQUILIBRIUM	$0.100 - x$		$+x$		$+x$
SOLUTION					

Entering the value of K_w obtained above, the value of K_b from Appendix 5, and the terms from the EQUILIBRIUM line into the equilibrium law yields

$$\frac{1.0 \times 10^{-14}}{1.8 \times 10^{-5}} = \frac{(x)(x)}{0.100 - x}$$

Assuming that $0.100 >> x$, we obtain

$$\frac{1.0 \times 10^{-14}}{1.8 \times 10^{-5}} = \frac{(x)(x)}{0.100}$$

which may be solved as

$$(5.6 \times 10^{-10}) = x^2$$

$$x = 7.4 \times 10^{-6}$$

Since x is much smaller than 0.100, the assumption was valid. We now use the value of x to calculate the values in the SOLUTION line of the table.

REACTION:	NH_4^+	$\rightleftharpoons$	NH_3	+	H^+
INIT. CONC.	0.100		0.00		0.00
CHANGE	$-x$		$+x$		$+x$
EQUILIBRIUM	$0.100 - x$		$+x$		$+x$
SOLUTION	**0.100**		$\mathbf{7.4 \times 10^{-6}}$		$\mathbf{7.4 \times 10^{-6}}$

The pH is then calculated:

$$pH = -\log(7.4 \times 10^{-6}) = 5.13$$

This result is consistent with the concept that NH_4^+ is a conjugate acid of NH_3 and that the solution must be acidic.

To summarize this calculation in a simpler form, we can write

$$[H^+] = \sqrt{\frac{K_w}{K_b} C_s} \quad \text{provided that} \quad C_s > 100 \frac{K_w}{K_b}$$

where K_b is the dissociation constant of the weak base used to produce the acidic salt.

For basic salts, calculations are performed in a very similar manner. For example, NaF is a basic salt since it is formed by reacting the strong base NaOH with the weak acid HF. We may ignore Na^+ while writing the hydrolysis equation for F^- as follows:

$$F^- \quad + \quad H_2O \quad \rightleftharpoons \quad HF \quad + \quad OH^-$$

The base dissociation constant is calculated by dividing K_w by the acid dissociation constant (K_a) of HF:

$$K_b = \frac{K_w}{K_a} = \frac{[HF][OH^-]}{[F^-]}$$

Questions may then be answered by using the procedures outlined above. The general solution can be simplified to

$$[OH^-] = \sqrt{\frac{K_w}{K_a} C_s} \quad \text{provided that} \quad C_s > 100 \frac{K_w}{K_a}$$

where C_s is the initial concentration of the salt and K_a is the acid dissociation constant of HF.

The salt KNO_3 does not change the pH of the solution since it is formed from the strong acid HNO_3 and the strong base KOH. The K^+ and NO_3^- ions are such weak Brönsted-Lowry acids and bases that they have no effect on the pH.

Two simple equations suffice for calculating the pH values of weak acid, weak base, acid salt, and basic salt solutions:

Weak acids and acid salts: $[H^+] = \sqrt{K_a C_a}$ (10.9)

Weak bases and basic salts: $[OH^-] = \sqrt{K_b C_b}$ (10.10)

For solutions of weak acids or weak bases, K_a and K_b are the acid and base dissociation constants. In salt solutions, K_a and K_b are the conjugate acid and base dissociation constants of the ions in the salt, which are calculated as K_w/K_b and K_w/K_a, respectively. Equations 10.9 and 10.10 work only when the initial concentration (C) is more than 100 times larger than the dissociation constant (K). Otherwise the problem must be solved using the quadratic equation as shown on pages 384–385.

Buffer Solutions

A mixture that contains a conjugate acid-base pair is known as a **buffer solution** because the pH changes by a relatively small amount if a strong acid or base is added. Buffer solutions, which are used to control pH levels in many biological and chemical reactions, are governed by the same equilibrium law as weak acids and weak bases.

Buffer Calculations

For instance, a buffer can be prepared by dissolving 0.200 mole of hydrofluoric acid, HF, and 0.100 mole of sodium fluoride, NaF, in water to make 1.00 liter of solution. The dissociation reaction of HF is

$$HF(aq) \rightleftharpoons F^-(aq) + H^+(aq)$$

and its equilibrium law is

$$K_a = \frac{[F^-][H^+]}{[HF]} = 6.8 \times 10^{-4}$$

(The value of $K = 6.8 \times 10^{-4}$ for HF is obtained from the table of ionization constants in Appendix 3.) An equilibrium table is set up, using the given information. For 0.100 M NaF, the concentration of F^- is determined as 0.100 M. The Na^+ in NaF is a spectator ion in the dissociation reaction and is not needed.

REACTION:	HF	$\rightleftharpoons$	F^-	+	H^+
INIT. CONC.	0.200		0.100		0.00
CHANGE	$-x$		$+x$		$+x$
EQUILIBRIUM	$0.200 - x$		$0.100 + x$		$+x$
SOLUTION					

Substituting the information from the EQUILIBRIUM line into the equilibrium law yields

$$K_a = \frac{(0.100 + x)(x)}{0.200 - x} = 6.8 \times 10^{-4}$$

Assuming that x is small compared to both 0.100 and 0.200, we can simplify the equation to

$$K_a = \frac{(0.100)\,(x)}{0.200} = 6.8 \times 10^{-4}$$

Solving this equation for x yields $x = 1.36 \times 10^{-3}$. This value of x satisfies the assumptions made, and the last line of the equilibrium table can be completed as shown,

REACTION:	HF	$\rightleftharpoons$	F^-	+	H^+
INIT. CONC.	0.200		0.100		0.00
CHANGE	$-x$		$+x$		$+x$
EQUILIBRIUM	$0.200 - x$		$0.100 + x$		$+x$
SOLUTION	**0.199**		**0.101**		**1.36×10^{-3}**

The last column in the table then gives us $[H^+] = 1.36 \times 10^{-3}$ M. The pH of the buffer solution is calculated as 2.87. For almost all buffer solutions, the assumptions hold true and this type of problem is solved by simply entering the given concentrations directly into the equilibrium law.

EXERCISE

Calculate the pH of each of the following buffer solutions:

(a) 0.250 M acetic acid and 0.150 M sodium acetate

(b) A solution of 10.0 grams each of formic acid and potassium formate dissolved in 1.00 liter of H_2O

(c) 0.0345 M ethylamine and 0.0965 M ethyl ammonium chloride

(d) 0.125 M hydrazine and 0.321 M hydrazine hydrochloride

Solutions

(a) $$K_a = \frac{[C_2H_3O_2^-][H^+]}{[HC_2H_3O_2]}$$

Substituting data gives

$$1.8 \times 10^{-3} = \frac{(0.150)[H^+]}{0.250}$$

$$[H^+] = 3.0 \times 10^{-3} \quad \text{and} \quad pH = 2.52$$

(b) Calculate 0.217 M formic acid and 0.417 M sodium formate. Then

$$K_a = \frac{[CHO_2^-][H^+]}{[HCHO_2]}$$

Substituting data gives

$$1.8 \times 10^{-4} = \frac{(0.417)[H^+]}{0.217}$$

$$[H^+] = 9.4 \times 10^{-5} \quad \text{and} \quad pH = 4.03$$

(c) $$K_b = \frac{[C_2H_5NH_3^+][OH^-]}{[C_2H_5NH_2]}$$

Substituting data gives

$$4.3 \times 10^{-4} = \frac{(0.0965)[OH^-]}{0.0345}$$

$$[OH^-] = 1.5 \times 10^{-4} \quad \text{and} \quad pOH = 3.82 \quad \text{and} \quad pH = 10.18$$

(d) $$K_b = \frac{[N_2H_5^+][OH^-]}{[N_2H_4]}$$

Substituting data gives

$$9.6 \times 10^{-7} = \frac{(0.321)[OH^-]}{0.125}$$

$$[OH^-] = 3.7 \times 10^{-7} \quad \text{and} \quad pOH = 6.43 \quad \text{and} \quad pH = 7.57$$

In (d), the hydroxide ion concentration is close to the hydroxide ion concentration of water and a rigorous calculation would include the 1×10^{-7} M hydroxide ions.

Shortcuts in Buffer Calculations

The equilibrium law used for buffers involves a ratio of the concentrations of the conjugate acid and conjugate base. For a ratio, we may use the moles of conjugate acid and moles of conjugate base instead of their molarities, thereby saving a step or two in the calculations.

In addition, if the concentrations of the conjugate acid and conjugate base are equal, their ratio is exactly 1.00. The result is that, with equal molarities or equal numbers of moles of conjugate acid and conjugate base, for a buffer made by mixing an equal number of moles of a weak acid and its conjugate base:

$$K_a = [H^+] \quad \text{and} \quad pK_a = pH$$

Also, for a buffer made from a weak base and its conjugate acid:

$$K_b = [OH^-] \quad \text{and} \quad pK_b = pOH$$

pH Changes in Buffers

Calculating the pH Change

Addition of a strong acid to a buffer will decrease the pH slightly, and addition of a strong base will increase the pH slightly. To determine the amount by which the pH changes when a strong acid or base is added to a buffer, we must calculate the change in concentration of the conjugate acid and conjugate base in the buffer.

In an acetate buffer, acetic acid, $HC_2H_3O_2$, is the conjugate acid, and the acetate ion, $C_2H_3O_2^-$, is the conjugate base. Addition of a strong acid, represented as H^+, results in this reaction:

$$C_2H_3O_2^- \quad + \quad H^+ \quad \rightarrow \quad HC_2H_3O_2$$

Addition of a strong acid increases the concentration of acetic acid and decreases the concentration of acetate ions. Similarly, addition of a strong base increases the acetate ion concentration and decreases the acetic acid concentration, as in the reaction

$$HC_2H_3O_2 \quad + \quad OH^- \quad \rightarrow \quad C_2H_3O_2^-$$

> Addition of a strong acid to a buffer always increases the conjugate acid and decreases the conjugate base of the buffer. Conversely, addition of a strong base increases the conjugate base and decreases the conjugate acid in the solution. As long as the strong acid or base is the limiting reactant, the solution will remain a buffer.

To calculate the effect of adding a strong acid or base to a buffer, the following steps are followed:

1. Determine either the molarity or the number of moles of conjugate acid and conjugate base in the original buffer.
2. Determine the amount of strong acid or base added in the same units as were used for the conjugate acid and base in Step 1.

3a. If a strong acid is added to the buffer, add the value from Step 2 to the conjugate acid and subtract it from the conjugate base.

3b. If a strong base is added to the buffer, add the value from Step 2 to the conjugate base and subtract it from the conjugate acid.

4. Substitute the new values for the conjugate acid and conjugate base into the equilibrium law, and calculate the pH.

EXERCISE

An acetate buffer is prepared with 0.250 M acetic acid and 0.100 M $NaC_2H_3O_2$. If 0.002 mole of solid NaOH is added to 100 milliliters of this buffer, calculate the change in pH of the buffer due to the addition of the NaOH.

Solution

To calculate the change in pH, the pH of the original buffer is needed, along with the final pH after the base is added. First, the dissociation reaction is written for acetic acid:

$$HC_2H_3O_2 \rightleftharpoons H^+ + C_2H_3O_2^-$$

Next the equilibrium law is written:

$$K_a = \frac{[H^+][C_2H_3O_2^-]}{[HC_2H_3O_2]}$$

The value for K_a and the acetate and acetic acid concentrations are substituted into the equilibrium law:

$$1.8 \times 10^{-5} = \frac{[H^+](0.100)}{0.250}$$

The hydrogen ion concentration is calculated as $[H^+] = 4.5 \times 10^{-5}$ M, and the pH is 4.35.

To calculate the pH of the buffer after the NaOH is added, we must first convert either the molarities of the conjugate acids and bases to moles or the moles of NaOH to molarity so that all of the units will be the same.

$$\text{Molarity of NaOH} = \frac{0.00200 \text{ mol NaOH}}{0.100 \text{ L}} = 0.0200 \text{ M}$$

Adding this value to the acetate concentration gives us 0.120 M $C_2H_3O_2^-$, and subtracting it from the acetic acid concentration yields 0.230 M $HC_2H_3O_2$. Substituting these new values into the equilibrium law, we have

$$1.8 \times 10^{-5} = \frac{[H^+]\ (0.120)}{0.230}$$

$$[H^+] = 3.45 \times 10^{-5} \text{ M}$$

$$\text{pH} = 4.46$$

There is an increase of 0.11 pH unit when the NaOH is added.

Buffer Capacity

Buffer capacity is defined as the number of moles of strong acid or strong base needed to change the pH of 1 liter of buffer by 1 pH unit. Given the initial concentrations of the conjugate acid and conjugate base, buffer capacity can be calculated. A quick estimate of buffer capacity is obtained by multiplying the sum of the molarities of the conjugate acid and conjugate base by 0.4:

$$\text{Approximate buffer capacity} = \text{Total molarity} \times 0.4$$

EXERCISE

Calculate the buffering capacity toward strong acid of the buffer in the previous exercise.

Solution

The pH of this buffer was found to be 4.35. The moles of strong acid needed to change the pH of 1 liter of this solution to 3.35 is the buffer capacity.

At pH 3.35 the ratio of conjugate acid $HC_2H_3O_2$ to conjugate base $C_2H_3O_2^-$ is calculated (pH 3.35 = 4.47×10^{-4}M H^+):

$$1.8 \times 10^{-5} = \frac{(4.47 \times 10^{-4})[C_2H_3O_2^-]}{[HC_2H_3O_2]}$$

$$0.040 = \frac{[C_2H_3O_2^-]}{[HC_2H_3O_2]}$$

From the previous example we also know that the total number of moles of conjugate acid and conjugate base in 1 L of buffer is 0.350 (0.100 mol of acetate ion + 0.250 mol of acetic acid):

$$0.350 \text{ mol} = [HC_2H_3O_2] + [C_2H_3O_2^-]$$

Solving the last two equations above, we obtain 0.337 mol of acetic acid and 0.013 mol of acetate ion. The moles of strong acid added are equal to the increase in the acetic acid concentration (0.337 − 0.250 = 0.087 mol) or the decrease in the acetate ion concentration (0.100 − 0.013 = 0.087 mol). The buffer capacity of this buffer is 0.087 mol of strong acid per liter.

Henderson-Hasselbach Equation

To derive the Henderson-Hasselbach equation, we write for a weak acid, HA:

$$HA \rightleftharpoons H^+ + A^-$$

The equilibrium law

$$K_a = \frac{[A^-][H^+]}{[HA]}$$

Taking the logarithm of both sides of this equation, we obtain

$$pK_a = pH - \log\left(\frac{[A^-]}{[HA]}\right)$$

Rearrangement yields

$$pH = pK_a + \log\left(\frac{[\text{conjugate base}]}{[\text{conjugate acid}]}\right)$$

This is known as the **Henderson-Hasselbach equation**. A similar derivation for weak bases results in

$$pOH = pK_b + \log\left(\frac{[\text{conjugate acid}]}{[\text{conjugate base}]}\right)$$

The Henderson-Hasselbach equation can be used to calculate the pH of a buffer mixture, or, alternatively, to calculate the ratio of conjugate acid to conjugate base when the pH of the buffer and pK_a are known.

The same calculations may be made directly from the equilibrium law, as shown on pages 406–409 on buffers.

Buffer Preparation

Preparation of a buffer starts with the selection of the desired pH. Then, a table of pK_a and pK_b values for weak acids and bases is consulted to find an appropriate conjugate acid-base pair to use. These two steps determine the ratio of the salt to the weak acid or base. Next, the moles of acid or base that need to be buffered are estimated. The total number of moles of the conjugate acid-base pair should be at least 20 times the amount of acid or base that needs to be buffered. The fourth step is to determine the volume of buffer solution that is needed. Finally, the method for preparing the buffer is selected.

There are three methods for preparing a buffer solution. In the first method the conjugate acid and conjugate base are measured and dissolved to the desired volume. In the second, the desired amount of conjugate acid is measured and then the appropriate amount of a strong base is added to convert some of the conjugate acid into its conjugate base. The third alternative is to measure the conjugate base and add the necessary amount of strong acid to convert some of the conjugate base into its conjugate acid.

Selecting the Buffer Materials

Assume that we want to prepare 100 milliliters of a buffer at pH 5.00 that will be capable of buffering 6.0×10^{-5} mole of strong acid for an enzyme experiment. We need to select an appropriate weak acid or weak base to use for this buffer. Our table gives only K_a values. Instead of converting all of them to pK values, we realize that we need a pK between 4.00 and 6.00, a range that is equivalent to a pK_a between 1.0×10^{-4} and 1.0×10^{-6}. The possible choices are H_3AsO_4, $HCHO_2$, $HC_2H_3O_2$, HF, HNO_2, and HN_3. Since this is a biological experiment, we have to be careful that the buffer has no adverse effects on the biochemicals to be used. Therefore it is wise to avoid H_3AsO_4 because of the arsenic, $HCHO_2$ and HF because of their toxicity, HNO_2 because of its redox properties, and HN_3 because it is generally not readily available. That leaves us with acetic acid, $HC_2H_3O_2$, which is a very popular buffer material.

Next, we need at least 20 times the 6.0×10^{-5} mole of strong acid expected, or 1.2×10^{-3} mole of conjugate acid plus conjugate base. This will be dissolved in 100 milliliters of solution, so the total molarity is 1.2×10^{-2} M. Since we already know that 100 milliliters of buffer is needed, we now consider the three methods of preparation.

Three Methods of Preparing Buffers

Dissolution of Conjugate Acid and Conjugate Base: The first method involves measuring the appropriate amounts of the conjugate acid and base into the desired volume of water. The acid we have chosen is acetic acid, and its conjugate base is the acetate ion. Acetate ions can be conveniently obtained from the salt sodium acetate. It was decided that the total of the conjugate acid and base should be

$$1.2 \times 10^{-2}\ \text{M} = [HC_2H_3O_2] + [C_2H_3O_2^-] \qquad (10.11)$$

The equilibrium law for the dissociation of acetic acid:

$$HC_2H_3O_2 \rightleftharpoons C_2H_3O_2^- + H^+$$

is

$$K_a = \frac{[C_2H_3O_2^-][H^+]}{[HC_2H_3O_2]}$$

Rearranging this equation gives

$$\frac{K_a}{[H^+]} = \frac{[C_2H_3O_2^-]}{[HC_2H_3O_2]} \qquad (10.12)$$

$$\frac{1.8 \times 10^{-5}}{1.0 \times 10^{-5}} = \frac{[C_2H_3O_2^-]}{[HC_2H_3O_2]}$$

$$1.8 = \frac{[C_2H_3O_2^-]}{[HC_2H_3O_2]}$$

Equations 10.11 and 10.12 are used to calculate the needed concentrations of acetic acid and acetate ions:

$$1.8[HC_2H_3O_2] = [C_2H_3O_2^-]$$

Substituting $1.8[HC_2H_3O_2]$ for $[C_2H_3O_2^-]$ in the first equation gives

$$1.2 \times 10^{-2} = 1.8[HC_2H_3O_2] + [HC_2H_3O_2]$$

$$1.2 \times 10^{-2} = 2.8[HC_2H_3O_2]$$

$$4.3 \times 10^{-3}\ M = [HC_2H_3O_2]$$

and then

$$1.8(4.3 \times 10^{-3}) = [C_2H_3O_2^-]$$

$$7.7 \times 10^{-3}\ M = [C_2H_3O_2^-]$$

Since the volume is 100 milliliters, we can calculate the grams of acetic acid as follows:

$$?\ g\ HC_2H_3O_2 = 100\ mL\ HC_2H_3O_2 \left(\frac{4.3 \times 10^{-3}\ mol\ HC_2H_3O_2}{1000\ mL\ HC_2H_3O_2}\right)\left(\frac{60.0\ g\ HC_2H_3O_2}{1\ mol\ HC_2H_3O_2}\right)$$

$$= 0.0258\ g\ HC_2H_3O_2$$

In a similar calculation the grams of sodium acetate can be determined since the molarity of sodium acetate is the same as the molarity of the acetate ion:

$$?\ g\ NaC_2H_3O_2 = 100\ mL\ NaC_2H_3O_2 \left(\frac{7.7 \times 10^{-3}\ mol\ NaC_2H_3O_2}{1000\ mL\ NaC_2H_3O_2}\right)\left(\frac{82.0\ g\ NaC_2H_3O_2}{1\ mol\ NaC_2H_3O_2}\right)$$

$$= 0.0631\ g\ NaC_2H_3O_2$$

To prepare this buffer, we weigh 0.0258 grams of $HC_2H_3O_2$ and 0.0631 grams of $NaC_2H_3O_2$ into a 100-milliliter volumetric flask, add water to the mark, and mix thoroughly.

Partial Neutralization of a Conjugate Acid: In the second method we have to obtain the same result by measuring only some acetic acid and neutralizing part of it with a base such as NaOH. A 0.100 M solution of NaOH can be used for this purpose. We calculate the number of grams of acetic acid needed from the 100 milliliters of solution and the total molarity of 1.2×10^{-2}:

$$?\ g\ HC_2H_3O_2 = 100mL\ HC_2H_3O_2 \left(\frac{1.2 \times 10^{-2}\ mol\ HC_2H_3O_2}{1000\ mL\ HC_2H_3O_2}\right)\left(\frac{60.0\ g\ HC_2H_3O_2}{1\ mol\ HC_2H_3O_2}\right)$$

$$= 0.0720\ g\ HC_2H_3O_2$$

The reaction of acetic acid with NaOH is as follows:

$$NaOH \quad + \quad HC_2H_3O_2 \quad \rightarrow \quad NaC_2H_3O_2 \quad + \quad H_2O$$

We can now calculate the volume of NaOH needed to produce 100 milliliters of 7.7×10^{-3} M sodium acetate solution:

$$? \text{ mL NaOH} = 100 \text{ mL } NaC_2H_3O_2 \left(\frac{7.7 \times 10^{-3} \text{ mol } NaC_2H_3O_2}{1000 \text{ mL } NaC_2H_3O_2}\right)\left(\frac{1 \text{ mol NaOH}}{1 \text{ mol } NaC_2H_3O_2}\right)$$
$$\left(\frac{1000 \text{ mL}}{0.100 \text{ mol}}\right)$$
$$= 7.7 \text{ mL NaOH}$$

To prepare this buffer, we measure 7.7 milliliters of 0.100 M NaOH and 0.0720 gram of $HC_2H_3O_2$ and dissolve them in enough water to make 100 milliliters of solution.

Partial Neutralization of a Conjugate Base: In the final method, we first calculate the number of grams of $NaC_2H_3O_2$ needed to make a 1.2×10^{-2} M solution, and then determine the volume of 0.0500 M HCl needed to convert some of this sodium acetate into the required amount of acetic acid:

$$? \text{ g } NaC_2H_3O_2 = 100 \text{ mL } NaC_2H_3O_2 \left(\frac{1.2 \times 10^{-2} \text{ mol } NaC_2H_3O_2}{1000 \text{ mL } NaC_2H_3O_2}\right)\left(\frac{82.0 \text{ g } NaC_2H_3O_2}{1 \text{ mol } NaC_2H_3O_2}\right)$$
$$= 0.0984 \text{ g } NaC_2H_3O_2$$

The reaction of sodium acetate with HCl is as follows:

$$HCl \quad + \quad NaC_2H_3O_2 \quad \rightarrow \quad HC_2H_3O_2 \quad + \quad NaCl$$

We now calculate the volume of HCl needed to make 100 mL of 4.3×10^{-3} M $HC_2H_3O_2$:

$$? \text{ mL HCl} = 100 \text{ mL } HC_2H_3O_2 \left(\frac{4.3 \times 10^{-3} \text{ mol } HC_2H_3O_2}{1000 \text{ mL } HC_2H_3O_2}\right)\left(\frac{1 \text{ mol HCl}}{1 \text{ mol } HC_2H_3O_2}\right)$$
$$\left(\frac{1000 \text{ mL HCl}}{0.0500 \text{ mol HCl}}\right)$$
$$= 8.6 \text{ mL HCl}$$

To prepare this buffer, we weigh 0.0984 gram of sodium acetate, measure 8.6 milliliters of 0.0500 M HCl, and dilute the mixture to make 100 milliliters of solution.

In preparing a buffer the chemist uses one of these three methods, usually the easiest one.

pH Values of Polyprotic Acids and Their Salts

Polyprotic acids dissociate in stepwise equilibria, each of which has its own dissociation constant. (See Appendix 4 for a list of polyprotic acid K_as.) Each dissociation constant is smaller than the preceding one ($K_{a1} > K_{a2} > K_{a3}$). When a polyprotic acid such as H_2A dissolves in water, the possible reactions are as follows:

$$H_2A \rightleftharpoons HA^- + H^+$$

$$HA^- \rightleftharpoons A^{2-} + H^+$$

Since K_{a2} is always less than K_{a1}, the second dissociation must occur to a smaller extent than the first one. In addition, the first dissociation produces hydrogen ions, and these ions further suppress the second dissociation according to Le Châtlier's principle. These two considerations, along with experimental evidence, demonstrate that the first dissociation step is the only important one in determining the $[H^+]$ and pH values of a solution of a polyprotic acid. The equation to use is

$$[H^+] = \sqrt{K_{a1}C_a}$$

where C_a is the initial concentration of the polyprotic acid.

EXERCISE

What is the pH value of each of the following solutions?

(a) 0.800 M H_3PO_4

(b) 0.0200 M H_2S

(c) 0.00400 M H_2CO_3

Answers

(a) 1.12 (b) 4.35 (c) 4.37

Similar reasoning is used for the hydrolysis of the fully deprotonated anion of a polyprotic acid such as A^{2-} in the second reaction above. The two hydrolysis reactions are

$$A^{2-} + H_2O \rightleftharpoons HA^- + OH^-$$

$$HA^- + H_2O \rightleftharpoons H_2A + OH^-$$

The first of these hydrolysis reactions has the larger equilibrium constant, and the OH^- formed in the first step suppresses the second step, again according to Le Châtlier's principle. The hydrolysis of a fully deprotonated anion of a polyprotic acid is calculated using this equation:

$$[OH^-] = \sqrt{\frac{K_w}{K_{a2}}C_{A^{2-}}}$$

EXERCISE

What is the pH of each of the following solutions?

(a) 3.00 M Na_3PO_4
(b) 0.0500 M Na_2CO_3

Answers
(a) 13.41 (b) 11.51

Anions of Polyprotic Acids

Anions of polyprotic acids, such as HA^-, are amphiprotic and can act as either conjugate acids or conjugate bases. For a salt that contains an amphiprotic anion, the hydrogen ion concentration is calculated as the square root of the product of the K_a values for that anion acting as a conjugate acid and as a conjugate base:

$$[H^+] = \sqrt{K_{a1}K_{a2}}$$

EXERCISE

Calculate the pH of a solution of each of the following ions:

(a) HSO_3^-
(b) $H_2PO_4^-$
(c) HPO_4^{2-}
(d) HCO_3^-
(e) HS^-

Answers
(a) 4.55 (c) 9.77 (e) 10.00
(b) 4.67 (d) 8.34

Buffers Made from Polyprotic Acids

Buffer solutions are often made using a conjugate acid and a conjugate base of a polyprotic acid. The pH of such a buffer is governed by the equilibrium law that contains both the conjugate acid and the conjugate base present in the solution. We also remember that the pH of a buffer solution must be within 1 pH unit of the pK_a of the acid.

As an illustration, we can consider the possible buffers that can be made by using phosphoric acid. The three dissociation steps for phosphoric acid are as follows:

$$\begin{array}{lclcll} H_3PO_4 & \rightleftharpoons & H_2PO_4^- & + & H^+ & pK_{a1} = 2.1 \\ H_2PO_4^- & \rightleftharpoons & HPO_4^{2-} & + & H^+ & pK_{a1} = 7.2 \\ HPO_4^{2-} & \rightleftharpoons & PO_4^{3-} & + & H^+ & pK_{a1} = 12.3 \end{array}$$

For phosphoric acid the first dissociation step involves H_3PO_4 and $H_2PO_4^-$, and a buffer with a pH in the range of 1.1 to 3.1 may be prepared from mixtures of these species. The second dissociation step involves $H_2PO_4^-$ and HPO_4^{2-}, and salts with these two ions are used to prepare a buffer with a pH in the range of 6.2 to 8.2. The HPO_4^{2-} and PO_4^{3-} anions involved in the third ionization step can be used to prepare a buffer with a pH in the range of 11.3 to 13.3.

TITRATION CURVES

Titration Curves for Neutralization Reactions

The titration technique is described on pages 352–356 along with the important calculations that apply to a wide variety of titrations. Acid-base titrations are often used to explain what occurs as an experiment is performed. There are four major points of interest during a titration experiment:

1. The start of the titration, where only sample is present.
2. The region where titrant is added up to the end point, where the solution contains a mixture of unreacted sample and products.
3. The end point, where all of the reactant has been converted into product.
4. The region after the end point, where the solution contains product and excess titrant.

For a neutralization reaction, the start of the titration involves a sample that is a pure acid or base and the pH is calculated as explained on pages 393–399. At the end point, the solution is the salt of the acid or base and the method of calculating the pH is described on pages 401–404. Between the start and the end point the solution is a mixture of a conjugate acid and its conjugate base. If a weak acid or base is involved, this is a buffer solution and the pH calculations are shown on pages 404–407. If only strong acids and bases are used, this region is unbuffered. After the end point, the pH depends on the excess titrant used.

The pH values during a titration may be calculated or measured in an experiment. A plot of pH versus the volume of titrant added is called a **titration curve**.

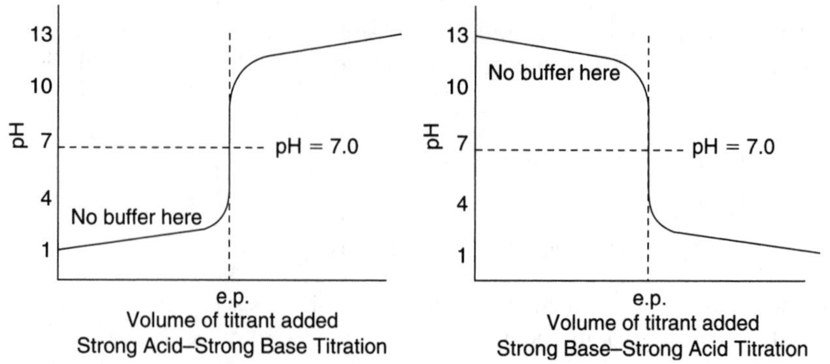

FIGURE 10.1. *Titration curves for a strong acid titrated with a strong base (left) and a strong base titrated with a strong acid (right). The region from the start to the end point (e.p.) is not buffered. The end point pH is always 7.0.*

The general shapes of strong acid–strong base titration curves are shown in Figure 10.1. Three important points about these curves should be noted. First, there is no buffer region since no weak acids or bases are present. Second, the end-point pH is always 7.0 because the products do not hydrolyze. Third, from the start the pH increases almost linearly until just before the end point.

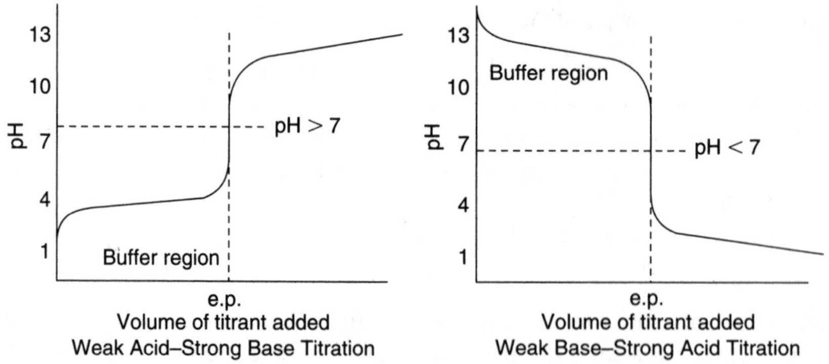

FIGURE 10.2. *Titration curves for a weak acid titrated with a strong base (left) and a weak base titrated with a strong acid (right). Between the start and the end point is the buffer region due to the presence of a conjugate acid-base mixture. The end point (e.p.) pH is not 7.0 because of hydrolysis of the products.*

As Figure 10.2 shows, weak acid–weak base titrations have a buffer region where a conjugate acid-base pair of moderate strength exists. In the middle of the buffer region the pH = pK_a. A large change in pH occurs just at the start

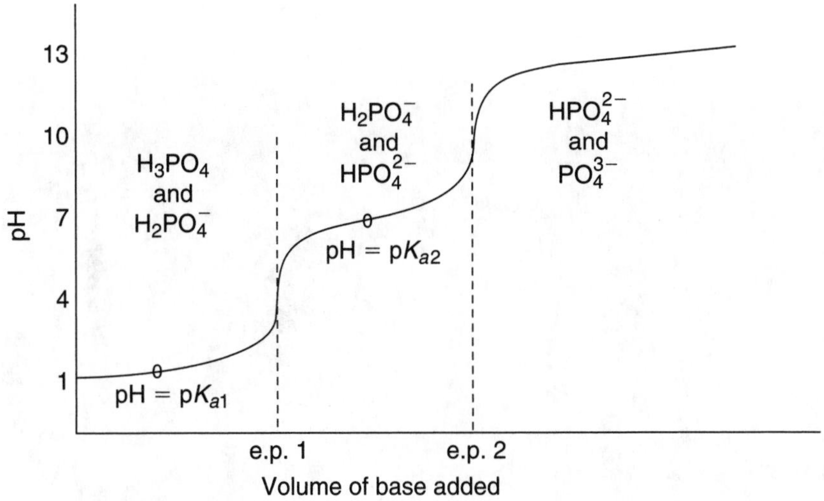

FIGURE 10.3. *Titration curve for the titration of phosphoric acid with a strong base.*

as the buffer mixture is formed, in contrast to the situation in strong acid–strong base titrations. At the end point the salts of weak acids and bases hydrolyze and the pH is not 7. Salts of weak acids hydrolyze to give basic solutions and $pH_{ep}>7$; the reverse is true for the salts of weak bases.

Titrations of polyprotic acids produce more than one end point because of the sequence of dissociation steps. The titration of phosphoric acid with a strong base is shown in Figure 10.3. Only two end points are observed since the third occurs at a pH that is too high to observe. At the start only pure H_3PO_4 is present. At each end point the acid is neutralized to the pure salts, $H_2PO_4^-$ and then HPO_4^{2-}; the pH values of these solutions are calculated using the equations on pages 414–417. Between the end points, in what are known as buffer regions, there is a mixture of a conjugate acid and its conjugate base, as shown. At the midpoint of each buffer region the pH is equal to the pK_a that governs that particular dissociation step. Figures 10.2 and 10.3 illustrate why and how a buffer can be made by partial neutralization of a weak acid. Partial neutralization results in a solution containing both a conjugate acid and conjugate base in the buffer regions of these curves

11
THERMODYNAMICS

ENERGY MEASUREMENT

Calorimeters for Energy Measurements

Specific Heat

Heat is often called the lowest form of energy. In this form it is easily measurable by determining the temperature changes caused by the heat released or absorbed in a chemical process. Heat energy was originally defined in terms of the calorie, which is the amount of heat needed to raise the temperature of 1 gram of pure water from 14.5 to 15.5°C. The joule is the S.I. unit for energy, and 1 calorie is equal to exactly 4.184 joules. By virtue of these definitions, 4.184 joules of energy is needed to raise the temperature of 1 gram of water by 1 Celsius degree. This quantity is known as the **specific heat** of water:

$$\text{Specific heat of water} = 4.184 \text{ J g}^{-1}\ ^\circ\text{C}^{-1}$$

Once the specific heat of water was defined, the specific heat of any other substance could be determined. One method is to immerse a hot object in a known quantity of water and to measure the temperature change that occurs. This method is illustrated in the next exercise.

As an interesting sidelight, in 1818 Dulong and Petit discovered that in most cases the specific heat multiplied by the atomic mass of a metal is equal to a constant. The law of Dulong and Petit can be written as follows:

$$\text{Specific heat} \times \text{molar mass} = 25 \text{ J mol}^{-1}\ ^\circ\text{C}^{-1}$$

This law helped confirm the molar masses of the metallic elements when disagreements occurred. In addition, specific heat is an intensive physical property of all elements and compounds and can be used to identify substances.

The definition of the specific heat of water allows chemists to determine the heat energy of any process by measuring the change in temperature of a known mass of water:

$$\text{heat energy} = (\text{Specific heat}) \times \text{ g } \times (\text{Mass}) \times (\text{Temperature change}) \quad (11.1)$$

or, in symbols:

$$q = \text{sp. ht.} \times \Delta T$$

Here q represents the heat energy in joules. The temperature change is always determined as the final temperature minus the initial temperature ($°C_{final} - °C_{initial}$). Equation 11.1 is applicable to any substance, not just water.

There is a distinct difference between heat energy and temperature, as Equation 11.1 indicates. Temperature is a measure of the average kinetic energy of a group of atoms, and a temperature change is the change in the average kinetic energy. Heat energy is produced when both the kinetic energy and the energy of attraction between atoms change in a chemical process.

EXERCISE

An insulated cup contains 75.0 grams of water at 24.00°C. A 26.00-gram sample of a metal at 85.25°C is added. The final temperature of the water and metal is 28.34°C.

1. What is the specific heat of the metal?
2. According to the law of Dulong and Petit, what is the approximate molar mass of the metal?
3. What is the apparent identity of the metal?

Solutions

1. The law of conservation of energy requires that the heat energy gained by the water must be exactly equal to the heat energy lost by the metal as it cools in the water. Mathematically, this is written as

$$+q_{water} = -q_{metal}$$

Using equation 11.1 we expand these terms to read

$$(\text{sp. heat } H_2O)(\text{g } H_2O)(\Delta T_{water}) = -(\text{sp. heat}_{metal})(\text{g metal})(\Delta T_{metal})$$

Entering the data yields

$$\begin{aligned}(4.184 \text{ J g}^{-1}\ °\text{C}^{-1})(75 \text{ g})(4.34\ °\text{C}) &= -(\text{sp. heat}_{metal})(26.0 \text{ g})(-56.91\ °\text{C}) \\ 1362 \text{ J} &= 1480 \text{ g } °\text{C}(\text{sp. heat}_{metal}) \\ \text{sp. heat}_{metal} &= 0.920 \text{ J g}^{-1}\ °\text{C}^{-1}\end{aligned}$$

2. Using the equation for the law of Dulong and Petit, we obtain

$$\begin{aligned}(0.920 \text{ J g}^{-1}\ °\text{C}^{-1})(\text{molar mass}) &= 25 \text{ J mol}^{-1}\ °\text{C}^{-1} \\ \text{molar mass} &= 27 \text{ g mol}^{-1}\end{aligned}$$

3. Aluminum is the only metal with a molar mass of 27 g mol^{-1}.

Calorimeters and Heat Measurements

A calorimeter is the device used to measure the heat energy produced by chemical reactions and physical changes. Figure 11.1 illustrates the design features of a calorimeter. The reaction vessel, usually made of metal, efficiently transfers heat to the rest of the apparatus, which includes a large mass of water in an insulated container that prevents heat from being lost to the surroundings. Also included is a stirrer and a very accurate thermometer.

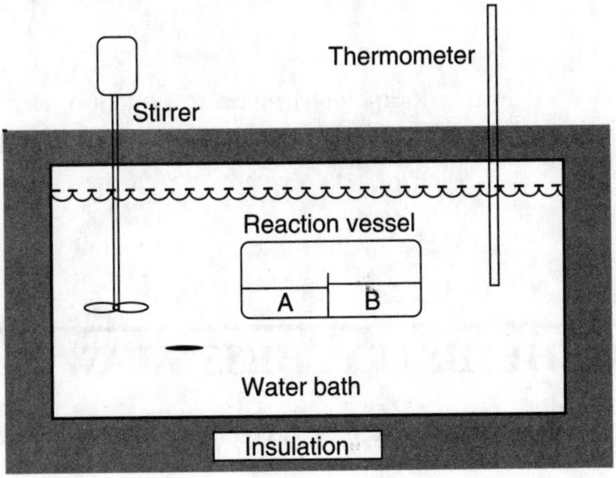

FIGURE 11.1. *Diagram of a calorimeter and its essential components. Reactants A and B may be mixed by rotating the reaction vessel. In a "bomb calorimeter" the reaction vessel is filled with reactant and O_2, which are then ignited with an electric spark.*

All parts of the calorimeter heat up or cool down as heat is released or absorbed in the chemical process. Water is the major part of the system; however, for the most accurate measurements, the specific heats and masses of the reaction vessel, the thermometer, the stirrer, and the container itself must be included in the calculation. The total heat capacity (C) of the calorimeter is defined as the sum of the products of the specific heat and the masses of all components of the calorimeter:

$$C\,(\text{J}\ ^\circ\text{C}^{-1}) = (\text{sp. ht.})_1(\text{mass})_1 + (\text{sp. ht.})_2(\text{mass})_2 + (\text{sp. ht.})_3(\text{mass})_3 + \ \ldots$$

The heat capacity is sometimes called the **calorimeter constant**.

The heat energy produced in a calorimeter is calculated as

$$q = (C)\,(^\circ\text{C}_{\text{final}} - {}^\circ\text{C}_{\text{initial}}) \qquad (11.2)$$

where C in Equation 11.2 replaces the sp. ht × g terms in Equation 11.1.

The most convenient method for determining the heat capacity (C) of a calorimeter is to calibrate the device using a reaction that will produce a known amount of heat and then calculate C from the observed temperature change.

EXERCISE

A calorimeter has a heat capacity of 1265 joules per degree Celsius. A reaction causes the temperature of the calorimeter to change from 22.34°C to 25.12°C. How many joules of heat were released in this process?

Solution

The heat released in a calorimeter is given in equation 11.2. We calculate ΔT as 25.12 − 22.34 = +2.78°C. Entering these data into the equation gives

$$q = 1265 \text{ J } {}^\circ\text{C}^{-1} (2.78\ {}^\circ\text{C})$$

$$= 3517 \text{ J of heat energy released}$$

THERMOCHEMISTRY: HESS'S LAW

Thermochemistry and Hess's Law

Hess's Law

Hess's law states that, whatever mathematical operations are performed on a chemical reaction, the same mathematical operations are also applied to the heat of reaction. Hess's law is summarized as follows:

1. If the coefficients of a chemical reaction are all multiplied by a constant, the heat of reaction (ΔH°_{rxn}) is multiplied by that same constant.
2. If two or more reactions are added to obtain an overall reaction, the heats of these reactions are also added to obtain the heat of the overall reaction.

Hess's law allows the chemist to measure ΔH°_{rxn} for several reactions and then to combine the reactions and their heats to obtain ΔH°_{rxn} for a completely different reaction.

For example, the burning of propane produces a large amount of heat. ($\Delta H^\circ_{rxn} = -2044$ kJ) It is an easy reaction to perform; propane is ignited in the presence of oxygen. Although the reverse reaction for the synthesis of 1 mole of propane from carbon dioxide and water is impossible to perform, it may be written as

$$3CO_2(g) \quad + \quad 4H_2O(g) \quad \rightarrow \quad CH_3CH_2CH_3(g) \quad + \quad 5O_2(g)$$

This reaction is the reverse of the combustion reaction for propane. Since reversing a reaction is the same as multiplying it by −1, the heat needed for this reaction is +2044 kilojoules. The reason is that $\Delta H°$ is a state function, depending only on the final and initial states of the system. Synthesis of 1 mole of propane has the same final and initial states as the combustion of 1 mole of propane; the only difference is the direction of the process. We must come to the conclusion that $\Delta H°$ has the same magnitude for these two reactions, but the values have different signs since the reactions go in opposite directions. Figure 11.2 illustrates this process.

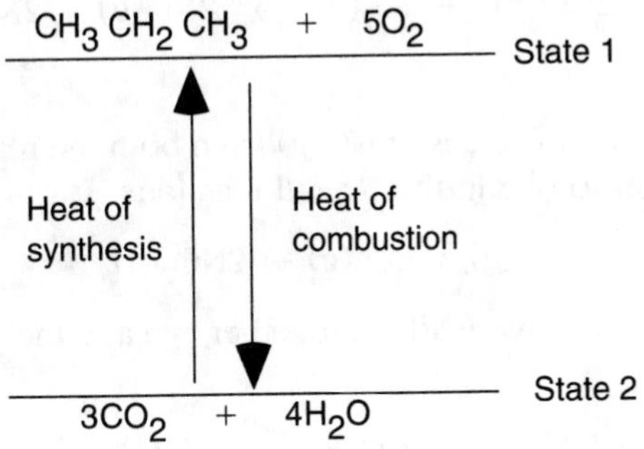

FIGURE 11.2. *Diagram illustrating that the heat of a reaction has the same magnitude whether the reaction is run in the forward or the reverse direction. The sign of the heat of reaction is positive in one direction and negative in the other.*

The general principle that

$$\Delta H°_{\text{forward } rxn} = -\Delta H°_{\text{reverse } rxn}$$

is an essential consequence of Hess's law.

For the combustion of propane, Hess's first rule tells us that multiplying the reaction by 2 will result in $\Delta H°$ being multiplied by 2:

$$2CH_3CH_2CH_3(g) + 10O_2(g) \rightarrow 6CO_2(g) + 8H_2O(g) \qquad (\Delta H = -4088 \text{ kJ})$$

If the reaction is multiplied by ½, $\Delta H°$ will also be multiplied by ½:

$$\tfrac{1}{2}CH_3CH_2CH_3(g) + 2.5O_2(g) \rightarrow 1.5CO_2(g) + 2H_2O(g) \qquad (\Delta H = -1011 \text{ kJ})$$

If the reaction is multiplied by 6, $\Delta H°$ will be multiplied by 6:

$$6CH_3CH_2CH_3(g) + 30O_2(g) \rightarrow 18CO_2(g) + 24H_2O(g) \qquad (\Delta H = -12{,}264 \text{ kJ})$$

Once $\Delta H°$ is multiplied by any coefficient, it is no longer a standard-state function and the superscript "°" is dropped as shown above.

The second of Hess's rules concerns the addition of chemical reactions. When chemical reactions are added, all of the reactants are combined as reactants of the overall reaction, and all of the products are combined as products of the overall reaction. The last step in adding reactions is the cancelation of any identical reactants and products in the overall reaction.

To add the two reactions below, we combine the reactants and products as follows:

$$\begin{array}{rcl} N_2(g) + O_2(g) & \rightarrow & 2NO(g) \\ 2NO(g) + O_2(g) & \rightarrow & 2NO_2(g) \\ \hline N_2(g) + 2NO(g) + 2O_2(g) & \rightarrow & 2NO(g) + 2NO_2(g) \end{array}$$

We then cancel two molecules of NO(g) from both the reactant and product sides of this equation to obtain the overall reaction:

$$N_2(g) + 2O_2(g) \rightarrow 2NO_2(g)$$

Since the two reactions were added together, we add their heats of reaction:

$$\begin{array}{lcll} N_2(g) + O_2(g) & \rightarrow & 2NO(g) & \Delta H_1^\circ = +180.5 \text{ kJ} \\ 2NO(g) + O_2(g) & \rightarrow & 2NO_2(g) & \Delta H_2^\circ = -114.1 \text{ kJ} \\ \hline N_2(g) + 2O_2(g) & \rightarrow & 2NO_2(g) & \Delta H^\circ_{overall} = +66.4 \text{ kJ} \end{array}$$

In this example, $\Delta H_1^\circ + \Delta H_2^\circ = \Delta H^\circ_{overall}$.

The heats of many reactions can be determined if the heats of combustion are known for all of the reactants and products. To illustrate a more complex combination of reactions, we will determine the heat of reaction for the synthesis of propane from carbon and hydrogen:

$$3C(s) + 4H_2(g) \rightarrow CH_3CH_2CH_3(g) \qquad (11.3)$$

using the combustion reactions of propane, hydrogen, and carbon. These combustion reactions, along with their ΔH° values, are taken from standard tables and are as follows:

$$\begin{array}{lll} CH_3CH_2CH_3(g) + 5O_2(g) \rightarrow 3CO_2(g) + 4H_2O(g) & & (\Delta H_1^\circ = -2{,}044 \text{ kJ}) \\ 2H_2(g) + O_2(g) \rightarrow 2H_2O(g) & & (\Delta H_2^\circ = -483.6 \text{ kJ}) \\ C(s) + O_2(g) \rightarrow CO_2(g) & & (\Delta H_3^\circ = -393.5 \text{ kJ}) \end{array}$$

If it is not immediately obvious how these three equations should be combined, some general principles offer guidance on how to approach the problem logically. To find which equation to start with, we:

1. Focus on the most complex molecules first.
2. Focus on atoms and molecules that occur in only one reaction.
3. Focus on atoms and molecules that are in the overall equation.

Using these principles, we see that the combustion of propane equation should be considered first. The first equation on the previous page contains 1 mole of propane, and so does equation 11.3, the overall equation. However, in the combustion reaction propane is a reactant and in the overall reaction it is a product. Consequently, the combustion equation on the previous page must be reversed by multiplying it by −1, which also changes the sign of $\Delta H°$:

$$3CO_2(g) + 4H_2O(g) \rightarrow CH_3CH_2CH_3(g) + 5O_2(g) \quad [\Delta H° = -2044 \text{ kJ} \times (-1)]$$

Once the first equation is established, we can use the same principles to select and manipulate the remaining reactions. If needed, an additional principle may be used:

4. Focus on finding atoms and molecules to cancel unneeded ones from already selected equations.

Using principle 3, we focus on the second (hydrogen) reaction on the previous page and note that it has $2H_2(g)$ in it and that we need $4H_2(g)$ in our overall equation. The hydrogens are reactants in both equations, and the second equation above needs to be multiplied by 2 so that we will have the correct number of $H_2(g)$ in the final reaction. Also, $\Delta H°$ must be multiplied by 2, giving

$$4H_2(g) \quad + \quad 2O_2(g) \quad \rightarrow \quad 4H_2O(g) \qquad (\Delta H = -483.6 \text{ kJ} \times 2)$$

Using principle 4, we note that the remaining (carbon) reaction above contains one $CO_2(g)$ as a product. We need to cancel three $CO_2(g)$ molecules from the two equations we have established. Since $CO_2(g)$ is a product in one equation and a reactant in the other, the $CO_2(g)$ molecules will cancel when the reactions are added. However, the third equation must be multiplied by 3 so that all three $CO_2(g)$ molecules will cancel:

$$3C(s) \quad + \quad 3O_2(g) \quad \rightarrow \quad 3CO_2(g) \qquad (\Delta H = -393.5 \text{ kJ} \times 3)$$

Adding the three equations gives

$$3C(s) + 3O_2(g) + 4H_2(g) + 3O_2(g) + 3CO_2(g) + 4H_2O(g)$$
$$\rightarrow 3CO_2(g) + 4H_2O(g) + 5O_2(g) + CH_3CH_2CH_3(g)$$

After we cancel the $3CO_2(g)$, $4H_2O(g)$, and $5O_2(g)$ molecules, the equation becomes

$$3C(s) \quad + \quad 4H_2(g) \quad \rightarrow \quad CH_3CH_2CH_3(g)$$

The heat of reaction is the sum of the three $\Delta H°$ values multiplied by the operations performed:

$$\begin{aligned}
\Delta H°\text{overall} &= \Delta H_1° \times (-1) && + \Delta H_2° \times 2 && + \Delta H_3° \times 3 \\
&= -2044 \text{ kJ} \times (-1) && + -483.6 \text{ kJ} \times 2 && + -393.5 \text{ kJ} \times 3 \\
&= +2044 \text{ kJ} && - \ 967.2 \text{ kJ} && - \ 1180.5 \text{ kJ} \\
&= -103.7 \text{ kJ}
\end{aligned}$$

The advantage of being able to perform some simple experiments to obtain data for complex, even impossible, reactions was recognized very quickly. During the energy crisis of the early 1970s the feasibility of producing alternative fuels was determined from thermochemical calculations.

EXERCISE

We may wish to synthesize methane from carbon and hydrogen:

$$C(s) \quad + \quad 2H_2(g) \quad \rightarrow \quad CH_4(g)$$

Based on the heat energy needed for this reaction and the heat of combustion for methane, is this a worthwhile effort? The standard heats of combustion in kilojoules, are as follows: $CH_4 = -890.3$, $H_2 = -571.8$, $C = -393.5$.

Solution

The three combustion reactions are written as follows:

$$CH_4(g) + 2O_2(g) \rightarrow CO_2(g) + 2H_2O(\ell) \qquad (\Delta H = -890.3 \text{ kJ})$$
$$2H_2(g) + O_2(g) \rightarrow 2H_2O(\ell) \qquad (\Delta H = -571.8 \text{ kJ})$$
$$C(s) + O_2(g) \rightarrow CO_2(g) \qquad (\Delta H = -393.5 \text{ kJ})$$

To obtain the desired reaction, we need to reverse the combustion of methane and add this reaction to the remaining reactions. The heat of this reaction is

$$\Delta H_{react} = (-890.3 \text{ kJ})(-1) + (-571.8 \text{ kJ}) + (-393.5 \text{ kJ})$$
$$= -75.0 \text{ kJ}$$

Ignoring all other factors, we find that the formation of CH_4 produces heat and the combustion produces a much greater amount of heat. On this basis, the synthesis of CH_4 is a feasible process. If the energy required to make $H_2(g)$ from water is included in the calculation, however, the energy gain is very small and the investment in such a project may be suspect.

This reaction, in which methane is formed from the elements carbon and hydrogen, is known as a formation reaction, and we determined the heat of formation.

Formation Reactions and Heats of Formation

In the preceding section we saw that the heats of many chemical reactions can be determined if the heats of combustion of all the reactants and products are known. For other types of reactions these heats can also be tabulated, but such tables would be very long and complex. Also, finding the appropriate reactions to combine would be a monumental task. Heats of formation allow chemists to tabulate thermochemical data in a short, easy-to-use arrangement.

A **formation reaction** is defined as a reaction in which the reactants are elements in their standard state at 25°C and 1 atmosphere of pressure and there is only 1 mole of product. Here are some examples of formation reactions:

$$Fe(s) + O_2(g) \rightarrow FeO(s)$$
$$2Fe(s) + \tfrac{3}{2}O_2(g) \rightarrow Fe_2O_3(s)$$
$$2K(s) + \tfrac{1}{2}H_2(g) + 2O_2(g) + P(s) \rightarrow K_2HPO_4(s)$$

Fractional coefficients may be used in formation reactions. Since there is always 1 mole of product, the heats of formation (ΔH_f°) are tabulated as the heat produced per mole of product. Since the reaction can easily be deduced if the product is known, a table of data needs to contain only the name or formula of the product and its corresponding ΔH_f°. Some heats of formation are given in the table in Appendix 2.

Heats of formation of the elements, whether they are molecules or atoms, are always zero. The reason is that the formation reaction for an element such as oxygen is defined as

$$O_2(g) \rightarrow O_2(g)$$

Since the oxygen is at 25°C and 1.00 atmosphere of pressure both as the product and as the reactant, the initial and final states of the oxygen are the same and their difference must be zero:

$$(\Delta H_f^\circ) \text{ (any element)} = 0$$

Heats of formation can be calculated from heats of combustion, as shown in the previous exercise.

The value of ΔH° for the combustion of propane:

$$CH_3CH_2CH_3(g) + 5O_2(g) \rightarrow 3CO_2(g) + 4H_2O(g)$$

can be calculated using the formation reactions and the tabulated heats of formation. We will need a formation reaction for each of the reactants and products. Elements are excluded, however, since, as stated above, their heats of formation are always zero. In this example, the formation reactions for propane, carbon dioxide and water are needed. The heats of formation are taken from Appendix 2.

$$3C(s) + 4H_2(g) \rightarrow CH_3CH_2CH_3(g) \qquad (\Delta H_f^\circ = -103.8 \text{ kJ mol}^{-1})$$
$$C(s) + O_2(g) \rightarrow CO_2(g) \qquad (\Delta H_f^\circ = -393.5 \text{ kJ mol}^{-1})$$
$$H_2(g) + \tfrac{1}{2}O_2(g) \rightarrow H_2O(g) \qquad (\Delta H_f^\circ = -241.8 \text{ kJ mol}^{-1})$$

The following operations are performed on these reactions so that they can be combined to yield the combustion reaction:

1. Reverse the propane formation reaction:

$$CH_3CH_2CH_3(g) \rightarrow 3C(s) + 4H_2(g) \qquad (\Delta H_f^\circ = -103.8 \text{ kJ mol}^{-1} \times -1 \text{ mol})$$

2. Multiply the carbon dioxide formation reaction by 3:

$$3C(s) + 3O_2(g) \rightarrow 3CO_2(g) \quad (\Delta H_f^\circ = -393.5 \text{ kJ mol}^{-1} \times 3 \text{ mol})$$

3. Multiply the $H_2O(g)$ formation reaction by 4:

$$4H_2(g) + 2O_2(g) \rightarrow 4H_2O(g) \quad (\Delta H_f^\circ = -241.8 \text{ kJ mol}^{-1} \times 4 \text{ mol})$$

After these three operations, the three reactions are added to obtain the combustion reaction:

$$CH_3CH_2CH_3(g) + 5O_2(g) \rightarrow 3CO_2(g) + 4H_2O(g)$$

The corresponding sum of the heats is the heat of reaction:

$$\begin{aligned}\Delta H^\circ &= (-103.8 \text{ kJ mol}^{-1})(-1 \text{ mol}) + (-393.5 \text{ kJ mol}^{-1})(3 \text{ mol}) \\ &\quad + (-241.8 \text{ kJ mol}^{-1})(4 \text{ mol}) \\ &= -2044 \text{ kJ}\end{aligned}$$

After several calculations using heats of formation, a pattern appears. The heat of any reaction will be the sum of the ($\Delta H_f^\circ \times \text{mol}_{\text{product}}$) values for all the products minus the sum of the ($\Delta H_f^\circ \times \text{mol}_{\text{reactant}}$) values for all the reactants. The terms $\text{mol}_{\text{product}}$ and $\text{mol}_{\text{reactant}}$ refer to the stoichiometric coefficients in the balanced equation for the reaction.

$$\Delta H^\circ = \Sigma(\Delta H_f^\circ \times \text{coeff})_{\text{products}} - \Sigma(\Delta H_f^\circ \times \text{coeff})_{\text{reactants}}$$

EXERCISE

Calculate the heat of combustion of CH_4. The heats of formation, in kilojoules per mole, are as follows: $CH_4 = -74.8$, $(CO_2) = -110.5$, and $H_2O(g) = -241.8$.

Solution

For the combustion of CH_4 the equation is

$$CH_4(g) + 2O_2(g) \rightarrow CO_2(g) + 2H_2O(g)$$

The heat of this reaction is calculated as

$$\begin{aligned}\Delta H^\circ_{\text{react}} &= [(-110.5 \text{ kJ mol}^{-1})(1 \text{ mol}) + (-241.8 \text{ kJ mol}^{-1})(2 \text{ mol})] \\ &\quad - [(-74.8 \text{ kJ mol}^{-1})(1) + (0.00 \text{ kJ mol}^{-1})(2 \text{ mol})] \\ &= -519.3 \text{ kJ}\end{aligned}$$

The value of 0.00 kJ mol^{-1} in this calculation represents the heat of formation of $O_2(g)$, which by definition, is zero.

ELECTROCHEMISTRY

Determining Concentrations by Using Galvanic Cells

The Nernst Equation

The **Nernst equation** suggests a method for using galvanic cell measurements to determine concentrations. A redox reaction that can be used for this purpose is

$$2Ag^+(ag) \quad + \quad Cu(s) \quad \rightarrow \quad 2Ag(s) \quad + \quad Cu^{2+}(aq)$$

The cell diagram for this reaction is as follows:

$$Cu\,|\,Cu^{2+}\|Ag^+\,|\,Ag$$

The cell voltage is represented as

$$E_{cell} = E_{cathode} - E_{anode}$$

Inserting the Nernst equations for $E_{cathode}$ and E_{anode}, we obtain

$$E_{cell} = \left(E^\circ_{Ag/Ag^+} - \frac{0.0591}{1}\log\frac{1}{[Ag^+]}\right) - \left(E^\circ_{Cu/Cu^{2+}} - \frac{0.0591}{2}\log\frac{1}{[Cu^{2+}]}\right) \quad (11.4)$$

If $[Ag^+]$ can be kept at a constant, known value, the value of E_{cell} determined in an experiment can be used to calculate $[Cu^{2+}]$. A saturated solution of AgCl will maintain the silver ion concentration at a constant 1.0×10^{-5} M. Since $E^\circ_{Ag/Ag^+} = +0.80$ volt and $E^\circ_{Cu/Cu^{2+}} = +0.34$ volt, equation 11.4 becomes

$$E_{cell} = 0.164 - \frac{0.0591}{2}\log\frac{1}{[Cu^{2+}]}$$

when all three constants are entered. If $E_{cell} = 0.128$ volt for a solution that contains an unknown concentration of Cu^{2+}, we can calculate the concentration as

$$0.128 = 0.164 - \frac{0.0591}{2}\log\frac{1}{[Cu^{2+}]}$$

$$-0.036 = -\frac{0.0591}{2}\log\frac{1}{[Cu^{2+}]}$$

$$+1.218 = \log\frac{1}{[Cu^{2+}]}$$

$$16.52 = \frac{1}{[Cu^{2+}]}$$

$$[Cu^{2+}] = 0.0605\text{ M } Cu^{2+}$$

The key to using galvanic cell measurements for determining concentrations is that all but one concentration in the Nernst equation must be known and must remain constant during the experiment. In the preceding analysis the silver ion concentration around the cathode was held constant. This constant electrode is also known as the **reference electrode**. The anode that monitored the Cu^{2+} ion concentration is called the **indicator electrode**.

pH Measurement and pH Electrodes

The concentration of hydrogen ions may be determined by using a galvanic cell in a manner similar to that for the determination of copper(II) ions illustrated on the previous page. **pH** is defined as the negative logarithm of the hydrogen ion concentration:

$$pH = -\log[H^+]$$

As in the preceding example, one of the electrodes in the galvanic cell must be a constant reference electrode. Two common reference electrodes for pH measurement are constructed using these half-reactions:

$$AgCl(s) \quad + \quad e^- \quad \rightarrow \quad Ag(s) \quad + \quad Cl^-$$

$$Hg_2Cl_2 \quad + \quad 2e^- \quad \rightarrow \quad 2Hg(l) \quad + \quad 2Cl^-$$

The silver-silver chloride reference electrode consists of a silver wire with a coating of insoluble silver chloride in a saturated solution of potassium chloride. The second half-reaction uses the compound calomel, Hg_2Cl_2, and the reference electrode is called a calomel electrode.

The indicator electrode for pH measurements is the **glass electrode**, which consists of a very thin glass membrane. On the inside of the glass electrode is a 1 M solution of hydrochloric acid. The outside of the glass membrane is in contact with the solution to be measured. When the glass membrane has been soaked in water for 24 hours, the glass surface becomes a hydrated gel. The hydrogen ions adsorb to this hydrated glass surface in proportion to the concentration of these ions in the solution. The 1 M HCl on the inside produces a constant positive charge on the inner membrane, while the solution to be measured produces a smaller positive charge on the outside of the membrane. The difference in charge is reflected as the electrode potential.

A galvanic cell used to measure pH follows this equation:

$$E_{cell} = E_{reference} + \text{constant} + 0.0591 \log [H^+] \qquad (11.5)$$

$$= C' - 0.0591 \text{ pH} \qquad (11.6)$$

The constant in equation 11.5 arises because of the variable properties of the glass electrode. It is combined with the potential of the reference electrode, which is also constant, to yield the constant C' in equation 11.6. In use, a pH electrode must be standardized by immersing it into a buffer solution with a known pH. Measuring E_{cell} under these conditions allows the calculation of C'. Next the electrode is rinsed with distilled water and immersed in the solution to be measured. The C' just determined and the E_{cell} value for the unknown solution allow calculation of the pH.

Glass is an excellent insulator and does not allow a large amount of electrical current to pass through it. Even the very thin glass membrane of the glass electrode is a good insulator. A special instrument, the pH meter, is needed when making measurements using the glass electrode. With a pH meter the calibration step is done by adjusting the meter reading to the pH of the buffer; in the measurement step the pH is read directly from the meter.

Standard Cell Voltages and Equilibrium

Equilibrium Constant

If the two electrodes of a galvanic cell are connected so that electrons flow freely, the reaction will proceed to equilibrium. At equilibrium $E_{cell} = 0$, and mathematical combination of the two Nernst equations gives us

$$\begin{aligned} E^\circ_{cell} &= E^\circ_{cathode} - E^\circ_{anode} \\ &= \frac{RT}{nF} \ln K_{eq} \\ &= \frac{0.0591}{n} \log K_{eq} \end{aligned}$$

where n is the total number of electrons transferred in the balanced redox equation (i.e., the number of electrons canceled in the final step of the ion-electron method), and the mathematical combination of the anode reaction quotient and cathode reaction quotient gives the equilibrium constant (K_{eq}).

The net result is that a table of standard reduction potentials (see page 211) also provides the information necessary to calculate the equilibrium constant for a redox reaction.

EXERCISE

Determine the equilibrium constant for each of the following reactions at 298 K:

(a) The single displacement of copper(II) by zinc metal

(b) The single displacement of H^+ by iron, forming Fe^{2+}

(c) The reduction of tin(IV) to tin(II) by iron(II) being oxidized to iron(III)

(d) The reduction of dichromate ions to chromium(III) ions by Mn^{2+} forming MnO_2

(e) The oxidation of AsO_3^{3-} to $H_2AsO_4^-$ by the reduction of iodine to iodide ions

Answers

In a previous exercise on page 212, the reactions and E°_{cell} values are determined.

(a) $Cu^{2+} + Zn \rightarrow Zn^{2+} + Cu$

$E^\circ_{cell} = +0.34\text{ V} - (-0.76\text{ V}) = +1.10\text{ V}$

Two electrons are transferred, and $K_{eq} = 1.7 \times 10^{37}$.

(b) $Fe + 2H^+ \rightarrow Fe^{2+} + H_2$

$E^\circ_{cell} = 0.00\text{ V} - (-0.44\text{ V}) = +0.44\text{ V}$

Two electrons are transferred, and $K_{eq} = 7.8 \times 10^{14}$.

(c) $Sn^{4+} + 2Fe^{2+} \rightarrow 2Fe^{3+} + Sn^{2+}$

$E^\circ_{cell} = +0.15\text{ V} - (+0.77\text{ V}) = -0.62\text{ V}$

Two electrons are transferred, and $K_{eq} = 1.0 \times 10^{-21}$.

(d) $2H^+ + Cr_2O_7^{2-} + 3Mn^{2+} \rightarrow 3MnO_2 + 2Cr^{3+} + H_2O$

$E^\circ_{cell} = +1.33\text{ V} - (+1.23\text{ V}) = +0.10\text{ V}$

Six electrons are transferred, and $K_{eq} = 1.4 \times 10^{10}$.

(e) $H_2O + AsO_3^{3-} + I_2 \rightarrow H_2AsO_4^- + 2I^-$

$E^\circ_{cell} = +0.54\text{ V} - (+0.58\text{ V}) = -0.04\text{ V}$

Two electrons are transferred, and $K_{eq} = 4.4 \times 10^{-2}$.

In all cases except (e), fairly small voltages translate into very large or very small equilibrium constants, depending on the sign of E°_{cell}.

Free Energy Change (ΔG°) and Standard Cell Voltage

The standard cell voltage is related to the equilibrium constant as follows:

$$E^\circ_{cell} = \frac{RT}{nF} \ln K_{eq}.$$

The standard free-energy change (ΔG°) is also related to the equilibrium constant:

$$\Delta G^\circ = -RT \ln K_{eq}$$

The relationship between ΔG° and E°_{cell} is as follows:

$$\Delta G^\circ = -nFE^\circ_{cell}$$

A major method for determining the standard free energy of a reaction uses the measurement of standard cell voltages.

EXERCISE

Determine the standard free-energy change for each of the following reactions:

(a) The single displacement of copper(II) by zinc metal
(b) The single displacement of H^+ by iron, forming Fe^{2+}
(c) The reduction of tin(IV) to tin(II) by Fe(II) being oxidized to Fe(III)
(d) The reduction of dichromate ions to chromium(III) ions by Mn^{2+}, forming MnO_2
(e) The oxidation of AsO_3^{3-} to $H_2AsO_4^-$ by the reduction of iodine to iodide ions.

Answers

In a previous exercise on page 434, the reactions and E°_{cell} values were determined.

(a) $Cu^{2+} + Zn \rightarrow Zn^{2+} + Cu$

$E^\circ_{cell} = +0.34\ V - (-0.76\ V) = +1.10\ V$

Two electrons are transferred, and $\Delta G^\circ = -212$ kJ.

(b) $Fe + 2H^+ \rightarrow Fe^{2+} + H_2$

$E^\circ_{cell} = 0.00\ V + (-0.44\ V) = +0.44\ V$
Two electrons are transferred, and $\Delta G^\circ = -84.9$ kJ.

(c) $Sn^{4+} + 2Fe^{2+} \rightarrow 2Fe^{3+} + Sn^{2+}$

$E^\circ_{cell} = +0.15\ V - (+0.77\ V) = -0.62\ V$

Two electrons are transferred, and $\Delta G^\circ = +120$ kJ.

(d) $2H^+ + Cr_2O_7^{2-} + 3Mn^{2+} \rightarrow 3MnO_2 + 2Cr^{3+} + H_2O$

$E^\circ_{cell} = +1.33\ V - (+1.23\ V) = +0.10$

Six electrons are transferred, and $\Delta G^\circ = -57.9$ kJ.

(e) $H_2O + AsO_3^{3-} + I_2 \rightarrow H_2AsO_4^- + 2I^-$

$E^\circ_{cell} = +0.54\ V - (+0.58\ V) = -0.04\ V$

Two electrons are transferred, and $\Delta G^\circ = +7.7$ kJ.

Practical Galvanic Cells

Galvanic cells can be used to convert chemical energy into electrical energy. Devices that do this are called **batteries**; several different types of batteries are in common use.

The **lead-acid battery** is the type used to start automobiles. The anode of this battery is lead, the cathode is lead coated with PbO_2. The half-reactions are as follows:

$$PbO_2(s) + 4H^+(aq) + SO_4^{2-}(aq) + 2e^- \rightarrow PbSO_4(s) + 2H_2O \quad \text{(cathode)}$$

$$Pb(s) + SO_4^{2-} \rightarrow PbSO_4(s) + 2e^- \quad \text{(anode)}$$

As the battery is discharged, the sulfate ions are used up and the electrodes become coated with lead sulfate. The reactions may be reversed in charging the battery. The reverse reactions regenerate the sulfate ion in solution and reduce the amount of lead sulfate contaminating the surface of the electrodes. Each pair of electrodes produces approximately 2 volts. Six pairs of electrodes are used in a 12-volt car battery.

The **alkaline battery** is the type most often used in flashlights, tape recorders, and TV remote controllers. The case of the battery is zinc, which is the anode. The cathode is a graphite rod inserted into a paste made of manganese dioxide, water, and potassium hydroxide. The half-reactions are as follows:

$$Zn(s) + 2OH^-(aq) \rightarrow ZnO(s) + H_2O + 2e^- \quad \text{(anode)}$$

$$2MnO_2(s) + H_2O + 2e^- \rightarrow Mn_2O_3(s) + 2OH^-(aq) \quad \text{(cathode)}$$

The total voltage is 1.54 volts. This battery is not rechargeable because the reactions cannot be reversed.

Other batteries of interest are the **nickel-cadmium** or **nicad** battery, the **mercury battery**, and the **silver oxide battery**. Fuel cells are also used to generate electricity. In a fuel cell hydrogen may be oxidized at an anode and oxygen reduced at a cathode to form water with the production of electricity. In comparison to gas-, oil-, or coal-powered generators, a fuel cell is approximately twice as efficient in converting chemical energy into electricity.

12
KINETICS

REACTION RATES

Experimental Rates of Reactions

The rate of a chemical reaction is the rate of change in concentration per unit time. It is expressed as the number of moles per liter that react each second; thus the units are always moles per liter per second (mol L^{-1} s^{-1}). Reaction rates are determined by measuring the concentration(s) of one or more of the chemicals involved in the reaction at different times during the course of the reaction.

Figure 12.1 illustrates the results of such measurements as a **kinetic curve** that is also called a concentration versus time curve. The kinetic curve on the left is obtained from five individual measurements of the concentration of a reaction product. These points are connected with a smooth line. The kinetic curve on the right may be obtained by using instrumentation that continuously monitors the concentration of the same product. Figure 12.1 illustrates the increase in product as a reaction occurs. Similarly, if the concentration of a reactant is measured, a decrease in concentration with time will be recorded.

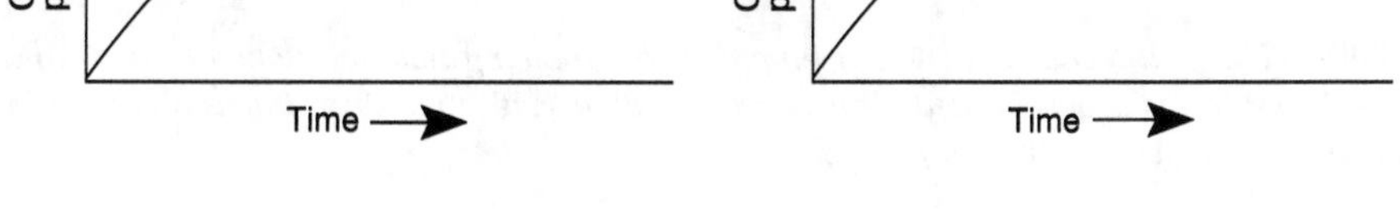

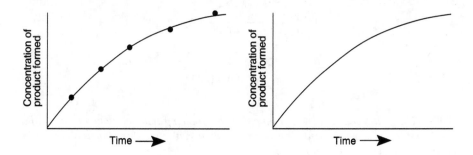

FIGURE 12.1. *Kinetic curves obtained by measuring the formation of a product of a chemical reaction. The curve on the left is obtained from five individual measurements connected with a smooth line. The curve on the right is automatically recorded using an instrument designed to monitor the concentration continuously.*

The rate of a chemical reaction is obtained from a kinetic curve by determining the slope of the curve at the desired point in time. The slope of a curve is determined in a three-step process.

1. Select the desired point on the curve, and draw a tangent to it.
2. Select two points on the tangent, and determine the concentrations (C) and times (t) corresponding to those points.
3. Use equation 12.1 to calculate the slope:

$$\text{Rate} = -\left(\frac{C_2 - C_1}{t_2 - t_1}\right) = -\frac{\Delta C}{\Delta t} \tag{12.1}$$

Figure 12.2 illustrates the construction of the tangents for the determination of the initial rate, $t = 0$, and the rate at 180 seconds, $t = 180$. The small triangles on the tangents connect point 1 and point 2 with a right triangle. The necessary values for ΔC ($C_2 - C_1$) and Δt ($t_2 - t_1$) are the lengths of the sides of these triangles.

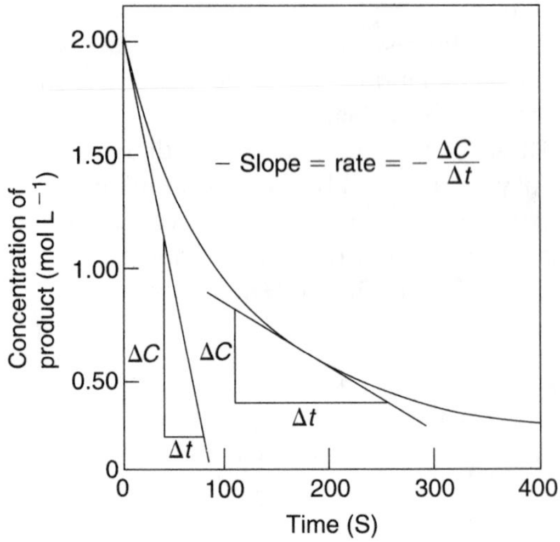

FIGURE 12.2. *Tangents drawn to a kinetic curve showing how the slope at $t = 0$ and $t = 180$ seconds is determined. This curve decreases with time since it was obtained by measuring the concentration of a reactant.*

In another, less desirable method to determine the rate of a chemical reaction, the concentration, (C) of one reactant or product is measured at only two different times. The rate is then calculated using equation 12.1. A two-point calculation like this one gives the average rate of the reaction over the time interval used.

Since the reactants disappear during a chemical reaction, the rate calculated by measuring a reactant will have a negative sign. If a product is measured, the calculated rate will have a positive sign. By convention, chemists always use a positive value for the rate of a reaction, whether it is the positive rate of the appearance of products or the negative rate of disappearance of reactants:

$$\text{Rate} = \frac{\Delta C_{\text{products}}}{\Delta t}$$

and

$$\text{Rate} = \frac{-\Delta C_{\text{reactants}}}{\Delta t}$$

The sign for the rate is implied by calling the rate either the rate of appearance (implied positive sign) of products or the rate of disappearance (implied negative sign) of reactants.

The value of a reaction rate depends on which reactant or product is measured. After the rate has been measured based on one component of the reaction, the rates of change of the other components can be calculated by a stoichiometric conversion. Consider the following reaction:

$$2C_2H_6 \quad + \quad 7O_2 \quad \rightarrow \quad 4CO_2 \quad + \quad 6H_2O$$

and assume that the rate of reaction was determined by measuring the CO_2 produced and was determined to be 2.50 moles CO_2 per liter per second. This can be converted to the rates of appearance or disappearance of all other chemical species by using the stoichiometric relationships in the chemical equation.

First, the problem is set up as a stoichiometry question. The equation below can be interpreted to read, What will the rate of consumption of C_2H_6 be if the rate of production of CO_2 is 2.50 moles per liter per second?

$$?\,\frac{\text{mol } C_2H_6}{\text{L s}} = \frac{2.50 \text{ mol } CO_2}{\text{L s}}$$

Since the denominators of the two ratios are the same, the problem involves only the conversion of moles of CO_2 into moles of C_2H_6. The chemical reaction gives us a conversion factor of $\left(\frac{2 \text{ mol } C_2H_6}{4 \text{ mol } CO_2}\right)$, which is used to obtain

$$?\,\frac{\text{mol } C_2H_6}{\text{L s}} = \frac{2.50 \text{ mol } CO_2}{\text{L s}}\left(\frac{2 \text{ mol } C_2H_6}{4 \text{ mol } CO_2}\right)$$

Since the mol CO_2 units cancel, leaving the units desired, the problem is finished and the answer can be calculated as

$$?\,\frac{\text{mol } C_2H_6}{\text{L s}} = \frac{1.25 \text{ mol } C_2H_6}{\text{L s}}$$

In a similar fashion we can calculate the corresponding rates based on O_2 and H_2O:

$$? \frac{\text{mol } O_2}{\text{L s}} = \frac{2.50 \text{ mol } CO_2}{\text{L s}} \left(\frac{7 \text{ mol } O_2}{4 \text{ mol } CO_2} \right)$$

$$= \frac{4.38 \text{ mol } O_2}{\text{L s}}$$

and

$$? \frac{\text{mol } H_2O}{\text{L s}} = \frac{2.50 \text{ mol } CO_2}{\text{L s}} \left(\frac{6 \text{ mol } H_2O}{4 \text{ mol } CO_2} \right)$$

$$= \frac{3.75 \text{ mol } H_2O}{\text{L s}}$$

It is apparent that any reaction rate must specify to which reactant or product the rate actually applies.

EXERCISES

1. Using Figure 12.2, determine the rates of reaction at 100 seconds and 200 seconds after the reaction starts.
2. In the reaction

$$N_2(g) \quad + \quad 3H_2(g) \quad \rightarrow \quad 2NH_3(g)$$

the rate of disappearance of $H_2(g)$ is found to be 4.8×10^{-2} mole per liter per second.
1. What is the rate at which $N_2(g)$ is reacting?
2. What is the rate at which $NH_3(g)$ is being produced under the same conditions?

Solutions

1. Drawing the tangents, we determine the slopes to be -6.5×10^{-3} and -2.8×10^{-3} mol L^{-1}/s^{-}. Since this curve represents the disappearance of a reactant, the rate is the negative of the slope, or 6.5×10^{-1} mol L^{-1}/s^{-1} at 100 s and 2.8×10^{-3} mol L^{-1}/s^{-1} at 200 s.

2. This problem requires the conversion of one rate into another. The initial setup is

$$? \frac{\text{mol } N_2}{\text{L s}} = \frac{4.8 \times 10^{-2} \text{ mol } H_2}{\text{L s}}$$

The next step uses the factor-label method for the conversion between $N_2(g)$ and $H_2(g)$:

$$? \frac{\text{mol } N_2}{\text{L s}} = \frac{4.8 \times 10^{-2} \text{ mol } H_2}{\text{L s}} \left(\frac{1 \text{ mol } N_2}{3 \text{ mol } H_2} \right)$$

$$= 1.6 \times 10^{-2} \text{ mol } N_2 \text{ L}^{-1} \text{ s}^{-1}$$

2. The same steps are used to determine the rate of production of $NH_3(g)$:

$$? \frac{\text{mol } NH_3}{\text{L s}} = \frac{4.8 \times 10^{-2} \text{ mol } H_2}{\text{L s}}$$

$$? \frac{\text{mol } N_2}{\text{L s}} = \frac{4.8 \times 10^{-2} \text{ mol } H_2}{\text{L s}} \left(\frac{2 \text{ mol } NH_3}{3 \text{ mol } H_2} \right)$$

$$= 3.2 \times 10^{-2} \text{ mol } NH_3 \text{ L}^{-1} \text{ s}^{-1}$$

RATE LAWS

Rate Laws and Concentration

The Rate Law Equation

The effect of concentration on a chemical reaction is expressed in the rate law. All rate laws start with the same form:

$$\text{Rate} = k[A]^x[B]^y[C]^z$$

In this equation k stands for the **rate constant**, and the square brackets indicate the concentrations of the reactants A, B, and C, which have exponents x, y, and z. These exponents are usually small whole numbers. In more complex rate laws they may be negative numbers or rational fractions. Exponents must be determined from laboratory experiments; they have no relationship to the stoichiometric coefficients of the balanced chemical equation.

Determination of Rate Laws

All rate laws must be determined by using the data from a group of well-planned experiments. By changing the concentration of one reactant while holding all other concentrations constant, we can determine whether the change we made has an effect on the rate and, if it does, we can calculate the exponent for the changed reactant in the rate law. **There is no theoretical way to predict the exponents of a rate law.**

An example of the methods used to determine rate laws can be shown using the reaction of peroxydisulfate, $S_2O_8^{2-}$, with iodide ions according to this equation:

$$S_2O_8^{2-} + 3I^- \rightarrow 2SO_4^{2-} + I_3^- \qquad (12.2)$$

Table 12.1 shows three experiments using this reaction. In experiments 1 and 2 the concentration of iodide ion is the same and the concentrations of peroxydisulfate ions are different. In experiments 2 and 3 the iodide ion concentration is changed while the peroxydisulfate concentration is held constant. The measured initial rate for each experiment is given in the last column.

TABLE 12.1. KINETIC DATA FOR THE PEROXYDISULFATE REACTION AT 20.0°C

Experiment Number	$[S_2O_8^{2-}]$ (mol L^{-1})	$[I^-]$ (mol L^{-1})	Initial Rate of Reaction (mol $L^{-1}s^{-1}$)
1	0.200	0.200	2.2×10^{-3}
2	0.400	0.200	4.4×10^{-3}
3	0.400	0.400	8.8×10^{-3}

Since there are only two reactants, the rate law must have the form

$$\text{Rate} = k[S_2O_8^{2-}]^x[I^-]^y \qquad (12.3)$$

where the exponents x and y need to be determined.

To determine the exponent for the peroxydisulfate ion, we focus attention on the change in reaction rate as the concentration of $S_2O_8^{2-}$ is changed in experiments 1 and 2. A rate law can be written for experiment 1 and another for experiment 2 by entering the data from the table into Equation 12.3:

$$\text{Rate}_1 = 2.2 \times 10^{-3} = k(0.200)^x(0.200)^y$$

$$\text{Rate}_2 = 4.4 \times 10^{-3} = k(0.400)^x(0.200)^y$$

The ratio of these two equations is then written (the calculations are usually easier if the larger numbers are used for the numerator) as follows:

$$\frac{\text{Rate}_2}{\text{Rate}_1} = \frac{4.4 \times 10^{-3}}{2.2 \times 10^{-3}} = \frac{k(0.400)^x(0.200)^y}{k(0.200)^x(0.200)^y}$$

The two k's and the two $(0.200)^y$ factors cancel to yield

$$\frac{\text{Rate}_2}{\text{Rate}_1} = \frac{4.4 \times 10^{-3}}{2.2 \times 10^{-3}} = \frac{(0.400)^x}{(0.200)^x}$$

Since both exponents are x, the right-hand term may be rewritten to give the equation

$$\frac{\text{Rate}_2}{\text{Rate}_1} = \frac{4.4 \times 10^{-3}}{2.2 \times 10^{-3}} = \left(\frac{0.400}{0.200}\right)^x$$

Solving this, we obtain

$$2.0 = 2.00^x$$
$$= 2.00^1$$

From this result we conclude that exponent $x = 1$. In a similar manner the value of exponent y is determined using experiments 2 and 3, where the concentration of $S_2O_8^{2-}$ is held constant and I^- is varied. The ratio of rate laws for experiments 2 and 3 is as follows:

$$\frac{\text{Rate}_3}{\text{Rate}_2} = \frac{8.8 \times 10^{-3}}{4.4 \times 10^{-3}} = \frac{k(0.400)^x(0.400)^y}{k(0.400)^x(0.200)^y}$$

Solving this gives

$$2.0 = 2.00^y$$
$$= 2.00^1$$

The value of exponent y is 1. Using the exponents determined in this manner, we write the rate law:

$$\text{Rate} = k[S_2O_8^{2-}]^1[I^-]^1 \quad \text{or} \quad k[S_2O_8^{2-}][I^-]$$

The exponents in the rate law are definitely not the same as the exponents in balanced Equation 12.2.

Once the rate law is known, the rate constant can be calculated by taking **any one** of the three experiments in Table 12.1, substituting the values into the rate equation, and solving for the rate constant (k). The rate constant should be the same no matter which experiment (1, 2, or 3) is chosen for this calculation. To verify this statement, we will calculate the rate constant for each experiment and express it in the proper units:

$$\text{Experiment 1: } k = \frac{2.2 \times 10^{-3} \text{ mol L}^{-1} \text{ s}^{-1}}{(0.200 \text{ mol L}^{-1})(0.200 \text{ mol L}^{-1})}$$
$$= 0.055 \text{ L mol}^{-1} \text{ s}^{-1}$$

$$\text{Experiment 2: } k = \frac{4.4 \times 10^{-3} \text{ mol L}^{-1} \text{ s}^{-1}}{(0.400 \text{ mol L}^{-1})(0.200 \text{ mol L}^{-1})}$$
$$= 0.055 \text{ L mol}^{-1} \text{ s}^{-1}$$

$$\text{Experiment 3: } k = \frac{8.8 \times 10^{-3} \text{ mol L}^{-1} \text{ s}^{-1}}{(0.400 \text{ mol L}^{-1})(0.400 \text{ mol L}^{-1})}$$
$$= 0.055 \text{ L mol}^{-1} \text{ s}^{-1}$$

We have demonstrated that an easy way to verify the rate law is to calculate the rate constant for each experiment. The rate constants should all agree within experimental error; if they do not agree, something is wrong with the rate law.

Chemists use the term **order of reaction** to indicate the exponents in the rate law. The reaction we have been considering is said to be **first order** with respect to peroxydisulfate and first order with respect to iodide ions. It is also said to be **second order** overall. Individual reactants have the same orders as their exponents. The order of the overall reaction is the sum of all of the exponents.

EXERCISES

1. Determine (a) the rate law and (b) the value of the rate constant, with its units, for this reaction:

$$A + 2B \rightarrow C + D$$

The data are given in the table below.

Experiment Number	[A] (mol L^{-1})	[B] (mol L^{-1})	Initial Rate of Reaction (mol L^{-1}s^{-1})
1	0.100	0.200	1.1×10^{-6}
2	0.100	0.600	9.9×10^{-6}
3	0.400	0.600	9.9×10^{-6}

2. What is the overall order of each of the following rate laws, and what are the units of the rate constant, k, in each of these rate laws?

(a) Rate = k[A][B][C]

(b) Rate = $k[X]^2[Y]^3$

(c) Rate = $k[M]^2[N]$

(d) Rate = k

(e) Rate = k[R]

Solutions

1. (a) The general form of the rate law will be as follows:

$$\text{Rate} = k[A]^x[B]^y$$

Taking the ratio of the rate laws for experiments 1 and 2 yields

$$\frac{\text{Rate}_2}{\text{Rate}_1} = \frac{k(0.100)^x(0.600)^y}{k(0.100)^x(0.200)^y}$$

Canceling the identical terms, k and $(0.100)^x$, on the right and entering the rates on the left, we have

$$\frac{9.9 \times 10^{-6}}{1.1 \times 10^{-6}} = \frac{(0.600)^y}{(0.200)^y}$$

$$9 = 3^y$$

Three squared is equal to 9, and therefore the exponent $y = 2$.

To determine the exponent x, the ratio of the rate laws for experiments 2 and 3 are used:

$$\frac{\text{Rate}_2}{\text{Rate}_3} = \frac{k(0.100)^x(0.600)^y}{k(0.400)^x(0.600)^y}$$

Entering the rates and canceling as before, we have

$$\frac{9.9 \times 10^{-6}}{9.9 \times 10^{-6}} = \frac{(0.100)^x}{(0.400)^x}$$

$$1.0 = 0.25^x$$

Any number raised to the zero power is equal to 1, and therefore $x = 0$. The rate law is

$$\text{Rate} = k[\text{A}]^0[\text{B}]^2 = k[\text{B}]^2$$

Since the concentration of A has no effect on the reaction rate, it is not part of the rate law.

(b) The value of the rate constant is determined by taking any of the three experiments, substituting the rate and concentrations into the rate law, and calculating k. Using the data from experiment 1 yields

$$1.1 \times 10^{-6} \text{ mol B L}^{-1} \text{ s}^{-1} = k(0.200 \text{ mol B L}^{-1} \text{ s}^{-1})^2$$

$$k = \frac{1.1 \times 10^{-6} \text{ mol B L}^{-1} \text{ s}^{-1}}{(0.200 \text{ mol B L}^{-1})^2}$$

$$= 2.75 \times 10^{-5} \text{ L mol}^{-1} \text{ s}^{-1}$$

2. (a) Order = 3; units are $\text{L}^2 \text{ mol}^{-2} \text{ s}^{-1}$
(b) Order = 5; units are $\text{L}^4 \text{ mol}^{-4} \text{ s}^{-1}$
(c) Order = 3; units are $\text{L}^2 \text{ mol}^{-2} \text{ s}^{-1}$
(d) Order = 0; units are $\text{mol L}^{-1} \text{ s}^{-1}$
(e) Order = 1; units are s^{-1}

Integrated Rate Laws

Zero-Order Reactions

A reaction that is **zero order** has a rate law in which the exponents of all of the reactants are zero. Since any number raised to the zero power is equal to 1 ($x^0 = 1$), the rate law is

$$\text{Rate} = k$$

This type of reaction does not depend on the concentration of any reactant. Its rate is equal to the rate constant, and the rate constant has the same units, (mol L^{-1} s^{-1}) as a rate. Catalytic reactions are often zero-order reactions. One catalyzed reaction is the decomposition of hydrogen peroxide in the presence of platinum metal:

$$2H_2O_2 \rightarrow 2H_2O + O_2$$

The concentration versus time plot for any zero-order reaction is a straight line, as shown in Figure 12.3.

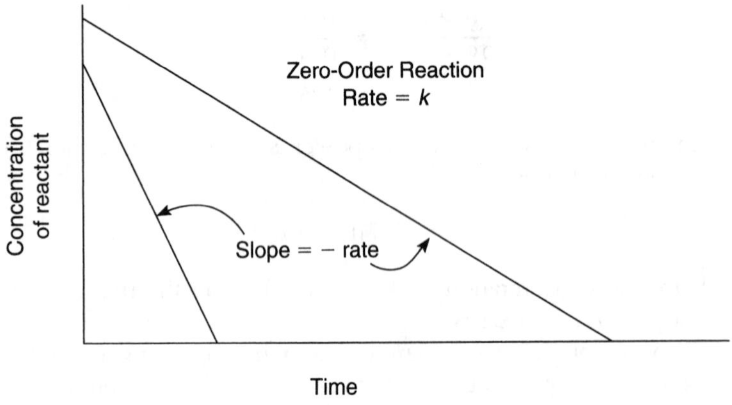

FIGURE 12.3. *Kinetic curves of two different zero-order reactions, illustrating the straight-line relationship between concentration and time.*

First-Order Reactions

Any rate law in which the sum of the exponents is 1 is a first-order reaction. There are a variety of ways in which a reaction can be first order, including the case where the exponents of two reactants are 0.5 each. In the most common case the concentration of one reactant, A, has an exponent of 1, as in the following rate law:

$$\text{Rate} = k[A]$$

As a first-order reaction progresses, the concentration of reactant A decreases and the rate decreases. When concentration of A decreases to half its original amount, the rate will be half the initial rate. The time required for this to occur is called the half-life, symbolized as $t_{1/2}$. In another half-life, the concentration will decrease by half again, to one-fourth of the initial concentration. Figure 12.4 illustrates the shape of the kinetic curve for all first-order reactions.

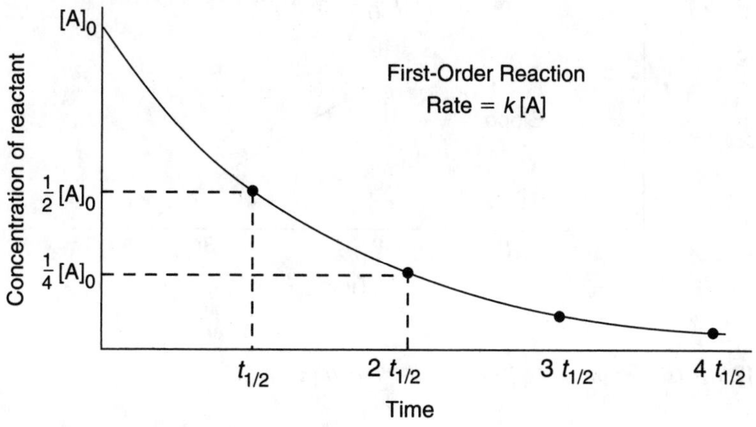

FIGURE 12.4. *Kinetic curve for all first-order reactions. The time at which the concentration is one-half the original concentration is the half-life,* $t_{1/2}$.

The Integrated Equation: Using elementary calculus, we can integrate the first-order rate equation to obtain an equation that allows us to calculate concentrations at any time after the reaction has started:

$$\ln \frac{[A]_0}{[A]_t} = kt$$

or

$$\ln [A]_0 - \ln [A]_t = kt$$

Each form of the integrated equation contains four variables: the time (t), the rate constant (k), an initial concentration of A ($[A]_0$), and the concentration at some time after the reaction has started ($[A]_t$). Using the integrated equation, chemists often plot the natural logarithm of the reactant concentration versus time in a graph, as shown in Figure 12.5. In this type of graph, only a first-order reaction will result in a straight line and the slope will be equal to $-k$.

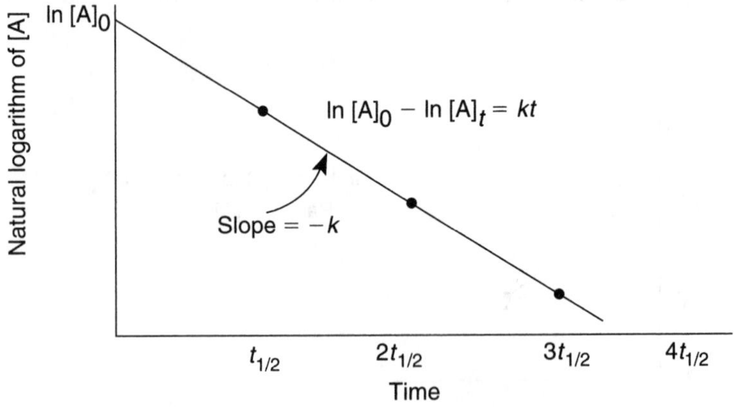

FIGURE 12.5. *Logarithm plot of a first-order reaction.*

EXERCISE

A certain-first order reaction has a rate constant of 4.5×10^{-3} reciprocal second (s^{-1}). millimolar (mM) sample will have reacted after 75.0 seconds?

Solution

Substitution of the data into the integrated rate equation yields

$$\ln 0.0500 - \ln [A]_t = (4.5 \times 10^{-3}\ s^{-1})(75.0\ s)$$
$$-2.995 - \ln [A]t = 0.3375$$
$$\ln [A]_t = -3.333$$
$$[A]_t = 0.0356\ M$$

This indicates that the sample is now 35.6 mM. The question asks how much of the sample has reacted; the answer is

$$50.0\ mM - 35.6\ mM = 14.4\ mM$$

Half-Life: As mentioned above, a first-order reaction has a unique property called the **half-life**, which is the time required for half of the reactant to be used up. When half of the reactant is used up, $[A]_t = 0.5[A]_0$. Substituting into the integrated rate equation yields

$$\ln \frac{0.5[A]_0}{[A]_0} = -kt_{1/2}$$

$$\ln 0.5 = -kt_{1/2}$$

$$-0.693 = -kt_{1/2}$$

$$t_{1/2} = \frac{0.693}{k}$$

where $t_{1/2}$ is the half-life of the reaction. The half-life may be used to quickly estimate the fraction of the starting material left after it has been allowed to react for a given number of half-lives.

EXERCISE

What percentage of a sample has reacted after six half-lives?

Solution

After each half-life, half of the sample present at the start of the half-life will be left. After one half-life, $\frac{1}{2}$ of the sample is left; after two half-lives, $\frac{1}{2} \times \frac{1}{2}$ or $\frac{1}{4}$ remains; after three half-lives, $\frac{1}{2} \times \frac{1}{2} \times \frac{1}{2}$ or $\frac{1}{8}$ remains. After six half-lives

$$\frac{1}{2} \times \frac{1}{2} \times \frac{1}{2} \times \frac{1}{2} \times \frac{1}{2} \times \frac{1}{2} = \frac{1}{64}$$

of the original is left, meaning that $\frac{63}{64}$ has reacted. The percentage that has reacted is calculated as

$$\frac{63}{64} \times 100 = 98.4\% \text{ reacted}$$

The most prominent first-order processes in chemistry are those associated with the radioactive decay of elements. The decay of radioactive elements follows first-order kinetics.

Second-Order Reactions

Two of the most common second-order rate laws are as follows:

$$\text{Rate} = k[A]^2 \qquad (12.4)$$

and

$$\text{Rate} = k[A]\,[B] \qquad (12.5)$$

The integrated form of the rate law in equation 12.4 is given in Equation 12.6. Equation 12.5 can be successfully integrated only in the special case where [A] = [B]. Then it becomes the same as Equation 12.4 and integrates to Equation 12.6:

$$\frac{1}{[A]_t} - \frac{1}{[A]_0} = kt \tag{12.6}$$

The half-life of the second-order reactions specified above may be determined by substituting $0.5[A]_o$ for $[A]_t$ in equation 12.6:

$$\frac{1}{0.5[A]_0} - \frac{1}{[A]_0} = kt_{1/2}$$

$$\frac{2}{[A]_0} - \frac{1}{[A]_0} = -kt_{1/2}$$

$$\frac{1}{[A]_0} = -kt_{1/2}$$

$$t_{1/2} = \frac{1}{k[A]_0}$$

The important information derived from these equations is that the half-life of second-order reactions depends on the starting concentrations of the reactants. In addition, a straight-line graph will be obtained by plotting 1/[A] versus time. The slope of the line will be equal to k, as shown in Figure 12.6.

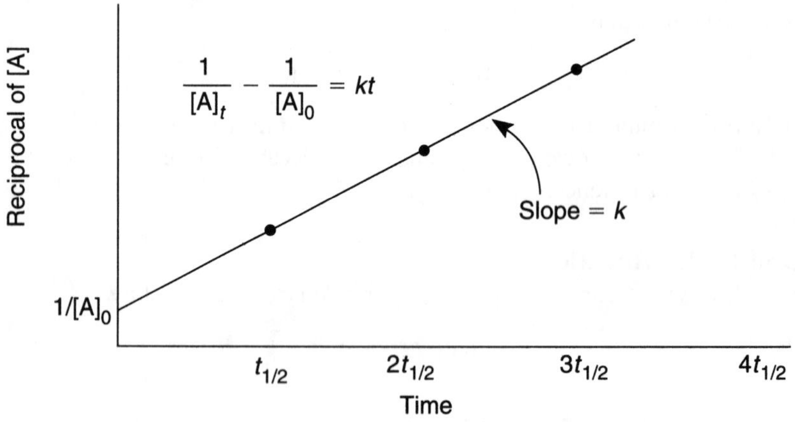

FIGURE 12.6. *Plot of 1/[A] versus time, illustrating the straight-line relationship for a second-order reaction.*

Rate Laws and Reverse Reactions

Factors Affecting Reaction Rates

The rate law for a reaction as determined from initial concentrations will describe a kinetic curve such as that shown in Figure 12.7 as long as no other significant reactions are occurring. Since the concentrations of the reactants decrease as the reaction progresses, we expect the rate of reaction to decrease, as is the case in Figure 12.7.

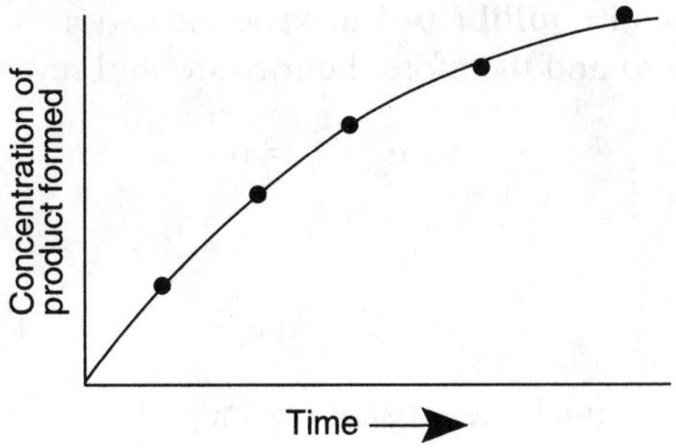

FIGURE 12.7. *A typical kinetic curve in which the increase in product is monitored as the reaction proceeds.*

In addition to the decrease in the concentration of reactants, the products may react with each other to make more reactants. This **reverse reaction** also slows the reaction rate.

The rate law for the forward reaction:

$$A + B \rightarrow C + D$$

is

$$\text{Rate}_f = k_f[A]^x[B]^y$$

where the subscript *f* indicates the forward reaction.

When the products C and D react to form the original reactants, the reverse reaction:

$$C + D \rightarrow A + B$$

occurs. Its rate law is written as

$$\text{Rate}_r = k_r[C]^z[D]^w$$

The overall rate of this reaction will be

$$\begin{aligned}\text{Rate}_{\text{overall}} &= \text{Rate}_f - \text{Rate}_r \qquad (12.7)\\ &= k_f[A]^x[B]^y - k_r[C]^z[D]^w\end{aligned}$$

Equation 12.7 indicates that, as products C and D increase in concentration, the reaction rate will decrease. Chemists use initial reaction rates to determine rate laws in order to avoid the complications that a reverse reaction may introduce into the results.

Chemical Equilibrium

When a chemical reaction stops, there are two possibilities. One is that the reaction goes all the way to completion and the reaction stops because the limiting reactant is used up. The other is that the reaction comes to chemical equilibrium. **Chemical equilibrium** may be viewed as a situation in which the overall rate is zero and therefore the forward and reverse rates are equal:

$$\text{Rate}_{\text{overall}} = 0$$

and therefore

$$\text{Rate}_f = \text{Rate}_r$$

If the two rates are equal, then, using the individual rate laws, we can write the relationship as

$$k_f[A]^x[B]^y = k_r[C]^z[D]^w$$

Rearranging this equation so that the constants are all on the same side gives

$$\frac{k_f}{k_r} = \frac{[C]^z[D]^w}{[A]^x[B]^y} \qquad (12.8)$$

This is known as the **equilibrium law**. The left side is the ratio of two constants that are combined into the equilibrium constant (K_{eq}). The right side is the ratio of the concentrations of the products divided by the concentrations of the reactants.

This derivation of the equilibrium constant clearly shows the nature of chemical equilibrium as a dynamic situation. Reaction is constantly occurring in both the forward and reverse directions at tremendous rates. Since these rates are equal, however, no change is observed in the concentrations of the reactants and products at equilibrium.

When determining ratio laws, it was emphasized that the exponents in the rate law are not necessarily equal to the coefficients in the chemical reaction except for elementary reactions. On pages 194–196 it was shown that in the equilibrium law the exponents are the coefficients found in the balanced chemical equation. Equation 12.8 seems to indicate that the exponents of the rate law should be equal to the reaction coefficients. The answer to this seeming inconsistency lies in a detailed understanding of reaction mechanisms that is beyond the scope of this book. We simply state that both principles are true.

> The coefficients of a chemical equation are the exponents used in the equilibrium law, but they are not necessarily the exponents in the rate laws.

It is important to realize that, although Equation 12.8 shows a relationship between the rate constants and the equilibrium constant, a single rate constant indicates nothing about the equilibrium constant. Similarly, since the equilibrium constant is a ratio of the rate constants, it indicates only the relative sizes of the forward and reverse rate constants. The equilibrium constant tells us nothing about the actual reaction rate.

ACTIVATION ENERGY

Reaction Rates: The Effect of Temperature

Almost instinctively, we add heat when we wish to increase the speed of a reaction. Higher temperatures increase the rates at which foods cook and solids dissolve in water. Cold-blooded animals such as snakes and lizards are almost immobile in cold temperatures; they sun themselves to warm up in order to hunt for food.

In 1889 Svante Arrhenius developed the equation, called the **Arrhenius equation** in his honor, for the relationship between the rate constant and temperature:

$$k = Ae^{-Ea/RT}$$

This equation uses the rate constant (k), the activation energy (E_a), the universal gas law constant (R = 8.314 J mol^{-1} K^{-1}), the Kelvin temperature (T), and a proportionality constant (A). When the natural logarithm of this equation is taken, the result is

$$\ln k = \frac{-E_a}{RT} + \ln A \qquad (12.9)$$

Equation 12.9 may be utilized in two ways to eliminate the necessity of knowing the value of A. First, a graph of ln k versus $1/T$ may be constructed after determining the rate constant at a variety of temperatures. This graph, shown in Figure 12.8, is often called an **Arrhenius plot**. The slope of the line is equal to $-E_a/R$. Arrhenius plots are used to determine the activation energy (E_a) and also to determine the rate constant at any desired temperature.

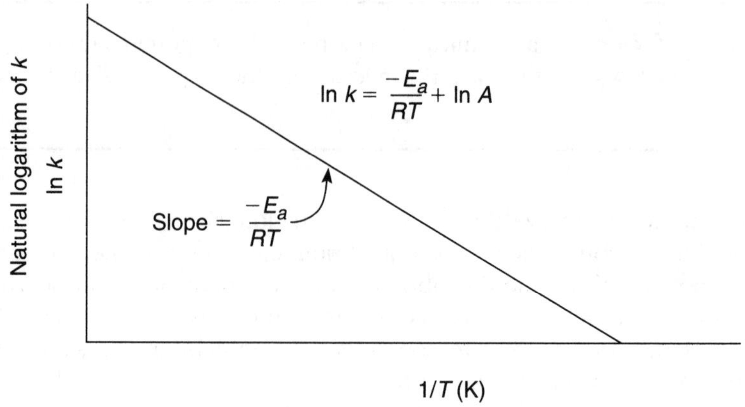

FIGURE 12.8. *An Arrhenius plot of ln k versus 1/T. The activation energy (E_a) is determined from the slope of the straight line.*

The second way to utilize Equation 12.9 is a two-point approach using only the rate constants determined at two different temperatures. First, an Arrhenius equation is written for the rate constant determined at each temperature, shown by the subscripts 1 and 2 in the equations below:

$$\ln k_1 = \frac{-E_a}{RT_1} + \ln A$$

$$\ln k_2 = \frac{-E_a}{RT_2} + \ln A$$

When the second equation is subtracted from the first, the result is

$$\ln k_1 - \ln k_2 = \frac{-E_a}{RT_1} - \frac{-E_a}{RT_2}$$

which is commonly rewritten as

$$\ln\left(\frac{k_1}{k_2}\right) = \frac{-E_a}{R}\left(\frac{1}{T_1} - \frac{1}{T_2}\right) \qquad (12.10)$$

Second, given four of the five variables, E_a, k_1, k_2, T_1, and T_2, in Equation 12.10, we can determine the fifth by substitution into the equation.

Equation 12.10 may also be written using the ratio of the rates rather than the rate constants:

$$\ln\left(\frac{\text{Rate}_1}{\text{Rate}_2}\right) = \frac{-E_a}{R}\left(\frac{1}{T_1} - \frac{1}{T_2}\right)$$

since the rate and the rate constant are directly proportional to each other as long as the concentrations are held constant.

Often the signs in Equation 12.10 become confused since the calculation is somewhat involved. It is always possible, however, to test the answer for reasonableness. Two principles must be remembered:

1. The larger rate constant (or rate) is always associated with the higher temperature.
2. The activation energy always has a positive sign.

EXERCISES

1. At 200 K the rate constant for a reaction is 3.5×10^{-3} s^{-1} and at 250 K the rate constant is 4.0×10^{-3} reciprocal second. What is the activation energy?
2. A common rule of thumb is that, near room temperature, the rate of a reaction will double with each 10°C rise in temperature. Estimate the activation energy needed for this "rule" to be true.
3. What is the rate of reaction at 450°C if the reaction rate is 6.75×10^{-6} mole per liter per second at 25°C? The activation energy was previously determined to be 35.5 kilojoules per mole.

Solutions

1. Using the given data, we can assign 200 K as T_1 and 3.5×10^{-3} as k_1. Similarly, 250 K will be T_2; then 4.0×10^{-3} must be k_2. These values are then substituted into the proper places in the equation, along with R = 8.3145 J mol^{-1} K^{-1}:

$$\ln\left(\frac{3.5 \times 10^{-3}}{4.0 \times 10^{-3}}\right) = \frac{-E_a}{8.3145 \text{ J mol}^{-1} \text{ K}^{-1}}\left(\frac{1}{200 \text{ K}} - \frac{1}{250 \text{ K}}\right)$$

$$-0.1335 = (-1.20 \times 10^{-4} \text{ mol J}^{-1})\, E_a$$

$$E_a = 1.11 \times 10^3 \text{ J mol}^{-1} = 1.11 \text{ kJ mol}^{-1}$$

2. Choose two temperatures 10°C apart, such as 290 K and 300 K, since they are close to room temperature, 298 K. The rates double, meaning that the rate constants must also double. We can assign rate constants of 1.00 to the 290 K temperature and 2.00 to the 300 K temperature. Now the problem is solved by entering the data into the Arrhenius equation.

$$\ln\left(\frac{1}{2}\right) = \frac{-E_a}{8.3145 \text{ J mol}^{-1} \text{ K}^{-1}}\left(\frac{1}{290 \text{ K}} - \frac{1}{300 \text{ K}}\right)$$

$$-0.693 = (-1.382 \times 10^{-5} \text{ mol J}^{-1})\, E_a$$

$$E_a = 5.01 \times 10^4 \text{ J mol}^{-1} = 50.1 \text{ kJ mol}^{-1}$$

Choosing other temperatures near 298 K will result in slightly different answers, but all will be close to 50 kJ mol^{-1}. Many common reactions have activation energies near this value.

3. Using the equation with rates, we enter the variables given, after converting kilojoules to joules, to obtain

$$\ln\left(\frac{\text{Rate}_1}{6.75 \times 10^{-6}}\right) = \frac{-35500 \text{ J mol}^{-1}}{8.3145 \text{ J mol}^{-1} \text{ K}^{-1}}\left(\frac{1}{723} - \frac{1}{298}\right)$$

Remembering that the logarithm of a ratio may be written as the logarithm of the numerator minus the logarithm of the denominator, we obtain

$$\begin{aligned} \ln \text{Rate}_1 - \ln(6.75 \times 10^{-6}) &= 8.42 \\ \ln \text{Rate}_1 - (-11.91) &= 8.42 \\ \ln \text{Rate}_1 &= -3.49 \\ \text{Rate}_1 &= 3.05 \times 10^{-2} \text{ mol L}^{-1} \text{ s}^{-1} \end{aligned}$$

As expected, the reaction rate is greater at a higher temperature.

MECHANISMS

Reaction Mechanisms

One of the most important uses of chemical kinetics involves deciphering the sequence of steps that lead to an observed chemical reaction. Chemists write chemical equations for reactions as a single step. However, most chemical reactions occur in a series of **elementary reactions** called a mechanism. All of the elementary reactions in a mechanism must add up to the **overall balanced equation**.

The Sequence of Elementary Reactions

In a mechanisms, the elementary reactions usually involve the collision of only two molecules. It is extremely rare that three molecules will collide simultaneously, and the simultaneous collision of more than three molecules is such a rare event that it is hardly ever considered. However, collisions between just two molecules occur millions of times each second. Therefore, the most probable elementary reaction is one where only two molecules collide. An important feature of the elementary reaction equation is that its coefficients are the exponents used in the rate law.

In a sequence of two or more elementary reactions, one reaction is always slower than all the rest. This slowest reaction determines the overall **rate of reaction** and is called the **rate-determining step** or the **rate-limiting step**.

Since chemists can measure only the overall reaction rate in the laboratory, they are in fact measuring the rate of the slowest, rate-determining elementary reaction in the mechanism. Thus the rate law determined for a reaction is directly related to the rate-determining step.

Using these principles, chemists experimentally determine the rate law for a chemical reaction and then postulate a series of elementary reactions based on the fact that one of the steps in the mechanism must obey the experimental rate law. When two or more mechanisms satisfy the rate law requirement, additional experiments must be done to decide which mechanism is correct.

To illustrate the procedure, we consider the reaction

$$H_2 + 2ICl \rightarrow I_2 + 2HCl$$

which has a rate law of

$$\text{Rate} = k[H_2][ICl]$$

A two-step mechanism that satisfies the rate law is as follows:

$$H_2 + ICl \rightarrow HI + HCl$$
$$HI + ICl \rightarrow I_2 + HCl$$

The two steps in this mechanism add up to the overall reaction. They also involve only two molecules as reactants. The rate law describes the first step of the mechanism, which is presumably the slow step. By reacting HI and ICl, chemists can easily demonstrate that the second step of the mechanism is a much faster reaction.

The HI in the above mechanism does not appear in the balanced equation. Any chemical species that is part of a mechanism but not of the balanced equation is called an **intermediate**. In this example, the intermediate HI is also a known compound that enabled the experimental verification of the mechanism. In other cases intermediates are very unstable, and often exotic, chemical species. Demonstrating the presence of an intermediate provides evidence to support one mechanism over another.

Considering the above example, we see that another mechanism might be considered. This is a three-step process:

$$H_2 + ICl \rightarrow HI + HCl$$
$$H_2 + ICl \rightarrow HI + HCl$$
$$2HI \rightarrow H_2 + I_2$$

When added, these three reactions give the same overall reaction. To decide if this mechanism is possible, scientists determined the rate of decomposition of the intermediate HI to H_2 and I_2 (the last step in the mechanism). This reaction was found to be much slower than the reaction between H_2 and ICl. It was concluded that this mechanism was not the correct one since the rate-determining step would have given a completely different rate law.

From this description we can see that a kinetic study of a reaction will determine the rate law for the slowest step in the mechanism. Possible mechanisms are proposed; the correct mechanism must have one step that obeys the observed rate law. If there appear to be several possible mechanisms, appropriate experiments must be devised to determine which mechanism is correct.

Determining Rate Laws for Elementary Reactions

The coefficients of the reactants in an elementary reaction are the exponents of the reactant concentrations in the rate law. Therefore the rate law for each step of a mechanism can be predicted directly. One complication is that, after the first step, most of the elementary reactions in a mechanism will have intermediates as reactants. To compare an experimental rate law with the rate laws of the elementary reactions, it will be necessary to convert a rate law that includes an intermediate into one that has only reactants.

For example, the reaction

$$2NO + O_2 \rightarrow 2NO_2$$

has a possible mechanism consisting of these two elementary reactions:

$$NO + O_2 \rightarrow NO_3$$
$$NO_3 + NO \rightarrow 2NO_2$$

The rate law for the first step is

$$\text{Rate} = k_1 [NO][O_2]$$

The rate law for the second step includes the intermediate NO_3:

$$\text{Rate} = k_2 [NO_3] [NO]$$

To eliminate the NO_3 and convert it into one of the reactants, we use the **steady-state assumption**. This assumption says that, if the second step is the rate-determining step, the first reaction must be relatively fast and reversible, meaning that the rate at which NO_3 is formed is equal to the rate at which it disappears:

$$NO_3 \text{ rate formation} = NO_3 \text{ rate disappearance}$$

The rate laws governing the forward and reverse reactions in the first step are as follows:

$$\text{Rate}_{\text{forward}} = k_f\,[\text{NO}]\,[\text{O}_2]$$

$$\text{Rate}_{\text{reverse}} = k_r\,[\text{NO}_3]$$

Since the forward and reverse rates are equal, we can write

$$k_f\,[\text{NO}]\,[\text{O}_2] = k_r\,[\text{NO}_3]$$

Solving for $[NO_3]$ gives

$$[\text{NO}_3] = \frac{k_f}{k_r}[\text{NO}]\,[\text{O}_2]$$

Substituting this result for the intermediate, NO_3, in the rate law, we obtain

$$\text{Rate} = k\frac{k_f}{k_r}[\text{NO}]\,[\text{O}_2]\,[\text{NO}]$$

Combining all of the k, k_f and k_r rate constants, we can write the rate law for the second step of the mechanism:

$$\text{Rate} = k[\text{NO}]^2[\text{O}_2] \tag{12.11}$$

EXERCISE

For the kinetics of the following reaction:

$$2\text{NO} + \text{O}_2 \rightarrow 2\text{NO}_2$$

two possible mechanisms are

$$2\text{NO} \rightarrow \text{N}_2\text{O}_2$$

$$\text{N}_2\text{O}_2 + \text{O}_2 \rightarrow 2\text{NO}_2$$

and

$$\text{NO} + \text{O}_2 \rightarrow \text{NO}_3$$

$$\text{NO}_3 + \text{NO} \rightarrow 2\text{NO}_2$$

Describe how to determine which mechanism is correct.

Solution

The rate laws for each of these pairs of elementary reactions can be determined. For the first mechanism they are

$$\text{Rate}_1 = k_1\,[\text{NO}]^2$$

$$\text{Rate}_2 = k[\text{N}_2\text{O}_2]\,[\text{O}_2]$$

Using the steady-state assumption to obtain the rate law of the second elementary reaction in terms of measurable reactants, we obtain

$$k_f[NO]^2 = k_r\,[N_2O_2]$$

and the rate law will be

$$Rate_2 = k_2[NO]^2[O_2]$$

For the second mechanism the rate laws are

$$Rate_1 = k[NO]\,[O_2]$$

$$Rate_2 = k[NO]\,^2[O_2]$$

The second rate law was derived previously as Equation 12.11.

From these rate laws we can see that, if the first step is the slow step, the two mechanisms give two distinctly different rate laws and the decision is clear cut. If, however, the second step is the slow step, both mechanisms yield the same rate law. In that case, to determine which mechanism is correct, additional experiments must be designed to learn which intermediate, NO_3 or N_2O_2, is formed during the reaction.

RADIOACTIVE DECAY

Rate of Radioactive Decay

Radioactive decay is a random event; we cannot predict when a single isotope will spontaneously decay. For a large group of radioactive atoms, however, we can apply the well-defined laws of kinetics. Radioactive decay is described as a first-order reaction process. The number of nuclei that disintegrate per second depends only on the number of radioactive nuclei in the sample. The equation for the number of radioactive disintegrations per second (dps) is as follows:

$$\text{Rate (dps)} = \text{(constant)(number of radioactive nuclei)}$$

$$= kN$$

This rate equation may be integrated by using simple calculus to obtain this expression:

$$\ln\left(\frac{N_0}{N_t}\right) = kt \tag{12.12}$$

where N_0 is the original number of radioactive atoms, N_t is the number of radioactive atoms left after t seconds have elapsed, k is the rate constant with units of reciprocal seconds (s^{-1}), and t is the time in seconds from the start of the experiment. Since the number of atoms in a sample is directly proportional to the mass of the sample, we may interpret N_0 and N_t as the masses of radioactive atoms at the start and at time t, respectively.

When half of the atoms have decayed, $N_0 = 2N_t$. Substituting $2N_t$ for N_0 in Equation 12.12 allows us to develop a relationship between the half-life ($t_{1/2}$) of an isotope and the specific rate constant (k):

$$\ln\left(\frac{2N_t}{N_t}\right) = kt_{1/2}$$

N_t cancels on the left side of this equation to give

$$\ln(2) = kt_{1/2}$$

The natural logarithm of 2 is 0.693, so

$$t_{1/2} = \frac{0.693}{k} \qquad (12.13)$$

In Equation 12.13 $t_{1/2}$ is the half-life of the isotope.

The half-life is an important concept. It means that one-half of a radioactive isotope present at the start of an experiment will decompose in a length of time equal to the half-life of that isotope. The bar graph in Figure 12.9 shows how the amount of radioactive material declines with each half-life. After one half-life half of the isotope is left; after two half-lives half of that half, or one quarter of the original amount, remains. With the next half-life half of what remained after the second half-life decays, and only one-eighth is left. We note that each of the bars in Figure 12.9 is one-half the size of the preceding bar.

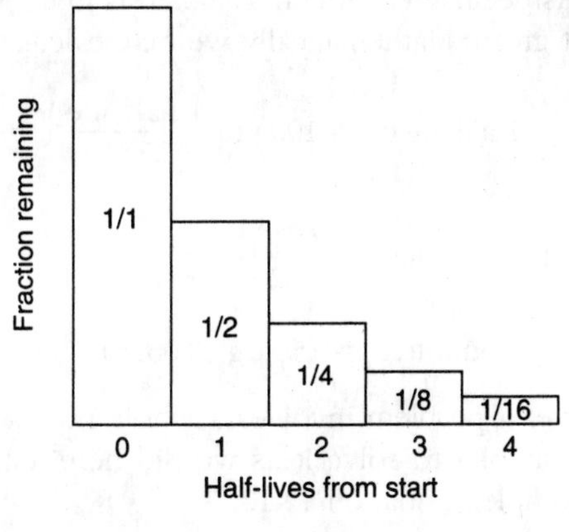

FIGURE 12.9. *A bar graph illustrating how half of a radioactive material decomposes with each half-life.*

We can calculate the fraction of the original sample left after a given number of half-lives by using Equation 12.14, where the fraction ½ is raised to the power equal to the number of half-lives:

$$\text{Fraction left} = \left(\frac{1}{2}\right)^{\text{number of half-lives}} \qquad (12.14)$$

The decimal value of each fraction multiplied by 100 gives us the percentage of the radioactive isotope remaining. The percentage that has decayed is obtained by subtracting the percentage remaining from 100.

It must be appreciated that only four half-lives are required for over 90 percent of a radioactive material to disintegrate (less than 10 percent is left). However, even a small sample of a radioactive isotope may contain upward of 10^{18} atoms. To reduce 10^{18} atoms to just one atom, 60 half-lives are required: four to disintegrate the first 90 percent of the isotope, and another 56 to disintegrate the rest. This fact has important implications for public policies concerning the disposal of radioactive waste materials.

Radioactive Decay Calculations

Calculations involving the disintegration of radioactive isotopes demand a thorough understanding of Equations 12.12, 12.13, and 12.14. Usually such calculations require a combination of two or more of these equations.

Example The half-life of ^{210}Pb is 25 years. If a sample starts with 50 micrograms of ^{210}Pb, how much is left after 100 years?

Solution: Since 25 years is one half-life, 100 years is equal to four half-lives for ^{210}Pb. Therefore one-sixteenth of the original amount will remain (see Figure 12.9). To calculate the amount remaining, 1⁄16 or its decimal equivalent (0.0625) is multiplied by the original amount given. Mathematically we may calculate as follows:

$$?\ \text{half-lives} = 100\ \text{yr}\left(\frac{1\ \text{half-life}}{25\ \text{yr}}\right) = 4\ \text{half-lives}$$

$$?\ \text{fraction left} = \left(\frac{1}{2}\right)^4 = 0.0625$$

$$?\ \text{amount left} = 50\ \mu\text{g}\ (0.0625) = 3.1\ \mu\text{g}$$

Whenever a problem involves a whole number of half-lives, it is usually simpler to solve it as we did here, rather than using the more complex Equation 12.12.

Example How many years will 50 micrograms of ^{210}Pb take to decrease to 5.0 micrograms? ($t_{1/2}$ = 25 years)

Solution: A quick calculation shows that 5.0 μg/50 μg is 0.1 or ⅒ of the original amount. From the previous discussion we know that after three half-lives (75 yr) ⅛ remains and that after four half-lives (100 yr) 1/16 remains. We may estimate that the answer should be between three and four half-lives (75 and 100 yr). If a more precise answer is needed, Equation 12.12 must be used:

$$\ln\left(\frac{N_0}{N_t}\right) = kt$$

Here N_0 is the original number of atoms of the isotope, and N_t is the number remaining after time t has elapsed. We can use the mass of the isotope in place of the number of atoms since mass is directly proportional to the number of atoms. Also, k is the rate constant for the radioactive disintegration, calculated from the equation:

$$t_{1/2} = \frac{0.693}{k}$$

Substitution yields

$$k = \frac{0.693}{25 \text{ yr}}$$
$$= 2.77 \times 10^{-2} \text{ yr}^{-1}$$

Entering our data into equation 12.12 gives

$$\ln\left(\frac{50 \ \mu\text{g}}{5.0 \ \mu\text{g}}\right) = (2.77 \times 10^{-2} \text{ yr}^{-1})\, t$$

$$\frac{2.30}{2.77 \times 10^{-2} \text{ yr}^{-1}} = t$$

$$t = 83 \text{ yr}$$

As expected, the result fits the preliminary estimate of somewhere between 75 and 100 years.

The importance of the effective use of good estimates cannot be overemphasized. At times an estimate allows us to select the correct answer in a multiple-choice problem. If not, the estimate serves as a check on our calculated answer.

Example Ninety-nine percent of a radioactive element disintegrates in 36.0 hours. What is the half-life of this isotope?

Solution: First, it must be realized that the problem tells us how much of the elements has disintegrated. Half-life calculations using our equations refer to the amount that is left (N_t). Subtracting

99% from 100%, we find that only 1% (or 1⁄100) is left after 36.0 hrs. Next we can estimate the half-life by continuing the sequence depicted in Figure 12.9. We find that 1⁄32 is left after five half-lives, 1⁄64 after six half-lives, and 1⁄128 after seven half-lives. Therefore our sample has undergone at least six, but not quite seven, half-lives. Six half-lives in 36.0 hrs means a half-life of 6 hrs. Seven half-lives in 36.0 hrs is approximately 5 hrs for a half-life. Our estimate of the half-life is between 5 and 6 hrs.

To calculate the exact half-life (if needed), we substitute the data into Equation 12.12:

$$\ln\left(\frac{N_0}{N_t}\right) = kt$$

$$\ln\left(\frac{100}{1}\right) = k(36.0\text{ hr})$$

$$4.605 = k(36.0\text{ hr})$$

$$k = \frac{4.605}{36.0} = 0.128\text{ hr}^{-1}$$

Then we use the Equation 12.13:

$$t_{1/2} = \frac{0.693}{k}$$

$$= \frac{0.693}{0.128\text{ hr}^{-1}}$$

$$= 5.41\text{ hr}$$

This answer agrees with our quick estimate. Often the estimate is sufficient to make an informed selection of the correct answer to a multiple-choice question. More important, it allows us to reject incorrect answers quickly.

Dating Archaeological Samples

Equation 12.12, the integrated rate equation for radioactive decay, can be used to calculate the time elapsed as long as N_0, N_t, and k are known. The rate constant can be obtained from equation 12.13 if the half-life ($t_{1/2}$) of an isotope is known. Therefore the problem of determining the ages of very old materials involves obtaining valid data for N_0 and N_t. Two examples will illustrate the different types of logic used.

Uranium-238 decays slowly through the uranium disintegration series to lead-206. The half-life of this conversion is 4.5×10^9 years. To determine the age of a rock, we need to determine the amount of uranium present when the rock solidified (N_0) and the amount of uranium in the rock today (N_t). The amount of ^{238}U present in the rock today can be measured. It is impossible, however, to go back billions of years to obtain a sample containing original amount of ^{238}U, which is needed to determine the value for N_0.

The solution to this dilemma lies in the fact that each atom of uranium ends up as an atom of lead. Therefore, if we take a rock sample and determine the number of atoms of ^{238}U and the number of atoms of ^{206}Pb it contains, we can say that

$$N_0 = \text{atoms}^{238}\text{U} + \text{atoms}^{206}\text{Pb}$$

and

$$N_t = \text{atoms}^{238}\text{U}$$

When we set up an experiment, we must consider all possibilities for error. Two obvious ones exist in this experiment. First, if the original rock contained some ^{206}Pb, our value for N_0 would be too high. Second, if some of the original ^{238}U did not end up as ^{206}Pb the value for N_0 would be too low. This could occur, for example, if randon gas, which is part of the uranium series, escaped from the rock before it decayed to the next solid isotope. Scientists working on this type of project evaluate these possibilities and make corrections or do additional experiments as needed.

In determining the ages of carbon-containing materials, the radioactive isotope carbon-14 is measured. ^{14}C is not a natural isotope; it is constantly formed in the upper atmosphere, where ^{14}N is bombarded with neutrons. This keeps the proportion of ^{14}C relatively constant in the biosphere. While alive, animals and plants maintain that same proportion of ^{14}C in their bodies because carbon is continuously recycled. When an organism dies, however, the ^{14}C is no longer replenished by the diet and the fraction of this isotope in dead organic matter decreases with time.

Understanding the process whereby ^{14}C enters living matter allows us to obtain reasonable measures for N_0 and N_t for our calculations. In this case we assume that the fraction of ^{14}C in the biosphere today is the same as it was in prehistoric times. With that assumption, we can write these equations:

$$N_0 = \frac{\text{living g } ^{14}\text{C}}{\text{living g total C}}$$

and

$$N_t = \frac{\text{ancient g } ^{14}\text{C}}{\text{ancient g total C}}$$

Now, with a value for the half-life of ^{14}C, we can solve Equation 12.12 to determine the age of an ancient sample.

In using radioactive dating techniques, the equation we solve involves the ratio N_0/N_t. This ratio is difficult to determine accurately if it is very close to 1.00 or if it is very large. If the ratio is close to 1.00, very few radioactive disintegrations have occurred. If the ratio is very large, most of the sample has disintegrated (the denominator is very small in the ratio N_0/N_t). A rule of thumb for radioactive isotope dating of materials is that the age of the sample should be 0.3 to 3 half-lives of the isotope used for the dating. Uranium's half-life is close to that of the age of the Earth and is appropriate for dating purposes. The half-life of ^{14}C is 5730 years. As a result, we may most reliably determine the ages of biological materials that range from 1700 to 17,000 years old. Radiocarbon dating would certainly not be appropriate for verifying the age of a bottle of wine with a label date of 1865.

GLOSSARY

Absolute uncertainty. The uncertainty of ±1 in the last digit of a measurement. If this uncertainty is different from ±1, it is written as part of the number; for example, 23.45 ± 0.05 indicates an uncertainty of ±5 in the last digit.

Absolute zero. The lowest possible temperature, 0.0 K or −273.16°C.

Absorbance (*A*). A measure of the amount of light absorbed by a chemical.

Absorptivity (*a*). A constant the value of which depends on the sample and the wavelength at which the measurement is made in spectroscopy.

Accuracy. The degree of closeness between a measured value and the true value.

Acid. Any substance that donates protons; Lewis acids are electron-pair acceptors.

Acid anhydride. The oxide of a nonmetal, which forms an acid when dissolved in water.

Acid dissociation constant (K_a). The value of the equilibrium law for the dissociation of a weak acid.

Activated complex. The structures of colliding molecules at the moment of collision, generally thought to be intermediate between the structures of the products and of the reactants.

Activation energy. The increase in potential energy, due to a molecular collision, necessary to convert a reactant into a product.

Activity series. A listing of elemental substances in the order of their abilities to be oxidized or reduced. This listing makes it possible to predict whether an element will cause the oxidation or the reduction of an ion of another element.

Addition reaction. The reaction in which a double bond opens to form two additional single bonds in a polymer.

Adhesive force. The attractive forces between two dissimilar substances.

Alcohol. An organic compound with an −OH group.

Aldehyde. An organic compound with a terminal −CHO group.

Alkali metals. The extremely reactive elements in Group 1 (first column) of the Periodic Table. They all have ns^1 electrons as valence electrons.

Alkaline earth metals. The very reactive elements in Group 2, (second column) of the Periodic Table. They all have ns^2 electrons for valence electrons.

Alkanes. Organic compounds with the general formula C_nH_{2n+2}.

Alkenes. Organic compounds with double bonds in their structures.

Alkyl group. A functional group that is an alkane in nature.

Alkynes. Organic compounds with triple bonds in their structures.

Allotrope(s). One or more distinct forms of an element; classification as an allotrope is based on structure and physical properties. For example, diamond and graphite are two allotropes of carbon.

Alpha particle. A helium nucleus.

Amines. Compounds, related to ammonia, in which one or more of the hydrogen atoms in ammonia have been replaced by organic functional groups.

Amino acid. An organic acid that contains both an amine and an acid functional group on adjacent carbon atoms.

Amorphous: A term meaning "without structure."

Amphiprotic (amphoteric). A term designating a substance that can act as both a conjugate acid and a conjugate base.

Amphoteric. *See* **Amphiprotic.**

Anhydride. The oxide of a metal or nonmetal that reacts with water to form an acid or a base, respectively.

Anion. An ion with a negative charge.

Aqueous. A term designating a system that involves water or a chemical mixture or solution having water as the solvent.

Arrhenius equation. The equation that relates temperature and activation energy to the rate constant.

Arrhenius theory. The theory that an acid increases hydrogen ion concentration when dissolved and that a base increases hydroxide ion concentration when dissolved.

Aryl group. A functional group that is aromatic in nature.

Atomic mass (*A*). The relative mass of an element compared to the mass of isotope carbon-12, which is defined as exactly 12.

Atomic number (*Z*). The number of an element in the Periodic Table; also, a number representing the number of protons in the nucleus of the atom.

Atomic orbital. The orbital structure of an element; also, an orbital within an element.

Atomic symbol. A one- or two-letter abbreviation of an element's name. Some symbols (e.g., Pb for lead) are derived from Latin names of the elements.

Atoms. The fundamental particles of chemistry; 112 are currently known and are arranged in an orderly manner in the Periodic Table.

Autopyrolysis constant of water (K_w). The value of the equilibrium law for the dissociation of water into H^+ and OH^-.

Avogadro's number. A quantity equal to 6.02×10^{23}.

Avogadro's principle. A statement of the direct relationship between the moles of a gas and the volume of that gas.

Axial atom. A term used to describe the position of an atom in a covalent molecule of the AX_5 or AX_6 basic structure. The axial atoms are on the vertical axis of the molecule in positions similar to the Earth's North and South Poles.

Azimuthal quantum number (ℓ). The quantum number that specifies the sublevel in which an electron is located; ℓ may be any number from 0 up to $n - 1$.

Balanced reaction. A chemical equation that has the smallest whole-number coefficients for the reactants and products that will result in the same number of atoms of each element on both sides of the arrow.

Barometer. A closed-end manometer used for measuring atmospheric pressure.

Base. Any compound that increases the hydroxide concentration of a solution or is a proton acceptor; Lewis bases are electron-pair donors.

Base anhydride. The oxide of a metal that forms a base when dissolved in water.

Base dissociation constant, (K_b). The value of the equilibrium law for the dissociation of a weak base.

Basic structure. One of five basic geometries—linear, triangular planar, tetrahedral, trigonal planar, octahedral—that a molecule may take.

Battery. A galvanic cell used to produce electricity for consumer items such as flashlights, portable radios, and heart pacemakers.

Beer's law. The law stating that the absorbance of a sample is the product of the absorptivity, optical path length, and sample concentration; $A = abc$.

Beta particle. An electron.

Bidentate ligand. A Lewis base that donates two pairs of electrons.

Binary acid. An acid that contains hydrogen and one other element in its formula.

Body-centered cubic (bcc). A term describing a cubic structure in which one atom is at each of the eight corners and one atom is in the center of the unit cell.

Bohr atom. The model of the atom developed by Neils Bohr. This model views electrons as circling the nucleus like a miniature solar system.

Boiling point, normal. The temperature at which the vapor pressure of a liquid is equal to 760 millimeters of mercury (1.00 atmosphere) of pressure; also, the temperature at which a gas condenses. Also called condensation point.

Boiling-point-elevation constant (k_b). The temperature increase of the boiling point per molal of solute particles.

Bond order. The average number of bonds per atom covalently bonded to a central atom.

Bonding electron pair. A pair of electrons that participate in a covalent bond.

Boyle's law. The law that states the inverse relationship between the volume and the pressure of a gas; $PV =$ constant.

Bragg equation. The equation that relates the atomic dimensions in a crystal to the angles at which monochromatic X rays will undergo constructive reinforcement.

Brönsted-Lowry theory. The theory that acids are proton donors and bases are proton acceptors.

Buckminsterfullerene. The allotrope of carbon that has the formula C_{60}.

Buffer capacity. The number of moles of strong acid or strong base required to change the pH of 1 liter of buffer by 1 pH unit.

Buffer solution. An aqueous solution containing a conjugate acid and its conjugate base in a molar ratio greater than 0.1 but less than 10.0.

Bumping. The violent boiling that occurs when a solution becomes superheated.

Buret. A tube approximately 1 centimeter in diameter that is used for measuring liquid volumes of 10–100 milliliters.

Calorimeter. An instrument used to determine heat energy.

Catalyst. A substance that speeds up the rate of a chemical reaction by providing an alternative reaction pathway with a lower activation energy.

Cation. An ion with a positive charge.

Cell voltage, (*E*). The voltage of a galvanic cell under non-standard-state conditions.

Centrifugation. The process of separating a solid from a liquid by spinning it rapidly to artificially increase the gravitational force.

Chain reaction. A nuclear reaction that produces more neutrons than were needed to initiate it, therefore causing more reactions than occurred in the preceding step.

Charles's law. The statement of the direct relationship between the temperature and the volume of a gas; $P/V =$ constant.

Chelate. A Lewis base that usually has more than one pair of electrons to donate.

Cis isomer. An isomer with substituents on the same side of the double bond. *See also* **Trans isomer.**

Clausius-Clapeyron equation. The equation that relates vapor pressure to heat of vaporization.

Closed system. A system in which mass cannot be lost to, or gained from, the surroundings. *See also* **Open system.**

Coefficient. A number placed in front of a chemical formula to represent the number of molecules of that substance that are included in an equation. This number multiplies the number of atoms in the formula unit.

Cohesive force. The sum of all the attractive forces in a pure substance.

Colligative property. Any one of several physical properties of a solution that changes depending on the amount of solute particles present in the solution.

Collision theory. The theory of kinetics that relates reaction rates to the frequency, energy, and orientation of molecules in collisions.

Complex. *See* **Complex ion.**

Complexation reaction. A reaction between a Lewis acid and a Lewis base.

Complexing agent. *See* **Lewis base.**

Complex ion. A combination of one or more compounds or anions with a metal ion; also called a complex.

Compound. A combination of two or more elements into a distinct substance with definite physical properties.

Concentrated. A qualitative term indicating a large amount of solute in a given amount of solvent.

Concentration. An expression of the amount of solute mixed with a given amount of solvent.

Concentration versus time curve. A graph of reactant or product concentration as a function of time.

Condensation point. *See* **boiling point.**

Condensation reaction. A polymerization reaction between an acid and either an alcohol or an amine.

Condensation. The conversion of a gas into a liquid.

Conjugate acid. Any substance that has a proton that may be donated to a base.

Conjugate acid-base pair. A pair of substances whose formulas differ by one H^+ ion.

Conjugate base. Any substance that can accept a proton.

Conjugated double bonds. A series of two or more double bonds; each separated by only one single bond in a molecule.

Cooling curve. A graph showing the changes that occur while a substance is cooling.

Coordinate covalent bond. The covalent bond that is formed between two atoms when one substance donates both electrons.

Covalent bond. The bond between two atoms that arises from the sharing of a pair of electrons.

Covalent compound. A chemical compound in which the atoms are held together with covalent bonds.

Covalent crystal. A crystal that consists of only one molecule. All atoms are joined to others with covalent bonds. Also called a network crystal.

Critical mass. The minimum mass of a fissile (fissionable) material needed to sustain a nuclear chain reaction.

Critical point. The temperature and pressure above which a gas cannot be condensed to a liquid.

Crystal lattice. The arrangement of atoms, ions, or molecules in a crystal structure.

Crystallization point. *See* **Melting point.**

Cyclotron. A device for accelerating charged particles, which are usually used to bombard targets in an effort to generate nuclear reactions.

Dalton's law of partial pressures. The law that the total pressure of a gas is the sum of the individual pressures of all gases in the mixture; $P_{total} = p_1 + p_2 + \ldots$

Decay. The spontaneous emission of a particle in a radioactive event.

ΔE. The energy change due to a chemical reaction. It is equal to the heat and work of the reaction or to the change in potential energy of the products compared to that of the reactants.

$\Delta G°$. The standard free-energy change for a reaction. The temperature is 298 K unless otherwise indicated.

$\Delta G_f°$. The standard free energy of formation, corresponding to the formation of 1 mole of product from its elements at 298 K.

ΔH. The enthalpy change occurring in a chemical process, Without the superscript "°" it indicates an extensive property often called the heat of reaction.

$\Delta H°$. The standard enthalpy change occurring in a reaction; refers to the heat produced or absorbed when the moles of reactants specified in the chemical reaction react at standard state.

$\Delta H_f°$. The standard heat of formation, which is the heat produced or absorbed when 1 mole of a product is formed.

ΔS. The change in entropy between the final state and the initial state in a chemical process. Without the superscript "°" it indicates an extensive property.

$\Delta S°$. The standard entropy change of a chemical process; an intensive property based on the number of moles of substance in the balanced chemical reaction.

Dependent variable. The variable that an experiment measures; its value depends on the value of the independent variable. *See also* **Independent variable.**

Derived structure. A molecular structure that is derived from one of the five basic structures.

Detergent. A chemical substance that has both polar and nonpolar properties and is soluble in both polar and nonpolar solvents.

Determinate error. An error associated with faulty instruments, calibrations, or techniques.

Dextrorotatory. A term describing an optical isomer that rotates polarized light to the right.

Diagonal relationship. A term describing the fact that some properties of atoms vary regularly from the lower left corner to the upper right corner of the Periodic Table; the diagonal line indicating the diagonal relationship. Electronegativity, ionization energy, and electron affinity are some of these diagonal relationships.

Diamond. An allotrope of carbon in which all carbon atoms have sp^3 hybridization.

Diatomic. A term describing a molecule that contains only two atoms (e.g., HCl. H_2).

Differentiating electron. The electron that differentiates one element from an adjacent element in the Periodic Table.

Diffraction. The combination of light waves that results in either constructive or destructive reinforcement.

Diffusion. The movement of molecules from one place to another by random motion.

Dilute. A qualitative term indicating a small amount of solute in a given amount of solvent.

Dimer. A substance composed of two identical molecular or ionic units.

Dipole. A polar molecule. The term *dipole* reminds us that a polar molecule has only two poles, one positive and one negative.

Direct relationship. A relationship between two variables whereby one must increase when the other increases.

Dispersion forces. *See* **London forces.**

Dissociation. The breakup of a molecular or ionic compound into ions. *See also* **Ionization.**

Double-replacement reaction. A chemical reaction in which the cation of one substance replaces the cation of a second substance. At the same time, the cation of the second substance replaces the cation of the first substance.

dsp^3 hybrid orbital. An orbital formed from one *s*, three *p*, and one *d* orbital. The electrons in these orbitals are all equal in energy. Structures are all related to the trigonal bipyramid.

d^2sp^3 hybrid orbital. An orbital formed from one *s*, three *p*, and two *d* orbitals. The electrons in the hybrid all have the same energy. Structures are all related to the octahedron.

Ductile. A term describing the property of being able to be drawn into wire forms.

Dynamic equilibrium. A state in which a chemical process is going in the forward direction at the same rate that it is going in the reverse direction and the concentrations of reactants and products remain constant. Equilibrium follows the kinetic period, in which reaction occurs and the concentrations of reactants and products change. Also called equilibrium.

EDTA. Ethylene diamine tetraacetic acid, a very useful complexing agent.

Effusion. The movement of molecules through a small hole from one container to another.

Electrode. A metal placed in a liquid to transfer electrons in a galvanic or electrolytic cell.

Electrolysis. The process of using electric current to reduce a chemical substance at the cathode and oxidize a chemical substance at the anode.

Electrolyte. A substance that dissociates completely into ions in solution.

Electrolytic cell. An arrangement of electrodes and a power source used to force nonspontaneous redox reactions to occur.

Electron. 1. The unit of negative charge in the atom; the diffuse electron cloud surrounds the dense nucleus. 2. One of the three particles, along with the proton and neutron, that make up an atom. An electron has a negative charge and virtually no mass in comparison to the neutron and proton. Electrons are arranged in an orderly fashion around the nucleus. There are as many electrons as neutrons in a neutral element.

Electron affinity. The energy released or absorbed in adding an electron to an atom.

Electron deficient. A term describing a Lewis structure that has less than an octet of electrons around one or more of its atoms.

Electronegativity. A measure of an atom's tendency to attract electrons. Fluorine has the highest, and francium the lowest, electronegativities in the Periodic Table.

Electroneutrality, law of. A statement of the fact that no chemical compound has a net charge. In addition, an element has no net charge.

Electronic configuration. A listing of the electrons within an atom based on the sublevels that are filled and the relative energies of these sublevels. For example, the electronic configuration for silicon is $1s^2$, $2s^2$, $2p^6$, $3s^2$, $3p^2$.

Electrostatic potential energy. The energy of attraction of two oppositely charged particles; the energy of repulsion of two like-charged particles.

Element. Any one of the 112 distinct particles, known as atoms, that are currently known. Each has distinct chemical and physical properties.

Elementary reaction. One reaction in a mechanism. It is usually bimolecular or unimolecular and its coefficients are the exponents in the rate law.

Empirical formula. The formula that gives only the simplest whole-number ratio of the atoms that make up the compound. *See also* **Molecular formula; Structural formula.**

End point. The point, where neither reactant is in excess, that marks the end of a titration experiment.

Endothermic. A term describing any process that absorbs heat from the surroundings. Endothermic processes cool the system.

Enthalpy. The heat content of a chemical substance. Enthalpy is related to the internal potential energy of the substance.

Entropy. A measure of the randomness of a chemical substance.

Enzyme. One of many naturally occurring catalysts found in biological materials.

Enzyme-substrate complex. The activated complex formed in an enzyme-catalyzed reaction.

Equatorial atom. A term used to describe the position of an atom in a covalent molecule of the AX_5 or AX_6 basic structure. The equatorial atoms are around the center of the molecule in positions similar to the Earth's equator.

Equilibrium. *See* **Dynamic equilibrium.**

Equilibrium constant (*K*). The numerical value of the equilibrium law. The only variable that has an effect on the equilibrium constant is temperature.

Equilibrium law. The basic equation governing chemical equilibrium. Each balanced chemical equation has its own equilibrium law.

Equilibrium table. A table of data used to summarize the numerical data and stoichiometric relationships of an equilibrium system.

Equivalent. An amount of a chemical substance determined by dividing its mass by its equivalent weight.

Equivalent weight. The mass of a compound, which loses or gains 1 mole of electrons in an oxidation-reduction reaction or the mass of a compound that functions or reacts with one mole of H^+ in an acid-base reaction.

Escape energy. The minimum kinetic energy needed to transform a liquid molecule into a gas.

Ether. An organic compound containing the $-C-O-C-$ functional group.

Eudiometer. A closed-end manometer.

Evaporation. The process whereby a liquid is transformed into a gas at a temperature below the boiling point.

Exact number. A number that has no uncertainty; exact numbers include stoichiometric coefficients, formula subscripts, and most defined quantities.

Excess reactant. Any reactant that is not completely consumed in a chemical reaction.

Exothermic. A term describing any process that gives off heat to the surroundings. Exothermic processes heat the system.

Extrinsic property. A physical or chemical property that varies in proportion to the amount of matter.

Face-Centered cubic (fcc). A term describing cubic structure in which atoms are at the corners and there is an atom on each cube face of the unit cell.

Factor label. A ratio used to convert a number with one set of units into the equivalent number with different units.

Faraday's constant (ℱ). The constant that expresses the relationship between the coulomb and moles of electrons; 96,485 coulombs = 1 mole of electrons.

Filtrate. The liquid that passes through a filter.

Filtration. The process of separating a solid from a liquid by passing the liquid through porous paper.

First law of thermodynamics. The stating that in any chemical or physical process all energy is conserved.

First-order reaction. A reaction with a rate law having exponents that add up to 1. The rate law is Rate = k[A].

Fissile. A term describing a nucleus that is capable of undergoing nuclear fission.

Fluorescence. A property of some atoms and molecules that allows them to absorb photons of light and then reemit them very rapidly, but with a different energy. The emitted light always has a lower energy, longer wavelength than the absorbed light. *See also* **Phosphorescence.**

Formal charge. The charge on an atom in a covalent compound, calculated by assuming that all bonding electrons are equally shared.

Formation reaction. A chemical reaction in which the reactants are elements at standard state and the product is 1 mole of one compound.

Formula. The representation of a chemical substance using chemical symbols and appropriate subscripts for the numbers of atoms and superscripts to represent charges if the substance is an ion.

Free radical. A molecule that contains an unpaired electron in its Lewis structure.

Freezing-point-depression constant (k_f). The temperature decrease in the freezing point per molal of solute particles.

Functional group. A group of atoms on an organic compound that represent a characteristic chemical entity.

Fused salt. *See* **Molten salt.**

Galvanic cell. The experimental setup used to convert chemical energy into electrical energy. All batteries are galvanic cells.

Gamma ray. A high-energy photon often emitted in a nuclear reaction.

Gas. A state of matter characterized by the ability to flow and to fill any container completely without regard to the amount of gas in the container.

Gay-Lussac's law. The law that expresses direct relationship between the temperature and the pressure of a gas; P/T = constant.

Gibbs free energy. The maximum energy from any chemical reaction.

Graham's law of effusion. The law that relates the rate at which a gas passes through a small hole to the mass of the molecule; $\sqrt{m_1/m_2} = \overline{v}_2/\overline{v}_1$

Gram-atomic mass. *See* **Atomic mass.**

Gram-molar mass. *See* **Molar mass.**

Graph. A pictorial method for presenting and evaluating experimental data.

Graphite. An allotrope of carbon in which carbon has sp^2 hybridization.

Gravitational potential energy. The attractive energy of any two masses toward each other.

Group. A column in the Periodic Table.

Half-life. The time required for half of the reactants to be consumed in a chemical reaction or a radioactive decay.

Halide. An organic compound with a halogen in its structure.

Halogen. One of the five elements in Group 17 of the Periodic Table. Halogens are reactive elements with ns^2, np^5 valence electron structures.

Halogenation. The addition of a halogen to a double or triple bond.

Heat capacity. The amount of heat energy that a system needs in order for its temperature to change by 1 Celsius degree.

Heating curve. A graph showing the changes that occur while a substance is heating.

Heat of combustion. The heat released when 1 mole of sample, usually an organic compound, is completely burned in oxygen to form CO_2 and H_2O.

Heat of fusion. The heat energy needed to convert a solid into a liquid. The units are either joules per gram or joules per mole.

Heat of reaction. The heat released or absorbed in a chemical reaction.

Heat of vaporization. The heat energy needed to convert a liquid into a gas. The units are either joules per gram or joules per mole.

Henderson-Hasselbach equation. A derivation of the equilibrium law obtained by taking the negative logarithm of the equilibrium law; $pH = pK_a + \log([\text{conjugate base}]/[\text{conjugate acid}])$.

Henry's law. The law that expresses the relationship between the solubility of a gas and its partial pressure.

Hess's law. The law that states that heats of reaction are additive when chemical reactions are added.

Hund's rule. The rule that all orbitals in a sublevel must fill with one electron before a second electron of opposite spin can be added to any orbital in that sublevel.

Hybrid orbital. An orbital constructed by combining electrons, usually from *s* and *p* orbitals, into a new orbital in which the electrons all have the same properties. Hybrid orbitals are designated as sp, sp^2, sp^3, dsp^3, and d^2sp^3.

Hydrate. A substance that contains a fixed number of water molecules. The water molecules are written separately from the chemical formula itself and are connected to it with a dot in the center of the line between the formula and the water molecules.

Hydrogenation. The addition of H_2 to a double or triple bond.

Hydrogen bonding. An extra strength dipole-dipole attractive force due to the large electronegativity difference between hydrogen and nitrogen, oxygen, or fluorine.

Hydrohalogenation. The addition of a hydrogen and a halogen to a double or triple bond by using a binary acid such as HF, HCl, HBr, or HI.

Hydrolysis reaction. The reaction of a substance, usually a conjugate acid or base, with water.

Ideal gas. A gas that obeys the ideal gas law; conceptually, a gas molecule with no volume and no attractive forces with other molecules.

Ideal gas law. The law that relates temperature, pressure, volume, and moles of gas; $PV = nRT$.

Independent variable. The variable in an experiment that is under the control of the experimenter. *See also* **Dependent variable.**

Indeterminate error. An error in estimating the uncertain digit in a measurement; also called random error.

Indicator. A chemical added to a titration experiment that changes color at the end point.

Indicator electrode. An electrode placed in a sample in order to measure the concentration of an ion in the sample.

Induced dipole. A dipole formed by the interaction of a nonpolar substance and either a polar substance or an instantaneous dipole.

Initial conditions. A quantitative description of a chemical system at the start of a reaction.

Inspection method. A method for balancing a chemical equation. The inspection involves counting the number of each atom on each side of the equation and then balancing by adding appropriate coefficients to the reactants and/or products.

Instantaneous dipole. A distortion of the electron cloud around an atom or molecule that gives the atom or molecule momentary polarity.

Intermediate. A substance that appears in the elementary reactions of a mechanism but not in the overall balanced equation.

Intermolecular forces. The attractive forces—dipole-dipole attractions, London forces, and hydrogen-bonding—between moles and atoms that allow them to condense into liquids and solidify into solids.

Internuclear axis. An imaginary line connecting the nuclei of two atoms.

Intrinsic property. A physical or chemical property that does not change with the amount of matter.

Inverse relationship. A relationship between two variables whereby one must increase if the other decreases.

Ion. An element that has lost or gained one or more electrons. *See also* **Polyatomic ion.**

Ion-electron method. A method for balancing more complex oxidation-reduction equations.

Ionic bond. The attraction of a negative anion for a positive cation.

Ionic compound. A chemical compound composed of negatively charged anions and positively charged cations. The unit is held together by the attraction of the positive charges toward the negative charges.

Ionic crystal. A crystal formed from cations and anions in which the main attractive force is the attraction of positive charges toward negative charges.

Ionization. Often confused with **dissociation.** The removal or addition of an electron from an atom or molecule.

Ionization energy. The energy required to remove an electron completely from an atom.

Isoelectronic. A term describing any two substances that have identical electronic configurations. These atoms may be ions or elements.

Isolated system. A system in which neither mass nor energy is transferred to or from the surroundings.

Isomers. Distinctly different compounds that have the same elemental composition.

Isotope. A form of an element with a specified number of protons, neutrons, and electrons.

IUPAC. The International Union of Pure and Applied Chemistry, which sets nomenclature standards.

***K*.** The equilibrium constant, often written as K_{eq}.

K_a. The acid dissociation constant, a special term denoting the equilibrium of a weak acid. A weak acid dissociation always has the form

$$HA + H_2O \rightleftharpoons A^- + H_3O^+$$

K_b. The base dissociation constant, a special term denoting the equilibrium of a weak base. A weak base dissociation always has the form

$$B + H_2O \rightleftharpoons BH^+ + OH^-$$

K_c. The equilibrium constant used when the reactants and products are specified as concentrations.

K_d. The dissociation constant, a special term used mainly to describe the dissociation of complex ions.

K_f. The formation constant, a special term used to describe the formation of complex ions.

K_p. The equilibrium constant used when the reactants and products are specified in terms of partial pressures.

K_{sp}. The solubility product, a special term denoting an equilibrium between a solid and its solution products.

K_w. The autopyrolysis constant of water, equal to 1.0×10^{-14} at 25°C.

Ketone. An organic compound with a nonterminal $-C{=}O$ group.

Kinetic curve. A graph of reactant or product concentration as a function of time.

Kinetic energy (KE). The energy that matter possesses because of its motion; $KE = \frac{1}{2}mv^2$.

Kinetic molecular theory. The theory of the motion of molecules in the gas phase, which explains gas pressure, effusion and diffusion rates, and the effect of temperature on the behavior of gases.

Law of Dulong and Petit. The law stating that the specific heat of a metal multiplied by its molar mass is equal to a constant of approximately 25 joules per molecule per Celsius degree.

Leveling effect. An expression of the fact that the strongest acid in water is the H^+ (H_3O^+) ion and the strongest base is the hydroxide ion, OH^-.

Levorotatory. A term describing an optical isomer that rotates polarized light to the left.

Lewis acid. Any substance that can accept electron pairs.

Lewis base. Any substance that can donate electron pairs; also called complexing agent; ligand; sequestering agent.

Lewis structure. A molecular structure based on the concept that all atoms try to achieve noble-gas electronic configurations by sharing electrons.

Lewis theory. The theory that acids are electron-pair acceptors and bases are electron-pair donors.

Ligand. *See* **Lewis base.**

Limiting reactant. The reactant that is completely consumed in a reaction, causing it to stop.

Limiting reagent. Same as **limiting reactant.**

Linear. A term referring to atoms aligned in a straight line; a three-atom arrangement with a 180° bond angle.

Liquid. A state of matter characterized by the ability to flow in order to fill any container from the bottom up.

Litmus paper. A type of pH paper using the indicator litmus, which is pink in acid and blue in base.

Lock and key. The analogy used to depict how an enzyme recognizes reactants on the basis of physical geometry as well as chemical characteristics.

London forces. The attractive forces from instantaneous dipoles. These forces are due to the fact that the electron clouds around atoms and molecules are not perfectly symmetrical at all times. Also called dispersion forces.

Lone pairs. Electron pairs in Lewis structures that are not used for bonding.

Magnetic quantum number (m_ℓ). The quantum number that specifies the orbital in which an electron is located and the orientation of the orbital in space; m_ℓ may be any number from -1 to $+1$, including 0.

Malleable. A term describing the property of being able to be hammered into new shapes.

Manometer. A device used to measure gas pressure.

Mass fraction (wt/wt). A concentration unit defined as the mass of one solute divided by the total mass of solution.

Mass-volume fraction (wt/vol). A concentration unit defined as the mass of one solute in a given volume of solution.

Mass. A quantity of matter.

Melting point, normal. The temperature at which a solid melts at 1.00 atmosphere of pressure; also, the temperature at which a liquid becomes a solid. Also called crystallization point.

Meniscus. The curved surface of a liquid in a tube or container.

Metal. A substance with characteristic properties of high electrical conductivity, malleability, and a metallic silver or yellow luster.

Metallic crystal. A crystal formed from the metals in the Periodic Table. Metallic crystals are malleable, ductile, and conduct electricity.

Metalloid. An element that has properties of both metals and nonmetals.

Metastable. A term describing a physical situation in which a material is stable unless disturbed.

Metric base unit. One of the seven basic units of measurement in the metric system. Complex units are combinations of base units.

Metric prefix. A symbol used with a metric base unit to represent a specific exponential value.

Michaelis-Menton equation. The rate equation that applies to enzyme-catalyzed reactions.

Mirror image. A description of stereoisomers in which the structure of one isomer is the reflection of the other in a mirror.

Molality (m). A concentration unit defined as the number of moles of solute dissolved in 1 kilogram of solvent.

Molarity (M). A concentration unit defined as the number of moles of solute in 1 liter of solution.

Molar mass. The sum of the gram-atomic masses of all the atoms in a chemical formula. Also called gram-molar mass; molecular mass.

Molar volume. The volume of 1 mole of gas, usually at standard temperature and pressure (STP).

Mole (mol). The quantity of any substance that contains 6.02×10^{23} units of that substance.

Mole fraction (X). A concentration unit defined as the number of moles of solute divided by the total number of moles in a solution.

Molecular crystals. A crystal formed from molecules. The attractive forces that hold molecular crystals together are London forces, dipole-dipole attractions, hydrogen-bonding, or a combination of these.

Molecular formula. The formula for a molecular or covalent substance, showing all of the atoms that comprise the molecular unit. A molecular formula may be simplified to an empirical formula if all of the subscripts can be divided by a common number.

Molecular mass. *See* **Molar mass.**

Molecular orbital. An orbital created by the pairing of electrons from different atoms. A molecular orbital encircles the atoms that are bonded together.

Molecule. A group of atoms bound together by covalent bonds.

Molten salt. A solid salt heated to the temperature at which it becomes a liquid; also called a fused salt.

Monochromatic. A term describing light that has a single wavelength.

Monodentate ligand. A Lewis base that donates one pair of electrons.

Monomer. One of the individual repeating units of a polymer.

Natural abundance. The percentage of an isotope of an element that is found in nature.

Network crystal. *See* **Covalent crystal.**

Neutralization reactions. A chemical reaction of an acid with a base.

Neutron. One of three particles, along with the electron and proton, that make up an atom. A neutron has no charge but has a mass approximately equal to that of a proton. Neutrons and protons make up the bulk of the atomic mass.

Noble gas. An element in Group 18 in the Periodic Table. Noble gases are unusually stable elements, and all have ns^2, np^6 valence electrons.

Nonbonding electron pair. A pair of electrons in a Lewis structure that is not shared with any other atoms.

Nonelectrolyte. A substance that does not dissociate at all in solution.

Nonmetal. An element that is not metallic. Nonmetals do not conduct electricity well and do not have a shiny metallic luster. They are located in the upper right portion of the Periodic Table.

Nonpolar. A term describing a bond or molecule that has its charge distributed evenly. Only diatomic elements from truly nonpolar bonds. Symmetrical molecules are nonpolar.

Normal. A term describing an organic compound in which all carbon atoms are arranged in one straight chain.

Normality (N). The number of equivalents of a substance dissolved in 1 liter of solution.

Nuclear charge. The number of positive charges in the nucleus of an atom. This is the same as the number of protons in the nucleus (Z) and is also the atomic number.

Nuclear fission. A radioactive decay process initiated by the absorption of a neutron; the result is that a large nucleus divides roughly in half.

Nuclear fusion. The combination of two nuclei to form a new atom.

Nuclear mass. The total mass of the nucleus; the sum of the masses of the protons and neutrons in the nucleus. Since electrons have virtually no mass, it is also the atomic mass (A) of an isotope.

Nuclear reactor. A device that uses a nuclear reaction to create heat energy for the purpose of generating electricity.

Nucleon. Either a proton or a neutron, both of which are fundamental particles of the nucleus.

Nucleus. The center of an atom, which contains the protons and neutrons. The nucleus is extremely dense and comprises a very small fraction of the atom's volume; the rest of the atom is empty space.

Octahedron. A geometric structure of six atoms covalently bound to a central atom. Each atom is 90° from any other.

Octet rule. A simple but effective rule stating that covalent molecules tend to have octets of electrons around each atom in their structures. These octets simulate the electronic configurations of the noble gases.

Open system. A system in which matter and energy can be exchanged with the surroundings. *See also* **Closed system.**

Optical isomer. A stereoisomer that rotates polarized light.

Optical path length (*b*). The thickness of a sample in a spectroscopic experiment.

Orbital. A region of space that may be occupied by, at most, two electrons. The shape of an orbital is defined by the sublevel it is in, and its orientation depends on its assigned quantum number, m_ℓ. Every orbital in a given sublevel must be filled with one electron before a second electron may fill that orbital.

Orbital diagram. A diagram in which boxes represent individual valence orbitals. Electrons are represented as arrows to show the spins of the electrons in each orbital.

Order of reaction. The sum of the exponents in a rate law.

Organic acid. An acid containing carbon and the −COOH functional group.

Organic compound. A compound composed of carbon and usually hydrogen.

Osmotic pressure. The pressure needed to stop the migration of solute through a semipermeable membrane.

Oxidation. The loss of electrons; also, the increase in oxidation number.

Oxidation number. A number assigned to atoms in compounds by assuming that all bonds are ionic.

Oxidizing agent. A substance that causes another to be oxidized; also, a substance that is reduced.

Oxoacid. An acid that contains hydrogen, oxygen, and another element in its formula, excluding organic acids.

Partial pressure. The pressure of a single gas in a mixture of gases.

Parts per billion (ppb). A unit of measurement similar in concept to percent, obtained by multiplying a fraction by 1 billion (10^6).

Parts per million (ppm). A unit of measurement similar in concept to percent, obtained by multiplying a fraction by 1 million (10^6).

Pauli exclusion principle. The requirement that no two electrons in an atom have the same set of four quantum numbers, n, ℓ, m_ℓ, and m_s.

Peptide bond. The bond formed from the reaction of an amine and an organic acid.

Percent (%). A unit of measurement defined as parts per hundred, obtained by multiplying a fraction by 100.

Period. A row in the Periodic Table.

Periodic Table. The table in which the elements are arranged in an orderly fashion that shows the relationships of their chemical and physical characteristics.

pH. The negative logarithm of the hydrogen ion concentration in a solution; pH = $-\log H^+$.

Phase diagram. A graph showing the relationship between temperature and pressure and the conversion of matter among its three states: solid, liquid, and gas.

pH indicator. A weak acid or a weak base whose conjugate acid and conjugate base have different colors. An indicator changes color at the end point of a titration.

pH meter. An electronic device used with pH electrodes to measure the pH of a solution.

pH paper. Paper with a pH indicator absorbed on it so that it changes color depending on the pH of the solution; pH paper is used to estimate pH.

Phosphorescence. A property of some atoms and molecules that allows them to absorb photons of light and reemit them seconds to hours later. The emitted light always has a lower energy, longer wavelength than the absorbed light.

Pi bond. A bond made from the sideways overlap of two *p* orbitals. The electron density of a pi bond lies outside the internuclear axis.

Pipet. A narrow tube calibrated for precise measurement of small volumes of liquids.

pK_w. The negative logarithm of the autopyrolysis constant of water, equal to 14.00.

Planar triangle. A geometric structure of four atoms, three bonded to a central atom, with 120° angles between the atoms, which are all in the same plane.

Pneumatic trough. An experimental setup for collecting gases by the displacement of water.

pOH. The negative logarithm of the hydroxide ion concentration in a solution; $pOH = -\log OH^-$.

Polar. A term describing the property of a covalent bond or molecule of having one end more positive than the other.

Polarizability. The tendency for an atom's electron cloud to be deformed so that polarity is created.

Polyatomic ion. An ion composed of more than one atom, with the atoms covalently bonded together. A polyatomic ion acts as a unit in most chemical reactions.

Polymer. A long-chain organic molecule with repeating units.

Polyprotic acid. An acid with two or more ionizable hydrogen atoms in its formula.

Positron. A positive electron.

Potential energy. The energy of matter, which may, under appropriate conditions, be converted into work.

Precipitate. 1. (v.) To cause the formation of a solid by a chemical reaction. 2. (n.) The solid formed as a result of a chemical reaction.

Precision. The degree of closeness of repeated measurements in a group to each other.

Pressure. The force per unit area; gas molecules exert this force by collisions with the container walls.

Principal quantum number (*n*). The quantum number that specifies the energy level of the atom in which an electron is located; *n* may be any integer from 1 to infinity.

Product. The result of a chemical reaction. Products are placed on the right-hand side of a chemical equation.

Proton. One of the three particles, along with the electron and the neutron, that make up the atom. The proton has a positive charge, equal in magnitude (but with the opposite sign) to the charge of the electron. The number of protons is equal to the atomic number (*Z*) of an element. Protons and neutrons make up the bulk of the mass of an atom.

Q. *See* **Reaction quotient.**

Qualitative. A term referring to a

description of a physical or chemical property without the use of numbers or equations.

Qualitative analysis. A logical sequence of experiments and observations used to determine the composition of a sample.

Quantitative. A term referring to a description of a physical or chemical property using numbers or equations.

Quantum number. One of four numbers—n, ℓ, m_l, m_s—used in the wave-mechanical model of the atom to describe an electron in an atom.

Racemic mixture. An equal molar mixture of L and D optical isomers.

Radioactive disintegration series. A sequence of radioactive disintegrations, in more than one step, from a heavy isotope to a lighter, stable isotope.

Radioactivity. The property that some unstable nuclei have of decaying spontaneously with the emission of a small particle and/or energy.

Radioisotope. A radioactive isotope of an element.

Random error. *See* **Indeterminate error.**

Randomness. A qualitative description of the disorder of the molecules in any sample.

Raoult's law. The law that expresses the relationship between the vapor pressure of a solution and the mole fraction of solute in that solution.

Rate constant (k). A constant in the direct relationship between the amount of reacting substance and the rate of the reaction.

Rate-determining step. The slowest reaction in a mechanism, which limits the overall rate of reaction; also called rate-limiting step.

Rate law. The mathematical relationship between reactant concentrations and reaction rate. The general form of a rate law is Rate = $k[A]^x[B]^y[C]^z$.

Rate-limiting step. *See* **Rate-determining step.**

Reactant. One of the starting materials in a chemical reaction. The reactants are placed on the left side of a chemical equation.

Reaction mechanism. The detailed series of elementary reactions that add up to the overall reaction. *See also* **Elementary reaction.**

Reaction profile. A plot of the potential energy of molecules as they collide in a reaction, illustrating the nature of the activation energy.

Reaction quotient (Q). The value of the ratio of the equilibrium law when a chemical system is not in a state of equilibrium. The value of Q in comparison to that of K indicates the direction of the reaction.

Reaction rate. The velocity, in moles per liter per second, at which reactants are converted into products in a chemical reaction.

Reagent blank. A solution used to set the zero point of a spectrophotometer.

Real gas. A gas in which there are attractive forces between the molecules, and the molecules have a finite volume.

Redox. A word coined to combine

the terms *reduction* and *oxidation*. It indicates that reduction and oxidation always occur together.

Reducing agent. A substance that causes another substance to be reduced; also, a substance that is oxidized.

Reduction. The gain of electrons; also, the decrease in oxidation number.

Reference electrode. An electrode in a galvanic cell, which has a constant potential since the concentrations of all reactants are kept constant.

Relative mass. A term indicating that the mass units in the Periodic Table are relative, without units, as compared to the mass of the carbon-12 isotope.

Relative uncertainty. The absolute uncertainty of a measurement divided by the value of the measurement.

Resonance structure. A Lewis structure that can be drawn in more than one equally probable way. The actual structure is a mixture of all possible resonance structures.

Reversible process. A chemical or physical process in which a substance can be changed from one state to another and then back to the original state.

S°. The standard entropy of 1 mole of a substance.

Saturated. A term describing an organic compound in which all carbon-carbon bonds are single bonds. Saturated compounds have the maximum number of hydrogen atoms; that is, they are saturated with hydrogen.

Saturated solution. A solution in which the maximum amount of solute is dissolved.

Scientific notation. A method of writing numbers in which the significant figures are numbers from 1 to 10, which are multiplied by 10 raised to the appropriate power to indicate the position of the decimal point.

Second law of thermodynamics. The law that states that in all physical and chemical processes the overall entropy of the universe must increase.

Second-order reaction. A reaction with a rate law having exponents adding up to 2. The rate law is either Rate = $k[A][B]$ or Rate = $[A]^2$.

Semipermeable membrane. A thin, solid material through which certain molecules can diffuse while others cannot; may be visualized as a barrier with small holes that allow only molecules of a certain size to pass through.

Sequestering agent. *See* **Lewis base.**

Shell. The old term for the principal energy level of an electron; the region in space where electrons are located around the nucleus of the atom. Energy levels are numbered starting with the energy level closest to the nucleus. The number of the principal energy level is also known as the principal quantum number.

Sigma bond. A covalent bond formed by the sharing of a pair of electrons. The electron pair is located along the internuclear axis between the two atoms that share it.

Significant figures. All the digits in a measurement except preceding zeros.

Simple cubic. A term describing a cubic structure with one atom in each of the eight corners of a unit cell.

Solid. A state of matter characterized by a rigid structure that retains its shape without a container.

Solubility. A property of a solute that refers to the maximum amount of that solute that can be dissolved in a solvent. This term can be a qualitative or quantitative description of a solute.

Solute. The substance—gas, liquid, or solid—that is dissolved in a solvent.

Solution. A uniform mixture of chemicals. In a solution it is impossible to distinguish separate solute and solvent particles.

Solvent. Typically, the liquid phase in which a gas, another liquid, or a solid is dissolved. In a mixture of two or more liquids the solvent is the liquid present in the largest amount.

***sp* Hybrid orbital.** An orbital constructed from an *s* and a *p* orbital; the resulting orbitals all have the same energy. Structures related to the *sp* hybrid orbital are linear.

sp^2 Hybrid orbital. An orbital constructed from an *s* and two *p* orbitals; the resulting orbitals all have the same energy. Structures related to the sp^2 hybrid orbital are triangular planar.

sp^3 Hybrid orbital. An orbital constructed from an *s* and three *p* orbitals. The resulting orbitals all have the same energy. Structures related to the sp^3 hybrid orbital are tetrahedral.

Specific heat. An intrinsic property of matter that describes the quantity of heat energy needed to raise the temperature of 1 gram of substance by 1 Celsius degree.

Spectrophotometer. An instrument for determining the amount of light absorbed by a sample.

Spin quantum number (m_s). The quantum number that specifies the spin of an electron as either $+\frac{1}{2}$ or $-\frac{1}{2}$. Two electrons in the same orbital must have opposite spins.

Spontaneous reaction. Any reaction that occurs without outside assistance; quantitatively, any reaction that has an equilibrium constant greater than 1.

Standard cell voltage, (E°_{cell}). The voltage of a galvanic cell when the system is at standard state; also the combination of two standard reduction potentials as

$$E^\circ_{cell} = E^\circ_{reduction} - E^\circ_{oxidation}$$

Standard reduction potential. The potential (voltage) of a reduction half-reaction at standard state.

Standard state. Defined temperature, pressure, and concentrations. In electrochemistry the standard state is 1.00 atmosphere pressure, 298 K or 25°C, and 1.00 molar concentrations for all soluble compounds. Solids and pure liquids are defined as 1.00 molar.

State function. A variable whose value depends only on the initial and final states of the system. State functions are ΔH, ΔS, ΔG, and ΔE.

Steady-state assumption. The assumption that, in evaluating rate constants for elementary reactions, the concentrations of intermediates may be mathematically eliminated by assuming that all prior fast steps are in chemical equilibrium.

Stereoisomers. Compounds that have the same formula and same bonding but differ in the geometric arrangement of the atoms.

Stereospecific. A term describing a chemical reaction that produces only one stereoisomer.

Stoichiometry. The quantitative relationships between chemical substances in a chemical equation.

Stopcock. The valve on the end of a buret.

Strong acid. An acid that dissociates completely when dissolved in water.

Strong base. A base that dissociates completely when dissolved in water.

Structural formula. A formula that shows the actual arrangement of atoms within a molecule and the bonds between the atoms.

Structural isomers. Compounds with the same formula but with the atoms bonded in different arrangements.

Sublevel. A subdivision of an energy level. Electrons in each principal energy level are localized in sublevels. Each sublevel has a distinct shape associated with it. Sublevels are numbered from zero up to 1 less than the number of the principal energy level. These sublevel numbers are the azimuthal quantum numbers, ℓ. Sublevels are also designated by the letters *s*, *p*, *d*, *f*.

Subscript. A number placed to the right of, and slightly below, the symbol for an element to represent the number of times that atom is present in the formula unit.

Substrate(s). The reactant(s) in an enzyme-catalyzed reaction.

Supercritical fluid. A gas at a temperature and pressure above the critical point. A supercritical fluid has properties of both a gas and a liquid.

Supersaturated solution. A metastable solution in which more than the maximum amount of solute is dissolved.

Supercooling. The property of some materials of being able to be cooled to temperatures below their melting points without solidifying.

Supernatant. The liquid remaining above a solid after centrifugation.

Surface tension. The added attractive force per molecule at the surface of a liquid. Surface tension causes liquids to assume shapes that minimize surface area.

Surfactant. A substance that lowers the surface tension of liquids.

Surroundings. All parts of the universe not included in the system being studied.

Symmetrical. A term describing a geometrical property whereby a structure may be rotated by some angle less than 360° and after rotation has the same configuration as before.

System. The portion of the universe that is under study.

TC ("To Contain"). A label on glassware indicating that the item is calibrated to contain the indicated volume.

TD ("To Deliver"). A label on

glassware indicating that the calibration is based on the volume delivered.

Teflon. The addition polymer of $CF_2{=}CF_2$ with extraordinary nonstick properties.

Tetrahedron. A geometric structure with four atoms bound to a central atom by covalent bonds. Each bond is equidistant from any other with a bond angle of 109°.

Thermodynamics. The study of energy changes in chemical and physical processes.

Titration: An experimental procedure for reacting two solutions in order to determine the quantity or concentration of one of the solutions.

Titration curve: A plot of pH versus the volume of titrant added to a sample.

Tracer. A radioactive element used to detect the movement of materials in a complex system.

Trans isomer. An isomer with substituents on opposite sides of a double bond. *See also* **Cis isomer.**

Transition element. An element having a *d* electron as the differentiating electron in its electronic configuration.

Transition-state theory. The reaction-rate theory that details the events and energy changes that occur as two molecules collide.

Trans-uranium element. Any of the 18 elements from atomic number 93 to 112.

Triangular bipyramid. A geometric structure with five atoms covalently bound to a central atom. Three atoms in the equatorial position are 120° from each other. Two additional atoms in the axial positions are 90° from the equatorial atoms.

Triple point. The temperature and pressure at which all three states of matter—solid, liquid, and gas—are in equilibrium.

Unit cell. The fundamental building block of crystals. An entire crystal is formed by repetitive stacking of the unit cells.

Universal gas constant (*R*). The constant needed to relate the temperature, pressure, volume, and moles of gas in the ideal gas law, $PV = nRT$.

Universe. The entirety of all matter and space that exist.

Unsaturated. A term describing any organic compound that contains one or more double or triple bonds in its structure.

Unsaturated solution. A solution in which the solute concentration is less than the maximum amount possible.

Vacuum distillation. The laboratory technique of vaporizing and condensing a liquid for the purpose of purification. Vacuum distillation is used to reduce the boiling points of heat-sensitive compounds.

Valence electrons. The outermost *s* and *p* electrons in an atom. The number and the arrangement of valence electrons define the chemical and physical properties of the atom.

Valence shell electron-pair repulsion (VSEPR) theory. A method of evaluating molecular structures by relating the number of bonding and nonbonding electron pairs on an atom to the geometrical structure of the atom.

Van der Waal's forces. *See* **London forces.**

Vapor pressure. The pressure developed by a liquid or solid in a closed container at a constant temperature.

Viscosity. The ability of a fluid to flow. The more easily a fluid flows, the lower is its viscosity.

Vital force theory. The theory now discredited, that all organic molecules must be formed in living matter.

Volume-volume fraction (vol/vol). A concentration unit defined as the volume of one liquid solute divided by the total volumes of the liquids mixed to prepare a solution.

Volumetric flask. A flask calibrated to contain a precise volume of liquid.

Weak acid. An acid that dissociates slightly when dissolved in water.

Weak electrolyte. A substance that partially dissociates into ions in solution.

Weight. The force developed by the gravitational attraction of two masses.

Weighted average. An average that depends on the abundance of the objects being averaged.

Wetting. The spreading of a liquid on a surface that occurs because the adhesive forces overcome the cohesive forces in the liquid.

X ray. The high-energy electromagnetic radiation emitted in nuclear decay events.

Zero-order reaction. A reaction in which the rate is independent of reactant concentration. The rate law is Rate = k.

APPENDICES

1. ELECTRONIC CONFIGURATIONS OF THE ELEMENTS

H	$1s^1$
He	$1s^2$
Li	$1s^2\ 2s^1$
Be	$1s^2\ 2s^2$
B	$1s^2\ 2s^2\ 2p^1$
C	$1s^2\ 2s^2\ 2p^2$
N	$1s^2\ 2s^2\ 2p^3$
O	$1s^2\ 2s^2\ 2p^4$
F	$1s^2\ 2s^2\ 2p^5$
Ne	$1s^2\ 2s^2\ 2p^6$
Na	$1s^2\ 2s^2\ 2p^6\ 3s^1$
Mg	$1s^2\ 2s^2\ 2p^6\ 3s^2$
Al	$1s^2\ 2s^2\ 2p^6\ 3s^2\ 3p^1$
Si	$1s^2\ 2s^2\ 2p^6\ 3s^2\ 3p^2$
P	$1s^2\ 2s^2\ 2p^6\ 3s^2\ 3p^3$
S	$1s^2\ 2s^2\ 2p^6\ 3s^2\ 3p^4$
Cl	$1s^2\ 2s^2\ 2p^6\ 3s^2\ 3p^5$
Ar	$1s^2\ 2s^2\ 2p^6\ 3s^2\ 3p^6$
K	$1s^2\ 2s^2\ 2p^6\ 3s^2\ 3p^6\ 4s^1$
Ca	$1s^2\ 2s^2\ 2p^6\ 3s^2\ 3p^6\ 4s^2$
Sc	$1s^2\ 2s^2\ 2p^6\ 3s^2\ 3p^6\ 4s^2\ 3d^1$
Ti	$1s^2\ 2s^2\ 2p^6\ 3s^2\ 3p^6\ 4s^2\ 3d^2$
V	$1s^2\ 2s^2\ 2p^6\ 3s^2\ 3p^6\ 4s^2\ 3d^3$
Cr	$1s^2\ 2s^2\ 2p^6\ 3s^2\ 3p^6\ 4s^1\ 3d^5$
Mn	$1s^2\ 2s^2\ 2p^6\ 3s^2\ 3p^6\ 4s^2\ 3d^5$
Fe	$1s^2\ 2s^2\ 2p^6\ 3s^2\ 3p^6\ 4s^2\ 3d^6$
Co	$1s^2\ 2s^2\ 2p^6\ 3s^2\ 3p^6\ 4s^2\ 3d^7$
Ni	$1s^2\ 2s^2\ 2p^6\ 3s^2\ 3p^6\ 4s^2\ 3d^8$
Cu	$1s^2\ 2s^2\ 2p^6\ 3s^2\ 3p^6\ 4s^1\ 3d^{10}$
Zn	$1s^2\ 2s^2\ 2p^6\ 3s^2\ 3p^6\ 4s^2\ 3d^{10}$
Ga	$1s^2\ 2s^2\ 2p^6\ 3s^2\ 3p^6\ 4s^2\ 3d^{10}\ 4p^1$
Ge	$1s^2\ 2s^2\ 2p^6\ 3s^2\ 3p^6\ 4s^2\ 3d^{10}\ 4p^2$
As	$1s^2\ 2s^2\ 2p^6\ 3s^2\ 3p^6\ 4s^2\ 3d^{10}\ 4p^3$
Se	$1s2\ 2s^2\ 2p^6\ 3s^2\ 3p^6\ 4s^2\ 3d^{10}\ 4p^4$
Br	$1s^2\ 2s^2\ 2p^6\ 3s^2\ 3p^6\ 4s^2\ 3d^{10}\ 4p^5$
Kr	$1s^2\ 2s^2\ 2p^6\ 3s^2\ 3p^6\ 4s^2\ 3d^{10}\ 4p^6$
Rb	$1s^2\ 2s^2\ 2p^6\ 3s^2\ 3p^6\ 4s^2\ 3d^{10}\ 4p^6\ 5s^1$
Sr	$1s^2\ 2s^2\ 2p^6\ 3s^2\ 3p^6\ 4s^2\ 3d^{10}\ 4p^6\ 5s^2$
Y	$1s^2\ 2s^2\ 2p^6\ 3s^2\ 3p^6\ 4s^2\ 3d^{10}\ 4p^6\ 5s^2\ 4d^1$
Zr	$1s^2\ 2s^2\ 2p^6\ 3s^2\ 3p^6\ 4s^2\ 3d^{10}\ 4p^6\ 5s^2\ 4d^2$
Nb	$1s^2\ 2s^2\ 2p^6\ 3s^2\ 3p^6\ 4s^2\ 3d^{10}\ 4p^6\ 5s^1\ 4d^4$
Mo	$1s^2\ 2s^2\ 2p^6\ 3s^2\ 3p^6\ 4s^2\ 3d^{10}\ 4p^6\ 5s^1\ 4d^5$
Tc	$1s^2\ 2s^2\ 2p^6\ 3s^2\ 3p^6\ 4s^2\ 3d^{10}\ 4p^6\ 5s^2\ 4d^5$
Ru	$1s^2\ 2s^2\ 2p^6\ 3s^2\ 3p^6\ 4s^2\ 3d^{10}\ 4p^6\ 5s^1\ 4d^7$
Rh	$1s^2\ 2s^2\ 2p^6\ 3s^2\ 3p^6\ 4s^2\ 3d^{10}\ 4p^6\ 5s^1\ 4d^8$
Pd	$1s^2\ 2s^2\ 2p^6\ 3s^2\ 3p^6\ 4s^2\ 3d^{10}\ 4p^6\ 4d^{10}$
Ag	$1s^2\ 2s^2\ 2p^6\ 3s^2\ 3p^6\ 4s^2\ 3d^{10}\ 4p^6\ 5s^1\ 4d^{10}$
Cd	$1s^2\ 2s^2\ 2p^6\ 3s^2\ 3p^6\ 4s^2\ 3d^{10}\ 4p^6\ 5s^2\ 4d^{10}$
In	$1s^2\ 2s^2\ 2p^6\ 3s^2\ 3p^6\ 4s^2\ 3d^{10}\ 4p^6\ 5s^2\ 4d^{10}\ 5p^1$

Sn	$1s^2$	$2s^2$	$2p^6$	$3s^2$	$3p^6$	$4s^2$	$3d^{10}$	$4p^6$	$5s^2$	$4d^{10}$	$5p^2$							
Sb	$1s^2$	$2s^2$	$2p^6$	$3s^2$	$3p^6$	$4s^2$	$3d^{10}$	$4p^6$	$5s^2$	$4d^{10}$	$5p^3$							
Te	$1s^2$	$2s^2$	$2p^6$	$3s^2$	$3p^6$	$4s^2$	$3d^{10}$	$4p^6$	$5s^2$	$4d^{10}$	$5p^4$							
I	$1s^2$	$2s^2$	$2p^6$	$3s^2$	$3p^6$	$4s^2$	$3d^{10}$	$4p^6$	$5s^2$	$4d^{10}$	$5p^5$							
Xe	$1s^2$	$2s^2$	$2p^6$	$3s^2$	$3p^6$	$4s^2$	$3d^{10}$	$4p^6$	$5s^2$	$4d^{10}$	$5p^6$							
Cs	$1s^2$	$2s^2$	$2p^6$	$3s^2$	$3p^6$	$4s^2$	$3d^{10}$	$4p^6$	$5s^2$	$4d^{10}$	$5p^6$	$6s^1$						
Ba	$1s^2$	$2s^2$	$2p^6$	$3s^2$	$3p^6$	$4s^2$	$3d^{10}$	$4p^6$	$5s^2$	$4d^{10}$	$5p^6$	$6s^2$						
La	$1s^2$	$2s^2$	$2p^6$	$3s^2$	$3p^6$	$4s^2$	$3d^{10}$	$4p^6$	$5s^2$	$4d^{10}$	$5p^6$	$6s^2$	$4f^1$					
Ce	$1s^2$	$2s^2$	$2p^6$	$3s^2$	$3p^6$	$4s^2$	$3d^{10}$	$4p^6$	$5s^2$	$4d^{10}$	$5p^6$	$6s^2$	$4f^2$					
Pr	$1s^2$	$2s^2$	$2p^6$	$3s^2$	$3p^6$	$4s^2$	$3d^{10}$	$4p^6$	$5s^2$	$4d^{10}$	$5p^6$	$6s^2$	$4f^3$					
Nd	$1s^2$	$2s^2$	$2p^6$	$3s^2$	$3p^6$	$4s^2$	$3d^{10}$	$4p^6$	$5s^2$	$4d^{10}$	$5p^6$	$6s^2$	$4f^4$					
Pm	$1s^2$	$2s^2$	$2p^6$	$3s^2$	$3p^6$	$4s^2$	$3d^{10}$	$4p^6$	$5s^2$	$4d^{10}$	$5p^6$	$6s^2$	$4f^5$					
Sm	$1s^2$	$2s^2$	$2p^6$	$3s^2$	$3p^6$	$4s^2$	$3d^{10}$	$4p^6$	$5s^2$	$4d^{10}$	$5p^6$	$6s^2$	$4f^6$					
Eu	$1s^2$	$2s^2$	$2p^6$	$3s^2$	$3p^6$	$4s^2$	$3d^{10}$	$4p^6$	$5s^2$	$4d^{10}$	$5p^6$	$6s^2$	$4f^7$					
Gd	$1s^2$	$2s^2$	$2p^6$	$3s^2$	$3p^6$	$4s^2$	$3d^{10}$	$4p^6$	$5s^2$	$4d^{10}$	$5p^6$	$6s^2$	$4f^7$	$5d^1$				
Tb	$1s^2$	$2s^2$	$2p^6$	$3s^2$	$3p^6$	$4s^2$	$3d^{10}$	$4p^6$	$5s^2$	$4d^{10}$	$5p^6$	$6s^2$	$4f^9$					
Dy	$1s^2$	$2s^2$	$2p^6$	$3s^2$	$3p^6$	$4s^2$	$3d^{10}$	$4p^6$	$5s^2$	$4d^{10}$	$5p^6$	$6s^2$	$4f^{10}$					
Ho	$1s^2$	$2s^2$	$2p^6$	$3s^2$	$3p^6$	$4s^2$	$3d^{10}$	$4p^6$	$5s^2$	$4d^{10}$	$5p^6$	$6s^2$	$4f^{11}$					
Er	$1s^2$	$2s^2$	$2p^6$	$3s^2$	$3p^6$	$4s^2$	$3d^{10}$	$4p^6$	$5s^2$	$4d^{10}$	$5p^6$	$6s^2$	$4f^{12}$					
Tm	$1s^2$	$2s^2$	$2p^6$	$3s^2$	$3p^6$	$4s^2$	$3d^{10}$	$4p^6$	$5s^2$	$4d^{10}$	$5p^6$	$6s^2$	$4f^{13}$					
Yb	1s2	$2s^2$	$2p^6$	$3s^2$	$3p^6$	$4s^2$	$3d^{10}$	$4p^6$	$5s^2$	$4d^{10}$	$5p^6$	$6s^2$	$4f^{14}$					
Lu	$1s^2$	$2s^2$	$2p^6$	$3s^2$	$3p^6$	$4s^2$	$3d^{10}$	$4p^6$	$5s^2$	$4d^{10}$	$5p^6$	$6s^2$	$4f^{14}$	$5d^1$				
Hf	$1s^2$	$2s^2$	$2p^6$	$3s^2$	$3p^6$	$4s^2$	$3d^{10}$	$4p^6$	$5s^2$	$4d^{10}$	$5p^6$	$6s^2$	$4f^{14}$	$5d^2$				
Ta	$1s^2$	$2s^2$	$2p^6$	$3s^2$	$3p^6$	$4s^2$	$3d^{10}$	$4p^6$	$5s^2$	$4d^{10}$	$5p^6$	$6s^2$	$4f^{14}$	$5d^3$				
W	$1s^2$	$2s^2$	$2p^6$	$3s^2$	$3p^6$	$4s^2$	$3d^{10}$	$4p^6$	$5s^2$	$4d^{10}$	$5p^6$	$6s^2$	$4f^{14}$	$5d^4$				
Re	$1s^2$	$2s^2$	$2p^6$	$3s^2$	$3p^6$	$4s^2$	$3d^{10}$	$4p^6$	$5s^2$	$4d^{10}$	$5p^6$	$6s^2$	$4f^{14}$	$5d^5$				
Os	$1s^2$	$2s^2$	$2p^6$	$3s^2$	$3p^6$	$4s^2$	$3d^{10}$	$4p^6$	$5s^2$	$4d^{10}$	$5p^6$	$6s^2$	$4f^{14}$	$5d^6$				
Ir	$1s^2$	$2s^2$	$2p^6$	$3s^2$	$3p^6$	$4s^2$	$3d^{10}$	$4p^6$	$5s^2$	$4d^{10}$	$5p^6$	$6s^2$	$4f^{14}$	$5d^7$				
Pt	$1s^2$	$2s^2$	$2p^6$	$3s^2$	$3p^6$	$4s^2$	$3d^{10}$	$4p^6$	$5s^2$	$4d^{10}$	$5p^6$	$6s^1$	$4f^{14}$	$5d^9$				
Au	$1s^2$	$2s^2$	$2p^6$	$3s^2$	$3p^6$	$4s^2$	$3d^{10}$	$4p^6$	$5s^2$	$4d^{10}$	$5p^6$	$6s^1$	$4f^{14}$	$5d^{10}$				
Hg	$1s^2$	$2s^2$	$2p^6$	$3s^2$	$3p^6$	$4s^2$	$3d^{10}$	$4p^6$	$5s^2$	$4d^{10}$	$5p^6$	$6s^2$	$4f^{14}$	$5d^{10}$				
Tl	$1s^2$	$2s^2$	$2p^6$	$3s^2$	$3p^6$	$4s^2$	$3d^{10}$	$4p^6$	$5s^2$	$4d^{10}$	$5p^6$	$6s^2$	$4f^{14}$	$5d^{10}$	$6p^1$			
Pb	$1s^2$	$2s^2$	$2p^6$	$3s^2$	$3p^6$	$4s^2$	$3d^{10}$	$4p^6$	$5s^2$	$4d^{10}$	$5p^6$	$6s^2$	$4f^{14}$	$5d^{10}$	$6p^2$			
Bi	$1s^2$	$2s^2$	$2p^6$	$3s^2$	$3p^6$	$4s^2$	$3d^{10}$	$4p^6$	$5s^2$	$4d^{10}$	$5p^6$	$6s^2$	$4f^{14}$	$5d^{10}$	$6p^3$			
Po	$1s^2$	$2s^2$	$2p^6$	$3s^2$	$3p^6$	$4s^2$	$3d^{10}$	$4p^6$	$5s^2$	$4d^{10}$	$5p^6$	$6s^2$	$4f^{14}$	$5d^{10}$	$6p^4$			
At	$1s^2$	$2s^2$	$2p^6$	$3s^2$	$3p^6$	$4s^2$	$3d^{10}$	$4p^6$	$5s^2$	$4d^{10}$	$5p^6$	$6s^2$	$4f^{14}$	$5d^{10}$	$6p^5$			
Rn	$1s^2$	$2s^2$	$2p^6$	$3s^2$	$3p^6$	$4s^2$	$3d^{10}$	$4p^6$	$5s^2$	$4d^{10}$	$5p^6$	$6s^2$	$4f^{14}$	$5d^{10}$	$6p^6$			
Fr	$1s^2$	$2s^2$	$2p^6$	$3s^2$	$3p^6$	$4s^2$	$3d^{10}$	$4p^6$	$5s^2$	$4d^{10}$	$5p^6$	$6s^2$	$4f^{14}$	$5d^{10}$	$6p^6$	$7s^1$		
Ra	$1s^2$	$2s^2$	$2p^6$	$3s^2$	$3p^6$	$4s^2$	$3d^{10}$	$4p^6$	$5s^2$	$4d^{10}$	$5p^6$	$6s^2$	$4f^{14}$	$5d^{10}$	$6p^6$	$7s^2$		
Ac	$1s^2$	$2s^2$	$2p^6$	$3s^2$	$3p^6$	$4s^2$	$3d^{10}$	$4p^6$	$5s^2$	$4d^{10}$	$5p^6$	$6s^2$	$4f^{14}$	$5d^{10}$	$6p^6$	$7s^2$	$6d^1$	
Th	$1s^2$	$2s^2$	$2p^6$	$3s^2$	$3p^6$	$4s^2$	$3d^{10}$	$4p^6$	$5s^2$	$4d^{10}$	$5p^6$	$6s^2$	$4f^{14}$	$5d^{10}$	$6p^6$	$7s^2$	$6d^2$	
Pa	$1s^2$	$2s^2$	$2p^6$	$3s^2$	$3p^6$	$4s^2$	$3d^{10}$	$4p^6$	$5s^2$	$4d^{10}$	$5p^6$	$6s^2$	$4f^{14}$	$5d^{10}$	$6p^6$	$7s^2$	$5f^2$	$6d^1$
U	$1s^2$	$2s^2$	$2p^6$	$3s^2$	$3p^6$	$4s^2$	$3d^{10}$	$4p^6$	$5s^2$	$4d^{10}$	$5p^6$	$6s^2$	$4f^{14}$	$5d^{10}$	$6p^6$	$7s^2$	$5f^3$	$6d^1$
Np	$1s^2$	$2s^2$	$2p^6$	$3s^2$	$3p^6$	$4s^2$	$3d^{10}$	$4p^6$	$5s^2$	$4d^{10}$	$5p^6$	$6s^2$	$4f^{14}$	$5d^{10}$	$6p^6$	$7s^2$	$5f^4$	$6d^1$
Pu	$1s^2$	$2s^2$	$2p^6$	$3s^2$	$3p^6$	$4s^2$	$3d^{10}$	$4p^6$	$5s^2$	$4d^{10}$	$5p^6$	$6s^2$	$4f^{14}$	$5d^{10}$	$6p^6$	$7s^2$	$5f^6$	
Am	$1s^2$	$2s^2$	$2p^6$	$3s^2$	$3p^6$	$4s^2$	$3d^{10}$	$4p^6$	$5s^2$	$4d^{10}$	$5p^6$	$6s^2$	$4f^{14}$	$5d^{10}$	$6p^6$	$7s^2$	$5f^7$	
Cm	$1s^2$	$2s^2$	$2p^6$	$3s^2$	$3p^6$	$4s^2$	$3d^{10}$	$4p^6$	$5s^2$	$4d^{10}$	$5p^6$	$6s^2$	$4f^{14}$	$5d^{10}$	$6p^6$	$7s^2$	$5f^7$	$6d^1$
Bk	$1s^2$	$2s^2$	$2p^6$	$3s^2$	$3p^6$	$4s^2$	$3d^{10}$	$4p^6$	$5s^2$	$4d^{10}$	$5p^6$	$6s^2$	$4f^{14}$	$5d^{10}$	$6p^6$	$7s^2$	$5f^9$	
Cf	$1s^2$	$2s^2$	$2p^6$	$3s^2$	$3p^6$	$4s^2$	$3d^{10}$	$4p^6$	$5s^2$	$4d^{10}$	$5p^6$	$6s^2$	$4f^{14}$	$5d^{10}$	$6p^6$	$7s^2$	$5f^{10}$	
Es	$1s^2$	$2s^2$	$2p^6$	$3s^2$	$3p^6$	$4s^2$	$3d^{10}$	$4p^6$	$5s^2$	$4d^{10}$	$5p^6$	$6s^2$	$4f^{14}$	$5d^{10}$	$6p^6$	$7s^2$	$5f^{11}$	
Fm	$1s^2$	$2s^2$	$2p^6$	$3s^2$	$3p^6$	$4s^2$	$3d^{10}$	$4p^6$	$5s^2$	$4d^{10}$	$5p^6$	$6s^2$	$4f^{14}$	$5d^{10}$	$6p^6$	$7s^2$	$5f^{12}$	
Md	$1s^2$	$2s^2$	$2p^6$	$3s^2$	$3p^6$	$4s^2$	$3d^{10}$	$4p^6$	$5s^2$	$4d^{10}$	$5p^6$	$6s^2$	$4f^{14}$	$5d^{10}$	$6p^6$	$7s^2$	$5f^{13}$	
No	$1s^2$	$2s^2$	$2p^6$	$3s^2$	$3p^6$	$4s^2$	$3d^{10}$	$4p^6$	$5s^2$	$4d^{10}$	$5p^6$	$6s^2$	$4f^{14}$	$5d^{10}$	$6p^6$	$7s^2$	$5f^{14}$	
Lr	$1s^2$	$2s^2$	$2p^6$	$3s^2$	$3p^6$	$4s^2$	$3d^{10}$	$4p^6$	$5s^2$	$4d^{10}$	$5p^6$	$6s^2$	$4f^{14}$	$5d^{10}$	$6p^6$	$7s^2$	$5f^{14}$	$6d^1$

2. THERMODYNAMIC DATA FOR SELECTED ELEMENTS, COMPOUNDS, AND IONS (25°C)

Substance	ΔH_f° (kJ/mol^{-1})	S° (J/mol^{-1} K^{-1})	ΔG_f° (kJ/mol^{-1})
Aluminum			
Al(s)	0	28.3	0
$AlCl_3$(s)	−704	110.7	−629
Al_2O_3(s)	−1676	51.0	−1576.4
$Al_2(SO_4)_3$(S)	−3441	239	−3100
Barium			
Ba(s)	0	66.9	0
$BaCO_3$(s)	−1219	112	−1139
$BaCl_2$(s)	−860.2	125	−810.8
$Ba(OH)_2$(s)	−998.22	−8	−875.3
$Ba(NO_3)_2$(s)	−992	214	−795
$BaSO_4$(s)	−1465	132	−1353
Bromine			
Br_2(l)	0	152.2	0
Br_2(g)	+30.9	245.4	3.11
HBr(g)	−36	198.5	53.1
Calcium			
Ca(s)	0	41.4	0
$CaCO_3$(s)	−1207	92.9	−1128.8
CaF_2(s)	−741	80.3	−1166
$CaCl_2$(s)	−795.8	114	−750.2
CaO(s)	−635.5	40	−604.2
$Ca(OH)_2$(s)	−986.6	76.1	896.76
$CaSO_4$(s)	−1433	107	−1320.3
Carbon			
C(s,graphite)	0	5.69	0
C(s,diamond)	+1.88	2.4	+2.9
CCl_4(l)	−134	214.4	−65.3
CO(g)	−110	197.9	−137.3
CO_2(g)	−394	213.6	−394.4
CO_2(aq)	−413.8	117.6	−385.98
H_2CO_3(aq)	−699.65	187.4	−623.08
HCO_3^-(aq)	−691.99	91.2	−586.77
CO_3^{2-}(aq)	−677.14	−56.9	−527.81
HCN(g)	+135.1	201.7	+124.7
CN^-(ag)	+150.6	94.1	+172.4
CH_4(g)	−74.9	186.2	−50.79
C_2H_2(g)	+227	200.8	+209
C_2H_4(g)	+51.9	219.8	+68.12
C_2H_6(g)	−84.5	229.5	−32.9
C_3H_8(g)	−104	269.9	−23
C_4H_{10}(g)	−126	310.2	−17.0
C_6H_6(l)	+49.0	173.3	+124.3
CH_3OH(l)	−238	126.8	−166.2
C_2H_5OH(l)	−278	161	−174.8

Substance	ΔH_f° (kJ/mol^{-1})	S° (J/mol^{-1} K^{-1})	ΔG_f° (kJ/mol^{-1})
$HCHO_2(g)$	−363	251	+335
$HC_2H_3O_2(l)$	−487.0	160	−392.5
$HCHO(g)$	−108.6	218.8	−102.5
$CH_3CHO(g)$	−167	250	−129
$(CH_3)_2CO(l)$	−248.1	200.4	−155.4
$C_6H_5CO_2H(s)$	−385.1	167.6	−245.3
Chlorine			
$Cl_2(g)$	0	223.0	0
$HCl(g)$	−92.5	186.7	−95.27
$HCl(ag)$	−167.2	56.5	−131.2
$HClO(aq)$	−131.3	106.8	−80.21
Chromium			
$Cr(s)$	0	23.8	0
$CrCl_2(s)$	−326	115	−282
$CrCl_3(s)$	−563.2	126	−493.7
Copper			
$Cu(s)$	0	33.15	0
$CuCl(s)$	−137.2	86.2	−119.87
$CuCl_2(s)$	−172	119	−131
$CuSO_4(s)$	−771.4	109	−661.8
Fluorine			
$F_2(g)$	0	202.7	0
$F^-(aq)$	−332.6	−13.8	−278.8
$HF(g)$	−271	173.5	−273
Hydrogen			
$H_2(g)$	0	130.6	0
$H_2O(\ell)$	−286	69.96	−237.2
$H_2O(g)$	−242	188.7	−228.6
Iron			
$Fe(s)$	0	27	0
$Fe_2O_3(s)$	−822.2	90.0	−741.0
$Fe_3O_4(s)$	−1118.4	146.4	−1015.4
Lead			
$Pb(s)$	0	64.8	0
$PbCl_2(s)$	−359.4	136	−314.1
$PbS(s)$	−100	91.2	−98.7
$PbSO_4(s)$	−920.1	149	−811.3
Lithium			
$Li(s)$	0	28.4	0
$LiF(s)$	−611.7	35.7	−583.3
$LiCl(s)$	−408	59.29	−383.7
Magnesium			
$Mg(s)$	0	32.5	0
$MgCO_3(s)$	−1113	65.7	−1029
$MgF_2(s)$	−1113	79.9	−1056
$MgCl_2(s)$	−641.8	89.5	−592.5

Substance	ΔH_f° (kJ/mol^{-1})	S° (J/mol^{-1} K^{-1})	ΔG_f° (kJ/mol^{-1})
MgO(s)	−601.7	26.9	−569.4
$Mg(OH)_2$(s)	−924.7	63.1	−833.9
Manganese			
Mn(s)	0	32.0	0
MnO_4^-(aq)	−542.7	191	−449.4
$KMnO_4$(s)	−813.4	171.71	−713.8
MnO_2(s)	−520.9	53.1	−466.1
Nitrogen			
N_2(g)	0	191.5	0
NH_3(g)	−46.0	192.5	−16.7
NH_4Cl(s)	−314.4	94.6	−203.9
NO(g)	+90.4	210.6	+86.69
NO_2(g)	+34	240.5	+51.84
N_2O(g)	+81.5	220.0	+103.6
HNO_3(l)	−174.1	155.6	−79.9
Oxygen			
O_2(g)	0	205.0	0
O_3(g)	+143	238.8	+163
OH^-(aq)	−230.0	−10.75	−157.24
Potassium			
K(s)	0	64.18	0
KF(s)	−567.3	66.6	−537.8
KCl(s)	−436.8	82.59	−408.3
KOH(s)	−424.8	78.9	−379.1
K_2SO_4(s)	−1433.7	176	−1316.4
Silver			
Ag(s)	0	42.55	0
AgCl(s)	−127.1	96.2	−109.8
$AgNO_3$(s)	−124	141	−32
Sodium			
Na(s)	0	51.0	0
NaF(s)	−571	51.5	−545
NaCl(s)	−413	72.38	−384.0
NaOH(s)	−426.8	64.18	−382
Na_2SO_4(s)	−1384.49	149.49	−1266.83
Sulfur			
S(s,rhombic)	0	31.8	0
SO_2(g)	−297	248	−300
SO_3(g)	−396	256	−370
H_2SO_4(aq)	−909.3	20.1	−744.5
SF_6(g)	−1209	292	−1105
Tin			
Sn(s,white)	0	51.6	0
$SnCl_4$(ℓ)	−511.3	258.6	−440.2
Zinc			
Zn(s)	0	41.6	0
$ZnCl_2$(s)	−415.1	111	−369.4

3. ACID (IONIZATION) CONSTANTS OF WEAK ACIDS

Monoprotic Acid	Name	K_a
HIO_3	Iodic acid	1.69×10^{-1}
HNO_2	Nitrous acid	7.1×10^{-4}
HF	Hydrofluoric acid	6.8×10^{-4}
$HCHO_2$	Formic acid	1.8×10^{-4}
$HC_3H_5O_3$	Lactic acid	1.38×10^{-4}
$HC_7H_5O_2$	Benzoic acid	6.28×10^{-5}
$HC_4H_7O_2$	Butanoic acid	1.52×10^{-5}
HN_3	Hydrazoic acid	1.8×10^{-5}
$HC_2H_3O_2$	Acetic acid	1.8×10^{-5}
$HC_3H_5O_2$	Propanoic acid	1.34×10^{-5}
$HOCl$	Hypochlorous acid	3.0×10^{-8}
HCN	Hydrocyanic acid	6.2×10^{-10}
HC_6H_5O	Phenol	1.3×10^{-10}
HOI	Hypoiodous acid	2.3×10^{-11}
H_2O_2	Hydrogen peroxide	1.8×10^{-12}

4. IONIZATION CONSTANTS OF POLYPROTIC ACIDS

Polyprotic Acid	Name	K_{a_1}	K_{a_2}	K_{a_3}
H_2SO_4	Sulfuric acid	Large	1.0×10^{-2}	
H_2CrO_4	Chromic acid	5.0	1.5×10^{-6}	
$H_2C_2O_4$	Oxalic acid	5.6×10^{-2}	5.4×10^{-5}	
H_3PO_3	Phosphorous acid	3×10^{-2}	1.6×10^{-7}	
H_2SO_3	Sulfurous acid	1.2×10^{-2}	6.6×10^{-8}	
H_2SeO_3	Selenous acid	4.5×10^{-3}	1.1×10^{-8}	
$H_2C_3H_2O_4$	Malonic acid	1.4×10^{-3}	2.0×10^{-6}	
$H_2C_8H_4O_4$	Phthalic acid	1.1×10^{-3}	3.9×10^{-6}	
$H_2C_4H_4O_6$	Tartaric acid	9.2×10^{-4}	4.3×10^{-5}	
H_2CO_3	Carbonic acid	4.5×10^{-7}	4.7×10^{-11}	
H_3PO_4	Phosphoric acid	7.1×10^{-3}	6.3×10^{-8}	4.5×10^{-13}
H_3AsO_4	Arsenic acid	5.6×10^{-3}	1.7×10^{-7}	4.0×10^{-12}
$H_3C_6H_5O_7$	Citric acid	7.1×10^{-4}	1.7×10^{-5}	6.3×10^{-6}

5. BASE (IONIZATION) CONSTANTS OF WEAK BASES

Weak Base	Name	K_b
$(CH_3)_2NH$	Dimethylamine	9.6×10^{-4}
CH_3NH_2	Methylamine	4.4×10^{-4}
$CH_3CH_2NH_2$	Ethylamine	4.3×10^{-4}
$(CH_3)_3N$	Trimethylamine	7.4×10^{-5}
NH_3	Ammonia	1.8×10^{-5}
N_2H_4	Hydrazine	9.6×10^{-7}
C_5H_5N	Pyridine	1.5×10^{-9}
$C_6H_5NH_2$	Aniline	4.1×10^{-10}

6. SOLUBILITY PRODUCT CONSTANTS

Salt	Dissolution Reaction	K_{sp}
Fluorides		
MgF_2	$MgF_2(s) \rightleftharpoons Mg^{2+}(aq) + 2F^-(aq)$	6.6×10^{-9}
CaF_2	$CaF_2(s) \rightleftharpoons Ca^{2+}(aq) + 2F^-(aq)$	3.9×10^{-11}
BaF_2	$BaF_2(s) \rightleftharpoons Ba^{2+}(aq) + 2F^-(aq)$	1.7×10^{-6}
PbF_2	$PbF_2(s) \rightleftharpoons Pb^{2+}(aq) + 2F^-(aq)$	3.6×10^{-8}
Chlorides		
$CuCl$	$CuCl(s) \rightleftharpoons Cu^+(aq) + Cl^-(aq)$	1.9×10^{-7}
$AgCl$	$AgCl(s) \rightleftharpoons Ag^+(aq) + Cl^-(aq)$	1.8×10^{-10}
$PbCl_2$	$PbCl_2(s) \rightleftharpoons Pb^{2+}(aq) + 2Cl^-(aq)$	1.7×10^{-5}
Bromides		
$CuBr$	$CuBr(s) \rightleftharpoons Cu^+(aq) + Br^-(aq)$	5×10^{-9}
$AgBr$	$AgBr(s) \rightleftharpoons Ag^+(aq) + Br^-(aq)$	5.0×10^{-13}
$PbBr_2$	$PbBr_2(s) \rightleftharpoons Pb^{2+}(aq) + 2Br^-(aq)$	2.1×10^{-6}
Iodides		
CuI	$CuI(s) \rightleftharpoons Cu^+(aq) + I^-(aq)$	1×10^{-12}
AgI	$AgI(s) \rightleftharpoons Ag^+(aq) + I^-(aq)$	8.3×10^{-17}
PbI_2	$PbI_2(s) \rightleftharpoons Pb^{2+}(aq) + 2I^-(aq)$	7.9×10^{-9}
Hydroxides		
$Mg(OH)_2$	$Mg(OH)_2(s) \rightleftharpoons Mg^{2+}(aq) + 2OH^-(aq)$	7.1×10^{-12}
$Ca(OH)_2$	$Ca(OH)_2(s) \rightleftharpoons Ca^{2+}(aq) + 2OH^-(aq)$	6.5×10^{-6}
$Mn(OH)_2$	$Mn(OH)_2(s) \rightleftharpoons Mn^{2+}(aq) + 2OH^-(aq)$	1.6×10^{-13}
$Fe(OH)_2$	$Fe(OH)_2(s) \rightleftharpoons Fe^{2+}(aq) + 2OH^-(aq)$	7.9×10^{-16}
$Fe(OH)_3$	$Fe(OH)_3(s) \rightleftharpoons Fe^{3+}(aq) + 3OH^-(aq)$	1.6×10^{-39}
$Ni(OH)_2$	$Ni(OH)_2(s) \rightleftharpoons Ni^{2+}(aq) + 2OH^-(aq)$	6×10^{-16}
$Cu(OH)_2$	$Cu(OH)_2(s) \rightleftharpoons Cu^{2+}(aq) + 2OH^-(aq)$	4.8×10^{-20}
$Cr(OH)_3$	$Cr(OH)_3(s) \rightleftharpoons Cr^{3+}(aq) + 3OH^-(aq)$	2×10^{-30}
$Zn(OH)_2$	$Zn(OH)_2(s) \rightleftharpoons Zn^{2+}(aq) + 2OH^-(aq)$	3.0×10^{-16}
$Cd(OH)_2$	$Cd(OH)_2(s) \rightleftharpoons Cd^{2+}(aq) + 2OH^-(aq)$	5.0×10^{-15}
Sulfites		
$CaSO_3$	$CaSO_3(s) \rightleftharpoons Ca^{2+}(aq) + SO_3^{2-}(aq)$	3×10^{-7}
$BaSO_3$	$BaSO_3(s) \rightleftharpoons Ba^{2+}(aq) + SO_3^{2-}(aq)$	8×10^{-7}
Sulfates		
$CaSO_4$	$CaSO_4(s) \rightleftharpoons Ca^{2+}(aq) + SO_4^{2-}(aq)$	2.4×10^{-5}
$BaSO_4$	$BaSO_4(s) \rightleftharpoons Ba^{2+}(aq) + SO_4^{2-}(aq)$	1.1×10^{-10}
Ag_2SO_4	$Ag_2SO_4(s) \rightleftharpoons 2Ag^+(aq) + SO_4^{2-}(aq)$	1.5×10^{-5}
$PbSO_4$	$PbSO_4(s) \rightleftharpoons Pb^{2+}(aq) + SO_4^{2-}(aq)$	6.3×10^{-7}
Chromates		
Ag_2CrO_4	$Ag_2CrO_4(s) \rightleftharpoons 2Ag^+(aq) + CrO_4^{2-}(aq)$	1.2×10^{-12}
Hg_2CrO_4	$Hg_2CrO_4(s) \rightleftharpoons Hg_2^{2+}(aq) + CrO_4^{2-}(aq)$	2.0×10^{-9}
$PbCrO_4$	$PbCrO_4(s) \rightleftharpoons Pb^{2+}(aq) + CrO_4^{2-}(aq)$	1.8×10^{-14}
Carbonates		
$MgCO_3$	$MgCO_3(s) \rightleftharpoons Mg^{2+}(aq) + CO_3^{2-}(aq)$	3.5×10^{-8}
$CaCO_3$	$CaCO_3(s) \rightleftharpoons Ca^{2+}(aq) + CO_3^{2-}(aq)$	4.5×10^{-9}
$SrCO_3$	$SrCO_3(s) \rightleftharpoons Sr^{2+}(aq) + CO_3^{2-}(aq)$	9.3×10^{-10}
$BaCO_3$	$BaCO_3(s) \rightleftharpoons Ba^{2+}(aq) + CO_3^{2-}(aq)$	5.0×10^{-9}
$MnCO_3$	$MnCO_3(s) \rightleftharpoons Mn^{2+}(aq) + CO_3^{2-}(aq)$	5.0×10^{-10}
$CuCO_3$	$CuCO_3(s) \rightleftharpoons Cu^{2+}(aq) + CO_3^{2-}(aq)$	2.3×10^{-10}
Ag_2CO_3	$Ag_2CO_3(s) \rightleftharpoons 2Ag^+(aq) + CO_3^{2-}(aq)$	8.1×10^{-12}
Hg_2CO_3	$Hg_2CO_3(s) \rightleftharpoons Hg_2^{2+}(aq) + CO_3^{2-}(aq)$	8.9×10^{-17}
$ZnCO_3$	$ZnCO_3(s) \rightleftharpoons Zn^{2+}(aq) + CO_3^{2-}(aq)$	1.0×10^{-10}
$PbCO_3$	$PbCO_3(s) \rightleftharpoons Pb^{2+}(aq) + CO_3^{2-}(aq)$	7.4×10^{-14}

INDEX